Introduction to

The Calculus of Variations and Control

with Modern Applications

CHAPMAN & HALL/CRC APPLIED MATHEMATICS AND NONLINEAR SCIENCE SERIES

Series Editor *H. T. Banks*

Published Titles

Advanced Differential Quadrature Methods, Zhi Zong and Yingyan Zhang
Computing with hp-ADAPTIVE FINITE ELEMENTS, Volume 1, One and Two Dimensional Elliptic and Maxwell Problems, Leszek Demkowicz
Computing with hp-ADAPTIVE FINITE ELEMENTS, Volume 2, Frontiers: Three Dimensional Elliptic and Maxwell Problems with Applications, Leszek Demkowicz, Jason Kurtz, David Pardo, Maciej Paszyński, Waldemar Rachowicz, and Adam Zdunek
CRC Standard Curves and Surfaces with Mathematica®*: Second Edition,* David H. von Seggern
Discovering Evolution Equations with Applications: Volume 1-Deterministic Equations, Mark A. McKibben
Discovering Evolution Equations with Applications: Volume 2-Stochastic Equations, Mark A. McKibben
Exact Solutions and Invariant Subspaces of Nonlinear Partial Differential Equations in Mechanics and Physics, Victor A. Galaktionov and Sergey R. Svirshchevskii
Fourier Series in Several Variables with Applications to Partial Differential Equations, Victor L. Shapiro
Geometric Sturmian Theory of Nonlinear Parabolic Equations and Applications, Victor A. Galaktionov
Green's Functions and Linear Differential Equations: Theory, Applications, and Computation, Prem K. Kythe
Group Inverses of M-Matrices and Their Applications, Stephen J. Kirkland and Michael Neumann
Introduction to Fuzzy Systems, Guanrong Chen and Trung Tat Pham
Introduction to non-Kerr Law Optical Solitons, Anjan Biswas and Swapan Konar
Introduction to Partial Differential Equations with MATLAB®, Matthew P. Coleman
Introduction to Quantum Control and Dynamics, Domenico D'Alessandro
Introduction to The Calculus of Variations and Control with Modern Applications, John A. Burns
Mathematical Methods in Physics and Engineering with Mathematica, Ferdinand F. Cap
Mathematical Theory of Quantum Computation, Goong Chen and Zijian Diao
Mathematics of Quantum Computation and Quantum Technology, Goong Chen, Louis Kauffman, and Samuel J. Lomonaco
Mixed Boundary Value Problems, Dean G. Duffy
Modeling and Control in Vibrational and Structural Dynamics, Peng-Fei Yao
Multi-Resolution Methods for Modeling and Control of Dynamical Systems, Puneet Singla and John L. Junkins
Nonlinear Optimal Control Theory, Leonard D. Berkovitz and Negash G. Medhin
Optimal Estimation of Dynamic Systems, Second Edition, John L. Crassidis and John L. Junkins
Quantum Computing Devices: Principles, Designs, and Analysis, Goong Chen, David A. Church, Berthold-Georg Englert, Carsten Henkel, Bernd Rohwedder, Marlan O. Scully, and M. Suhail Zubairy
A Shock-Fitting Primer, Manuel D. Salas
Stochastic Partial Differential Equations, Pao-Liu Chow

CHAPMAN & HALL/CRC APPLIED MATHEMATICS
AND NONLINEAR SCIENCE SERIES

Introduction to The Calculus of Variations and Control with Modern Applications

John A. Burns
Virginia Tech
Blacksburg, Virginia, USA

CRC Press
Taylor & Francis Group
Boca Raton London New York

CRC Press is an imprint of the
Taylor & Francis Group, an **informa** business
A CHAPMAN & HALL BOOK

CRC Press
Taylor & Francis Group
6000 Broken Sound Parkway NW, Suite 300
Boca Raton, FL 33487-2742

First issued in paperback 2019

ISBN-13: 978-1-4665-7139-6 (hbk)
ISBN-13: 978-0-367-37955-1 (pbk)

Visit the Taylor & Francis Web site at
http://www.taylorandfrancis.com

and the CRC Press Web site at
http://www.crcpress.com

Contents

Preface

It is fair to say that variational calculus had its beginnings in the 17th century when many of the mathematical giants of that time were focused on solving "calculus of variations" problems. In modern terminology, these early problems in the calculus of variations may be formulated as optimization problems over infinite dimensional spaces of functions. Although this might seem to be a very specialized area, many of the mathematical ideas that were developed to analyze such optimization problems provided the foundations of many areas of modern mathematics. The roots of functional analysis, optimal control, mechanics and the modern theory of partial differential equations can all be traced back to the classical calculus of variations. In addition to its historical connections to many branches of modern mathematics, variational calculus has applications to a wide range of current problems in engineering and science. In particular, variational calculus provides the mathematical framework for developing and analyzing finite element methods. Thus, variational calculus plays a central role in modern scientific computing.

Note that the word "modern" appears five times in the previous paragraph. This is no accident. Too often the calculus of variations is thought of as an old area of classical mathematics with little or no relevance to modern mathematics and applications. This is far from true. However, during the first half of the 20th century, most mathematicians in the United States focused on the intricacies of the mathematics and ignored many of the exciting new (modern) applications of variational calculus. This point was not lost on E. J. McShane who made many fundamental contributions to this area. In a 1978 lecture on the history of calculus of variations and control theory (see [133] and [134]), McShane explained why his fundamental papers on the classical Bolza problem in the calculus of variations "... burst on the mathematical world with the éclat of a butterfly's hiccough." McShane observed:

> *The problem of Bolza was the most general of the single-integral problems of the calculus of variations. Its mastery gave us the power to answer many deep and complicated questions that no one was asking. The whole subject was introverted. We who were working in it were striving to advance the theory of the calculus of variations as an end in itself, without attention to its relation with other fields of activity.*

In the same lecture, McShane provided the one reason why Pontryagin and his followers lead the development of optimal control theory:

> *In my mind, the greatest difference between the Russian approach and ours was in mental attitude. Pontryagin and his students encountered some problems in engineering and in economics that urgently asked for answers. They answered the questions, and in the process they incidentally introduced new and important ideas into the calculus of variations. I think it is excusable that none of us in this room found answers in the 1930's for questions that were not asked until the 1950's. But I for one am regretful that when the questions arose, I did not notice them. Like most mathematicians in the United States, I was not paying attention to the problems of engineers.*

The importance of applications as noted by McShane is still valid today. Optimal control is a vibrant and important offshoot of the classical calculus of variations. The development of a modern framework for the analysis and control of partial differential equations was developed by J. L. Lions and is based on variational theory (see [123], [128], [127] and [125]). Moreover, variational approaches are essential in dealing with problems in stochastic control [21] and differential games [20]. Perhaps one of the most important current applications of variational theory is to modern computational science. The 1972 book by Aubin [12] was in some sense ahead of its time. This book uses variational theory to develop a very general framework for constructing numerical approximation of elliptic boundary problems. Finite element methods produce numerical algorithms that are based on variational (weak) solutions to partial differential equations and provide a powerful approach to simulating a wide variety of physical systems. It is interesting to note that recent advances in computational science have come about because many people in this research community have "paid attention to applications". Thus, this is a case where focusing on a good application can lead to exciting new mathematics and pave the way for major breakthroughs in computational algorithms.

The main goal of this book is to provide an introduction to the calculus of variations and (finite dimensional) optimal control theory with modern applications. The book is based on lecture notes that provided the raw reading material for a course I have taught at Virginia Tech for the past thirty five years. However, the examples and motivating applications have changed and evolved over the years. The object of the course is to introduce the main ideas in a completely rigorous fashion, but to keep the content at a level that is accessible by first year graduate students in engineering and science. For example, we focus mostly on function spaces of piecewise continuous and piecewise smooth functions and thereby avoid measure theory. This is sufficient for the variational calculus and the simplest problem in optimal control. In Part I we develop the calculus of variations and provide complete proofs of the main results. Detailed proofs are given, not for the sake of proving theorems, but because the ideas behind these proofs are fundamental to the development of modern optimization and control theory. Indeed, many of these ideas provide the foundations to all of modern applied and computational mathematics including: functional analysis, distribution theory, the theory of partial differential equations, optimization and finite element methods.

In Part II we focus on optimal control problems and show how optimal control is a natural extension of the classical calculus of variations to more complex problems. Although the proof of the Maximum Principle was a tour de force in applied mathematics, the basic Maximum Principle for the simplest problem is given without proof. A complete (rigorous) proof is not given for two reasons. First, from the author's teaching experience I have found that the time spent to develop and present a proof of the Maximum Principle adds very little to the understanding of key ideas. Basically there are two approaches to proving the Maximum Principle. One approach is based on functional analysis techniques and would require that the student have a working knowledge of Lebesgue integration and measure theory. This approach is typical in the more mathematical treatments such as found in [100] and [119]. Although there is a second approach that only uses advanced calculus and geometric ideas, the complete proof is rather lengthy. This approach can be found in several excellent references such as [18], [101] and [120] and will not be repeated here. However, the basic Maximum Principle is used to rigorously develop necessary conditions for more complex problems and for optimal control of linear systems. The author feels that going through a rigorous development for extensions of the simplest optimal control problem provides the student with the basic mathematical tools to attack new problems of this type in their particular application area.

During the past fifty years a huge number of texts have been published on calculus of variations, optimization, design and control. Clearly this book can not capture this entire body of work. A major objective of this book is to provide the fundamental background required to develop necessary conditions that are the starting points for theoretical and numerical approaches to variational and control problems. Although we focus on necessary conditions, we present some classical sufficient conditions and discuss the importance of distinguishing between the two. In all cases the emphasis is on understanding the basic ideas and their mathematical development so that students leave the course with mathematical tools that allow them to attack new problems of this type. After a thorough treatment of the simplest problems in the Calculus of Variations and Optimal Control, we walk through the process of moving from the simplest problem to more complex problems to help the student see how one might begin to modify the basic optimality conditions to address more difficult problems. This is important since we stress the point:

> *It is impossible to solve all problems in a course and when the student moves on to the "real working world" there will be new applications and problems to be solved. This book provides the ideas and methodologies that might be used as a starting point to address yet unsolved problems.*

It is assumed that the reader has a basic background in differential equations and advanced calculus. The notes focus on the fundamental ideas that are needed to rigorously develop necessary and sufficient conditions and to present cases where these ideas have impact on other mathematical areas and applications. In Part I we provide complete proofs of the main theorems on necessary and sufficient conditions. One goal is to make sure the student has a clear understanding of the difference between necessary conditions and sufficient conditions and when to use them. Although this may seem like a trivial issue to mathematicians, the author has found that some students in other disciplines have trouble distinguishing between the two which can lead to mistakes. Moreover, since very little "advanced mathematics" is assumed, the initial proofs are very detailed and in a few cases these details are repeated to emphasize the important ideas which have applications to a variety of problems.

In order to keep the book at a reasonable length and to keep the mathematical requirements at the advanced calculus level, we have clearly omitted many important topics. For example, multi-integral problems are not discussed and "direct methods" that require more advanced mathematics such as Sobolev Space theory are also missing. The interested reader is referred to [86] for an elementary introduction to these topics and more advanced treatments may be found in [64], [65], [89] and [186].

Suggested References and Texts

The best textbooks to supplement this book are George Ewing's book, *Calculus of Variations with Applications* [77] and Leitmann's book, *The Calculus of Variations and Optimal Control* [120]. Also, the books by Joseph Z. Ben-Asher [18] and Lee and Markus [119] provide nice introductions to optimal control. Other books that the reader might find useful are [65], [122] and [135]. Finally, the following texts are excellent and will be cited often so that the interested reader can dive more deeply into individual topics.

- B. Anderson and J. Moore [6]
- A. Bensoussan, G. Da Prato, M. Delfour and S. Mitter [22] and [23]
- L. D. Berkovitz [24]
- G. Bliss [29]
- O. Bolza [31]
- A. Bryson and Y. Ho [46]
- C. Caratheodory [54]
- F. Clarke [57]
- R. Courant [59]
- R. Curtain and H. Zwart [63]
- A. Forsyth [80]
- I. Gelfand and S. Fomin [87]
- L. Graves [95]
- H. Hermes and J. LaSalle [100]
- M. Hestenes [101] and [102]
- A. Isidori [105]
- T. Kailath [110]
- H. Knobloch, A. Isidori and D. Flockerzi [114]
- H. Kwakernaak and R. Sivan [115]
- B. Lee and L. Markus [119]
- J. L. Lions [123] and [125]
- D. Luenberger [131]
- L. Neustadt [144]
- D. Russell [157]
- H. Sagan [158]
- D. Smith [166] and [167]
- J. Troutman [180]
- F. Wan [183]
- L. Young [186]

Disclaimer: *This book is based on "raw" lecture notes developed over the years and the notes were often updated to include new applications or eliminate old ones. Although this process helped the author find typos and errors, it also meant the introduction of new typos and errors. Please feel free to send the author a list of typos, corrections and any suggestions that might improve the book for future classes.*

Acknowledgments

I would like to acknowledge all my students who for the past thirty five years have provided feedback on the raw notes that were the basis of this book. Also, I thank my long time colleagues Gene Cliff and Terry Herdman who have provided valuable input, guidance and inspiration over the past three decades. This constant feedback helped me to update the subject matter as the years progressed and to find both typos and technical errors in the material. However, as noted above the constant updating of the material means that even this version is sure to have some typos. This is not the fault of my students and colleagues, so I wish to apologize in advance for not catching all these errors. I especially wish to thank my student Mr. Boris Kramer for a careful reading of the current version. I also thank the reviewers for their comments and valuable suggestions for improving the manuscript, the acquisitions editor Bob Stern at Taylor & Francis for his help and support and the series editor H. T. Banks for encouraging this project. Most importantly, I wish to thank my family for their support and understanding, particularly my wife Gail. They graciously gave up many nights of family time while this manuscript was being written.

Part I

Calculus of Variations

Chapter 1

Historical Notes on the Calculus of Variations

It is widely quoted that the calculus of variations (as a mathematical subject) had its beginnings in 1686 when Newton proposed and solved a problem of finding a body of revolution that produces minimum drag when placed in a flow. The problem and its solution were given in his 1687 Philosophiae Naturalis Principia Mathematica (Mathematical Principles of Natural Philosophy).

A second milestone occurred in 1696 when Johann (John) Bernoulli proposed the brachistochrone problem as a mathematical challenge problem. In 1697 his brother Jacob (James) Bernoulli published his solution and proposed a more general isoperimetric problem. In addition to the Bernoulli brothers, Newton, Leibniz and L'Hôpital also gave correct solutions to the brachistochrone problem. This is clearly a precise time in history where a "new" field of mathematics was born.

Between 1696 and 1900 a large core of mathematical giants worked in this area and the book by Herman H. Goldstine [91] provides a detailed treatment of this body of work. In particular, John and James Bernoulli, Leonhard Euler, Isaac Newton, Joseph-Louis Lagrange, Gottfried Wilhelm von Leibniz, Adrien Marie Legendre, Carl G. J. Jacobi and Karl Wilhelm Theodor Weierstrass were among the main contributors in this field. Other important contributions during this period were made by Paul du Bois-Reymond,

Johann Peter Gustav Lejeune Dirichlet, William Rowan Hamilton and Pierre-Louis Moreau de Maupertuis. At the end of the 19th century and into the first half of the 20th century, the single integral problem in the calculus of variations was expanded and refined by David Hilbert (Germany), Leonida Tonelli (Italy), Oskar Bolza (Germany and U.S.A.) and the "Chicago school" including G. A. Bliss, L. M. Graves, M. R. Hestenes, E.J. McShane, and W. T. Reid. Around 1950 the basic problem of minimizing an integral subject to differential equation constraints became a major problem of interest because of various military applications in the USA and the USSR. These problems required the treatment of "hard constraints" which were basically ignored in the classical calculus of variations and led to the theory of optimal control.

The history of the development of optimal control is less precise and the subject of varying opinions. The paper "300 *Years of Optimal Control: From the Brachystochrone to the Maximum Principle*" by Sussmann and Willems [172] clearly states that optimal control was born in 1667. Although everyone agrees that optimal control is an extension of the classical calculus of variations, others ([149], [47]) suggest that optimal control theory had its beginning around 1950 with the "discovery" of the Maximum Principle by various groups.

The road from the "classical" calculus of variations to "modern" optimal control theory is certainly not linear and it can honestly be argued that optimal control theory had its beginnings with the solution to the brachistochrone problem in 1697 as suggested in [172]. However, two important steps in moving from classical variational approaches to modern control theory occurred between 1924 and 1933. In L. M. Graves' 1924 dissertation [93] he treated the derivative as an independent function and hence distinguished between state and control variables. In 1926 C. Carathéodory gave the first formulation of the classical Weierstrass necessary condition in terms of a Hamiltonian [53] which, as noted in [172], is the "first fork in the road" towards modern control theory. Finally, in 1933 L. M. Graves [94] gave a control formulation of the classical Weierstrass condition for a Bolza type problem. These ideas are key to understanding the power of modern optimal control methods. The papers [149], [154] and [172] provide a nice historical summary of these results and their impact on modern optimal control.

Clearly, everyone agrees to some level that the classical calculus of variations is a starting point for modern optimal control theory. However, what is often lost in this historical discussion is that the calculus of variations also laid the foundations for the creation of other "new" fields in both applied and pure mathematics. Modern functional analysis, the theory of distributions, Hamiltonian mechanics, infinite (and finite dimensional) optimization and the modern theory of partial differential equations all trace their roots to the classical calculus of variations. Perhaps even more relevant today is the role that variational theory plays in modern scientific computing.

A key theme in this book is that all these problems fall within the purview of optimization. Although the theoretical issues first appear to be no more difficult than those that occur in finite dimensional optimization problems, there are major differences between infinite and finite dimensional optimization. Moreover, the computational challenges are different and a direct reduction to a finite dimensional problem through approximation is not always the best approach. For example, information about the form and properties of a solution to an infinite dimensional optimization problem can be lost if one introduces approximation too early in the problem solution. The paper [113] by C. T. Kelley and E. W. Sachs provides an excellent example of how the theory of infinite dimensional optimization can yield improved numerical algorithms that could not be obtained from finite dimensional theory alone.

We specifically mention that Lucien W. Neustadt's book on optimization [144] contains the most complete presentation of general necessary conditions to date. This book provides necessary conditions for optimization problems in topological vector spaces. Although a rather deep and broad mathematical background is required to follow the development in this book, anyone seriously considering research in infinite dimensional optimization, control theory and their applications should be aware of this body of work.

Most early problems in the calculus of variation were motivated by applications in the physical sciences. During the past century, calculus of variations was key to the development of game theory, existence theory for partial differential equations and convergence theory for finite element methods for numerical approximation of partial differential equations. Modern applications in optimal control, especially of systems governed by partial differential equations (PDEs), and computational science has sparked a renewed interest in variational theory and infinite dimensional optimization. The finite element method is based on minimizing certain functionals over spaces of functions and PDE optimization and control problems have applications ranging from fluid flow control to large space structures to the design and control of energy efficient buildings. All these applications lead to problems that require infinite dimensional optimization. In this book we focus on necessary conditions for the classical calculus of variations and then provide a short introduction to modern optimal control and the Maximal Principle.

1.1 Some Typical Problems

The term "calculus of variations" originally referred to problems involving the minimization of a functional defined by an integral

$$J(x(\cdot)) = \int_{t_0}^{t_1} f\left(s, x\left(s\right), \dot{x}\left(s\right)\right) ds \tag{1.1}$$

over a suitable function space. Most of the problems we consider will have an integral cost functions of the form (1.1). However, because of significant developments during the past century, we will expand this classical definition of the calculus of variations to include problems that now fall under the topic of optimal control theory. To set the stage, we begin with some standard problems.

1.1.1 Queen Dido's Problem

As noted above, the calculus of variations as a mathematical subject has its beginnings in 1696, when John Bernoulli suggested the brachistochrone problem. Although folklore presents Queen Dido's problem as one of the first problems in the calculus of variations, real scientific and mathematical investigations into such problems probably started with Galileo and Newton. In any case Queen Dido's problem provides a nice illustrative example to start the discussion.

In Roman mythology, Dido was the Queen of Carthage (modern-day Tunisia). She was the daughter of a king of Tyre. After her brother Pygmalion murdered her husband, she fled to Libya, where she founded and ruled Carthage. The legend has it that she was told that she could rule all the land around the coast that she could "cover with the hide of a cow". Being very clever, she cut the hide into one continuous thin string and used the string to outline the area she would rule. Thus, Queen Dido's problem is to find the maximum area that can be encompassed with a string of fixed length.

1.1.2 The Brachistochrone Problem

Galileo discovered that the cycloid is a brachistochrone curve, that is to say it is the curve between two points that is covered in the least time by a body rolling down the curve under the action of constant gravity. In 1602 Galileo provided a geometrical demonstration that the arc of circumference is the brachistochrone path. The apparatus is a wooden frame, mounted on two feet with adjusting screws and carrying a cycloidal groove. Parallel to this is a straight groove, the inclination of which may be varied by inserting wooden pins in the brass-reinforced openings made just underneath the cycloid. A lever with two small clamps allows the release of two balls along the two channels simultaneously. The ball that slides down the cycloidal groove arrives at the point of intersection of the two channels in less time than the ball rolling along an inclined plane. This device can still be seen in Museo Galileo at the Institute and Museum of the History of Science in Florence, Italy. The mathematical formulation of the problem of finding the curve of least descent time, the brachistochrone, was proposed by

John Bernoulli in 1696. In 1697 the problem was solved by John, his brother James, Newton and others. The solution curve was shown to be a cycloid and Jakob's solution contained the basic ideas leading to the theory of the calculus of variations.

1.1.3 Shape Optimization

As noted above, in 1686 Newton proposed the problem of finding a body of revolution (nose cone) that produces minimum drag when placed in a flow. In modern terms, the problem is to find a shape of the nose cone to minimize the drag with the constraint that this shape is defined by a surface of revolution. From a historical point of view this is one of the first mathematical formulations of a "shape optimization" problem. It is also interesting to note that because of Newton's choice of model for the aerodynamic forces, his solution is not accurate for subsonic flows (see [136], [137] and page 52 in [46]). On the other hand, Newton's assumption is fairly accurate for a hypersonic flow which is important today. In any case, this problem was an important milestone in the calculus of variations.

Today Newton might have considered the problem of finding a shape of a body attached to the front of a jet engine in order to produce a flow that matches as well as possible a given flow into the jet (a forebody simulator). The idea is to use a smaller forebody so that it can be placed in a wind tunnel and the engine tested. This problem had its beginnings in 1995 and is based on a joint research effort between the Air Force's Arnold Engineering Design Center (AEDC) and The Interdisciplinary Center for Applied Mathematics (ICAM) at Virginia Tech. The goal of the initial project was to help develop a practical computational algorithm for designing test facilities needed in the free-jet test program. At the start of the project, the main bottleneck was the time required to compute cost function gradients used in an optimization loop. Researchers at ICAM attacked this problem by using the appropriate variational equations to guide the development of efficient computational algorithms. This initial idea has since been

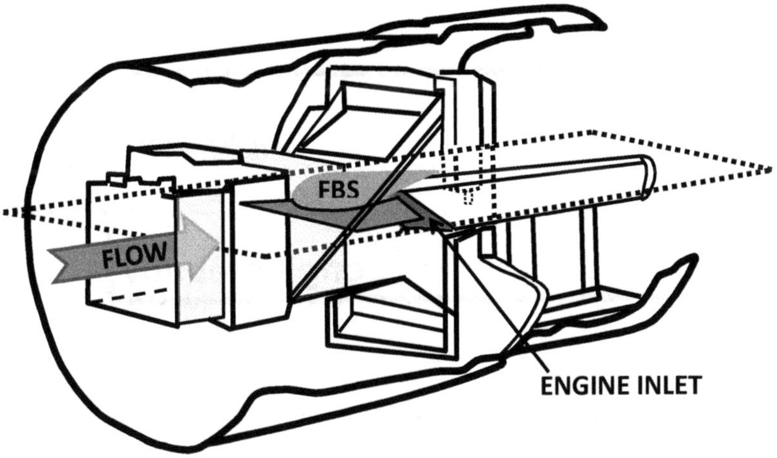

Figure 1.1: The AEDC Test Section

refined and has now evolved into a practical methodology known as the Continuous Sensitivity Equation Method (CSEM) for optimal design.

The wind tunnel is large enough to hold the engine and a smaller "forebody simulator" (FBS), but not large enough to hold the entire front of the airplane. The objective is to determine the shape of the forebody simulator and the inflow conditions (MACH number, angle, etc.) so that the flow going into the engine inlet matches (as well as possible) a given flow. This given flow is the flow that would be present if the jet were in free flight. This data can be generated by flight test or full 3D simulation.

Consider a 2D version of the problem. The green sheet represents a cut through the engine reference plane and leads to the following 2D problem. The goal is to find a shape Γ (that is constrained to be 1/2 the length of the long forebody) and an inflow mach number M_0 to match the flow entering the engine inlet generated by the long forebody. This shorter curve is called the "forebody simulator". This problem is much more complex than Newton's minimum drag problem because it is not assumed that the forebody is a surface of revolution. Although both problems are concerned with finding optimal shapes (or curves), the forebody simulator problem is still a challenge for mathematics and numer-

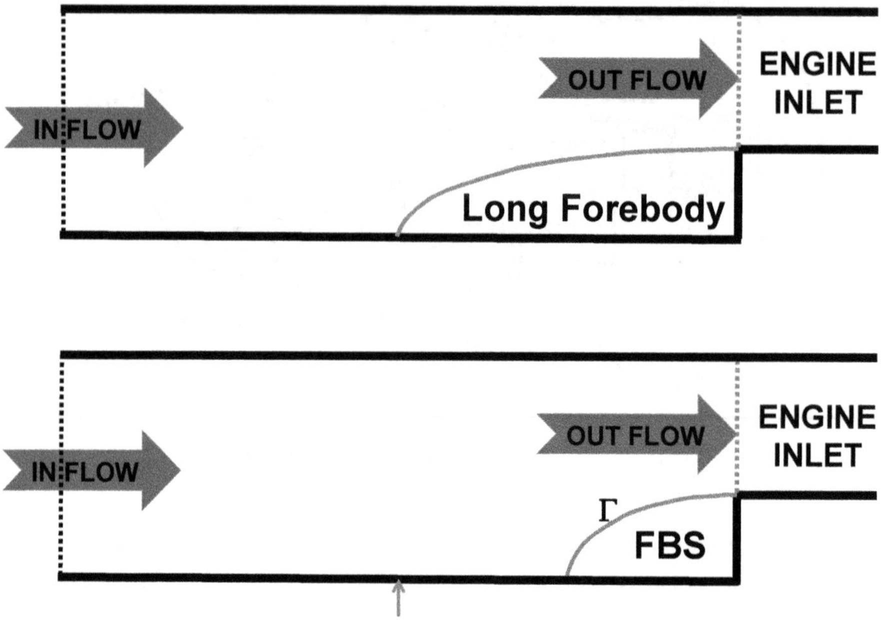

Figure 1.2: The 2D Forebody Simulator Problem

ical computations. Given data generated over the long forebody, the goal is to find a shorter (maybe "fatter") forebody simulator to optimally match the "real data" at the engine inlet.

Modern "optimal design" is one area that has its roots in the calculus of variations. However, the computer has changed both the range of problems that are possible to solve and the type of methods used to solve these problems. For more on the analysis, design and optimization of modern engineering systems see [8], [9], [13], [19], [38], [34], [32], [33], [35], [40], [41], [36], [37], [43], [42], [67], [68], [82], [83], [90], [98], [99], [108], [109], [111], [142], [150], [163], [168], [174], [177] and [178].

1.2 Some Important Dates and People

The following dates provide some insight into the history and the people who contributed to the development of calculus of variations and its modern realization in Optimal Control.

1600-1900

- 1630 - Galileo Galilei (1564-1642) formulated the brachistochrone problem.
- 1686 - Isaac Newton (1642-1727) proposed and gave solution to the surface of revolution of minimum drag when the body is moved through a fluid.
- 1696 - John Bernoulli (1667-1748) proposed the brachistochrone problem as a challenge to all mathematicians, but to his brother James Bernoulli (1654-1705) in particular. The problem was also solved by Newton, Leibniz (1646-1716), L'Hôpital (1661-1704), as well as both Bernoulli brothers.
- 1697 - James Bernoulli (1654-1705) published his solution and proposed a more general isoperimetric problem.
- 1724 - Jacopo Francesco Riccati (1676-1754) wrote his famous paper on the Riccati Equation. James Bernoulli had also worked on this equation.
- 1744 - Leonard Euler (1707-1783) extended James Bernoulli's methods (geometric, analytical) to general problems, and discovered the "Euler Equation". He also derived Newton's second law from "the principle of least action".
- 1760 - Joseph Louis Lagrange (1736-1813) first used the term "calculus of variations" to describe the methods he used in his work.
- 1762 and 1770 - Lagrange devised an analytic method for general constrained problems. He also indicated that mechanics and variational theory are connected, and he introduced the notation $y(x) + \delta y(x)$. This symbol, $\delta y(\cdot)$, was called the variation of $y(\cdot)$.

- 1786 - Lagrange published his necessary condition for a minimum of Newton's problem.
- 1786 - Adrien-Marie Legendre (1752-1833) studied the second variation $\delta^2 J$.
- 1788 - Lagrange showed that a large part of Newtonian dynamics could be derived from the principle of least action for "conservative" systems.
- 1788 - Lagrange showed that a curve could satisfy the Euler-Lagrange equation for Newton's problem and not minimize the functional. He used a proof essentially equivalent to Weierstrass's condition.
- 1835 - Sir William Rowan Hamilton (1805-1865) expanded the principle of least action to "Hamilton's Principle" of stationary action.
- 1837 - Karl Gustav Jacob Jacobi (1804-1851) used some of Legendre's ideas to construct Jacobi's (second order) necessary condition.
- 1842 - Jacobi gave an example to show that the principle of least action does not hold in general.
- 1879 - Karl Theodor Wilhelm Weierstrass (1815-1897) gave his necessary condition for a strong local minimum.
- 1879 - Paul David Gustav Du Bois-Reymond (1831-1889) gave proof of the fundamental lemma. In addition to providing a correct proof of the FLCV, in 1873 he gave an example of a continuous function with divergent Fourier series at every point. The term "integral equation" is also due to Du Bois-Reymond.
- 1898 - Adolf Kneser (1862-1930) defined focal point.

1900-1965

- 1900 - David Hilbert (1862-1943) gave a derivation of the "Hilbert Invariant Integral".
- 1904 - Hilbert gave his famous existence proof for the simplest problem.
- 1913 - Oskar Bolza (1857-1942) stated the problem of Bolza. The problem of Bolza is the forerunner of modern control problem.
- 1933 - Lawrence M. Graves (1896-1973) transformed the problem of Lagrange into a control theory formulation, and proved a "maximum principle" for normal problems. Between 1958 and 1962, V. G. Boltyanskii, R. Gamkrelidze, E. F. Mischenko and L. S. Pontryagin established the Pontryagin Maximum Principle for more general optimal control problems.
- 1937 - Lawrence C. Young (1905-2000) introduced generalized curves and relaxed controls.
- 1940 - Edward James McShane (1904-1989) established the existence of a relaxed control and proved that a generalized curve was real curve under certain convexity conditions.
- 1950 - Magnus R. Hestenes (1906-1991) formulated the first optimal control problems, and gave the maximum principle first published in a Rand Report.
- 1952 - Donald Wayne Bushaw (1926-2012) gave a mathematical solution to a simple time optimal control problem by assuming the bang-bang principle.
- 1959 - Joesph P. LaSalle (1916-1983) gave the first proof of (LaSalle's) Bang-Bang Principle. He also extended the classical Lyapunov theory to the LaSalle Invariance Principle.

- 1959 - Richard F. Bellman (1920-1984) developed the dynamic programming principle of optimality for control problems.

- 1962 - Lev Semenovich Pontryagin (1908-1988) derived the Pontryagin Maximum Principle along with V. G. Boltyanskii, R. V. Gamkrelidze and E. F. Mishchenko.

Other Important Players

- Gilbert Ames Bliss (1876-1951) - Bliss's main work was on the calculus of variations and he produced a major book, *Lectures on the Calculus of Variations*, on the topic in 1946. As a consequence of Bliss's results a substantial simplification of the transformation theories of Clebsch and Weierstrass was achieved. His interest in the calculus of variations came through two sources, firstly from lecture notes of Weierstrass's 1879 course of which he had a copy, and secondly, from the inspiring lectures by Bolza which Bliss attended. Bliss received his doctorate in 1900 for a dissertation *The Geodesic Lines on the Anchor Ring* which was supervised by Bolza. Then he was appointed as an instructor at the University of Minnesota in 1900. He left Minnesota in 1902 to spend a year in Göttingen where he interacted with Klein, Hilbert, Minkowski, Zermelo, Schmidt, Max Abraham, and Carathéodory.

- Constantin Carathéodory (1873-1950) - Carathéodory made significant contributions to the calculus of variations, the theory of point set measure, and the theory of functions of a real variable. He added important results to the relationship between first order partial differential equations and the calculus of variations.

- Jean Gaston Darboux (1842-1917) - Darboux studied the problem of finding the shortest path between two points on a surface and defined a Darboux point.

- Ernst Friedrich Ferdinand Zermelo (1871-1953) - His doctorate was completed in 1894 when the University of Berlin awarded him the degree for a dissertation *Untersuchungen zur Variationsrechnung* which followed the Weierstrass approach to the calculus of variations. In this thesis he extended Weierstrass' method for the extrema of integrals over a class of curves to the case of integrands depending on derivatives of arbitrarily high order. He also introduced the notion of a "neighborhood" in the space of curves.

- George M. Ewing (1907-1990) - Ewing's 1959 book, *Calculus of Variations with Applications* remains one of the best introductions to the classical theory.

- Christian Gustav Adolph Mayer (1839-1907) - Mayer focused on the principle of least action and is credited with formulating variational problems in "Mayer form" where the cost functional is given in terms of end conditions.

- Harold Calvin Marston Morse (1892-1977) - Morse developed "variational theory in the large" and applied this theory to problems in mathematical physics. He built his "Morse Theory" on the classical results in the calculus developed by Hilbert.

- William T. Reid (1907-1977) - Reid's work in the calculus of variations combined Sturm-Liouville theory with variational theory to study second order necessary and sufficient conditions. Reid established a generalization of Gronwall's Inequality which is known as the Gronwall-Reid-Bellman Inequality. The Reid Prize awarded by SIAM for contributions to Differential Equations and Control Theory is named after W. T. Reid.

- Frederick A. Valentine (1911-2002) - Valentine attended the University of Chicago where he received his Ph.D. in mathematics in 1937. His dissertation was entitled "The Problem of Lagrange with Differential Inequalities as Added Side Conditions" and was written under the direction of Bliss. Most of his work was in the area of convexity and his book *Convex Sets* [182] is a classic.

- Vladimir Grigorevich Boltyanskii (1925-) - Boltyanskii was one of the four authors of the book *The Mathematical Theory of Optimal Processes* and was awarded the Lenin Prize for the work presented in that book on optimal control.

- Revaz Valer'yanovich Gamkrelidze (1927-) - Gamkrelidze was one of the four authors of the book *The Mathematical Theory of Optimal Processes* and was awarded the Lenin Prize for the work presented in that book on optimal control.

- Evgenii Frolovich Mishchenko (1922-2010) - Mishchenko was one of the four authors of the book *The Mathematical Theory of Optimal Processes* and was awarded the Lenin Prize for the work presented in that book on optimal control.

- Lev Semenovich Pontryagin (1908-1988) - Pontryagin was 14 when an accident left him blind. The article [4] describes how Pontryagin's mother, Tat'yana Andreevna Pontryagin took complete responsibility for seeing that her son was educated and successful. As noted in [4]: *For many years she worked, in effect, as Pontryagin's secretary, reading scientific works aloud to him, writing in the formulas in his manuscripts, correcting his work and so on. In order to do this she had, in particular, to learn to read foreign languages.* Pontryagin's early work was on problems in topology and algebra. In the early 1950's Pontryagin began to study applied mathematics, differential equations and control theory. In 1961 he published *The Mathematical Theory of Optimal Processes* with his students V. G. Boltyanskii, R. V. Gamkrelidze and E. F. Mishchenko.

Chapter 2

Introduction and Preliminaries

In this chapter we discuss several problems that will serve as models and examples for the theory to follow. We first provide rough overviews of some classical problems to give a preliminary indication of the type of mathematical concepts needed to formulate these problems as mathematical optimization problems. We introduce some notation, discuss various classes of functions and briefly review some topics from calculus, advanced calculus, and differential equations. Finally, we close this chapter with precise mathematical statements of some of these model problems.

2.1 Motivating Problems

In order to motivate the topic and to set the stage for future applications, we provide a brief description of some problems that are important from both historical and scientific points of view.

2.1.1 Problem 1: The Brachistochrone Problem

This problem is perhaps the first problem presented in every endeavor of this type. It is called the brachistochrone problem and it

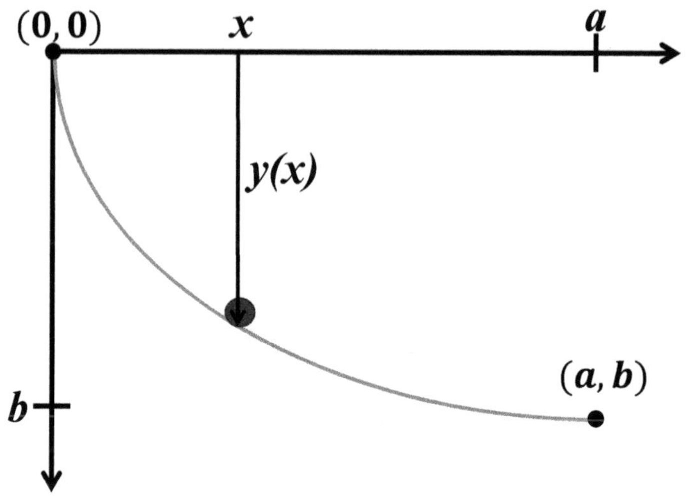

Figure 2.1: The Brachistochrone Problem

was first considered by Galileo Galilei (1564-1642). The mathematical formulation of the problem was first given by John Bernoulli in 1696, and is often quoted as the beginning of the classical calculus of variations. Suppose that $P_1 = [0\ 0]^T$ and $P_2 = [a\ b]^T$ are two points in the plane with P_1 "higher" than P_2 (see Figure 2.1).

Suppose that we slide a (frictionless) bead down a wire that connects the two points. We are interested in finding the shape of the wire down which the bead slides from P_1 to P_2 in minimum time. At first glance one might think that the wire should be the straight line between P_1 and P_2, but when we solve this problem it will be seen that this is not the case. The mathematical formulation of this problem will be derived below. But first, we describe other "typical problems" in the calculus of variations.

2.1.2 Problem 2: The River Crossing Problem

Another "minimum time" problem is the so called river crossing problem. We assume that the river of a fixed width of one mile has

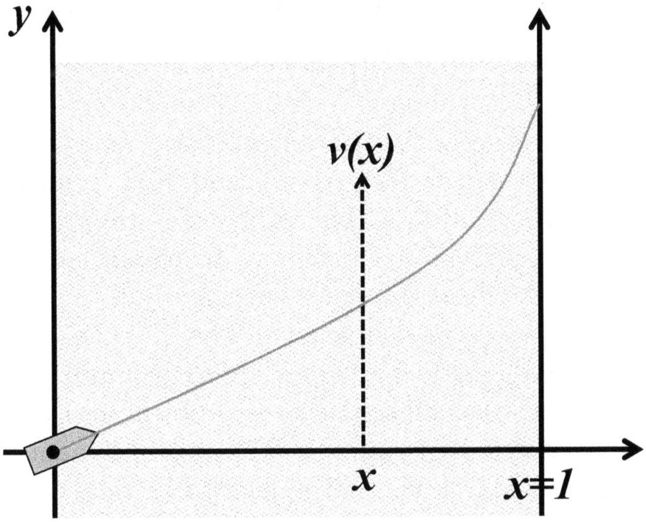

Figure 2.2: The River Crossing Problem

parallel banks and we let the y-axis be the left bank (see Figure 2.2). The current is directed downstream along the y-axis, and the velocity $v(\cdot)$ of the current depends only on x, i.e. $v = v(x)$ is the velocity of the current at a distance of x feet from the left bank. Given $v(x)$ and the assumption that the boat travels at a constant velocity (relative to the surrounding water), we wish to find the "steering angle" of the boat that will move the boat from the point $[0\ 0]^T$ to the right bank in minimum time. Note that we are not concerned with where along the right bank the boat lands. The goal is to find the path of the boat that starts at a prescribed point and reaches the opposite bank in minimum time. This problem is similar to the brachistochrone problem, except that landing site is not prescribed so that the downstream location is free. Thus, this is an example of a "free endpoint problem". Intuition would seem to imply that the shape of the minimum time crossing path is not dependent on the starting point. In particular, if the boat were to start at $[0\ y_0]^T$ with $y_0 > 0$ rather than $[0\ 0]^T$, then the shape (graph) of the minimum time crossing path would look the same, only shifted downstream by an additional distance of y_0.

2.1.3 Problem 3: The Double Pendulum

Consider an idealized double pendulum with masses m_1 and m_2 attached to weightless rods of lengths ℓ_1 and ℓ_2 as shown below. Suppose that we observe the pendulum at two times t_0 and t_1 and note the angles $\alpha_1(t_0)$, $\alpha_2(t_0)$, $\alpha_1(t_1)$, and $\alpha_2(t_1)$ and/or angular velocities $\dot{\alpha}_i(t_j)$, $i, j = 0, 1$. The problem is to write down the governing equations for the pendulum. In particular, we wish to find a system of differential equations in $\alpha_1(t)$ and $\alpha_2(t)$ that describe the motion of the double pendulum. If you recall Newton's laws of motion and a little mechanics, then the problem is not very difficult. However, we shall see that the problem may be solved by using what is known as Hamilton's Principle in dynamics.

Although Problem 3 does not appear to be related to the first two "optimization problems", it happens that in order to apply Hamilton's Principle one must first be able to solve problems similar to each of these.

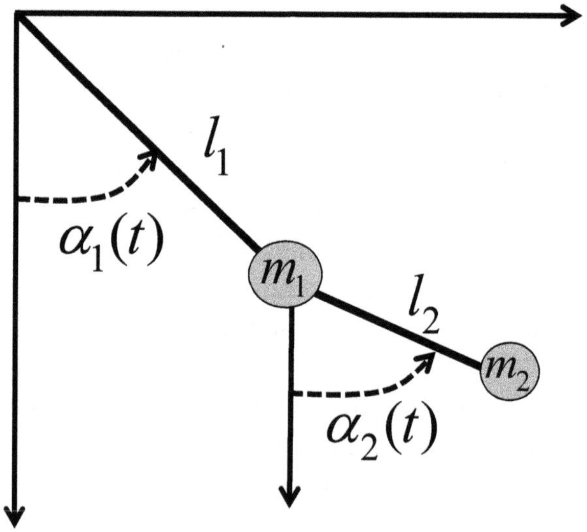

Figure 2.3: The Double Pendulum Problem

2.1.4 Problem 4: The Rocket Sled Problem

The orbit transfer problem cannot be solved by the classical techniques of the calculus of variations. The more modern theory of optimal control is required to address problems of this type. Another very simple example that falls under the category of optimal control is the so-called rocket car problem. This example will illustrate some of the elementary mathematics that are needed in order to solve the problems we have presented earlier. Consider a rocket sled illustrated in Figure 2.4 below. We assume the car is on a frictionless track, that the sled is of mass m, and that it is controlled by two rocket engines thrusting in opposite directions.

It is assumed that the mass m is so large that the weight of the fuel is negligible compared to m. Also, the thrusting force of the rockets is bounded by a maximum of 1. Let $x(t)$ denote the position of the sled with respect to the reference R, at time t. Given that at time $t = 0$, the sled is at an initial position x_0 with initial velocity v_0, we wish to find a thrust force action that will move the sled to position R and stop it there, and we want to accomplish this transfer in minimum time.

We derive a simple mathematical formulation of this problem. Let $x(t)$ be as in the figure and let $u(t)$ denote the thrusting force due to the rockets at time t. Newton's second law may be written as

$$m\ddot{x}(t) = u(t),$$

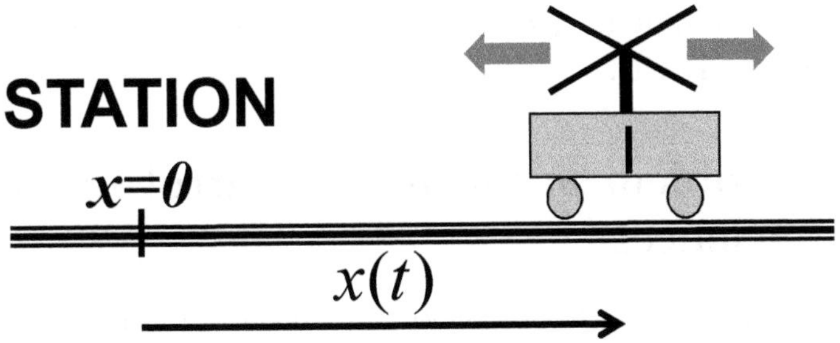

Figure 2.4: The Rocket Sled Control Problem

or equivalently,

$$\ddot{x}(t) = \left(\frac{1}{m}\right) u(t). \tag{2.1}$$

The initial conditions are given by

$$x(0) = x_0, \quad \dot{x}(0) = v_0. \tag{2.2}$$

The fact that the thrust force is bounded is written as the constraint

$$|u(t)| \le 1. \tag{2.3}$$

The problem we wish to solve may be restated as the following optimal control problem.

> Find a control function $u^*(t)$ satisfying the constraint (2.3) such that if $x^*(t)$ satisfies (2.1)-(2.2), then $x^*(t^*) = 0 = \dot{x}^*(t^*)$ for some time t^*. Moreover, if $\overline{u}(t)$ is any other control function satisfying (2.3) with $\overline{x}(t)$ satisfying (2.1)-(2.2) and $\overline{x}(\overline{t}) = 0 = \overline{x}(\overline{t})$ for some $\overline{t}$, then $t^* \le \overline{t}$.

Although this problem seems to be the easiest to understand and to formulate, it turns out to be more difficult to solve than the first three problems. In fact, this problem falls outside of the classical calculus of variations and to solve it, one must use the modern theory of optimal control. The fundamental new ingredient is that the control function $u(t)$ satisfies the "hard constraint" $|u(t)| \le 1$. In particular, $u(t)$ can take values on the boundary of the interval $[-1, +1]$.

2.1.5 Problem 5: Optimal Control in the Life Sciences

Although many motivating problems in the classical calculus of variations and modern optimal control have their roots in the physical sciences and engineering, new applications to the life sciences is a very active area of current research. Applications to cancer treatment and infectious diseases present new challenges

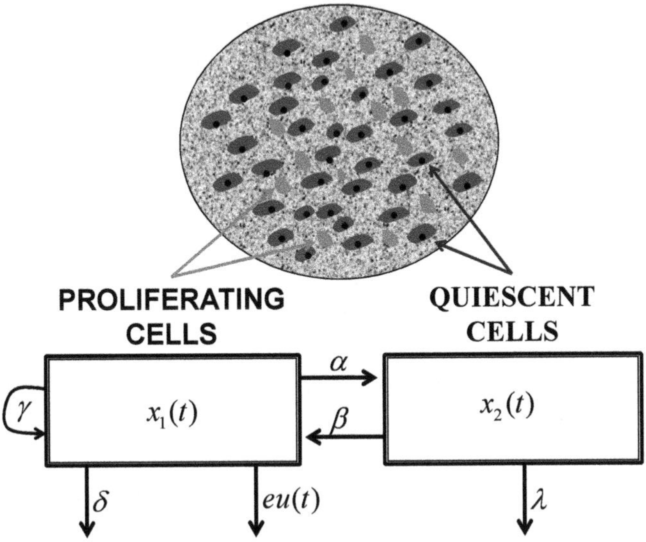

Figure 2.5: Cancer Control Problem

and opportunities for control (see [1], [14], [26], [62], [74], [106], [117], [118], [141], [140], [145], [146], [148], [165], [173]). The following optimal chemotherapy treatment problem may be found in the paper by Fister and Panetta [148]. The cancer cells are divided into two types. The proliferating cells are in a drug-sensitive phase and quiescent cells are in a drug-resistant phase (see Figure 2.5). The growth dynamics are given by the system

$$\dot{x}_1(t) = (\gamma - \delta - \alpha - eu(t))\, x_1(t) + \beta x_2(t)$$
$$\dot{x}_2(t) = \alpha x_1(t) - (\lambda + \beta)\, x_2(t)$$

with initial data

$$x_1(0) = x_{1,0} \text{ and } x_2(0) = x_{2,0}.$$

Here, $x_1(\cdot)$ is the cell mass of proliferating cancer cells and $x_1(\cdot)$ is the cell mass of quiescent cells in the bone marrow. The parameters are all constant. Here γ is the cycling cells' growth rate, α is the transition rate from proliferating to resting, δ is the natural cell death rate, β is the transition rate from resting to proliferating, λ is cell differentiation (where mature bone marrow cells leave the bone marrow and enter the blood stream as various types of blood cells) and e is the effectiveness of the treatment. The function $u(\cdot)$ is the control describing the effects of the chemotherapeutic treatment which has impact only on the proliferating cells.

The control function is assumed to be $PWC(0,T)$ and satisfies $0 \leq u(t) \leq 1$. The cost function is defined by

$$\tilde{J}(u(\cdot)) = -\int_0^T \left\{ \frac{b}{2}(1 - u(s))^2 - q(x_1(t) + x_2(t)) \right\} ds$$

where b and q are weighting parameters and the goal is to maximize $\tilde{J}(\cdot)$ on the set

$$\Theta = \{u(\cdot) \in PWC(0,T) : u(t) \in [0,1]\}.$$

This cost function is selected so that one can give as much drug as possible while not excessively destroying the bone marrow. The weighting parameters b and q are selected depending on the importance of the terms. Notice the negative sign in front of the integral so that maximizing $\tilde{J}(\cdot)$ is equivalent to minimizing

$$J(u(\cdot)) = \int_0^T \left\{ \frac{b}{2}(1 - u(s))^2 - q(x_1(t) + x_2(t)) \right\} ds$$

on Θ.

We note that similar problems occur in the control of HIV (see [1] and [14]) and the models vary from application to application. In addition, the models of cell growth have become more complex and more realistic leading to more complex control problems. As indicated by the current references [14], [26], [62], [74], [117], [118], [146] and [165], this is a very active area of research.

2.1.6 Problem 6: Numerical Solutions of Boundary Value Problems

The finite element method is a powerful computational method for solving various differential equations. Variational theory and the calculus of variations provide the mathematical framework required to develop the method and to provide a rigorous numerical analysis of the convergence of the method. In order to describe method and to illustrate the key ideas, we start with a simple two-point boundary problem.

Given an integrable function $f(\cdot)$, find a twice differentiable function $x(\cdot)$ such that $x(\cdot)$ satisfies the differential equation

$$-\ddot{x}(t) + x(t) = f(t), \quad 0 < t < 1, \tag{2.4}$$

subject to the Dirichlet boundary conditions

$$x(0) = 0, \quad x(1) = 0. \tag{2.5}$$

Observe that it is assumed that $x(\cdot)$ has two derivatives since $\ddot{x}(\cdot)$ appears in the equation. The goal here is to find a numerical solution of this two-point boundary value problem by approximating $x(\cdot)$ with a continuous piecewise linear function $x_N(\cdot)$ as shown in Figure 2.6 below. The idea is to divide the interval $[0, 1]$ into $N + 1$ subintervals (called *elements*) with nodes $0 = \hat{t}_0 < \hat{t}_1 < \hat{t}_2 < \ldots < \hat{t}_{N-1} < \hat{t}_N < \hat{t}_{N+1} = 1$ and construct the

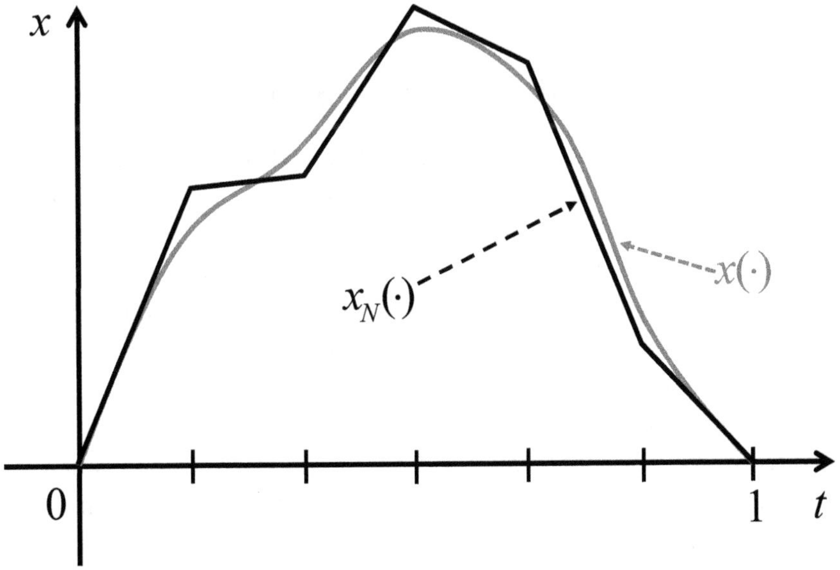

Figure 2.6: A Piecewise Linear Approximation

approximation $x_N(\cdot)$ so that it is continuous on all of $[0, 1]$ and linear between the nodes. It is clear that such a continuous piecewise linear approximating function $x_N(\cdot)$ may not be differentiable at the nodes and hence can not satisfy the differential equation (2.4). The finite element method is based on developing a weak (or variational) form of the two-point boundary value problem and using this formulation to devise a computational method for constructing the approximating solution $x_N(\cdot)$. We will work through this process in Section 2.4.5 below.

2.2 Mathematical Background

In order to provide a mathematical formulation of the above problems, we must first introduce some notation and discuss various classes of functions. At this point it is assumed that the reader is familiar with the basics of advanced calculus and elementary differential equations. However, we briefly review some topics from calculus, advanced calculus, and differential equations. Since this is a review of background material, we will not go into much detail on any topic.

2.2.1 A Short Review and Some Notation

In mathematical writings employing consistent and careful notation helps improve the understanding of the theory and reduces common errors in the application of the theory to problems. However, as is typical in mathematical texts it is sometimes useful to abuse notation as long as the precise meaning is made clear in the presentation. Every attempt is made to use precise notation and when notation is abused, we point out the abuse to keep the exposition as clear as possible. These observations are especially important for the notation used to describe functions since understanding functions is key to the development of the material in this book. In particular, it is essential to distinguish between a function, its name and the value of the function at a point in its domain.

Remark 2.1 *Let X and Y be two sets and assume that $F(\cdot)$ is a function with domain in X and range in Y. Observe that we use the symbol $F(\cdot)$ to denote the function rather than using the "**name**" of the function F. For example, if x is a real number and one writes $F(x) = x^2$, then the corresponding function $F(\cdot)$ has domain and range contained in the set of real numbers and for each x in the domain of $F(\cdot)$, the value of $F(\cdot)$ at x is given by $F(x) = x^2$. Thus, we distinguish between the **function** $F(\cdot)$, its **name** F and its **value at a specific point** $F(x)$. At first glance these distinctions may seem like "overkill". However, using the same symbol for a function and its name is often a major source of confusion for students and leads to a lack of understanding of the theory and its application.*

Let $F(\cdot)$ be a function with domain $\mathcal{D} \subseteq X$ and range $\mathcal{R} \subseteq X$. To emphasize the relationship between the function $F(\cdot)$, its domain and its range we write $\mathcal{D}(F) = \mathcal{D}$ and $\mathcal{R}(F) = \mathcal{R}$. This notation merely states the obvious that the **names of the domain and range** of a function $F(\cdot)$ should be attached to the name F of the function $F(\cdot)$. We denote this by writing

$$F : \mathcal{D}(F) \subseteq X \to Y.$$

If the domain of $F(\cdot)$ is all of X, i.e. if $\mathcal{D}(F) = X$, then we write

$$F : X \to Y.$$

For completeness, recall that the range of a function $F(\cdot)$ is given by

$$\mathcal{R}(F) = \{y \in Y \mid y = F(x), x \in \mathcal{D}(F)\}.$$

In elementary calculus the sets X and Y are often intervals of the form $[a, b]$, $(a, b]$, $[a, b)$ or (a, b). In advanced calculus X and Y may be n-dimensional Euclidean spaces. In the calculus of variations and optimal control X and Y are most often spaces of functions. The following notation is rather standard and will be used throughout the book.

- The space of real numbers is denoted by $\mathbb{R}$ and the space of complex numbers by $\mathbb{C}$, respectively. However, in some cases we may use the notation $\mathbb{R}^1$ to emphasize the fact that the real line is also the state space for a 1-dimensional system.

- Real n-dimensional Euclidean space, denoted by $\mathbb{R}^n$, is the space of n-dimensional real column vectors

$$\mathbb{R}^n = \left\{ \boldsymbol{x} = \begin{bmatrix} x_1 \\ x_2 \\ \vdots \\ x_n \end{bmatrix} : x_i \in \mathbb{R} \right\}.$$

Likewise, complex n-dimensional Euclidean space, denoted by $\mathbb{C}^n$, is the space of n-dimensional complex column vectors

$$\mathbb{C}^n = \left\{ \boldsymbol{z} = \begin{bmatrix} z_1 \\ z_2 \\ \vdots \\ z_n \end{bmatrix} : z_i \in \mathbb{C} \right\}.$$

Remark 2.2 *Note that we use boldface letters for vectors so that if one writes $\boldsymbol{x} \in \mathbb{R}^n$, then one knows that $\boldsymbol{x}$ is a vector with n real components. Also, in special cases we will abuse notation and use the symbols $\mathbb{R}^1$ for $\mathbb{R}$ and $\mathbb{C}^1$ for $\mathbb{C}$, respectively.*

- If $x \in \mathbb{R}$, then the absolute value of x is denoted by $|x|$ and if $z = x + iy \in \mathbb{C}$, then the complex modulus is denoted by $|z| = \sqrt{x^2 + y^2}$. The complex conjugate of $z = x + iy$ is $\bar{z} = x - iy$.

- The transpose of a column vector $\begin{bmatrix} z_1 \\ z_2 \\ \vdots \\ z_n \end{bmatrix}$ is a row vector

given by

$$\begin{bmatrix} z_1 \\ z_2 \\ \vdots \\ z_n \end{bmatrix}^T \triangleq \begin{bmatrix} z_1 & z_2 & \cdots & z_n \end{bmatrix}$$

and conversely,

$$\begin{bmatrix} z_1 & z_2 & \cdots & z_n \end{bmatrix}^T = \begin{bmatrix} z_1 \\ z_2 \\ \vdots \\ z_n \end{bmatrix}.$$

- If $\boldsymbol{x} = \begin{bmatrix} x_1 & x_2 & \cdots & x_n \end{bmatrix}^T \in \mathbb{R}^n$, then the *Euclidean (or 2) norm* of $\boldsymbol{x}$ is given by

$$\|\boldsymbol{x}\| = \|\boldsymbol{x}\|_2 = \sqrt{\sum_{i=1}^{n} |x_i|^2}$$

and likewise, if $\boldsymbol{z} = \begin{bmatrix} z_1 & z_2 & \cdots & z_n \end{bmatrix}^T \in \mathbb{C}^n$, then *the Euclidean (or 2) norm* of $\boldsymbol{z}$ is given by

$$\|\boldsymbol{z}\| = \|\boldsymbol{z}\|_2 = \sqrt{\sum_{i=1}^{n} |z_i|^2}.$$

- If $\boldsymbol{x} = \begin{bmatrix} x_1 & x_2 & \cdots & x_n \end{bmatrix}^T \in \mathbb{R}^n$ and $1 \leq p < +\infty$, then the *p-norm* of $\boldsymbol{x}$ is given by

$$\|\boldsymbol{x}\|_p = \sqrt[p]{\sum_{i=1}^{n} |x_i|^p} = \left[\sum_{i=1}^{n} |x_i|^p\right]^{1/p}$$

and likewise, if $\boldsymbol{z} = \begin{bmatrix} z_1 & z_2 & \cdots & z_n \end{bmatrix}^T \in \mathbb{C}^n$, then the *p-norm* of $\boldsymbol{z}$ is given by

$$\|\boldsymbol{z}\|_p = \sqrt[p]{\sum_{i=1}^{n} |z_i|^p} = \left[\sum_{i=1}^{n} |z_i|^p\right]^{1/p}.$$

The norms above are special cases of the general concept of a norm which measures the magnitude of a vector and provides a mechanism to define distance between vectors. In general, we have the following definition.

Definition 2.1 *A* ***norm*** *on* $\mathbb{R}^n$ *(or* $\mathbb{C}^n$*) is a function* $\|\cdot\| : \mathbb{R}^n$ *(or* $\mathbb{C}^n$*)* $\to \mathbb{R}$ *satisfying:*

1. $\|\boldsymbol{x}\| \geq 0$ and $\|\boldsymbol{x}\| = 0$ if and only if $\boldsymbol{x} = \boldsymbol{0}$.
2. $\|\alpha \cdot \boldsymbol{x}\| = |\alpha| \cdot \|\boldsymbol{x}\|$, $\alpha \in \mathbb{R}$ (or $\mathbb{C}$) and $\boldsymbol{x} \in \mathbb{R}^n$ (or $\mathbb{C}^n$).
3. $\|\boldsymbol{x} + \boldsymbol{y}\| \leq \|\boldsymbol{x}\| + \|\boldsymbol{y}\|$, for $\boldsymbol{x}, \boldsymbol{y} \in \mathbb{R}^n$ (or $\mathbb{C}^n$) (the triangle inequality).

Let $\|\cdot\| : \mathbb{R}^n \to \mathbb{R}$ be a norm on $\mathbb{R}^n$. If $\hat{\boldsymbol{x}} \in \mathbb{R}^n$ and $\delta > 0$, then the (open) δ-neighborhood of $\hat{\boldsymbol{x}}$ is the open ball of radius δ centered at $\hat{\boldsymbol{x}}$ given by

$$U(\hat{\boldsymbol{x}}, \delta) = \{\boldsymbol{x} \in \mathbb{R}^n \colon \|\boldsymbol{x} - \hat{\boldsymbol{x}}\| < \delta\}.$$

On the real line, $U(\hat{x}, \delta)$ is the open interval centered at $\hat{x}$ with radius δ. We sometimes abbreviate the δ -neighborhood by δ-nbd.

If $f : \mathcal{D}(f) \subseteq \mathbb{R}^n \longrightarrow \mathbb{R}$ is a real valued function of n real variables and $\boldsymbol{x} \in \mathcal{D}(f)$, then we often write $f(\boldsymbol{x}) = f(x_1, x_2, ..., x_n)$ rather than

$$f(\boldsymbol{x}) = f\left(\begin{bmatrix} x_1 \\ x_2 \\ \vdots \\ x_n \end{bmatrix}\right) = f\left(\begin{bmatrix} x_1 & x_2 & \cdots & x_n \end{bmatrix}^T\right).$$

Remark 2.3 *The use of* $f(\boldsymbol{x}) = f(x_1, x_2, ..., x_n)$ *rather than* $f(\boldsymbol{x}) = f([x_1, x_2, ..., x_n]^T)$ *is standard "abuse of notation" and should cause little misunderstanding in the material.*

We shall use various notations for partial derivatives. For example, we use subscripts for partial derivatives such as

$$f_{x_i}(\boldsymbol{x}), = f_{x_i}(x_1, x_2, ..., x_n) = \frac{\partial f(x_1, x_2, ..., x_n)}{\partial x_i} = \frac{\partial f(\boldsymbol{x})}{\partial x_i}$$

and

$$f_{xy}(x, y) = \frac{\partial^2 f(x, y)}{\partial x \partial y}.$$

If $\boldsymbol{x} : I \to \mathbb{R}^n$ is a vector-valued function with domain defined by an interval $I \subset \mathbb{R}$, then it follows that there exist n real valued functions $x_i(\cdot)$, $i = 1, 2, \ldots, n$, such that

$$\boldsymbol{x}(t) = \begin{bmatrix} x_1(t) \\ x_2(t) \\ \vdots \\ x_n(t) \end{bmatrix} = \begin{bmatrix} x_1(t) & x_2(t) & \cdots & x_n(t) \end{bmatrix}^T .$$

We use the standard definitions of continuity and differentiability for real valued functions. Assume $x : [t_0, t_1] \longrightarrow \mathbb{R}$ is a real valued function. If $t = \hat{t} \in [t_0, t_1)$, then the right-hand limit of $x(\cdot)$ at $t = \hat{t}$ is defined by

$$x\left(\hat{t}^+\right) = \lim_{t \to \hat{t}+} [x(t)],$$

provided that this limit exist (it could be infinite). Likewise, if $t = \hat{t} \in (t_0, t_1]$, then the left-hand limit of $x(\cdot)$ at $t = \hat{t}$ is defined by

$$x\left(\hat{t}^-\right) = \lim_{t \to \hat{t}^-} [x(t)],$$

provided that this limit exist (it could be infinite).

If $x : [t_0, t_1] \longrightarrow \mathbb{R}$ is a real valued function which is differentiable at $t = \hat{t}$, then $\dot{x}(\hat{t})$, $x'(\hat{t})$, and $\frac{dx(\hat{t})}{dt}$ all denote the derivative of $x(\cdot)$ at $t = \hat{t}$. If $t = t_0$, then $\dot{x}(t_0)$ denotes the right-hand derivative $\dot{x}^+(t_0)$ defined by

$$\frac{d^+x(t_0)}{dt} = \dot{x}^+(t_0) = \lim_{t \to t_0^+} \left[\frac{x(t) - x(t_0)}{t - t_0}\right]$$

and if $t = t_1$, then $\dot{x}(t_1)$ denotes the left-hand derivative $\dot{x}^-(t_0)$ defined by

$$\frac{d^-x(t_1)}{dt} = \dot{x}^-(t_1) = \lim_{t \to t_1^-} \left[\frac{x(t) - x(t_1)}{t - t_1}\right],$$

provided these limits exist and are finite.

Remark 2.4 *It is important to note that even if* $\frac{d^+x(\hat{t})}{dt}$ *and* $\dot{x}\left(\hat{t}^+\right)$ *both exist at a point* $\hat{t}$, *they may not be the same so that in general*

$$\dot{x}^+(\hat{t}) = \frac{d^+x(\hat{t})}{dt} \neq \dot{x}\left(\hat{t}^+\right).$$

Clearly, the same is true for the left-hand derivatives and limits. In words, the one-sided derivative of a function at a point is not the one-sided limit of the derivative at that point.

Example 2.1 *Let* $x : [-1, 1] \longrightarrow \mathbb{R}$, *be defined by*

$$x(t) = \begin{cases} t^2 \sin(1/t), & t > 0, \\ 0 & t = 0, \\ t^2 & t < 0. \end{cases}$$

Computing the right-hand derivative at $\hat{t} = 0$, *it follows that*

$$\begin{aligned} \frac{d^+}{dt}x(0) &= \dot{x}^+(0) = \lim_{t\to 0+}\left[\frac{x(t) - x(0)}{t}\right] = \lim_{t\to 0+}\left[\frac{t^2\sin(1/t) - 0}{t}\right] \\ &= \lim_{t\to 0+}\left[t\sin(1/t)\right] = 0 \end{aligned}$$

exists and is finite. On the other hand, if $0 < t \leq 1$, *then* $\dot{x}(t)$ *exists and*

$$\dot{x}(t) = 2t\sin(1/t) - \cos(1/t)$$

and

$$\lim_{t\to 0+}\left[2t\sin(1/t) - \cos(1/t)\right]$$

does not exist. Hence,

$$\frac{d^+}{dt}x(0) = \dot{x}^+(0) \neq \dot{x}(0^+).$$

However, it is true that

$$\frac{d^-}{dt}x(0) = \dot{x}^-(0) \neq \dot{x}(0^-) = 0.$$

Although, as the previous example illustrates, one can not interchange the limit process for one-sided derivatives, there are important classes of functions for which this is true. Consider the following example.

Example 2.2 *If* $x(t) = |t|$, *then*

$$\frac{d^+x(0)}{dt} = \lim_{t \to 0^+} \frac{x(t) - x(0)}{t - 0} = \lim_{t \to 0^+} \frac{|t|}{t} = 1$$

and

$$\frac{d^-x(0)}{dt} = \lim_{t \to 0^-} \frac{x(t) - x(0)}{t - 0} = \lim_{t \to 0^-} \frac{|t|}{t} = \lim_{t \to 0^-} \frac{-t}{t} = -1.$$

Note also that

$$\frac{d^+x(0)}{dt} = \lim_{t \to 0^+} \dot{x}(t) = 1$$

and

$$\frac{d^-x(0)}{dt} = \lim_{t \to 0^-} \dot{x}(t) = -1$$

are the left and right-hand limits of $\dot{x}(\cdot)$ *at* 0. *Note that* $x(t) = |t|$ *is differentiable at all points except* $t = 0$.

Definition 2.2 *Let* I *denote an interval and assume that* $x : I \subseteq \mathbb{R}^1 \longrightarrow \mathbb{R}^1$ *is a real valued function. For a given integer* $k \geq 1$, *we say that* $x(\cdot)$ *is* C^k *on* I *if* $x(\cdot)$ *and all its derivatives of order* k *exist and are continuous at all points* $t \in I$. *We say that* $x(\cdot)$ *is* C^0 *on* I *if* $x(\cdot)$ *is continuous and* $x(\cdot)$ *is* C^∞ *on* I *if* $x(\cdot)$ *is* C^k *for all* $k \geq 0$. *If* $x(\cdot)$ *is* C^∞ *on* I, *then we call* $x(\cdot)$ *a* ***smooth function****. Thus, we define the function spaces* $C^k(I) = C^k(I;\mathbb{R}^1)$ *by*

$$C^k(I) = C^k(I;\mathbb{R}^1) = \left\{x : I \subseteq \mathbb{R}^1 \longrightarrow \mathbb{R}^1 : x(\cdot) \ \text{ is } C^k \text{ on } I\right\}.$$

Definition 2.3 *Let* I *denote an interval and assume that* $\boldsymbol{x} : I \to \mathbb{R}^n$ *is a vector-valued function. We say that the function* $\boldsymbol{x}(\cdot) = \left[\begin{array}{cccc} x_1(\cdot) & x_2(\cdot) & \cdots & x_n(\cdot) \end{array}\right]^T$ *is* ***continuous at*** $\hat{\mathbf{t}}$ *if, for each* $\epsilon > 0$, *there is a* $\delta > 0$ *such that if* $t \in I$ *and* $0 < |t - \hat{t}| < \delta$, *then*

$$\left\|\boldsymbol{x}(t) - \boldsymbol{x}(\hat{t})\right\| < \epsilon.$$

The function $\boldsymbol{x}(\cdot)$ *is said to be a* ***continuous function*** *if it is continuous at every point in its domain* I.

Definition 2.4 *A function* $\boldsymbol{x} : I \to \mathbb{R}^n$ *is* ***differentiable at*** $\hat{\boldsymbol{t}}$ *if for each* $i = 1, 2, \ldots, n$, *the scalar function* $x_i(\cdot)$ *is differentiable at* $\hat{t}$ *and we define* $\dot{\boldsymbol{x}}(\hat{t})$ *by*

$$\dot{\boldsymbol{x}}(\hat{t}) \triangleq \begin{bmatrix} \dot{x}_1(\hat{t}) & \dot{x}_2(\hat{t}) & \cdots & \dot{x}_n(\hat{t}) \end{bmatrix}^T.$$

The right-hand and left-hand derivatives at $t = t_0$ *and* $t = t_1$ *are defined as above. A function* $\boldsymbol{x}(\cdot)$ *is said to be a* ***differentiable function*** *if it is differentiable at every point in its domain* I.

Remark 2.5 *Observe that the definitions given above imply that a vector-valued function* $\boldsymbol{x} : I \to \mathbb{R}^n$ *is continuous at a point* $\hat{t}$ *if and only if all of the component functions* $x_i : I \to \mathbb{R}^1$, $i = 1, 2, ..., n$, *are continuous at* $\hat{t}$. *Likewise, the vector-valued function* $\boldsymbol{x} : I \to \mathbb{R}^n$ *is differentiable at the point* $\hat{t}$ *if and only if all of the component functions* $x_i : I \to \mathbb{R}^1$, $i = 1, 2, ..., n$, *are differentiable at* $\hat{t}$.

Definition 2.5 *Let* I *denote an interval and assume that* $\boldsymbol{x} : I \subseteq \mathbb{R}^1 \longrightarrow \mathbb{R}^n$ *is a vector valued function. For a given integer* $k \geq 1$, *we say that* $x(\cdot)$ *is* C^k *on* I *if* $x_i(\cdot)$ *is* C^k *on* I *for all* $i = 1, 2, ..., n$. *We define the function spaces* $C^k(I) = C^k(I; \mathbb{R}^n)$ *by*

$$C^k(I) = C^k(I; \mathbb{R}^n) = \left\{\boldsymbol{x} : I \subseteq \mathbb{R}^1 \longrightarrow \mathbb{R}^n : \boldsymbol{x}(\cdot) \text{ is } C^k \text{ on } I\right\}.$$

Note that we use the same notation $C^k(I)$ *for* $C^k(I) = C^k(I; \mathbb{R}^1)$ *and* $C^k(I) = C^k(I; \mathbb{R}^n)$. *This should cause no confusion since a statement like* $\boldsymbol{x}(\cdot) \in C^k(I)$ *clearly implies that* $C^k(I) = C^k(I; \mathbb{R}^n)$ *because* $\boldsymbol{x}(\cdot)$ *is boldfaced, and hence, a vector valued function.*

Remark 2.6 *Although the above definition of a derivative is sufficient for the initial introduction here, this definition must be revisited when we move to more general functions. In particular, the concept of a "derivative" is best presented in terms of linear approximations of a non-linear function.*

2.2.2 A Review of One Dimensional Optimization

Here we consider the simple optimization problem that one "solves" in a first calculus course. Let $f : I \to \mathbb{R}$ be a differentiable function on the interval I (I may be open, closed, infinite, or of the form $(a, b]$, *etc.*). We are interested in finding minimizing points for $f(\cdot)$. We remind the reader of the following definitions.

Definition 2.6 *Given $f : I \to \mathbb{R}$, we say that x^* provides a* ***local minimum*** *for $f(\cdot)$ on I (or x^* is a* ***local minimizer*** *for $f(\cdot)$ on I) if*

1. *$x^* \in I$ and*

2. *there is a $\delta > 0$ such that*
$$f(x^*) \leq f(x)$$
for all $x \in U(x^*, \delta) \cap I$.

If in addition,
$$f(x^*) < f(x)$$
for all $x \in U(x^*, \delta) \cap I$ with $x \neq x^*$, then we say that x^* provides a ***proper local minimum*** *for $f(\cdot)$ on I. If $x^* \in I$ is such that*
$$f(x^*) \leq f(x)$$
for all $x \in I$, then x^* is said to provide a ***global minimum*** *for $f(\cdot)$ on I.*

In theory, there is always one way to "find" global minimizers. Simply pick $x_1 \in I$ and "test" x_1 to see if $f(x_1) \leq f(x)$ for all $x \in I$. If so, then x_1 is a global minimizer, and if not, there is a x_2 with $f(x_2) < f(x_1)$. Now test x_2 to see if $f(x_2) \leq f(x)$ for all $x \in I$, $x \neq x_1$. If so, x_2 is the minimizer, and if not, we can find x_3 so that $f(x_3) < f(x_2) < f(x_1)$. Continuing this process generates sequence of points $\{x_k : k = 1, 2, 3, \ldots\}$ satisfying
$$\ldots f(x_{k+1}) < f(x_k) < \ldots < f(x_3) < f(x_2) < f(x_1).$$

Such a sequence is called a minimizing sequence and under certain circumstances the sequence (or a subsequence) will converge to a point x^* and if $x^* \in I$ it will be a minimizer. The above "direct procedure" may be impossible to accomplish in finite time. Therefore, we need some way to reduce the number of possible candidates for x^*. We can do this by applying elementary necessary conditions from calculus.

Clearly, if x^* is a global minimizer, then x^* is a local minimizer. We shall try to find all the local minima of $f(\cdot)$ by using standard necessary conditions. The following theorem can be found in almost any calculus book.

Theorem 2.1 *Suppose that $f : I \to R$ is continuous, where I is an interval with endpoints $a < b$ and assume that $x^* \in I$ is a local minimizer for $f(\cdot)$ on I. If $\frac{df(x^*)}{dx}$ exists, then*

$$\frac{df\left(x^*\right)}{dx} = 0, \ \ if \ a < x_* < b, \tag{2.6}$$

or

$$\frac{df\left(x^*\right)}{dx} \geq 0, \ \ if \ x^* = a, \tag{2.7}$$

or

$$\frac{df\left(x^*\right)}{dx} \leq 0, \ \ if \, x^* = b. \tag{2.8}$$

In addition, if $\frac{d^2f(x^*)}{dx^2} = f''(x^*)$ exists, then

$$f''(x^*) \geq 0. \tag{2.9}$$

It is important to emphasize that **Theorem** 2.1 is only a necessary condition. The theorem says that if x^* is a local minimizer, then x^* must satisfy (2.6), (2.7) or (2.8). It does not imply that points that satisfy (2.6)-(2.8) are local minimizers. Necessary conditions like **Theorem** 2.1 can be used to reduce the number of possible candidates that must be tested to see if they provide a minimum to $f(\cdot)$. Consider the following example that illustrates the above theorem.

Example 2.3 *Let* $I = [-2, 3)$ *and define* $f(x) = x^4 - 6x^2$. *Let us try to find the local (maximum and) minimum. In order to apply* ***Theorem*** *2.1 we compute the derivatives*

$$f'(x) = 4x^3 - 12x$$

and

$$f''(x) = 12x^2 - 12.$$

If x^* *is a local minimum and* $-2 < x^* < 3$, *then*

$$f'(x^*) = 4[x^*]^3 - 12[x^*] = 4x^*\left([x^*]^2 - 3\right) = 0.$$

Solving this equation, we find that x^* *could be either* $0, \sqrt{3}, -\sqrt{3}$. *Checking (2.9) at* $x^* = 0$, *yields*

$$f''(0) = 12 \cdot 0 - 12 = -12 < 0$$

so that condition (2.9) fails and $x^* = 0$ *is not a local minimizer. Also, checking condition (2.7) at* $x^* = -2$, *yields*

$$f'(-2) = -32 + 24 = -8 < 0$$

so, $x^* = -2$ *is not a candidate. Thus, an application of* ***Theorem*** *2.1 has reduced the process to that of "testing" the two points* $\sqrt{3}$ *and*$-\sqrt{3}$. *The second derivative test (2.9) yields*

$$f''(\pm\sqrt{3}) = 12[\pm\sqrt{3}]^2 - 12 = 12(3 - 1) = 24 > 0$$

so that $x^* = -\sqrt{3}$, *and* $x^* = +\sqrt{3}$ *are possible local minimum. This is clear from the graph, but what one really needs are sufficient conditions.*

Theorem 2.2 *Suppose that* $f(\cdot)$ *is as in* ***Theorem 2.1****. If* $x^* \in (a, b)$, $f'(x^*) = 0$ *and* $f''(x^*) > 0$, *then* x^* *provides a local minimum for* $f(\cdot)$ *on* $[a, b]$.

We close this section with a proof of **Theorem** 2.1. The proof is very simple, but the idea behind the proof happens to be one of the key ideas in much of optimization.

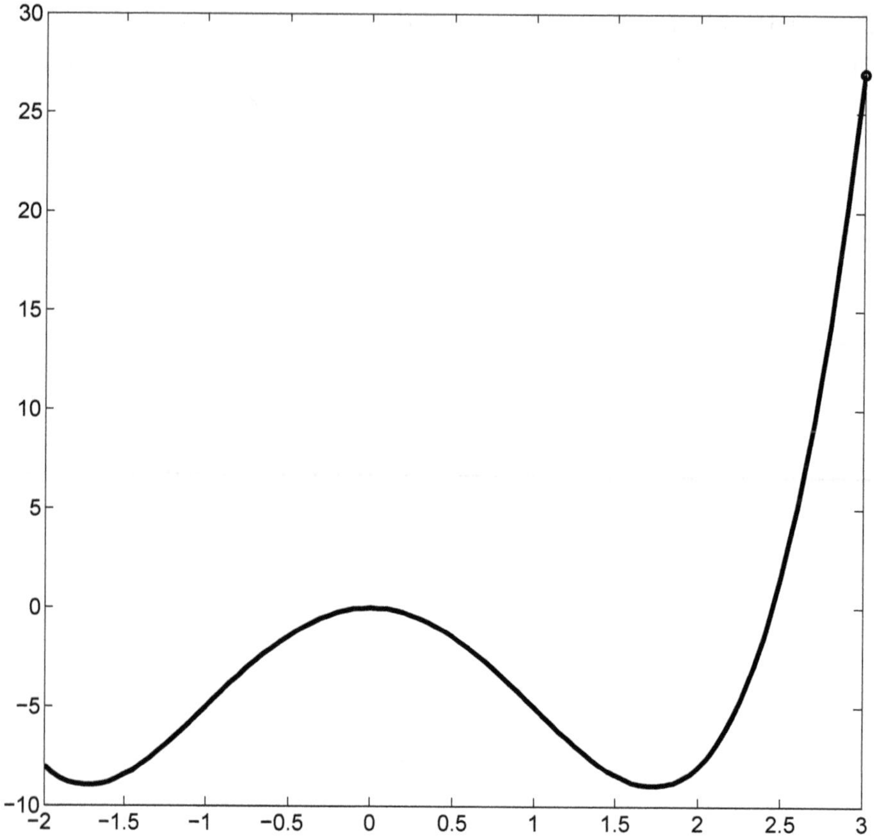

Figure 2.7: Plot of the Cost Function

Proof of Theorem 2.1: First assume that $a < x^* < b$, is a local minimizer. This means that there is a $\delta > 0$ such that

$$f(x^*) \le f(x)$$

for all $x \in U(x^*, \delta) \cap I$. Let $\delta_1 = \frac{1}{2}\min\{\delta, b - x^*, x^* - a\} > 0$ and note that the open interval $U(x^*, \delta_1)$ is contained in I. In particular, if $-\delta_1 < \varepsilon < \delta_1$, then the "variation" $x^* + \varepsilon \in U(x^*, \delta_1) \subset U(x^*, \delta) \cap I$. Hence x^* satisfies

$$f(x^*) \le f(x^* + \varepsilon),$$

or equivalently,

$$0 \le f(x^* + \varepsilon) - f(x^*). \tag{2.10}$$

Observe that this inequality holds for positive and negative ε. Dividing both sides of (2.10) by $\varepsilon > 0$, yields the inequality

$$0 \le \frac{f(x^* + \varepsilon) - f(x^*)}{\varepsilon},$$

and passing to the limit as $\varepsilon \to 0^+$ it follows that

$$0 \le f'(x^*). \tag{2.11}$$

Likewise, dividing both sides of (2.10) by $\varepsilon < 0$, yields the reverse inequality

$$0 \ge \frac{f(x^* + \varepsilon) - f(x^*)}{\varepsilon},$$

and passing to the limit as $\varepsilon \to 0^-$ it follows that

$$0 \ge f'(x^*). \tag{2.12}$$

Combining (2.11) with (2.12) it follows that

$$f'(x^*) = 0$$

and we have established (2.6).

Consider now the case where $x^* = a$. Let $\delta_1 = \frac{1}{2}\min\{\delta, b - x^*\} > 0$ and note that if $0 < \varepsilon < \delta_1$, then the "variation" $x^* + \varepsilon = a + \varepsilon \in U(x^*, \delta_1) \cap I \subset U(a, \delta) \cap I$. Hence $x^* = a$ satisfies

$$f(a) \le f(a + \varepsilon),$$

or equivalently,

$$0 \le f(a + \varepsilon) - f(a).$$

Observe that this inequality holds for $0 < \varepsilon < \delta_1$. Dividing both sides of this inequality by $\varepsilon > 0$, yields

$$0 \le \frac{f(a + \varepsilon) - f(a)}{\varepsilon},$$

and passing to the limit as $\varepsilon \to 0^+$ it follows that

$$0 \le f'(a).$$

This completes the proof of (2.7). The case $x^* = b$ is completely analogous. □

Remark 2.7 *The important idea in the above proof is that when* $a < x^* < b$*, the variations* $x^* + \varepsilon$ *belong to* $U(x^*, \delta) \cap I$ *for both positive and negative values of* ε*. In particular, one can "approach"* x^* *from both directions and still be inside* $U(x^*, \delta) \cap I$*. However, when* $x^* = a$*, the variations* $x^* + \varepsilon$ *belong to* $U(x^*, \delta) \cap I$ *only when* ε *is positive. Thus, one can "approach"* x^* *only from the right and still remain in* $U(x^*, \delta) \cap I$*. This simple observation is central to much of what we do in deriving necessary conditions for the simplest problem in the calculus of variations.*

Remark 2.8 *It is extremely important to understand that care must be exercised when applying necessary conditions. A necessary condition usually assumes the existence of of an optimizer. If an optimizer does not exist, then the necessary condition is vacuous. Even worse, one can draw the incorrect conclusion by applying the necessary condition. Perron's Paradox (see [186]) provides a very simple example to illustrate the danger of applying necessary conditions to a problem with no solution. Let*

$$\Phi = \{N : N \textit{is a positive integer}\} \tag{2.13}$$

and define

$$J(N) = N. \tag{2.14}$$

Assume that $\hat{N} \in \Phi$ *maximizes* $J(\cdot)$ *on the set* Φ*, i.e. that* $\hat{N}$ *is the largest positive integer. Thus,* $\hat{N} \geq 1$ *which implies that* $\hat{N}^2 \geq \hat{N}$*. However, since* $\hat{N}^2$ *is a positive integer and* $\hat{N}$ *is the largest positive integer, it follows that* $\hat{N} \geq \hat{N}^2$*. Consequently,* $\hat{N}^2 \leq \hat{N} \leq \hat{N}^2$ *which implies that* $\hat{N} \leq 1 \leq \hat{N}$ *so that* $\hat{N} = 1$*. Therefore, if one assumes that the optimization problem (2.13) - (2.14) has a solution, then one can (correctly) prove that the largest positive integer is* $\hat{N} = 1$*. Of course the issue is a point of logic, where a false assumption can be used to prove a false conclusion. If one assumes an optimizer exists and it does not, then necessary conditions can be used to produce incorrect answers. Unlike the simple Perron Paradox, it is often difficult to establish the existence of solutions to calculus of variation and optimal control problems.* ***One should take this remark as a warning when applying necessary conditions to such problems.***

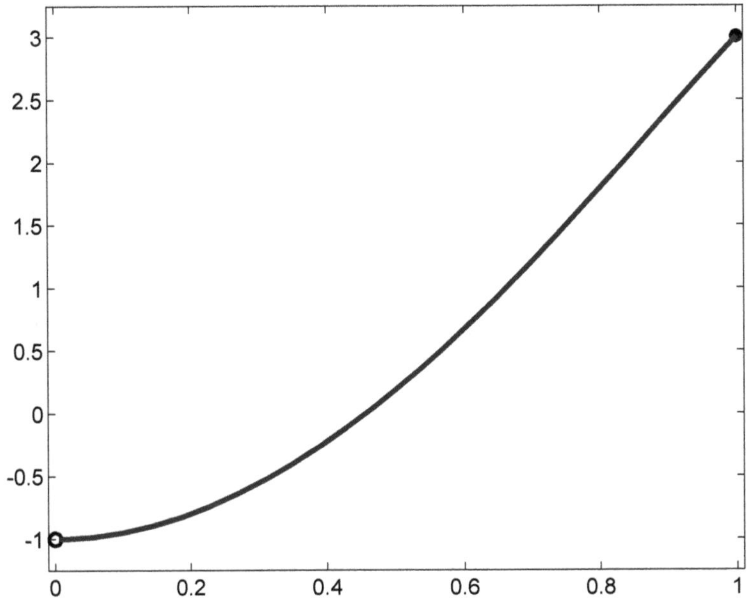

Figure 2.8: Plot of the Cost Function $f(\cdot)$ on $(0,1]$

Example 2.4 *Let* $I = (0,1]$ *and define* $f(x) = 5x^2 - x^4 - 1$. *The plot of* $f(\cdot)$ *on the interval* $(0,1]$ *is shown in Figure 2.8. It is also clear that* $f(\cdot)$ *does not have a global nor local minimizer on the* $(0,1]$ *since* $0 \notin (0,1]$. *However, if one "relaxes" (expands) the problem and considers the problem of minimizing* $f(\cdot)$ *on the closure of* I *given by* $\bar{I} = [0,1]$, *then* $\bar{x}^* = 0 \in (0,1]$ *solves the "relaxed problem". In addition, there exist a sequence of points* $x_k \in I = (0,1]$ *satisfying*

$$x_k \longrightarrow \bar{x}^* = 0$$

and

$$f(x_k) \longrightarrow f(\bar{x}^*) = -1.$$

This process of relaxing the optimization problem by expanding the set of admissible points is an important idea and leads to the concepts of generalized curves and relaxed controllers. The key point is that by relaxing the problem one obtains existence of a minimizer in a larger set. But equally important, it can be shown that the solution to the relaxed problem can be approximated by points in the original constraint set.

This general idea also has profound applications to the theory of nonlinear partial differential equations. In particular, a "viscosity solution" is an extension of the classical concept of what is meant by a "solution" to certain partial differential equations. The basic idea is to define a viscosity solution as a limit of (classical) solutions that are parameterized by parameter that tends to zero. This method produces existence of (weak) solution and these solutions are close related to the concept of Young's "generalized curves". The references [24], [61], [60], [75] and [186] provide an introduction to these ideas.

2.2.3 Lagrange Multiplier Theorems

The constraint set for the general optimization problem is often defined in terms of equality (or inequality) constraints. The isoperimetric problem in the calculus of variations is such a problem. In this section assume there are two vector spaces $\boldsymbol{Z}$ and $\boldsymbol{Y}$ and two functions

$$\boldsymbol{J} : \boldsymbol{D}(\boldsymbol{J}) \subseteq \boldsymbol{Z} \longrightarrow \mathbb{R}^1 \tag{2.15}$$

and

$$\boldsymbol{G} : \boldsymbol{D}(\boldsymbol{G}) \subseteq \boldsymbol{Z} \longrightarrow \boldsymbol{Y}. \tag{2.16}$$

The function $\boldsymbol{J}(\cdot)$ is called the *cost function* and $\boldsymbol{G}(\cdot)$ is called the *constraint function.* Define the constraint set

$$\boldsymbol{\Theta}_G \subseteq \boldsymbol{Z}$$

by

$$\boldsymbol{\Theta}_G = \{\boldsymbol{z} \in \boldsymbol{D}(\boldsymbol{G}) : \boldsymbol{G}(\boldsymbol{z}) = \boldsymbol{0} \in \boldsymbol{Y}\} \subset \boldsymbol{D}(\boldsymbol{G}). \tag{2.17}$$

The *Equality Constrained Optimization Problem* is defined to be:

Find an element $\boldsymbol{z}^* \in \boldsymbol{\Theta}_G \cap \boldsymbol{D}(\boldsymbol{J})$ such that

$$\boldsymbol{J}(\boldsymbol{z}^*) \leq \boldsymbol{J}(\boldsymbol{z})$$

for all $\boldsymbol{z} \in \boldsymbol{\Theta}_G \cap \boldsymbol{D}(\boldsymbol{J})$.

Observe that since $\boldsymbol{\Theta}_G \subset \boldsymbol{D}(\boldsymbol{G})$, it follows that $\boldsymbol{\Theta}_G \cap \boldsymbol{D}(\boldsymbol{J}) \subset \boldsymbol{D}(\boldsymbol{G}) \cap \boldsymbol{D}(\boldsymbol{J})$. Therefore, the equality constrained optimization problem is equivalent to finding $\boldsymbol{z}^* \in \boldsymbol{D}(\boldsymbol{G}) \cap \boldsymbol{D}(\boldsymbol{J})$ such that $\boldsymbol{z}^*$ minimizes $\boldsymbol{J}(\boldsymbol{z})$ subject to $\boldsymbol{G}(\boldsymbol{z}) = \boldsymbol{0} \in \boldsymbol{Y}$. We first discuss special cases and then move to the more abstract versions.

Lagrange Multiplier Theorem in $\mathbb{R}^n$

We consider the finite dimensional constrained optimization problem in n variables and m equality constraints. For the sake of simplicity of presentation, we assume that the cost function

$$\boldsymbol{J} : \mathbb{R}^n \longrightarrow \mathbb{R}^1$$

has domain equal to all $\mathbb{R}^n$, i.e. $\boldsymbol{D}(\boldsymbol{J}) = \mathbb{R}^n$. Also, we assume that the constraint function

$$\boldsymbol{G} : \mathbb{R}^n \longrightarrow \mathbb{R}^m$$

has domain equal to all $\mathbb{R}^n$, i.e. $\boldsymbol{D}(\boldsymbol{G}) = \mathbb{R}^n$ and $m < n$. In particular, there are m real-valued functions

$$\boldsymbol{g}_i : \mathbb{R}^n \longrightarrow \mathbb{R}^1, i = 1, 2, \ldots, m,$$

such that

$$\boldsymbol{G}(\boldsymbol{z}) = \begin{bmatrix} \boldsymbol{g}_1(\boldsymbol{z}) \\ \boldsymbol{g}_2(\boldsymbol{z}) \\ \vdots \\ \boldsymbol{g}_m(\boldsymbol{z}) \end{bmatrix},$$

where $\boldsymbol{z} = \begin{bmatrix} x_1 & x_2 & \cdots & x_n \end{bmatrix}^T \in \mathbb{R}^n$. We assume that all the functions $g_i(\cdot)$, $i = 1, 2, ..., n$, are C^1 real valued functions of the n real variables $\boldsymbol{z} = \begin{bmatrix} x_1 & x_2 & \cdots & x_n \end{bmatrix}^T \in \mathbb{R}^n$ so that the gradients

$$\nabla \boldsymbol{J}(\boldsymbol{z}) = \begin{bmatrix} \frac{\partial \boldsymbol{J}(\boldsymbol{z})}{\partial x_1} \\ \frac{\partial \boldsymbol{J}(\boldsymbol{z})}{\partial x_2} \\ \vdots \\ \frac{\partial \boldsymbol{J}(\boldsymbol{z})}{\partial x_n} \end{bmatrix} \quad \text{and} \quad \nabla \boldsymbol{g}_i(\boldsymbol{z}) = \begin{bmatrix} \frac{\partial\, \boldsymbol{g}_i(\boldsymbol{z})}{\partial x_1} \\ \frac{\partial\, \boldsymbol{g}_i(\boldsymbol{z})}{\partial x_2} \\ \vdots \\ \frac{\partial\, \boldsymbol{g}_i(\boldsymbol{z})}{\partial x_n} \end{bmatrix}$$

exist and are continuous on $\mathbb{R}^n$. The (equality) constrained minimization problem is to minimize $\boldsymbol{J}(\boldsymbol{z})$ subject to $\boldsymbol{G}(\boldsymbol{z}) = \boldsymbol{0} \in \mathbb{R}^m$.

Define the Lagrangian $L : \mathbb{R}^1 \times \mathbb{R}^m \times \mathbb{R}^n \to \mathbb{R}^1$ by

$$L(\lambda_0, \boldsymbol{\lambda}, \boldsymbol{z}) \triangleq \lambda_0 \boldsymbol{J}(\boldsymbol{z}) + \sum_{i=1}^{m} \lambda_i \boldsymbol{g}_i(\boldsymbol{z}), \tag{2.18}$$

where $\boldsymbol{\lambda} = \begin{bmatrix} \lambda_1 & \lambda_2 & \cdots & \lambda_m \end{bmatrix}^T \in \mathbb{R}^m$. Observe that $L(\lambda_0, \boldsymbol{\lambda}, \boldsymbol{z})$ can be written as

$$L(\lambda_0, \boldsymbol{\lambda}, \boldsymbol{z}) = \lambda_0 \boldsymbol{J}(\boldsymbol{z}) + \langle \boldsymbol{\lambda}, \boldsymbol{G}(\boldsymbol{z}) \rangle_m$$

where $\langle \cdot, \cdot \rangle_m$ is the standard inner product on $\mathbb{R}^m$. Also, the gradient $\nabla_{\boldsymbol{z}} L(\lambda_0, \boldsymbol{\lambda}, \boldsymbol{z})$ is given by

$$\nabla_{\boldsymbol{z}} L(\lambda_0, \boldsymbol{\lambda}, \boldsymbol{z}) = \lambda_0 \nabla \boldsymbol{J}(\boldsymbol{z}) + \sum_{i=1}^{m} \lambda_i \nabla \boldsymbol{g}_i(\boldsymbol{z}). \tag{2.19}$$

Finally, we shall need the Jacobian of $\boldsymbol{G}(\cdot)$ which is the $m \times n$ matrix defined by

$$\mathbb{J}\boldsymbol{G}(\boldsymbol{z}) \triangleq \begin{bmatrix} \frac{\partial\, g_1(\boldsymbol{z})}{\partial x_1} & \frac{\partial\, g_1(\boldsymbol{z})}{\partial x_2} & \cdots & \frac{\partial\, g_1(\boldsymbol{z})}{\partial x_n} \\ \frac{\partial\, g_2(\boldsymbol{z})}{\partial x_1} & \frac{\partial\, g_2(\boldsymbol{z})}{\partial x_2} & \cdots & \frac{\partial\, g_2(\boldsymbol{z})}{\partial x_n} \\ \vdots & \vdots & \ddots & \vdots \\ \frac{\partial\, g_m(\boldsymbol{z})}{\partial x_1} & \frac{\partial\, g_m(\boldsymbol{z})}{\partial x_2} & \cdots & \frac{\partial\, g_m(\boldsymbol{z})}{\partial x_n} \end{bmatrix}, \tag{2.20}$$

which can be written as

$$\mathbb{J}\boldsymbol{G}(\boldsymbol{z}) = \begin{bmatrix} \nabla \boldsymbol{g}_1(\boldsymbol{z}) & \nabla \boldsymbol{g}_2(\boldsymbol{z}) & \cdots & \nabla \boldsymbol{g}_m(\boldsymbol{z}) \end{bmatrix}^T.$$

As above we set

$$\boldsymbol{\Theta_G} = \{\boldsymbol{z} \in \mathbb{R}^n : \boldsymbol{G}(\boldsymbol{z}) = \boldsymbol{0} \in \mathbb{R}^m\}$$

and state the **Lagrange Multiplier Theorem** for this problem.

Theorem 2.3 (Multiplier Theorem for the n-D Problem) *If $\boldsymbol{z}^* \in \boldsymbol{\Theta_G}$ minimizes $\boldsymbol{J}(\cdot)$ on $\boldsymbol{\Theta_G}$ then there exists a constant λ_0^* and vector $\boldsymbol{\lambda}^* = \begin{bmatrix} \lambda_1^* & \lambda_2^* & \cdots & \lambda_m^* \end{bmatrix}^T \in \mathbb{R}^m$ such that* *(i)* $|\lambda_0^*| + \|\boldsymbol{\lambda}^*\| \neq 0$ *and*

(ii) the minimizer $\boldsymbol{z}^*$ *satisfies*

$$\nabla_{\boldsymbol{z}} \mathrm{L}(\lambda_0^*, \boldsymbol{\lambda}^*, \boldsymbol{z}^*) = \lambda_0^* \nabla \boldsymbol{J}(\boldsymbol{z}^*) + \sum_{i=1}^{m} \lambda_i^* \nabla \boldsymbol{g}_i(\boldsymbol{z}^*) = 0. \quad (2.21)$$

(iii) If in addition the gradients $\nabla \boldsymbol{g}_1(\boldsymbol{z}^*), \nabla \boldsymbol{g}_2(\boldsymbol{z}^*), \ldots, \nabla \boldsymbol{g}_m(\boldsymbol{z}^*)$ *are linearly independent, i.e. the Jacobian* $\mathbb{J}\boldsymbol{G}(\boldsymbol{z}) = \begin{bmatrix} \nabla \boldsymbol{g}_1(\boldsymbol{z}^*) & \nabla \boldsymbol{g}_2(\boldsymbol{z}^*) & \cdots & \nabla \boldsymbol{g}_m(\boldsymbol{z}^*) \end{bmatrix}^T$ *has maximal rank* m, *then* λ_0^* *is not zero.*

Remark 2.9 *Note that condition* (iii) *implies that the linear operator* $\mathcal{T} : \mathbb{R}^n \to \mathbb{R}^m$ *defined by*

$$\mathcal{T}\boldsymbol{h} = [\mathbb{J}\boldsymbol{G}(\boldsymbol{z}^*)]\boldsymbol{h}$$

is onto all of $\mathbb{R}^m$. *This form of condition* (iii) *is the key to more abstract forms of the Lagrange Multiplier Theorem. We say that the minimizer* $\boldsymbol{z}^*$ *is a* normal minimizer *if* $\lambda_0^* \neq 0$.

Example 2.5 *(A Normal Problem) Consider the problem in* $\mathbb{R}^2$ *of minimizing*

$$\boldsymbol{J}(x, y) = x^2 + y^2$$

subject to the single $(m = 1)$ *equality constraint*

$$\boldsymbol{G}(x, y) = y - x^2 - 1 = 0.$$

The gradients of $\boldsymbol{J}(\cdot)$ *and* $\boldsymbol{G}(\cdot)$ *are given by*

$$\nabla \boldsymbol{J}(x, y) = \begin{bmatrix} 2x & 2y \end{bmatrix}^T$$

and

$$\nabla \boldsymbol{G}(x, y) = \begin{bmatrix} -2x & 1 \end{bmatrix}^T,$$

respectively.

Assume that $\boldsymbol{z}^* = \begin{bmatrix} x^* & y^* \end{bmatrix}^T$ *minimizes* $\boldsymbol{J}(\cdot, \cdot)$ *subject to* $\boldsymbol{G}(x, y) = 0$. *The Lagrange Multiplier Theorem above implies that there exist* λ_0^* *and* $\lambda_1^* \in \mathbb{R}^1$ *such that*

$$\lambda_0^* \nabla \boldsymbol{J}(x^*, y^*) + \lambda_1^* \nabla \boldsymbol{G}(x^*, y^*) = 0 \quad (2.22)$$

and $|\lambda_0^*| + |\lambda_1^*| \neq 0$. *Observe that if* $\lambda_0^* = 0$, *then it follows from (2.22) that*

$$\lambda_1^* \nabla \boldsymbol{G}(x^*, y^*) = \lambda_1^* \left[\begin{array}{cc} -2x^* & 1 \end{array}\right]^T = \left[\begin{array}{cc} -\lambda_1^* 2x^* & \lambda_1^* \end{array}\right]^T = 0,$$

which implies that $\lambda_1^* = 0$. *Since* $\lambda_0^* = 0$ *this would imply that* $|\lambda_0^*| + |\lambda_1^*| = 0$ *which contradicts the theorem. Therefore,* $\lambda_0^* \neq 0$ *and if we define* $\lambda = \lambda_1^*/\lambda_0^*$, *then (2.22) is equivalent to*

$$\nabla \boldsymbol{J}(x^*, y^*) + \lambda \nabla \boldsymbol{G}(x^*, y^*) = 0. \tag{2.23}$$

Consequently, the Lagrange Multiplier Theorem yields the following system

$$2x^* - \lambda 2x^* = 2x^*(1 - \lambda) = 0,$$

$$2y^* + \lambda = 0,$$

$$g(x^*, y^*) = y^* - [x^*]^2 - 1 = 0.$$

Therefore, either $\lambda = 1$ *or* $x^* = 0$. *However, if* $\lambda = 1$, *then it would follow that* $y^* = -\lambda/2 = -1/2$ *and* $-1/2 = y^* = [x^*]^2 + 1 > 0$ *which is impossible. The only solution to the system is* $x^* = 0$, $y^* = 1$ *and* $\lambda = -2$.

The previous example was normal in the sense that one could show that $\lambda_0^* \neq 0$ and hence the Lagrange Multiplier Theorem resulted in three (nonlinear) equations to be solved for for three unknowns x^*, y^* and λ. When the problem is not normal this is not always the case. This is easily illustrated by the following example.

Example 2.6 (A Non-Normal Problem) *Consider the problem of minimizing*

$$\boldsymbol{J}(x, y) = x^2 + y^2$$

subject to the equality constraint

$$\boldsymbol{G}(x, y) = x^2 - (y - 1)^3 = 0.$$

The gradients of $\boldsymbol{J}(\cdot)$ *and* $\boldsymbol{G}(\cdot)$ *are given by*

$$\nabla \boldsymbol{J}(x, y) = \left[\begin{array}{cc} 2x & 2y \end{array}\right]^T$$

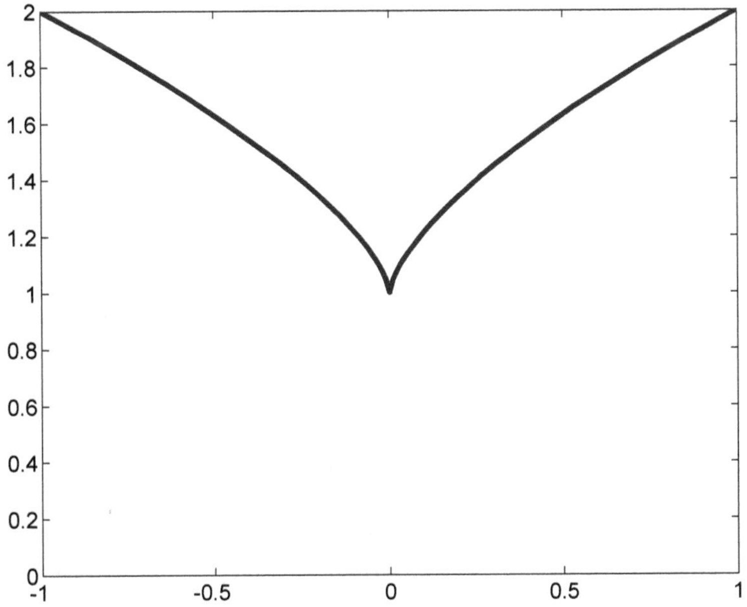

Figure 2.9: A Non-Normal Problem

and

$$\nabla \boldsymbol{G}(x,y) = \begin{bmatrix} 2x & -3(y-1)^2 \end{bmatrix}^T,$$

respectively.

It is clear from Figure 2.9 that the minimizer is given by $\boldsymbol{z}^* = \begin{bmatrix} x^* & y^* \end{bmatrix}^T = \begin{bmatrix} 0 & 1 \end{bmatrix}^T$. *The Lagrange Multiplier Theorem implies that there exist* λ_0^* *and* λ_1^* *such that*

$$\lambda_0^* \nabla \boldsymbol{J}(x^*, y^*) + \lambda_1^* \nabla \boldsymbol{G}(x^*, y^*) = 0 \tag{2.24}$$

and $|\lambda_0^*| + |\lambda_1^*| \neq 0$.

If $\lambda_0^* \neq 0$ *and if we define* $\lambda = \lambda_1^*/\lambda_0^*$, *then (2.24) is equivalent to*

$$\nabla \boldsymbol{J}(x^*, y^*) + \lambda \nabla \boldsymbol{G}(x^*, y^*) = 0. \tag{2.25}$$

Using the fact that $\boldsymbol{z}^* = \begin{bmatrix} x^* & y^* \end{bmatrix}^T = \begin{bmatrix} 0 & 1 \end{bmatrix}^T$ *is the minimizer,*

(2.24) and (2.25) imply that

$$\nabla \boldsymbol{J}(0,1) + \lambda \nabla \boldsymbol{G}(0,1) = 0,$$

or equivalently, that

$$\begin{bmatrix} 0 & 1 \end{bmatrix}^T + \lambda \begin{bmatrix} 0 & 0 \end{bmatrix}^T = \begin{bmatrix} 0 & 0 \end{bmatrix}^T.$$

Clearly, $\begin{bmatrix} 0 & 1 \end{bmatrix}^T \neq \begin{bmatrix} 0 & 0 \end{bmatrix}^T$ *so that* $\lambda_0^* = 0$ *is the only possible choice for* λ_0^* *and* λ_1^* *can be any non-zero value. The reason that the only choice for* λ_0^* *is* $\lambda_0^* = 0$ *is because the gradient of* $\boldsymbol{G}(\cdot)$ *at* $\boldsymbol{z}^* = \begin{bmatrix} x^* & y^* \end{bmatrix}^T = \begin{bmatrix} 0 & 1 \end{bmatrix}^T$ *is* $\nabla \boldsymbol{G}(0,1) = \begin{bmatrix} 0 & 0 \end{bmatrix}^T = \mathbf{0}$.

Proofs of the 2D Lagrange Multiplier Theorem

We shall present two proofs of the **Lagrange Multiplier Theorem 2.3** in the special case where $n = 2$. The first proof is based on a variational method and can be easily modified to deal with isoperimetric problems in the calculus of variations discussed in Chapter 7 below. The second proof is geometric and relies on a separation result. Both proofs can be extended to a very general setting and yield a Lagrange Multiplier Theorem in abstract spaces.

The first proof makes use of the Inverse Mapping Theorem from advanced calculus. The Inverse Mapping Theorem provides conditions that ensure a function will have an inverse (at least locally). Let $T : \mathcal{O} \subset \mathbb{R}^2 \to \mathbb{R}^2$ be a function from an open set $\mathcal{O}$ into the plane defined by

$$T(\alpha, \beta) = [p(\alpha, \beta) \ q(\alpha, \beta)]^T$$

where $p(\alpha, \beta)$ and $q(\alpha, \beta)$ are smooth functions. Assume that $[\hat{\alpha} \ \ \hat{\beta}]^T \in \mathcal{O}$ and

$$T(\hat{\alpha}, \hat{\beta}) = [p(\hat{\alpha}, \hat{\beta}) \ q(\hat{\alpha}, \hat{\beta})] = [\hat{p} \ \ \hat{q}]^T.$$

Roughly speaking, the Inverse Mapping Theorem implies that if the Jacobian matrix at $[\hat{\alpha} \ \ \hat{\beta}]^T$ is non-singular (i.e. invertible), then there is a neighborhood $\mathcal{U}$ of $[\hat{\alpha} \ \hat{\beta}]^T$ and an open neighborhood $\mathcal{V}$

of $T(\hat{\alpha}, \hat{\beta}) = [\hat{p}\ \ \hat{q}]^T$ so that $T(\alpha, \beta)$ restricted to $\mathcal{U}$ is a one-to-one and onto mapping from $\mathcal{U}$ to $\mathcal{V}$ with a continuous inverse function.

Recall that the Jacobian matrix at $[\hat{\alpha}\ \hat{\beta}]^T$ is given by

$$\mathbb{J}T(\hat{\alpha}, \hat{\beta}) = \begin{bmatrix} \frac{\partial p(\alpha,\beta)}{\partial \alpha} & \frac{\partial p(\alpha,\beta)}{\partial \beta} \\ \frac{\partial q(\alpha,\beta)}{\partial \alpha} & \frac{\partial q(\alpha,\beta)}{\partial \beta} \end{bmatrix}_{[\alpha\ \beta]^T = [\hat{\alpha}\ \hat{\beta}]^T} = \begin{bmatrix} \frac{\partial p(\hat{\alpha},\hat{\beta})}{\partial \alpha} & \frac{\partial p(\hat{\alpha},\hat{\beta})}{\partial \beta} \\ \frac{\partial q(\hat{\alpha},\hat{\beta})}{\partial \alpha} & \frac{\partial q(\hat{\alpha},\hat{\beta})}{\partial \beta} \end{bmatrix}, \tag{2.26}$$

and $\mathbb{J}T(\hat{\alpha}, \hat{\beta})$ in one-to-one and onto (i.e. non-singular) if and only if the determinant

$$\det \mathbb{J}T(\hat{\alpha}, \hat{\beta}) = det \begin{bmatrix} \frac{\partial p(\hat{\alpha},\hat{\beta})}{\partial \alpha} & \frac{\partial p(\hat{\alpha},\hat{\beta})}{\partial \beta} \\ \frac{\partial q(\hat{\alpha},\hat{\beta})}{\partial \alpha} & \frac{\partial q(\hat{\alpha},\hat{\beta})}{\partial \beta} \end{bmatrix} \neq 0.$$

The following version of the Inverse Mapping Theorem follows from Theorem 41.8 on page 381 in Bartle's book [15].

Theorem 2.4 (Inverse Function Theorem) *Let $T : \mathcal{O} \subset \mathbb{R}^2 \to \mathbb{R}^2$ be a C^1 function from the open set $\mathcal{O}$ into the plane defined by*

$$T(\alpha, \beta) = [p(\alpha, \beta)\ q(\alpha, \beta)]^T$$

where $p(\alpha, \beta)$ and $q(\alpha, \beta)$ are smooth functions. Assume that $[\hat{\alpha}\ \hat{\beta}]^T \in \mathcal{O}$ with

$$T(\hat{\alpha}, \hat{\beta}) = [p(\hat{\alpha}, \hat{\beta})\ q(\hat{\alpha}, \hat{\beta})] = [\hat{p}\ \hat{q}]^T$$

and that the Jacobian at $[\hat{\alpha}\ \ \hat{\beta}]^T$, $\mathbb{J}T(\hat{\alpha}, \hat{\beta})$ is non-singular. Then there are open neighborhoods $\mathcal{U}$ of $[\hat{\alpha}\ \ \hat{\beta}]^T$ and $\mathcal{V}$ of $T(\hat{\alpha}, \hat{\beta}) = [\hat{p}\ \ \hat{q}]^T$ such that $T(\alpha, \beta)$ restricted to $\mathcal{U}$ is a one-to-one and onto mapping from $\mathcal{U}$ onto $\mathcal{V}$. Moreover, if

$$\mathcal{T}(\alpha, \beta) \triangleq T(\alpha, \beta)|_{\mathcal{U}}$$

denotes the restriction of $T(\alpha, \beta)$ to $\mathcal{U}$, then $\mathcal{T}(\alpha, \beta) : \mathcal{U} \to \mathcal{V}$ has a continuous inverse $\mathcal{T}^{-1}(p, q) : \mathcal{V} \to \mathcal{U}$ belonging to C^1,

$$[\alpha\ \ \beta]^T = [\alpha(p, q)\ \ \beta(p, q)]^T = \mathcal{T}^{-1}(p, q),$$

and

$$\mathbb{J}[\mathcal{T}^{-1}(p, q)] = [\mathbb{J}T(\alpha(p, q), \beta(p, q))]^{-1} = [\mathbb{J}T(\mathcal{T}(p, q))]^{-1}$$

for all $[p\ q]^T \in \mathcal{V}$.

Proof of Theorem 2.3 for the 2D Case:

Assume $\boldsymbol{z}^* = [\, x^* \;\; y^*]^T \in \mathbb{R}^2$ minimizes

$$\boldsymbol{J}(x, y)$$

subject to

$$\boldsymbol{G}(x, y) = \boldsymbol{g}(x, y) = 0,$$

where

$$\boldsymbol{J} : \mathbb{R}^2 \longrightarrow \mathbb{R}^1$$

and

$$\boldsymbol{G} : \mathbb{R}^2 \longrightarrow \mathbb{R}^1.$$

Since there is only one constraint (i.e. $m = 1$), the Jacobian

$$\begin{aligned} \mathbb{J}\boldsymbol{G}(\boldsymbol{z}^*) &= \left[\; \nabla \boldsymbol{g}_1(\boldsymbol{z}^*) \quad \nabla \boldsymbol{g}_2(\boldsymbol{z}^*) \quad \cdots \quad \nabla \boldsymbol{g}_m(\boldsymbol{z}^*) \;\right]^T \\ &= [\nabla \boldsymbol{g}(\boldsymbol{z}^*)]^T = \left[\; \boldsymbol{g}_x(x^*, y^*) \quad \boldsymbol{g}_y(x^*, y^*) \;\right]^T \end{aligned}$$

has maximal rank $m = 1$, if and only if $\nabla \boldsymbol{g}(\boldsymbol{z}^*) \neq [0 \;\; 0]^T$.

First consider the case where $\boldsymbol{z}^* = [\, x^* \;\; y^*]^T$ satisfies

$$\nabla \boldsymbol{G}(x^*, y^*) = \nabla \boldsymbol{g}(x^*, y^*) = \begin{bmatrix} \boldsymbol{g}_x(x^*, y^*) \\ \boldsymbol{g}_y(x^*, y^*) \end{bmatrix} = \begin{bmatrix} 0 \\ 0 \end{bmatrix}.$$

In this case set $\lambda_0^* = 0$ and $\lambda_1^* = 1$. It follows that $|\lambda_0^*| + |\lambda_1^*| = 1 \neq 0$ and

$$\begin{aligned} \nabla_{\boldsymbol{z}} L(\lambda_0^*, \boldsymbol{\lambda}^*, \boldsymbol{z}^*) &= \lambda_0^* \nabla \boldsymbol{J}(\boldsymbol{z}^*) + \boldsymbol{\lambda}^* \nabla \boldsymbol{g}(\boldsymbol{z}^*) \\ &= 0 \nabla \boldsymbol{J}(\boldsymbol{z}^*) + 1 \nabla \boldsymbol{g}(\boldsymbol{z}^*) = 0. \end{aligned}$$

Hence,

$$\lambda_0^* \nabla \boldsymbol{J}(\boldsymbol{z}^*) + \lambda_1^* \nabla \boldsymbol{g}(\boldsymbol{z}^*) = 0$$

and the theorem is clearly true.

Now consider the case where $\boldsymbol{z}^* = [\, x^* \;\; y^*]^T$ satisfies

$$\nabla \boldsymbol{G}(x^*, y^*) = \nabla \boldsymbol{g}(x^*, y^*) = \begin{bmatrix} \boldsymbol{g}_x(x^*, y^*) \\ \boldsymbol{g}_y(x^*, y^*) \end{bmatrix} \neq \begin{bmatrix} 0 \\ 0 \end{bmatrix}.$$

In particular, at least one of the partial derivatives is not zero. Without loss of generality we assume $\boldsymbol{g}_x(x^*, y^*) \neq 0$ and define

$$\lambda_0^* = \boldsymbol{g}_x(x^*, y^*) \neq 0$$

and

$$\lambda_1^* = -\boldsymbol{J}_x(x^*, y^*).$$

We now show that $\boldsymbol{z}^* = [x^* \;\; y^*]^T$ satisfies

$$\lambda_0^* \nabla \boldsymbol{J}(\boldsymbol{z}^*) + \lambda_1^* \nabla \boldsymbol{g}(\boldsymbol{z}^*) = 0.$$

Observe that

$$\lambda_0^* \nabla \boldsymbol{J}(\boldsymbol{z}^*) + \lambda_1^* \nabla \boldsymbol{g}(\boldsymbol{z}^*) = \boldsymbol{g}_x(\boldsymbol{z}^*) \nabla \boldsymbol{J}(\boldsymbol{z}^*) - \boldsymbol{J}_x(\boldsymbol{z}^*) \nabla \boldsymbol{g}(\boldsymbol{z}^*),$$

so that

$$\lambda_0^* \boldsymbol{J}_x(\boldsymbol{z}^*) + \lambda_1^* \boldsymbol{g}_x(\boldsymbol{z}^*) = \boldsymbol{g}_x(\boldsymbol{z}^*) \boldsymbol{J}_x(\boldsymbol{z}^*) - \boldsymbol{J}_x(\boldsymbol{z}^*) \boldsymbol{g}_x(\boldsymbol{z}^*) = 0 \tag{2.27}$$

and

$$\begin{aligned} \lambda_0^* \boldsymbol{J}_y(\boldsymbol{z}^*) + \lambda_1^* \boldsymbol{g}_y(\boldsymbol{z}^*) &= \boldsymbol{g}_x(\boldsymbol{z}^*) \boldsymbol{J}_y(\boldsymbol{z}^*) \\ &\quad - \boldsymbol{J}_x(\boldsymbol{z}^*) \boldsymbol{g}_y(\boldsymbol{z}^*) \\ &= \det \begin{bmatrix} \boldsymbol{g}_x(\boldsymbol{z}^*) & \boldsymbol{J}_x(\boldsymbol{z}^*) \\ \boldsymbol{g}_y(\boldsymbol{z}^*) & \boldsymbol{J}_y(\boldsymbol{z}^*) \end{bmatrix}. \end{aligned}$$

Therefore, to establish that $\boldsymbol{z}^* = [\, x^* \;\; y^*]^T$ satisfies

$$\lambda_0^* \nabla \boldsymbol{J}(\boldsymbol{z}^*) + \lambda_1^* \nabla \boldsymbol{g}(\boldsymbol{z}^*) = \boldsymbol{g}_x(\boldsymbol{z}^*) \nabla \boldsymbol{J}(\boldsymbol{z}^*) - \boldsymbol{J}_x(\boldsymbol{z}^*) \nabla \boldsymbol{g}(\boldsymbol{z}^*) = 0,$$

we must show that

$$\det \begin{bmatrix} \boldsymbol{g}_x(\boldsymbol{z}^*) & \boldsymbol{J}_x(\boldsymbol{z}^*) \\ \boldsymbol{g}_y(\boldsymbol{z}^*) & \boldsymbol{J}_y(\boldsymbol{z}^*) \end{bmatrix} = \det \begin{bmatrix} \boldsymbol{J}_x(\boldsymbol{z}^*) & \boldsymbol{J}_y(\boldsymbol{z}^*) \\ \boldsymbol{g}_x(\boldsymbol{z}^*) & \boldsymbol{g}_y(\boldsymbol{z}^*) \end{bmatrix} = 0. \tag{2.28}$$

This is accomplished by applying the **Inverse Mapping Theorem** 2.4 above.

Define $T : \mathbb{R}^2 \to \mathbb{R}^2$ by

$$T(\alpha, \beta) = [(p(\alpha, \beta) \;\; q(\alpha, \beta)]^T, \tag{2.29}$$

where

$$p(\alpha, \beta) = \boldsymbol{J}(x^* + \alpha, y^* + \beta) \tag{2.30}$$

and

$$q(\alpha, \beta) = \boldsymbol{G}\left(x^* + \alpha, y^* + \beta\right), \tag{2.31}$$

respectively.

Observe that $T(\alpha, \beta)$ maps the open set $\mathbb{R}^2$ to $\mathbb{R}^2$, is defined by (2.29) - (2.31) with $[\hat{\alpha} \;\; \hat{\beta}]^T = [0 \;\; 0]^T$ and

$$T(\hat{\alpha}, \hat{\beta}) = T(0,0) = [\boldsymbol{J}(x^*, y^*) \;\; \boldsymbol{G}(x^*, y^*)]^T = [\boldsymbol{J}(\boldsymbol{z}^*) \;\; 0]^T = [\hat{p} \;\; 0]^T.$$

The Jacobian of $T(\alpha, \beta)$ at $[\hat{\alpha} \;\; \hat{\beta}]^T = [0 \;\; 0]^T$ is given by

$$\begin{bmatrix} \frac{\partial p(0,0)}{\partial \alpha} & \frac{\partial p(0,0)}{\partial \beta} \\ \frac{\partial q(0,0)}{\partial \alpha} & \frac{\partial q(0,0)}{\partial \beta} \end{bmatrix} = \begin{bmatrix} \boldsymbol{J}_x\left(x^*, y^*\right) & \boldsymbol{J}_y\left(x^*, y^*\right) \\ \boldsymbol{g}_x(x^*, y^*) & \boldsymbol{g}_y(x^*, y^*) \end{bmatrix}.$$

Assume that (2.28) is **not true**. This assumption implies that

$$\det \begin{bmatrix} \frac{\partial p(0,0)}{\partial \alpha} & \frac{\partial p(0,0)}{\partial \beta} \\ \frac{\partial q(0,0)}{\partial \alpha} & \frac{\partial q(0,0)}{\partial \beta} \end{bmatrix} = \det \begin{bmatrix} \boldsymbol{J}_x\left(x^*, y^*\right) & \boldsymbol{J}_y\left(x^*, y^*\right) \\ \boldsymbol{g}_x(x^*, y^*) & \boldsymbol{g}_y(x^*, y^*) \end{bmatrix} \neq 0$$

and the Jacobian of $T(\alpha, \beta)$ is non-singular at $[\hat{\alpha} \;\; \hat{\beta}]^T = [0 \;\; 0]^T$ so we may apply the **Theorem** 2.4. In particular, (see Figure 2.10) there is a neighborhood $\mathcal{U} = \left\{[\alpha \;\; \beta]^T : \sqrt{\alpha^2 + \beta^2} < \gamma\right\}$ of $[0 \;\; 0]^T$ and a neighborhood $\mathcal{V}$ of $[\boldsymbol{J}(x^*, y^*) \;\; 0]^T = [\hat{p} \;\; 0]^T$ such that the restriction of $T(\alpha, \beta)$ to $\mathcal{U}$, $\mathcal{T}(\alpha, \beta) : \mathcal{U} \rightarrow \mathcal{V}$, has a continuous inverse $\mathcal{T}^{-1}(p, q) : \mathcal{V} \rightarrow \mathcal{U}$ belonging to C^1.

Let $[\tilde{p} \;\; 0]^T \in \mathcal{V}$ be any point with $\tilde{p} < \boldsymbol{J}(x^*, y^*)$ and let $[\tilde{\alpha} \;\; \tilde{\beta}]^T = \mathcal{T}^{-1}(\tilde{p}, 0) \in \mathcal{U}$. Observe that

$$\boldsymbol{J}(x^* + \tilde{\alpha}, y^* + \tilde{\beta}) = p(\tilde{\alpha}, \tilde{\beta}) = \tilde{p} < \boldsymbol{J}(x^*, y^*) = \boldsymbol{J}(\boldsymbol{z}^*) \tag{2.32}$$

and

$$\boldsymbol{G}(x^* + \tilde{\alpha}, y^* + \tilde{\beta}) = q(\tilde{\alpha}, \tilde{\beta}) = 0. \tag{2.33}$$

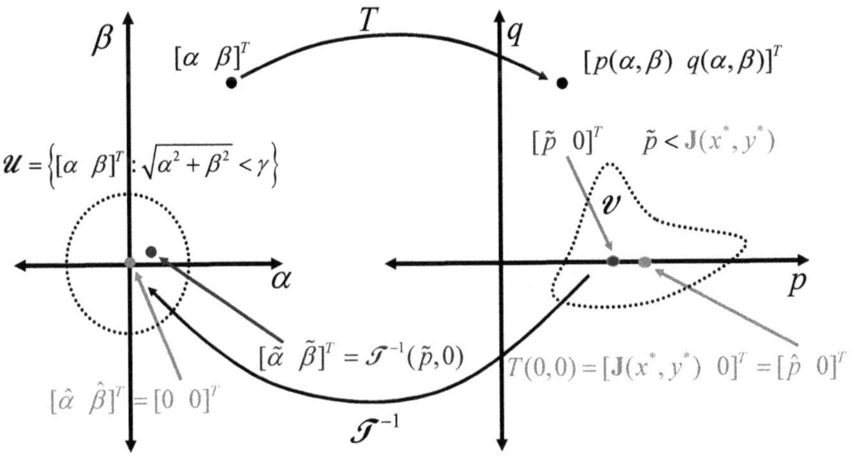

Figure 2.10: 2D Lagrange Multiplier Proof

Thus, the vector $\tilde{\boldsymbol{z}} = [x^* + \tilde{\alpha}\ \ y^*{+}\tilde{\beta}]^T$ satisfies

$$\boldsymbol{G}(\tilde{\boldsymbol{z}}) = \boldsymbol{G}(x^* + \tilde{\alpha}, y^*{+}\tilde{\beta}) = q(\tilde{\alpha}, \tilde{\beta}) = 0$$

and

$$\boldsymbol{J}(\tilde{\boldsymbol{z}}) = \boldsymbol{J}(x^* + \tilde{\alpha}, y^*{+}\tilde{\beta}) = p(\tilde{\alpha}, \tilde{\beta}) = \tilde{p} < \boldsymbol{J}(x^*, y^*) = \boldsymbol{J}(\boldsymbol{z}^*)$$

which contradicts the assumption that $\boldsymbol{z}^* = [x^*\ \ y^*]^T$ minimizes $\boldsymbol{J}\,(x, y)$ subject to $\boldsymbol{G}\,(x, y) = 0$.

Therefore, it follows that

$$\det \begin{bmatrix} \frac{\partial p(0,0)}{\partial \alpha} & \frac{\partial p(0,0)}{\partial \beta} \\ \frac{\partial q(0,0)}{\partial \alpha} & \frac{\partial q(0,0)}{\partial \beta} \end{bmatrix} = \det \begin{bmatrix} \boldsymbol{J}_x\,(x^*, y^*) & \boldsymbol{J}_y\,(x^*, y^*) \\ \boldsymbol{g}_x(x^*, y^*) & \boldsymbol{g}_y(x^*, y^*) \end{bmatrix} \neq 0$$

must be false and hence

$$\begin{aligned} &\boldsymbol{g}_x\,(x^*, y^*)\,\boldsymbol{J}_y\,(x^*, y^*) - \boldsymbol{J}_x\,(x^*, y^*)\,\boldsymbol{g}_y(x^*, y^*) \\ &= -\det \begin{bmatrix} \boldsymbol{J}_x\,(x^*, y^*) & \boldsymbol{J}_y\,(x^*, y^*) \\ \boldsymbol{g}_x(x^*, y^*) & \boldsymbol{g}_y(x^*, y^*) \end{bmatrix} \\ &= 0. \end{aligned} \tag{2.34}$$

Consequently, (2.27) and (2.34) together imply that

$$\lambda_0^* \nabla \boldsymbol{J}(\boldsymbol{z}^*) + \lambda_1^* \nabla \boldsymbol{g}(\boldsymbol{z}^*) = \boldsymbol{g}_x(\boldsymbol{z}^*) \nabla \boldsymbol{J}(\boldsymbol{z}^*) - \boldsymbol{J}_x(\boldsymbol{z}^*) \nabla \boldsymbol{g}(\boldsymbol{z}^*) = \boldsymbol{0}$$

and $\lambda_0^* = \boldsymbol{g}_x(x^*, y^*) \neq 0$ which completes the proof. □

We also outline another proof that is geometric in nature. The details can be found in Hestenes' book [102].

Second Proof of Theorem 2.3 for the 2D Case:

Assume $\boldsymbol{z}^* = [x^* \;\; y^*]^T \in \mathbb{R}^2$ minimizes

$$\boldsymbol{J}(x, y)$$

subject to

$$\boldsymbol{G}(x, y) = \boldsymbol{g}(x, y) = 0.$$

Observe that if $\nabla \boldsymbol{g}(\boldsymbol{z}^*) = \boldsymbol{0} = [0 \;\; 0]^T$, then $\lambda_0^* = 0$ and $\lambda^* = 1$ produces

$$\nabla_{\boldsymbol{z}} L(\lambda_0^*, \boldsymbol{\lambda}^*, \boldsymbol{z}^*) = \lambda_0^* \nabla \boldsymbol{J}(\boldsymbol{z}^*) + \lambda^* \nabla \boldsymbol{g}(\boldsymbol{z}^*) = 0$$

and hence we need only consider the case where $\nabla \boldsymbol{g}(\boldsymbol{z}^*) \neq \boldsymbol{0}$. Also, in the trivial case when $\nabla \boldsymbol{J}(\boldsymbol{z}^*) = \boldsymbol{0}$ one can set $\lambda_0^* = 1$ and $\lambda^* = 0$ so that without loss of generality we can consider the case where $\nabla \boldsymbol{g}(\boldsymbol{z}^*) \neq \boldsymbol{0}$ and $\nabla \boldsymbol{J}(\boldsymbol{z}^*) \neq \boldsymbol{0}$. Under this assumption, the condition that

$$\lambda_0^* \nabla \boldsymbol{J}(\boldsymbol{z}^*) + \lambda^* \nabla \boldsymbol{g}(\boldsymbol{z}^*) = 0$$

with $\lambda_0^* \neq 0$ is equivalent to the existence of a $\bar{\lambda}$ such that

$$\nabla \boldsymbol{J}(\boldsymbol{z}^*) + \bar{\lambda} \nabla \boldsymbol{g}(\boldsymbol{z}^*) = 0,$$

where $\bar{\lambda} = (\lambda^*/\lambda_0^*)$. In particular, the nonzero gradients $\nabla \boldsymbol{g}(\boldsymbol{z}^*)$ and $\nabla \boldsymbol{J}(\boldsymbol{z}^*)$ must be collinear.

To establish this we define the level set L^* by

$$L^* \triangleq \left\{ [x \;\; y]^T : \boldsymbol{J}(x, y) = m^* = \boldsymbol{J}(x^*, y^*) \right\}$$

and the constraint set C by

$$C = \left\{ [x \;\; y]^T : \boldsymbol{G}(x, y) = 0 \right\},$$

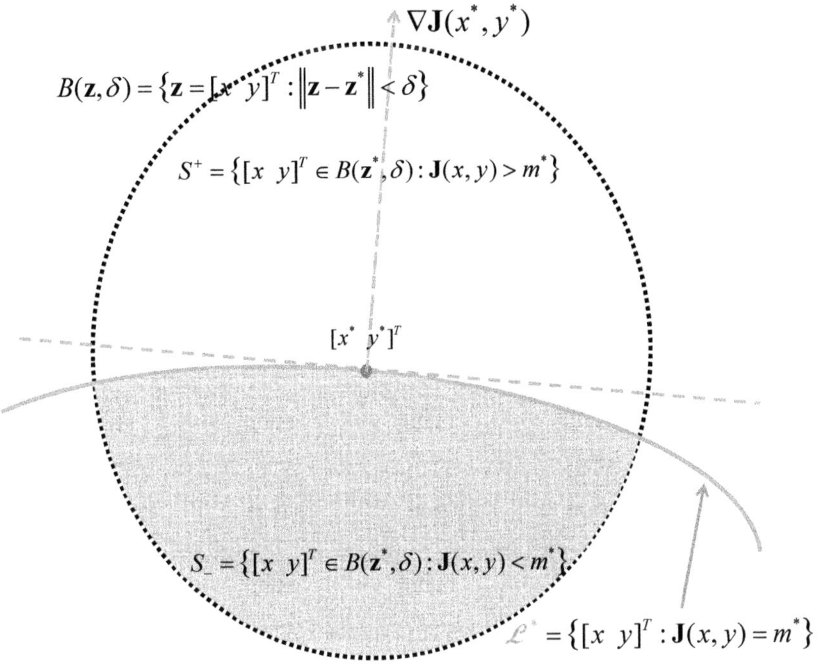

Figure 2.11: The Level Set

respectively. Thus, in a sufficiently small ball $B(\boldsymbol{z}^*, \delta)$ about $\boldsymbol{z}^* = [x^* \;\; y^*]^T$ the level set separates the ball $B(z^*, \delta)$ into two sets

$$S_- = \left\{[x \quad y]^T \in B(\boldsymbol{z}^*, \delta) : \boldsymbol{J}(x, y) < m^* = \boldsymbol{J}(x^*, y^*)\right\}$$

and

$$S^+ = \left\{[x \quad y]^T \in B(\boldsymbol{z}^*, \delta) : \boldsymbol{J}(x, y) > m^* = \boldsymbol{J}(x^*, y^*)\right\},$$

respectively (see Figure 2.11).

If we assume that $\nabla \boldsymbol{g}(\boldsymbol{z}^*)$ and $\nabla \boldsymbol{J}(\boldsymbol{z}^*)$ are not collinear, then as shown in Figure 2.12 the support line to S_- at $\boldsymbol{z}^* = [\, x^* \;\; y^*]^T$ (i.e. the line orthogonal to $\nabla \boldsymbol{J}(\boldsymbol{z}^*)$) must cross the line orthogonal to $\nabla \boldsymbol{g}(\boldsymbol{z}^*)$ at $\boldsymbol{z}^* = [\, x^* \;\; y^*]^T$. In particular, the constraint set C must intersect $S_- = \left\{[x \quad y]^T \in B(z^*, \delta) : \boldsymbol{J}(x, y) < m^*\right\}$ and there is a point $[\tilde{x} \quad \tilde{y}]^T \in C \cap S_-$ (see Figure 2.13). However,

$$C \cap S_- = \left\{[x \quad y]^T : \boldsymbol{G}(x, y) = 0 \text{ and } \boldsymbol{J}(x, y) < m^* = \boldsymbol{J}(x^*, y^*)\right\}$$

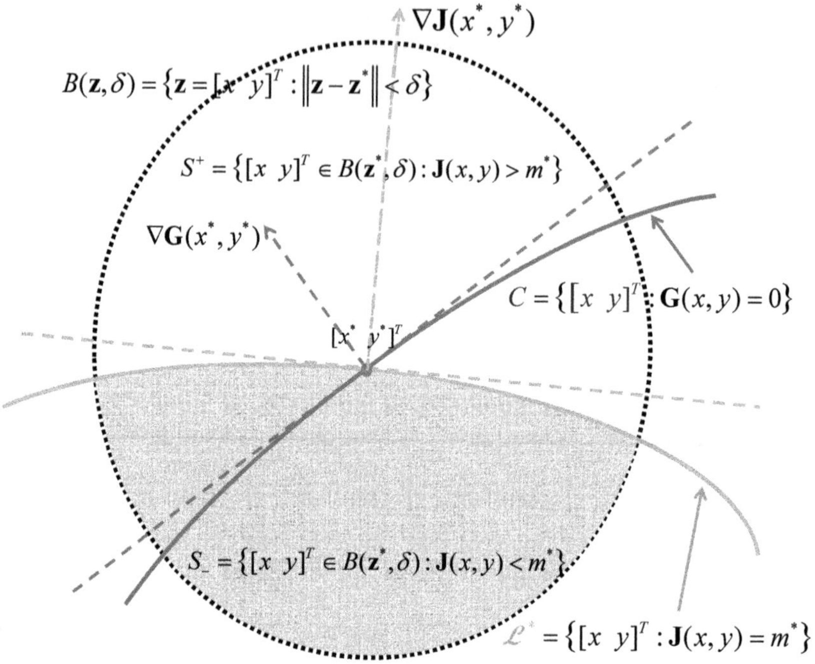

Figure 2.12: Non-collinear Gradients

so that $[\tilde{x} \ \ \tilde{y}]^T \in C \cap S_-$ satisfies

$$\boldsymbol{G}(\tilde{x}, \tilde{y}) = 0$$

and

$$\boldsymbol{J}(\tilde{x}, \tilde{y}) < m^* = \boldsymbol{J}(x^*, y^*).$$

Therefore, $[x^* \ \ y^*]^T$ is not a local minimizer of $\boldsymbol{J}(x, y)$ on the set

$$C = \left\{ [x \ \ y]^T : \boldsymbol{G}(x, y) = 0 \right\}$$

and hence can not minimize $\boldsymbol{J}(x, y)$ subject to $\boldsymbol{G}(x, y) = 0$. Consequently, $\nabla \boldsymbol{g}(\boldsymbol{z}^*)$ and $\nabla \boldsymbol{J}(\boldsymbol{z}^*)$ must be collinear and this completes the proof. □

We note that this geometric proof depends on knowing that the line orthogonal to the gradient $\nabla \boldsymbol{J}(\boldsymbol{z}^*)$ is a support plane for $S_- = \left\{ [x \ \ y]^T \in B(\boldsymbol{z}^*, \delta) : \boldsymbol{J}(x, y) < m^* \right\}$ at $\boldsymbol{z}^* = [\, x^* \ \ y^*]^T$. This

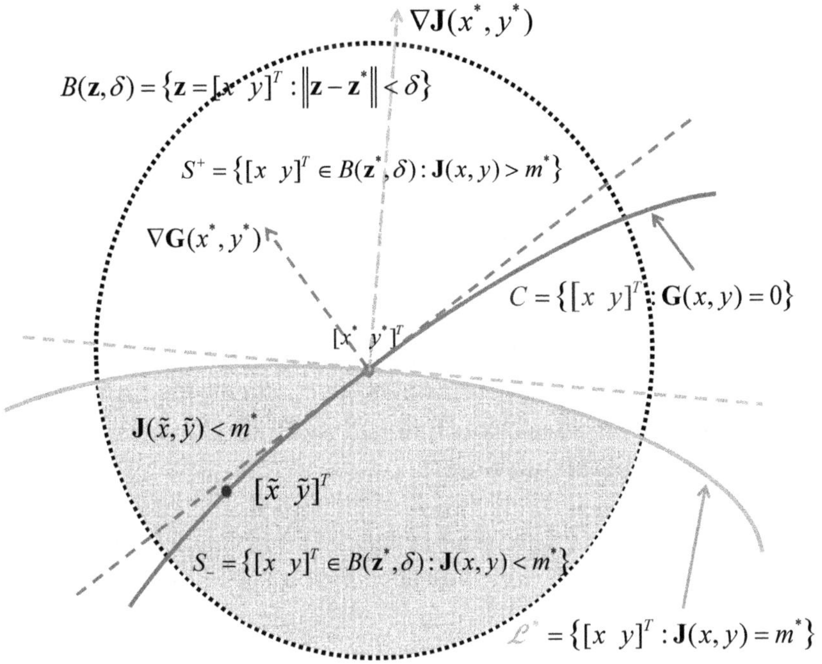

Figure 2.13: Contradiction to Assumption

"theme" will be important when we derive the simplest Maximum Principle for the time optimal control problem.

2.3 Function Spaces

Problems in the calculus of variations and optimal control involve finding functions that minimize some functional over a set of prescribed (admissible) functions. Therefore, we need to specify the precise space of functions that will be admissible. We start with the basic real valued piecewise continuous functions defined on an interval $I = [t_0, t_1]$.

Definition 2.7 *Let $I = [t_0, t_1]$ be a closed interval. Then $x : I \to \mathbb{R}$ is said to be* ***piecewise continuous*** *(PWC) on $[t_0, t_1]$ if:*

- *The function $x(\cdot)$ is bounded on I.*

- *The right-hand limit* $x\left(\hat{t}^{+}\right) = \lim_{t\to\hat{t}^{+}}\left[x(t)\right]$ *exists (is finite) for all* $\hat{t} \in [t_0, t_1)$.
- *The left-hand limit* $x\left(\hat{t}^{-}\right) = \lim_{t\to\hat{t}^{-}}\left[x(t)\right]$ *exists (is finite) for all* $\hat{t} \in (t_0, t_1]$.
- *There is a* ***finite*** *partition of* $[t_0, t_1]$, $t_0 = \hat{t}_0 < \hat{t}_1 < \hat{t}_2 < \ldots < \hat{t}_{p-1} < \hat{t}_p = t_1$, *such that* $x(\cdot)$ *is continuous on each open subinterval* $\left(\hat{t}_{i-1}, \hat{t}_i\right)$.

Note that if $x(\cdot)$ is defined and bounded on $[t_0, t_1]$, $\hat{t} \in [t_0, t_1]$ and $\lim_{t\to\hat{t}^{\pm}}\left[x(t)\right]$ exists (even one-sided), then this limit must be finite. Thus, for piecewise continuous functions the one-sided limits exist and are finite at all points.

Example 2.7 *If* $x(\cdot)$ *is defined as in Figure 2.14, then* $x(\cdot)$ *is piecewise continuous. Note however that* $x(\cdot)$ *is not continuous.*

Example 2.8 *The function shown in Figure 2.15 is not piecewise continuous since it is not bounded on* $[t_0, t_1)$.

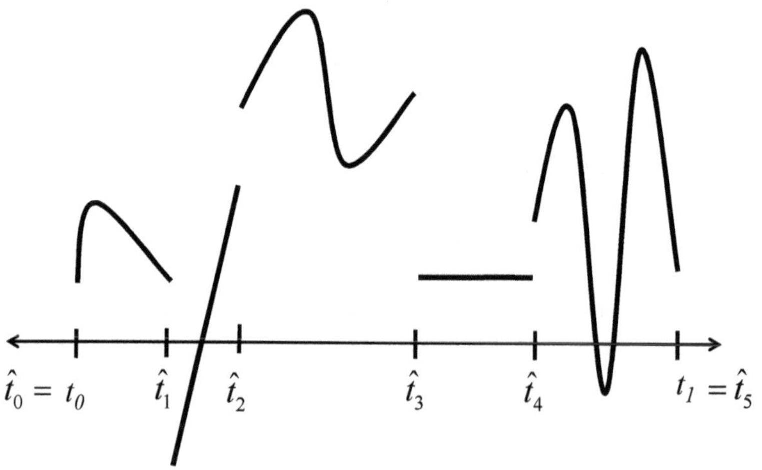

Figure 2.14: A Piecewise Continuous Function

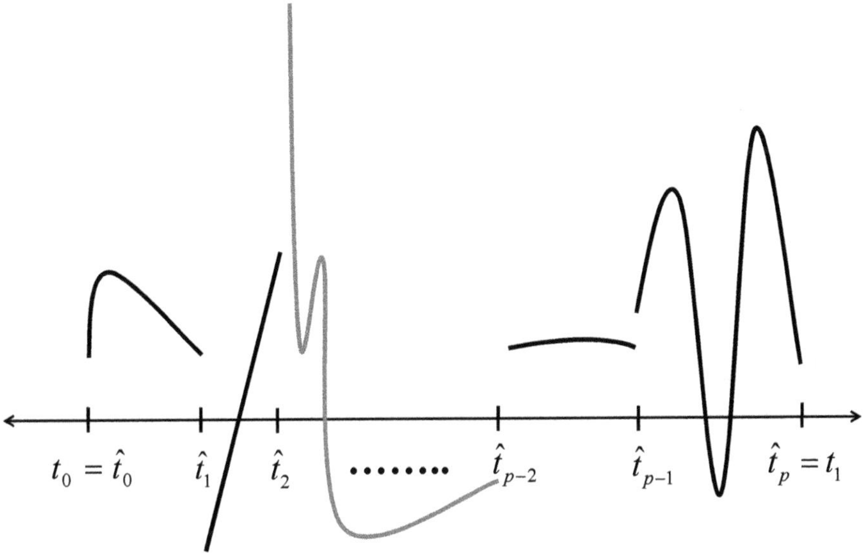

Figure 2.15: A Non-Piecewise Continuous Function

Example 2.9 *Let* $t_N = \frac{1}{2^{N-1}}$*, for* $N = 1, 2, 3, \cdots$ *and define* $\tilde{x}$: $[0,1] \longrightarrow \mathbb{R}$ *by*

$$\tilde{x}(t) = \frac{1}{2^{N-1}}, \quad \frac{1}{2^N} < t \leq \frac{1}{2^{N-1}}$$

and set $\tilde{x}(0) = 0$ *(see Figure 2.16). Note that* $\tilde{x}(\cdot)$ *is defined and bounded on* $[0,1]$ *and given any* $\hat{t} \in [0,1]$ *it follows that the limits* $\lim_{t \to \hat{t}^{\pm}} [\tilde{x}(t)]$ *exist and are finite. Clearly,* $\tilde{x}(\cdot)$ *is bounded, but there is no* ***finite*** *partition of* $[0,1]$ *such that* $\tilde{x}(\cdot)$ *is continuous on this partition. Therefore,* $\tilde{x}(\cdot)$ *is not piecewise continuous since it has an infinite number of discontinuous jumps.*

Definition 2.8 *A function* $x : [t_0, t_1] \to \mathbb{R}$ *is called* ***piecewise smooth*** *(PWS) on* $[t_0, t_1]$ *if*

(i) $x(\cdot)$ *is continuous on* $[t_0, t_1]$ *and*

(ii) there exists a piecewise continuous function $g(\cdot)$ *and a con-*

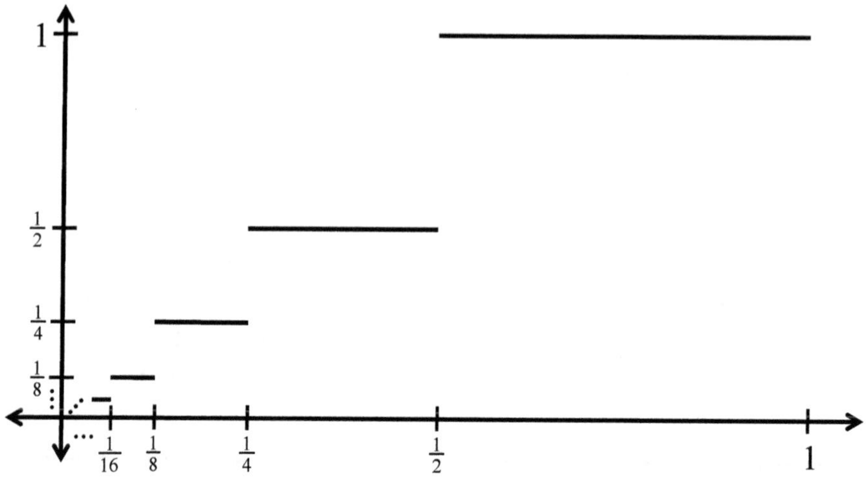

Figure 2.16: A Bounded Non-Piecewise Continuous Function

stant c such that for all $t \in [t_0, t_1]$

$$x(t) = c + \int_{t_0}^{t} g(s)\, ds.$$

Note that if $x : [t_0, t_1] \to \mathbb{R}$ is PWS on $[t_0, t_1]$ and there are two PWC functions $g_i(\cdot)$ and constants c_i, $i = 1, 2$ such that for all $t \in [t_0, t_1]$

$$x(t) = c_1 + \int_{t_0}^{t} g_1(s)\, ds$$

and

$$x(t) = c_2 + \int_{t_0}^{t} g_2(s)\, ds,$$

then $c_1 = x(t_0) = c_2$. Also,

$$\int_{t_0}^{t} (g_1(s) - g_2(s)) ds = 0$$

for all $t \in [t_0, t_1]$. It is straightforward to show that $g_1(t) = g_2(t)$ except at a finite number of points,

$$g_1(t^+) = g_2(t^+)$$

and

$$g_1(t^-) = g_2(t^-)$$

for all $t \in (t_0, t_1)$. Hence, $g_1(t)$ and $g_2(t)$ can only have discontinuities at the same discrete points $\hat{t}_i$ and between these points $g_1(t) = g_2(t)$ if $t \neq \hat{t}_i$.

If $x : [t_0, t_1] \to \mathbb{R}$ is PWS on $[t_0, t_1]$, then one could redefine $x : [t_0, t_1] \to \mathbb{R}$ at a finite number of points, but the resulting function would no longer be continuous. In particular, if $\hat{x}(\cdot)$ and $x(\cdot)$ are equal except at a finite number of points and $\hat{x}(\cdot)$ is continuous, then $\hat{x}(\cdot)$ is uniquely defined.

Thus, piecewise smooth functions are continuous and $\dot{x}(t) = g(t)$ exists except at a finite number of points. In addition, at the points $\hat{t}_i, i = 1, 2, ..., p-1$ the right-hand and left-hand derivatives exist and are given by

$$\frac{d^+x(\hat{t}_i)}{dt} = \dot{x}^+(\hat{t}_i) = g\left(\hat{t}_i^+\right) = \dot{x}(\hat{t}_i^+) \tag{2.35}$$

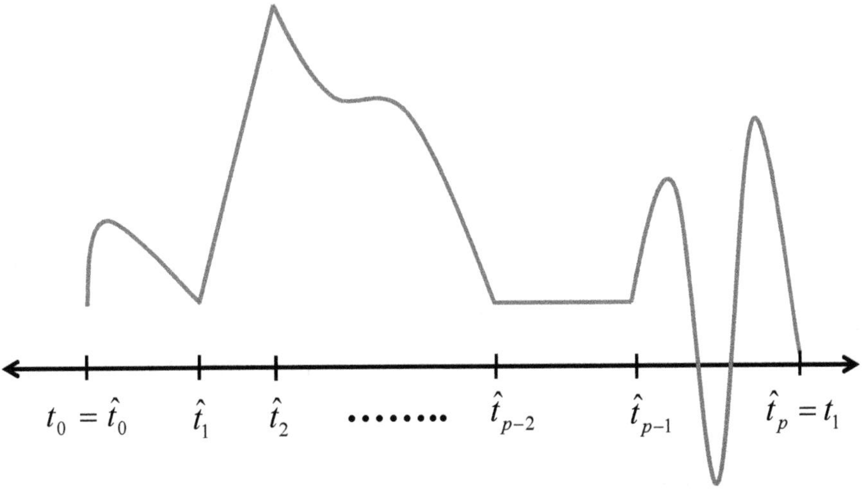

Figure 2.17: A Piecewise Smooth Function

and

$$\frac{d^- x(\hat{t}_i)}{dt} = \dot{x}^- \left(\hat{t}_i\right) = g\left(\hat{t}_i^-\right) = \dot{x}\left(\hat{t}_i^-\right), \tag{2.36}$$

respectively. Moreover, at the endpoints t_0 and t_1 it follows that

$$\frac{d^+ x(t_0)}{dt} = \dot{x}^+(t_0) = g\left(t_0^+\right) = \dot{x}(t_0^+) \tag{2.37}$$

and

$$\frac{d^- x(t_1)}{dt} = \dot{x}^- \left(t_1\right) = g\left(t_1^-\right) = \dot{x}\left(t_1^-\right), \tag{2.38}$$

respectively.

Definition 2.9 *If $x(\cdot)$ is piecewise smooth on $[t_0, t_1]$ and $t_0 < \hat{t} < t_1$ is such that $\dot{x}\left(\hat{t}^+\right) \neq \dot{x}\left(\hat{t}^-\right)$, then we say that $x(\cdot)$ has a* ***corner*** *at $\hat{t}$. From the remarks above, it follows that $\dot{x}\left(\hat{t}^+\right) = \lim_{t\to\hat{t}^+}[\dot{x}(t)] = g\left(\hat{t}^+\right)$ and $\dot{x}\left(\hat{t}^-\right) = \lim_{t\to\hat{t}^-}[\dot{x}(t)] = g\left(\hat{t}^-\right)$ always exist and are finite since $\dot{x}(t) = g(t)$ except at a finite number of points and $g(\cdot)$ is piecewise continuous.*

Before discussing specific spaces of functions, it is worthwhile to discuss what one means when we say that two functions are "equal" on a fixed interval $[t_0, t_1]$. For example, the functions $x(\cdot)$ and $z(\cdot)$ plotted in Figure 2.18 below are not equal at each point. However, for all "practical purposes" (like integration) they are essentially the same functions. Their values are the same except at a finite number of points. If $x : [t_0, t_1] \longrightarrow \mathbb{R}$ and $z : [t_0, t_1] \longrightarrow \mathbb{R}$ are two functions such that $x(t) = z(t)$ *except at an finite number of points* we shall write

$$x(\cdot) = z(\cdot) \quad e.f. \tag{2.39}$$

and, unless otherwise noted, we will rarely distinguish between $x(\cdot)$ and $z(\cdot)$. The functions $x(t)$ and $z(t)$ in Figure 2.18 are equal $e.f.$

We denote the space of all real-valued piecewise continuous functions defined on $[t_0, t_1]$ by $PWC(t_0, t_1)$ and the space of all real-valued piecewise smooth functions defined on $[t_0, t_1]$ by $PWS(t_0, t_1)$.

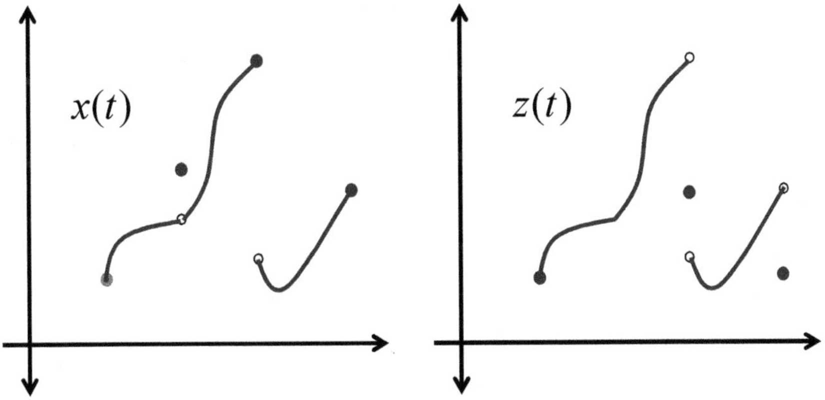

Figure 2.18: The Functions $x(\cdot)$ and $z(\cdot)$ are Equal *e.f.*

Remark 2.10 ***Important Remark on Notation:*** *Recall that each* $z(\cdot) \in PWS(t_0, t_1)$ *is continuous. Therefore, if* $x(\cdot) \in PWS(t_0, t_1)$ *and* $z(\cdot) \in PWS(t_0, t_1)$*, then* $x(\cdot) = z(\cdot)$ *e.f. if and only if* $x(t) = z(t)$ *for all* $t \in [t_0, t_1]$*. Also note that if* $x(\cdot) \in PWC(t_0, t_1)$*,* $z(\cdot) \in PWS(t_0, t_1)$ *and* $x(\cdot) = z(\cdot)$ *e.f., then the left and right limits of* $x(\cdot)$ *are equal at all points of* (t_0, t_1)*. In particular,*

$$x(t^+) = z(t^+) = z(t^-) = x(t^-).$$

Clearly, $z(\cdot)$ *is the only continuous function satisfying* $x(\cdot) = z(\cdot)$ *e.f. and since* $x(t^+) = z(t^+) = z(t^-) = x(t^-)$ *we shall make no distinction between* $x(\cdot)$ *and its* continuous representation $z(\cdot)$. ***Thus, for future reference we shall always use the equivalent continuous representation*** $z(\cdot)$ ***of*** $x(\cdot)$ ***when it exists and not distinguish between*** $x(\cdot)$ ***and*** $z(\cdot)$*.*

If $[t_0, +\infty)$ is a semi-infinite interval, then we say that $x(\cdot) \in PWS(t_0, +\infty)$ if $x(\cdot) \in PWS(t_0, T)$ for all $t_0 < T < +\infty$. Thus, we denote the space of all real-valued piecewise continuous functions defined on $[t_0, +\infty)$ by $PWC(t_0, +\infty)$ and the space of all real-valued piecewise smooth functions defined on $[t_0, +\infty)$ by $PWS(t_0, +\infty)$.

2.3.1 Distances between Functions

We shall need to consider what is meant by two functions "being close" in some sense. Although there are many possible definitions of "the distance" between $x(\cdot)$ and $z(\cdot) \in PWS(t_0, t_1)$, we shall consider only two specific metrics.

Definition 2.10 *If $x(\cdot)$ and $z(\cdot) \in PWS(t_0, t_1)$, then the d_0* ***distance between*** *two piecewise smooth functions $x(\cdot)$ and $z(\cdot)$ with domain $[t_0, t_1]$ is defined by*

$$d_0(x(\cdot), z(\cdot)) \triangleq \sup_{t_0 \le t \le t_1} \{|x(t) - z(t)|\}. \tag{2.40}$$

In this case we can define a norm on $PWS(t_0, t_1)$ by

$$\|x(\cdot)\|_0 = \sup_{t_0 \le t \le t_1} \{|x(t)|\},$$

and note that

$$d_0(x(\cdot), z(\cdot)) = \|x(\cdot) - z(\cdot)\|_0 .$$

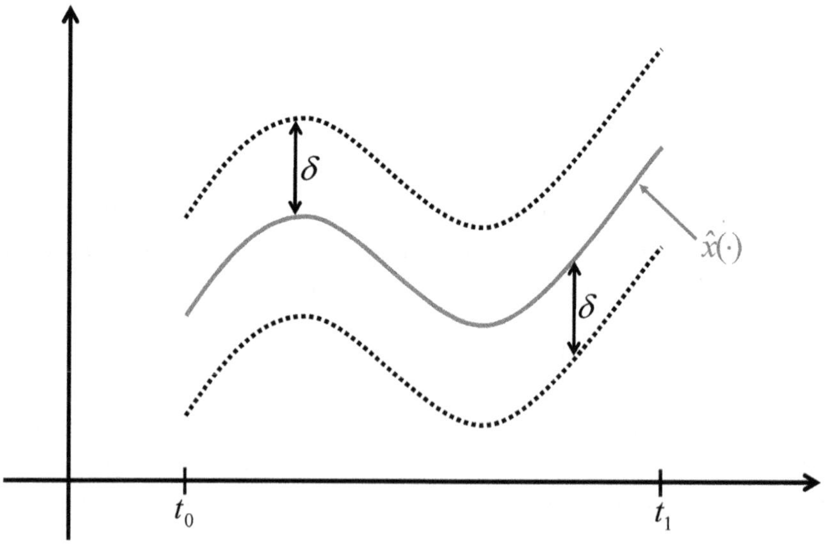

Figure 2.19: A $U_0(\hat{x}(\cdot), \delta)$-neighborhood of $\hat{x}(\cdot)$

Definition 2.11 *If $\hat{x}(\cdot) \in PWS(t_0, t_1)$ and $\delta > 0$, the $U_0(\hat{x}(\cdot), \delta)$-* ***neighborhood (or Strong Neighborhood) of*** *$\hat{x}(\cdot)$ is defined to be the open ball*

$$U_0(\hat{x}(\cdot), \delta) = \{x(\cdot) \in PWS(t_0, t_1) \colon d_0(\hat{x}(\cdot), x(\cdot)) < \delta\}.$$

It is easy to visualize what such neighborhoods look like. Given $\hat{x}(\cdot)$ and $\delta > 0$, the $U_0(\hat{x}(\cdot), \delta)$-neighborhood of $\hat{x}(\cdot)$ is the set of all $x(\cdot) \in PWS(t_0, t_1)$ with graphs in a tube of radius δ about the graph of $\hat{x}(\cdot)$ (see the figure below).

As we see from above, $x(\cdot)$ and $z(\cdot) \in PWS(t_0, t_1)$ are "close" in the d_0 metric if their graphs are close. However, the derivatives can be greatly different. For example the two functions shown in the Figure 2.20 below have very different derivatives.

In order to "fix this problem" we need a different metric. Recall that if $x(\cdot)$ and $z(\cdot) \in PWS(t_0, t_1)$, then their derivatives $\dot{x}(\cdot)$ and $\dot{z}(\cdot)$ are PWC on I. In particular, it follows that:

- $\dot{x}(\cdot)$ and $\dot{z}(\cdot)$ are bounded on I.
- $\dot{x}(t^+)$ and $\dot{z}(t^+)$ exist (are finite) on $[t_0, t_1)$.
- $\dot{x}(t^-)$ and $\dot{z}(t^-)$ exist (are finite) on $(t_0, t_1]$.
- There is a (finite) partition of $[t_0, t_1]$, say $t_0 = \hat{t}_0 < \hat{t}_1 < \hat{t}_2 < \ldots < \hat{t}_{p-1} < \hat{t}_p = t_1$, such that both $\dot{x}(t)$ and $\dot{z}(t)$ exist and are continuous (and bounded) on each open subinterval $(\hat{t}_{i-1}, \hat{t}_i)$. We can now define a weaker notion of distance.

Definition 2.12 *The d_1* ***distance between*** *two piecewise smooth functions $x(\cdot)$ and $z(\cdot)$ with domain $[t_0, t_1]$ is defined by*

$$\begin{aligned} d_1(x(\cdot), z(\cdot)) = &\sup\{|x(t) - z(t)| : t_0 \le t \le t_1\} \\ &+ \sup\{|\dot{x}(t) - \dot{z}(t)| : t_0 \le t \le t_1, \quad t \ne \hat{t}_i\}. \end{aligned} \tag{2.41}$$

Definition 2.13 *If $\hat{x}(\cdot) \in PWS(t_0, t_1)$ and $\delta > 0$, the $U_1(\hat{x}(\cdot), \delta)$-* ***neighborhood (or weak neighborhood) of*** *$\hat{x}(\cdot)$ is defined to be the open ball*

$$U_1(\hat{x}(\cdot), \delta) = \{x(\cdot) \in PWS(t_0, t_1) \colon d_1(\hat{x}(\cdot), x(\cdot)) < \delta\}.$$

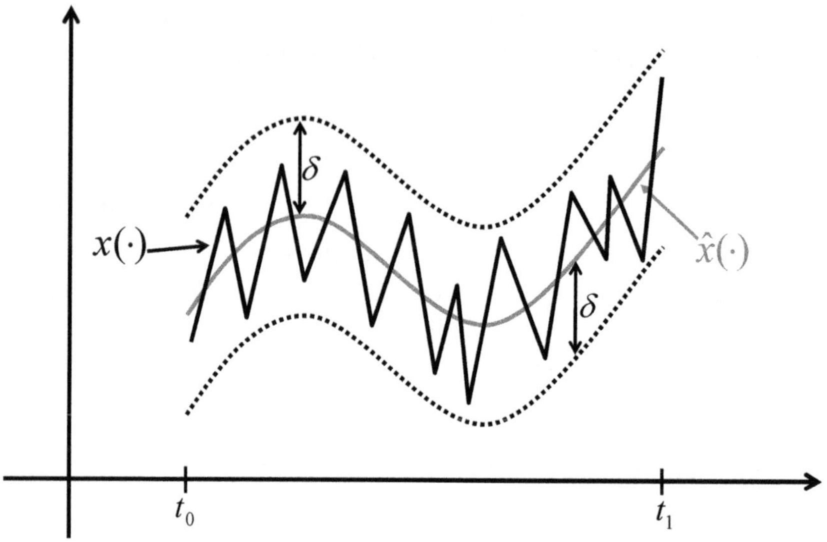

Figure 2.20: A Non-smooth Function in a $U_0(\hat{x}(\cdot), \delta)$-neighborhood of $\hat{x}(\cdot)$

Remark 2.11 *Note that the d_1 distance between $x(\cdot)$ and $z(\cdot)$ is given by*

$$d_1(x(\cdot), z(\cdot)) = d_0(x(\cdot), z(\cdot)) + \sup\{|\dot{x}(t) - \dot{z}(t)| : t_0 \le t \le t_1, \; t \ne \hat{t}_i\}. \tag{2.42}$$

If $x(\cdot)$ and $z(\cdot) \in PWS(t_0, t_1)$ and $d_1(x(\cdot), z(\cdot)) = 0$, then $x(t) = z(t)$ for all $t \in [t_0, t_1]$ and $\dot{x}(t) = \dot{z}(t)$ e.f. Also, since

$$d_0(x(\cdot), z(\cdot)) \le d_1(x(\cdot), z(\cdot)),$$

it follows that if $d_1(x(\cdot), z(\cdot)) < \delta$, then $d_0(x(\cdot), z(\cdot)) < \delta$. It is important to note that this inequality implies that

$$U_1(\hat{x}(\cdot), \delta) \subset U_0(\hat{x}(\cdot), \delta) \subset PWS(t_0, t_1), \tag{2.43}$$

so that the $U_1(\hat{x}(\cdot), \delta)$-neighborhood $U_1(\hat{x}(\cdot), \delta)$ is smaller than the $U_0(\hat{x}(\cdot), \delta)$-neighborhood $U_0(\hat{x}(\cdot), \delta)$ (see Figure 2.21).

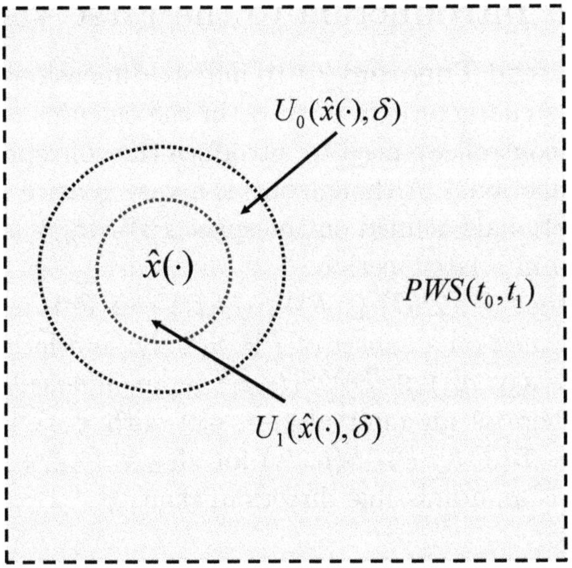

Figure 2.21: A Comparison of $U_0(\hat{x}(\cdot),\delta)$ and $U_1(\hat{x}(\cdot),\delta)$ - neighborhoods of $\hat{x}(\cdot)$

Two functions $x(\cdot)$ and $z(\cdot)$ are "close" in the d_0 sense if their graphs are within δ of each other at all points. It is helpful (although not quite accurate) to think of two functions $x(\cdot)$ and $z(\cdot)$ as "close" in the d_1 sense if their graphs and the graphs of $\dot{x}(\cdot)$ and $\dot{z}(\cdot)$ are within δ of each other except at a finite number of points.

Remark 2.12 ***Another Remark on Notation:*** *If $x(\cdot)$ is a piecewise smooth function, then $\dot{x}(t)$ is defined except perhaps at a finite number of points say, $t_0 = \hat{t}_0 < \hat{t}_1 < \hat{t}_2 < \ldots < \hat{t}_{p-1} < \hat{t}_p = t_1$. However, at these points the left and right derivatives exist and are given by $\dot{x}(\hat{t}_i^-)$ and $\dot{x}(\hat{t}_i^+)$, respectively. In order to keep notation at a minimum, when we use $\dot{x}(t)$ in any expression, we mean that this expression holds for all points t where $\dot{x}(t)$ exists. Also, recall that at corners $\hat{t}$ where $\dot{x}(\hat{t})$ do not exist,* ***both*** *$\dot{x}(\hat{t}^-)$* ***and*** *$\dot{x}(\hat{t}^+)$ exist and are finite. We shall treat conditions at corners as they arise in the development of the material.*

2.3.2 An Introduction to the First Variation

In order to extend **Theorem** 2.1 to infinite dimensional optimization problems characteristic of those in the calculus of variations and optimal control, we need to introduce the concept of a "Variation" of a functional. Although this is a very general concept, we focus on functionals defined on the space $PWS(t_0, t_1)$ and extend the definition in a later section.

Assume that $J : \mathcal{D}(J) \subseteq PWS(t_0, t_1) \longrightarrow \mathbb{R}$ is a real-valued functional defined on a subset of $PWS(t_0, t_1)$ and let $\hat{x}(\cdot) \in \mathcal{D}(J)$ be given. We say $\eta(\cdot) \in PWS(t_0, t_1)$ is an *admissible direction for* $\hat{x}(\cdot)$ if there is an interval $(-\hat{\varepsilon}, +\hat{\varepsilon})$ with $\hat{\varepsilon} > 0$ such that $\hat{x}(\cdot) + \varepsilon\eta(\cdot) \in \mathcal{D}(J) \subseteq PWS(t_0, t_1)$ for all $\varepsilon \in (-\hat{\varepsilon}, +\hat{\varepsilon})$. If $\eta(\cdot) \in PWS(t_0, t_1)$ is an admissible direction, then

$$F(\varepsilon) = J(\hat{x}(\cdot) + \varepsilon\eta(\cdot)) \tag{2.44}$$

defines a real-valued function of the real variable ε on the interval $(-\hat{\varepsilon}, +\hat{\varepsilon})$. In this case we have the following definition.

Definition 2.14 *If* $\hat{x}(\cdot) \in PWS(t_0, t_1)$, $\eta(\cdot) \in PWS(t_0, t_1)$ *is an admissible direction for* $\hat{x}(\cdot)$ *and* $F(\cdot) : (-\hat{\varepsilon}, +\hat{\varepsilon}) \longrightarrow \mathbb{R}$ *has a derivative at* $\varepsilon = 0$, *then the* ***first variation of*** $J(\cdot)$ ***at*** $\hat{x}(\cdot)$ ***in the direction of*** $\eta(\cdot)$ *is denoted by* $\delta J(\hat{x}(\cdot); \eta(\cdot))$ *and is defined by*

$$\delta J(\hat{x}(\cdot); \eta(\cdot)) = \frac{d}{d\varepsilon}F(\varepsilon)\bigg|_{\varepsilon=0} = \frac{d}{d\varepsilon}\left[J(\hat{x}(\cdot) + \varepsilon\eta(\cdot))\right]\bigg|_{\varepsilon=0}. \tag{2.45}$$

Likewise, we say $\eta(\cdot) \in PWS(t_0, t_1)$ is a *right admissible direction* for $\hat{x}(\cdot)$ if there is an $\hat{\varepsilon} > 0$ such that that $\hat{x}(\cdot) + \varepsilon\eta(\cdot) \in \mathcal{D}(J) \subseteq PWS(t_0, t_1)$ for all $\varepsilon \in [0, +\hat{\varepsilon})$. If $\eta(\cdot) \in PWS(t_0, t_1)$ is a right admissible direction, then $F(\varepsilon) = J(\hat{x}(\cdot) + \varepsilon\eta(\cdot))$ defines a real-valued function of the real variable ε on the interval $[0, +\hat{\varepsilon})$. If the right-hand derivative

$$\frac{d^+}{d\varepsilon}F(\varepsilon)\bigg|_{\varepsilon=0} = \frac{d^+}{d\varepsilon}\left[J(\hat{x}(\cdot) + \varepsilon\eta(\cdot))\right]\bigg|_{\varepsilon=0} \tag{2.46}$$

exists, then

$$\delta^+ J(\hat{x}(\cdot);\eta(\cdot)) = \left.\frac{d^+}{d\varepsilon}F(\varepsilon)\right|_{\varepsilon=0} = \left.\frac{d^+}{d\varepsilon}\left[J(\hat{x}(\cdot)+\varepsilon\eta(\cdot))\right]\right|_{\varepsilon=0}$$

is the **right-hand first variation of $J(\cdot)$ at $\hat{x}(\cdot)$ in the direction of $\eta(\cdot)$**. We will abuse notation and write

$$\delta J(\hat{x}(\cdot);\eta(\cdot)) = \delta^+ J(\hat{x}(\cdot);\eta(\cdot))$$

even when $\eta(\cdot) \in PWS(t_0,t_1)$ is only a *right admissible direction* for $\hat{x}(\cdot)$.

If $\hat{x}(\cdot) \in PWS(t_0,t_1)$, $\eta(\cdot) \in PWS(t_0,t_1)$ is an admissible direction for $\hat{x}(\cdot)$ and $F(\cdot) : (-\hat{\varepsilon}, +\hat{\varepsilon}) \longrightarrow \mathbb{R}$ has a second derivative at $\varepsilon = 0$, then the **second variation of $J(\cdot)$ at $\hat{x}(\cdot)$ in the direction of $\eta(\cdot)$** is denoted by $\delta^2 J(\hat{x}(\cdot);\eta(\cdot))$ and is defined by

$$\delta^2 J(\hat{x}(\cdot);\eta(\cdot)) = \left.\frac{d^2}{d\varepsilon^2}F(\varepsilon)\right|_{\varepsilon=0} = \left.\frac{d^2}{d\varepsilon^2}\left[J(\hat{x}(\cdot)+\varepsilon\eta(\cdot))\right]\right|_{\varepsilon=0}. \quad (2.47)$$

2.4 Mathematical Formulation of Problems

In this section we return to some of the motivating problems outlined at the beginning of this chapter. Now that we have carefully defined specific classes of functions, it is possible to develop precise mathematical formulations of these problems.

2.4.1 The Brachistochrone Problem

We consider the problem where the particle starts at $(0,0)$ and slides down a wire to the point $[a \;\; b]^T$ and there is no friction as again illustrated in Figure 2.22. Let g be the acceleration due to gravity and m be the mass of the bead. The velocity of the bead when it is located at $[x \;\; y]^T$ is denoted by $v(x)$. The kinetic energy of the bead is given by

$$T = \frac{1}{2}mv^2(x)$$

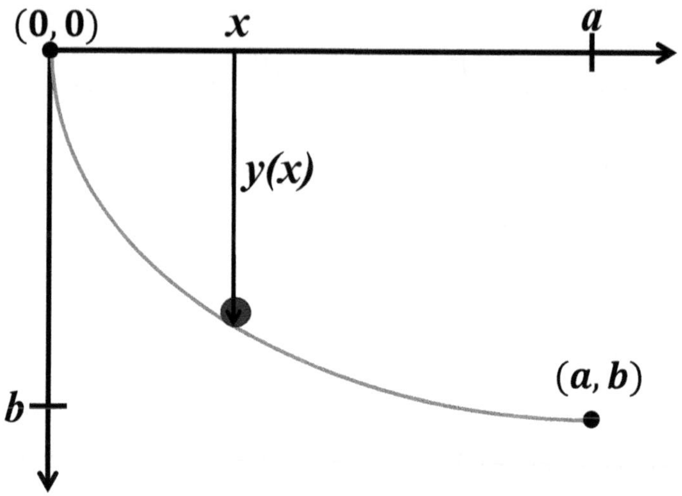

Figure 2.22: Mathematical Formulation of The Brachistochrone Problem

and the potential energy is given by

$$V = mg\left(-y\left(x\right)\right).$$

The total energy of the bead is

$$T + V$$

and conservation of energy requires that $T + V$ remains constant at all times. Since at $x = 0$, the kinetic energy is 0 (i.e. $v\left(0\right) = 0$) and the potential energy is 0 (i.e. $y\left(0\right) = 0$), we have that

$$T = -V,$$

or equivalently,

$$\frac{1}{2}mv^2 = mgy.$$

Now, we use some elementary calculus. The length of the path described by the graph of $y\left(x\right)$ is given by

$$s\left(x\right) = \int_0^x \sqrt{1 + \left[y'\left(\tau\right)\right]^2} d\tau.$$

The velocity is given by

$$v(x) = \frac{d}{dt}s(x) = \frac{d}{dx}s(x)\frac{dx}{dt} = \frac{\sqrt{1 + [y'(x)]^2}}{dt}dx.$$

However, $v(x) = \sqrt{2gy(x)}$ and hence we obtain

$$dt = \frac{\sqrt{1 + [y'(x)]^2}}{\sqrt{2gy(x)}}dx.$$

Integration gives the time of travel from $[0 \quad 0]^T$ to $[x \quad y(x)]^T$ as the integral

$$t(x) = \int_0^x \frac{\sqrt{1 + [y'(x)]^2}}{\sqrt{2gy(x)}}dx.$$

Hence, the time required for the bead to slide from $[0 \quad 0]^T$ to $[a \quad b]^T$ is given by

$$J(y(\cdot)) = t(a) = \int_0^a \frac{\sqrt{1 + [y'(x)]^2}}{\sqrt{2gy(x)}}dx. \tag{2.48}$$

Thus, we have derived a formula (2.48) for the time it takes the bead to slide down a given curve $y(\cdot)$. Observe that the time (2.48) depends only on the function $y(\cdot)$ and is independent of the mass of the bead. We can now state the mathematical formulation of the brachistochrone problem.

Among the set of all smooth functions $y : [0, a] \to \mathbb{R}$ satisfying

$$y(0) = 0, \quad y(a) = b,$$

find the function that minimizes

$$J(y(\cdot)) = \int_0^a \frac{\sqrt{1 + [y'(x)]^2}}{\sqrt{2gy(x)}}dx. \tag{2.49}$$

Note that in the derivation of (2.49) we assumed that $y(x)$ is smooth (i.e., that $y(x)$ and $y'(x)$ are continuous). Although we never explicitly assumed that $y(x) \geq 0$, this is needed to make the terms under the square root sign non-negative. Also, observe that at $x = 0$, $y(0) = 0$, so that the integrand is singular. All of these "little details" are not so important for the brachistochrone problem, but can become important in other cases.

2.4.2 The Minimal Surface of Revolution Problem

Consider the problem of generating a surface by revolving a continuous curve about the x-axis. Suppose we are given two points $P_0 = [a_0 \ \ b_0]^T$ and $P_1 = [a_1 \ \ b_1]^T$ in the plane. These two points are joined by a curve which is the graph of the function $y(\cdot)$. We assume that $y(\cdot)$ is smooth and we generate a surface by rotating the curve about the x-axis (see Figure 2.23 below). What shape should $y(\cdot)$ be so that this surface of revolution has minimum surface area?

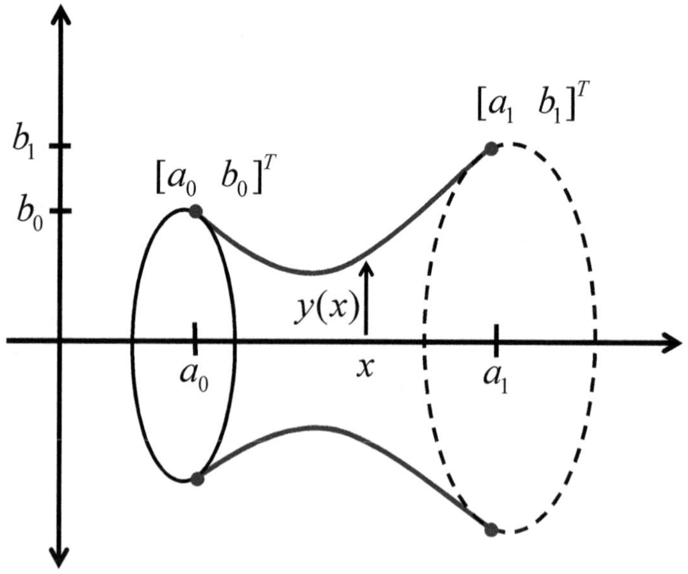

Figure 2.23: Minimal Surface of Revolution Problem

Again, we return to elementary calculus and recall the formula for surface area. If $y(\cdot) \geq 0$, then the surface area is given by

$$S = 2\pi \int_{a_0}^{a_1} y(x) \sqrt{1 + [y'(x)]^2} dx.$$

We are seeking, among the set of all smooth functions $y : [a_1, a_2] \to \mathbb{R}$ satisfying

$$y(a_0) = b_0, \quad y(a_1) = b_1,$$

a function $y^*(\cdot)$ that minimizes

$$J(y(\cdot)) = S = 2\pi \int_{a_0}^{a_1} y(x) \sqrt{1 + [y'(x)]^2} dx. \tag{2.50}$$

At this point you should have noticed that the brachistochrone problem and the problem of minimal surface of revolution are "almost" identical in form. Both problems involve finding a function, with fixed endpoints, that minimizes an integral among all sufficiently smooth functions that pass through these same endpoints.

2.4.3 The River Crossing Problem

Consider the river crossing problem discussed in section 2.1.2 and as illustrated in Figure 2.2. If the path of the boat is represented by the graph of the function $x(t)$, then it can be shown that the time it takes for the boat to cross the river is given by

$$time = \int_0^1 \frac{\sqrt{c^2(1 + [\dot{x}(s)]^2) - [v(s)]^2} - v(s)\dot{x}(s)}{c^2 - [v(s)]^2} ds, \tag{2.51}$$

where c is the constant speed of the boat in still water and we assume that $c^2 - [v(t)]^2 > 0$. The point of departure is given by $t_1 = 0$, $x_1 = 0$ so that

$$x(0) = 0. \tag{2.52}$$

Thus, the river crossing problem is to find a smooth function $x^*: [0,1] \longrightarrow \mathbb{R}$ that minimizes

$$J(x(\cdot)) = \int_0^1 \frac{\sqrt{c^2(1+[\dot{x}(s)]^2) - [v(s)]^2} - v(s)\dot{x}(s)}{c^2 - [v(s)]^2} ds \qquad (2.53)$$

among all smooth functions satisfying (2.52). Observe that unlike the brachistochrone problem and the problem of minimal surface of revolution, there is no specified condition at $x = 1$. Thus, this is a simple example of the classical free-endpoint problem.

2.4.4 The Rocket Sled Problem

Consider the rocket sled described in Section 2.1.4 above. We assume the sled is on a frictionless track, the sled is of mass m, it is controlled by two rocket engines thrusting in opposite directions and that the thrusting force of the rockets is bounded by a maximum of 1 so that $|u(t)| \leq 1$. Let $x(t)$ denote the displacement of the sled from the reference point R, at time t. Given that at time $t = 0$, the sled is at an initial position x_0 with initial velocity v_0, we wish to find a thrust force action that will move the sled to position R and stop it there, and we want to accomplish this transfer in minimal time. Newton's second law may be written as

$$m\ddot{x}(t) = u(t),$$

where $x(t)$ is the displacement from the base station at time t, m is the mass of the sled and $u(t)$ is the thrust. We assume that at time $t = 0$ the initial data is given by the initial displacement

$$x(0) = x_0,$$

and initial velocity

$$\dot{x}(0) = v_0.$$

The time optimal control problem is to find a *control* $u^*(t)$ that transfers the system from $[x_0 \;\; v]^T$ to $[0 \;\; 0]^T$ in minimal time, given that $|u(t)| \leq 1$.

To formulate the problem we write this as a first order system by defining

$$x_1(t) = x(t) = \text{position at time } t,$$
$$x_2(t) = \dot{x}(t) = \text{velocity at time } t,$$

with initial data

$$x_1(0) = x_{1,0} \triangleq x_0,$$
$$x_2(0) = x_{2,0} \triangleq v_0.$$

The system becomes

$$\frac{d}{dt}\begin{bmatrix} x_1(t) \\ x_2(t) \end{bmatrix} = \begin{bmatrix} 0 & 1 \\ 0 & 0 \end{bmatrix}\begin{bmatrix} x_1(t) \\ x_2(t) \end{bmatrix} + \begin{bmatrix} 0 \\ 1/m \end{bmatrix} u(t), \tag{2.54}$$

or

$$\dot{\boldsymbol{x}}(t) = A\boldsymbol{x}(t) + Bu(t), \quad \boldsymbol{x}(0) = \boldsymbol{x}_0 \tag{2.55}$$

where the matrices A and B are defined by

$$A = \begin{bmatrix} 0 & 1 \\ 0 & 0 \end{bmatrix} \text{ and } B = \begin{bmatrix} 0 \\ 1/m \end{bmatrix},$$

respectively. Here, $\boldsymbol{x}_0 = \begin{bmatrix} x_{1,0} & x_{2,0} \end{bmatrix}^T$ and $\boldsymbol{x}(t) = \begin{bmatrix} x_1(t) & x_2(t) \end{bmatrix}^T$ is the trajectory in the plane $\mathbb{R}^2$. Given a control $u(\cdot)$ we let $\boldsymbol{x}(t; u(\cdot))$ denote the solution of the initial value problem (2.55) with control input $u(\cdot)$.

We say that a control $u(t)$ *steers* $\boldsymbol{x}_0 = \begin{bmatrix} x_{1,0} & x_{2,0} \end{bmatrix}^T$ to a state $\boldsymbol{x}_1 = \begin{bmatrix} x_{1,1} & x_{2,1} \end{bmatrix}^T$ in time t_1 if there is a solution $\boldsymbol{x}(t; u(\cdot))$ to (2.55) satisfying

$$\boldsymbol{x}(0; u(\cdot)) = \boldsymbol{x}_0 \tag{2.56}$$

and

$$\boldsymbol{x}(t_1; u(\cdot)) = \boldsymbol{x}_1 \tag{2.57}$$

for some finite time $t_1 > 0$. The time optimal problem is to find the control $u^*(\cdot)$ such that

$$|u^*(t)| \leq 1,$$

and $u^*(\cdot)$ steers $\boldsymbol{x}_0 = \begin{bmatrix} x_{1,0} & x_{2,0} \end{bmatrix}^T$ to $\boldsymbol{x}_1 = \begin{bmatrix} x_{1,1} & x_{2,1} \end{bmatrix}^T$ in minimal time.

Observe that here the final time t_1 is not fixed and the cost function is very simple. In particular, the cost is the final time and this can be represented by the integral

$$J(\boldsymbol{x}(\cdot), u(\cdot)) = t_1 = \int_0^{t_1} 1 ds. \tag{2.58}$$

This time optimal control problem can be "solved" by several means, including a Lagrange Multiplier method. Also, since this particular problem is rather simple, one can use a geometric analysis of the problem and the optimal controller can be synthesized by means of a switching locus. We will present this solution in Chapter 9 because the basic ideas are useful in understanding the maximum principle and how one might derive this necessary condition.

2.4.5 The Finite Element Method

In its purest form the finite element method is intimately connected to variational calculus, optimization theory and Galerkin approximation methods. The books by Strang and Fix (see [170, 171]) provide an excellent introduction to the finite element method. The references [44], [96], [98], and [179] provide examples where the finite element method has been applied to a variety of areas in science and engineering.

The finite element method and its extensions are among the most powerful computational tools for solving complex ordinary and partial differential equations. We shall illustrate how one formulates the finite element approximation for the simple two-point boundary value problem described by (2.4)-(2.5) above. Later we shall see how the calculus of variations provides the mathematical framework and theory necessary to analyze the convergence of the method.

Here we focus on the application of the finite element method to a simple example defined by the two-point boundary value problem

$$-\ddot{x}(t) + x(t) = f(t), \quad 0 < t < 1, \tag{2.59}$$

subject to the Dirichlet boundary conditions

$$x(0) = 0, \quad x(1) = 0. \tag{2.60}$$

A *strong* (or classical) solution is a twice differentiable function $x(\cdot)$ that satisfies (2.59)-(2.60) at every value $0 \le t \le 1$. We are interested in developing a numerical algorithm for approximating solutions.

As noted above, in order to construct the finite element approximation one divides the interval $(0, 1)$ into N subintervals (called *elements*) with nodes $0 = \hat{t}_0 < \hat{t}_1 < \hat{t}_2 < \ldots < \hat{t}_{N-1} < \hat{t}_N < \hat{t}_{N+1} = 1$ and constructs the approximation $x_N(\cdot)$ so that it is continuous on all of $[0, 1]$ and linear between the nodes. Since continuous piecewise linear approximating functions $x_N(\cdot)$ are not typically differentiable at the nodes, it is not possible to insert this approximation directly into equation (2.59). In particular, the piecewise smooth function $x_N(\cdot)$ has only a piecewise continuous derivative $\dot{x}_N(\cdot)$ and hence $\ddot{x}_N(\cdot)$ does not exist. In order to deal with this lack of smoothness, we must define the concept of weak solutions.

Before introducing this approximation, one derives a weak (or variational) form of this equation. If $x(\cdot)$ is a solution to (2.59) - (2.60), then multiplying both sides of (2.59) by any function $\eta(\cdot)$ yields

$$[-\ddot{x}(t) + x(t)]\eta(t) = f(t)\eta(t), \quad 0 < t < 1. \tag{2.61}$$

If one assumes that $\eta(\cdot)$ is piecewise continuous, then integrating both sides of (2.61) implies that

$$\int_0^1 [-\ddot{x}(t) + x(t)]\eta(t)dt = \int_0^1 f(t)\eta(t)dt,$$

or equivalently,

$$\int_0^1 -\ddot{x}(t)\eta(t)dt + \int_0^1 x(t)\eta(t)dt = \int_0^1 f(t)\eta(t)dt, \tag{2.62}$$

for any function $\eta(\cdot) \in PWC(0,1)$. If $\eta(\cdot)$ is piecewise smooth, then one can use integration by parts on the first term in (2.62) which implies that

$$-\dot{x}(t)\eta(t)|_{t=0}^{t=1} + \int_0^1 \dot{x}(t)\dot{\eta}(t)dt + \int_0^1 x(t)\eta(t)dt = \int_0^1 f(t)\eta(t)dt, \tag{2.63}$$

for any function $\eta(\cdot) \in PWS(0,1)$. If in addition $\eta(\cdot) \in PWS(0,1)$ satisfies the boundary conditions (2.60), then the boundary terms in (2.63) are zero so that

$$\int_0^1 \dot{x}(t)\dot{\eta}(t)dt + \int_0^1 x(t)\eta(t)dt = \int_0^1 f(t)\eta(t)dt, \tag{2.64}$$

for all $\eta(\cdot) \in PWS(0,1)$ satisfying

$$\eta(0) = 0, \quad \eta(1) = 0. \tag{2.65}$$

What we have shown is that if $x(\cdot)$ is a solution of the two-point boundary value problem (2.59)-(2.60), then $x(\cdot)$ satisfies (2.64) for all $\eta(\cdot) \in PWS(0,1)$ that satisfy the Dirichlet boundary conditions (2.65). Observe that the equation (2.64) does not involve the second derivative of the function $x(\cdot)$ and equation (2.64) makes sense as long as $x(\cdot)$ is just piecewise smooth on $(0,1)$.

Let $PWS_0(0,1)$ denote the space of all functions $z(\cdot) \in PWS(0,1)$ satisfying $z(0) = 0$ and $z(1) = 0$. In particular,

$$PWS_0(0,1) = \{z(\cdot) \in PWS(0,1) : z(0) = 0, \quad z(1) = 0\}. \tag{2.66}$$

We now define what we mean by a weak solution of the boundary value problem (2.59)-(2.60).

Definition 2.15 (Weak Solution) *We say that the function $x(\cdot)$ is a **weak solution** of the two-point boundary value problem (2.59)-(2.60), if:*
(1) $x(\cdot) \in PWS_0(0,1)$,
(2) $x(\cdot)$ satisfies (2.64) for all $\eta(\cdot) \in PWS_0(0,1)$.

Observe that we have shown that a solution (in the strong sense) to the two-point boundary value problem (2.59)-(2.60) is always a weak solution. We shall show later that the Fundamental Lemma of the Calculus of Variations can be used to prove that a weak solution is also a strong solution. Therefore, for this one dimensional problem, weak and strong solutions are the same. Now we return to the issue of approximating this solution.

We begin by dividing the interval $[0,1]$ into $N+1$ subintervals (called *elements*) of length $\Delta = 1/(N+1)$ with nodes $0 = \hat{t}_0 < \hat{t}_1 < \hat{t}_2 < \ldots < \hat{t}_{N-1} < \hat{t}_N < \hat{t}_{N+1} = 1$, where for $i = 0, 1, 2, \ldots, N, N+1$, $\hat{t}_i = i \cdot \Delta$. For $i = 0, 1, 2, \ldots, N, N+1$, define the *hat functions* $h_i(\cdot)$ on $[0,1]$ by

$$h_0(t) = \begin{cases} (\hat{t}_1 - t)/\Delta, & 0 \le t \le \hat{t}_1 \\ 0, & \hat{t}_1 \le t \le 1 \end{cases},$$

$$h_{N+1}(t) = \begin{cases} (t - \hat{t}_N)/\Delta, & \hat{t}_N \le t \le 1 \\ 0, & 0 \le t \le \hat{t}_N \end{cases}, \tag{2.67}$$

$$h_i(t) = \begin{cases} (t - \hat{t}_{i-1})/\Delta, & \hat{t}_{i-1} \le t \le \hat{t}_i \\ (\hat{t}_{i+1} - t)/\Delta, & \hat{t}_i \le t \le \hat{t}_{i+1} \\ 0, & t \notin (\hat{t}_{i-1}, \hat{t}_{i+1}) \end{cases}, \text{ for } i = 1, 2, \ldots, N.$$

Plots of these functions are shown in Figures 2.24, 2.25 and 2.26 below.

These hat functions provide a basis for all continuous piecewise linear functions with (possible) corners at the nodes $\hat{t}_1 < \hat{t}_2 < \ldots < \hat{t}_{N-1} < \hat{t}_N$. Therefore, any continuous piecewise linear function $y_N(\cdot)$ with corners only at these nodes can be written as

$$y_N(t) = \sum_{i=0}^{N+1} y_i h_i(t), \tag{2.68}$$

where the numbers y_i determine the value of $y_N(t)$ at $\hat{t}_i$. In particular, $y_N(\hat{t}_i) = y_i$ and in order to form the function $y_N(\cdot)$ one must provide the coefficients y_i for $i = 0, 1, 2, \ldots, N, N+1$. Moreover, if $y_N(t)$ is assumed to satisfy the Dirichlet boundary conditions (2.60), then $y_N(\hat{t}_0) = y_N(0) = y_0 = 0$ and $y_N(\hat{t}_{N+1}) = y_N(1) = y_{N+1} = 0$ and $y_N(\cdot)$ can be written as

$$y_N(t) = \sum_{i=1}^{N} y_i h_i(t). \tag{2.69}$$

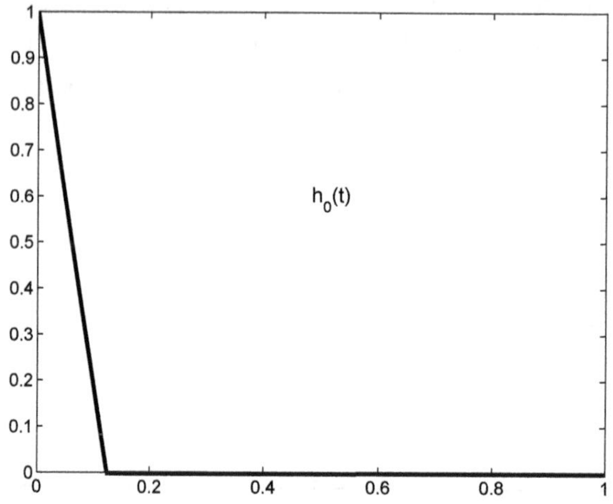

Figure 2.24: The Hat Function $h_0(\cdot)$

If we seek an approximate continuous piecewise linear solution $x_N(\cdot)$ to the weak form of the two-point value problem (2.59)-

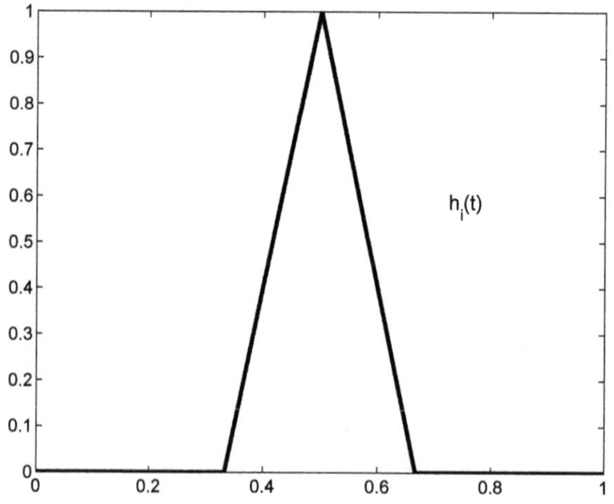

Figure 2.25: The Hat Function $h_i(\cdot)$

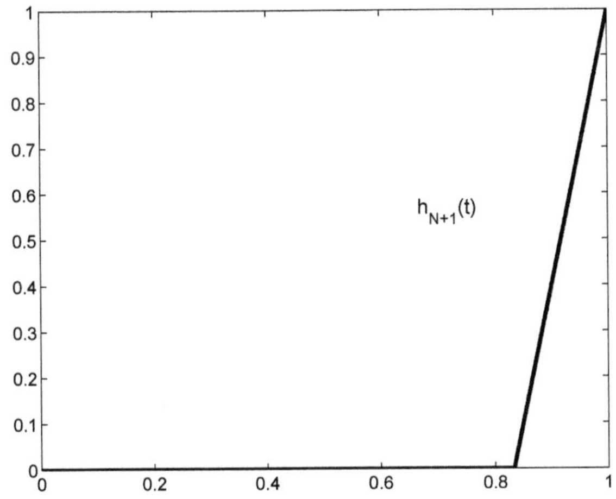

Figure 2.26: The Hat Function $h_{N+1}(\cdot)$

(2.60), then $x_N(\cdot)$ must have the form

$$x_N(t) = \sum_{i=1}^{N} x_i h_i(t) \tag{2.70}$$

and we need only to "compute" the values x_i for $i = 1, 2, \ldots, N$. We do this by substituting $x_N(t) = \sum_{i=1}^{N} x_i h_i(t)$ into the weak form of the equation given by (2.64). In particular, $x_N(t)$ is assumed to satisfy

$$\int_0^1 \dot{x}_N(t)\dot{\eta}(t)dt + \int_0^1 x_N(t)\eta(t)dt = \int_0^1 f(t)\eta(t)dt, \tag{2.71}$$

for all $\eta(\cdot) \in PWS_0(0,1)$.

Substituting $x_N(t) = \sum_{i=1}^{N} x_i h_i(t)$ into the weak equation (2.71) yields

$$\int_0^1 \left(\sum_{i=1}^N x_i \dot{h}_i(t)\right) \dot{\eta}(t)dt + \int_0^1 \left(\sum_{i=1}^N x_i h_i(t)\right) \eta(t)dt$$
$$= \int_0^1 f(t)\eta(t)dt,$$

for all $\eta(\cdot) \in PWS_0(0,1)$. This equation can be written as

$$\sum_{i=1}^N x_i \left(\int_0^1 \dot{h}_i(t)\dot{\eta}(t)dt\right) + \sum_{i=1}^N x_i \left(\int_0^1 h_i(t)\eta(t)dt\right)$$
$$= \int_0^1 f(t)\eta(t)dt, \tag{2.72}$$

for all $\eta(\cdot) \in PWS_0(0,1)$. In order to use the variational equation to compute the coefficients x_i for $i = 1, 2, \ldots, N$, we note that each basis function $h_i(\cdot)$ belongs to $PWS_0(0,1)$ for $i = 1, 2, \ldots, N$. Therefore, setting $\eta(\cdot) = h_j(t) \in PWS_0(0,1)$ for each index $j = 1, 2, \ldots, N$, yields N equations

$$\sum_{i=1}^N x_i \left(\int_0^1 \dot{h}_i(t)\dot{h}_j(t)dt\right) + \sum_{i=1}^N x_i \left(\int_0^1 h_i(t)h_j(t)dt\right)$$
$$= \int_0^1 f(t)h_j(t)dt. \tag{2.73}$$

Define the $N \times N$ *mass matrix* $\boldsymbol{M} = \boldsymbol{M}_N$ by

$$\boldsymbol{M} = \boldsymbol{M}_N = [m_{i,j}]_{i,j=1,2,\ldots,N},$$

where the entries $m_{i,j}$ of $\boldsymbol{M}_N$ are given by the integrals

$$m_{i,j} = \int_0^1 h_j(t)h_i(t)dt.$$

Likewise, define the $N \times N$ *stiffness matrix* $\boldsymbol{K} = \boldsymbol{K}_N$ by

$$\boldsymbol{K} = \boldsymbol{K}_N = [k_{i,j}]_{i,j=1,2,\ldots,N},$$

where the entries $k_{i,j}$ of $\boldsymbol{K}_N$ are given by the integrals

$$k_{i,j} = \int_0^1 \dot{h}_j(t)\dot{h}_i(t)dt.$$

Finally, let $\boldsymbol{f}_N$ be the $N \times 1$ (column) vector defined by

$$\boldsymbol{f}_N = \begin{bmatrix} f_1 & f_2 & \cdots & f_N \end{bmatrix}^T,$$

where entries f_j of $\boldsymbol{f}_N$ are given by the integrals

$$f_j = \int_0^1 h_j(t)f(t)dt.$$

If $\boldsymbol{x}$ is the solution vector

$$\boldsymbol{x} = \begin{bmatrix} x_1 \\ x_2 \\ \vdots \\ x_N \end{bmatrix},$$

of (2.73), then $\boldsymbol{x}$ satisfies the matrix system

$$\boldsymbol{K}_N\boldsymbol{x} + \boldsymbol{M}_N\boldsymbol{x} = \boldsymbol{f}_N. \tag{2.74}$$

Defining

$$\boldsymbol{L}_N = \boldsymbol{K}_N + \boldsymbol{M}_N$$

yields the algebraic equation

$$\boldsymbol{L}_N\boldsymbol{x} = \boldsymbol{f}_N \tag{2.75}$$

which must be solved to find x_i for $i = 1, 2, \ldots, N$. This approach provides a mechanism for finding the (finite element) approximation $x_N(t) = \sum_{i=1}^{N} x_i h_i(t)$. In particular, one solves the matrix equation (2.75) for $\boldsymbol{x} = \begin{bmatrix} x_1 & x_2 & \cdots & x_N \end{bmatrix}^T$ and then constructs the piecewise smooth *finite element* approximate function $x_N(t) = \sum_{i=1}^{N} x_i h_i(t)$.

There are many issues that need to be addressed in order to prove that this method actually produces accurate approximations of the solution to the two-point boundary value problem (2.59) - (2.60). What one might expect (hope) is that as $N \to +\infty$ the functions $x_N(\cdot)$ converge to $x(\cdot)$ in some sense. We shall return to this issue later, but we note here that this process breaks down if the matrix $\boldsymbol{L}_N$ is not invertible. In particular, one needs to prove that $\boldsymbol{L}_N$ is non-singular.

For this problem one can show that the mass and stiffness matrices are given by

$$\boldsymbol{M}_N = \frac{\Delta}{6}\begin{bmatrix} 4 & 1 & 0 & 0 & \cdots & 0 & 0 \\ 1 & 4 & 1 & 0 & \cdots & 0 & 0 \\ 0 & 1 & 4 & 1 & \cdots & 0 & 0 \\ 0 & 0 & 1 & 4 & \cdots & 0 & 0 \\ \vdots & \vdots & \vdots & \vdots & \ddots & \vdots & \vdots \\ 0 & 0 & \cdots & \cdots & 1 & 4 & 1 \\ 0 & 0 & \cdots & \cdots & 0 & 1 & 4 \end{bmatrix}$$

and

$$\boldsymbol{K}_N = \frac{1}{\Delta}\begin{bmatrix} 2 & -1 & 0 & 0 & \cdots & 0 & 0 \\ -1 & 2 & -1 & 0 & \cdots & 0 & 0 \\ 0 & -1 & 2 & -1 & \cdots & 0 & 0 \\ 0 & 0 & -1 & 2 & \cdots & 0 & 0 \\ \vdots & \vdots & \vdots & \vdots & \ddots & \vdots & \vdots \\ 0 & 0 & \cdots & \cdots & -1 & 2 & -1 \\ 0 & 0 & \cdots & \cdots & 0 & -1 & 2 \end{bmatrix},$$

respectively. Thus, $\boldsymbol{L}_N = \boldsymbol{K}_N + \boldsymbol{M}_N$ is given by

$$\boldsymbol{L}_N = \begin{bmatrix} \frac{4\Delta}{6}+\frac{2}{\Delta} & \frac{\Delta}{6}-\frac{1}{\Delta} & 0 & 0 & \cdots & 0 & 0 \\ \frac{\Delta}{6}-\frac{1}{\Delta} & \frac{4\Delta}{6}+\frac{2}{\Delta} & \frac{\Delta}{6}-\frac{1}{\Delta} & 0 & \cdots & 0 & 0 \\ 0 & \frac{\Delta}{6}-\frac{1}{\Delta} & \frac{4\Delta}{6}+\frac{2}{\Delta} & \frac{\Delta}{6}-\frac{1}{\Delta} & \cdots & 0 & 0 \\ 0 & 0 & \frac{\Delta}{6}-\frac{1}{\Delta} & \frac{4\Delta}{6}+\frac{2}{\Delta} & \cdots & 0 & 0 \\ \vdots & \vdots & \vdots & \vdots & \ddots & \vdots & \vdots \\ 0 & 0 & \cdots & \cdots & \frac{\Delta}{6}-\frac{1}{\Delta} & \frac{4\Delta}{6}+\frac{2}{\Delta} & \frac{\Delta}{6}-\frac{1}{\Delta} \\ 0 & 0 & \cdots & \cdots & 0 & \frac{\Delta}{6}-\frac{1}{\Delta} & \frac{4\Delta}{6}+\frac{2}{\Delta} \end{bmatrix}$$

and one can show that $\boldsymbol{L}_N$ is a symmetric positive definite matrix. Thus, $\boldsymbol{L}_N$ is non-singular (see [92]).

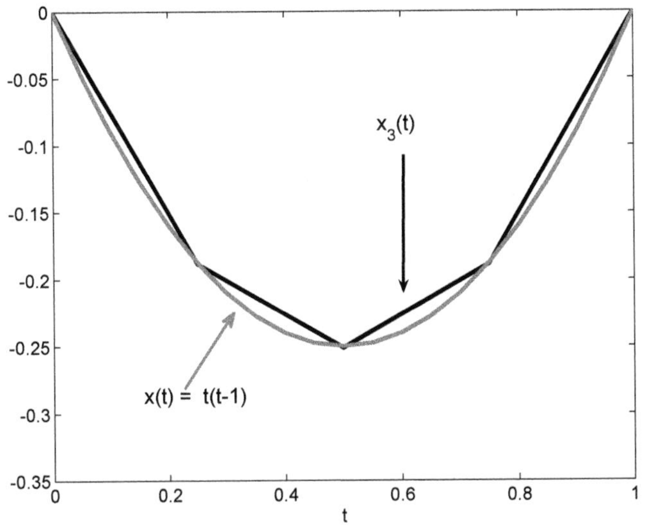

Figure 2.27: $N = 3$ Finite Element Approximation

Consider the case where the function $f(\cdot)$ is given by

$$f(t) = t^2 - t - 2.$$

It is easy to show (by direct substitution) that

$$x(t) = t(t-1)$$

is the solution to the boundary value problem (2.59)-(2.60). In Figures 2.27 and 2.28 below, we compare the finite element solutions for $N = 3$ and $N = 7$. Clearly, the approximating solutions converge very rapidly.

Although it might seem like the problem discussed above has nothing to do with classical problems such as the brachistochrone and minimal surface area problems, this is not true. In fact, we shall show later that finding solutions to the two-point boundary problem (2.59) - (2.60) is equivalent to solving the following simplest problem in the calculus of variations. Find $x^*(\cdot) \in PWS(0,1)$ to minimize the integral

$$J(x(\cdot)) = \frac{1}{2}\int_0^1 \left\{ [\dot{x}(s)]^2 + [x(s)]^2 - 2x(s)f(s) \right\} ds, \qquad (2.76)$$

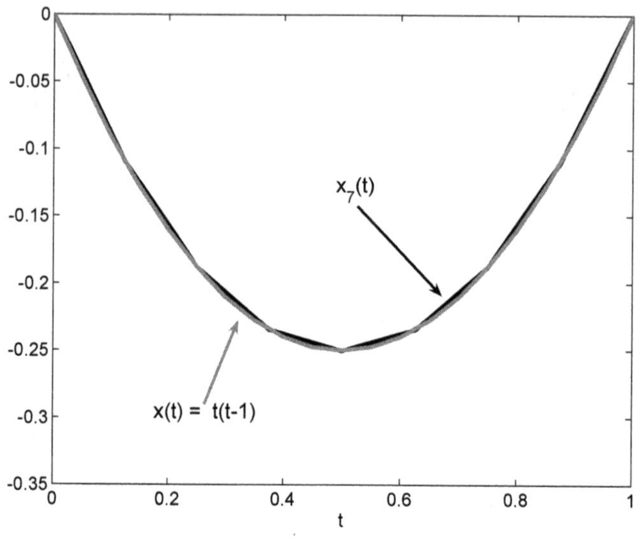

Figure 2.28: $N = 7$ Finite Element Approximation

subject to the end conditions

$$x(0) = 0 \text{ and } x(1) = 0. \tag{2.77}$$

Again, this is a problem of finding a function passing through two given points that minimize an integral where $f(t, x, u)$ is given by

$$f(t, x, u) = \frac{1}{2}\left\{u^2 + x^2 - 2xf(t)\right\}.$$

In fact, the three problems above are examples of the so called Simplest Problem in the Calculus of Variations (SPCV) to be discussed in Chapter 3.

2.5 Problem Set for Chapter 2

Problem 2.1 *Let $f : [0, 2] \to \mathbb{R}$ be defined by $f(x) = xe^{-x}$. Find all local minimizers for $f(\cdot)$ on $[0, 2]$.*

Problem 2.2 *Let $f : (0, 2] \to \mathbb{R}$ be defined by $f(x) = xe^{-x}$. Find all local minimizers for $f(\cdot)$ on $(0, 2]$.*

Problem 2.3 *Let* $f : [-1,+1] \to \mathbb{R}$ *be defined by* $f(x) = |x|$. *Find all local minimizers for* $f(\cdot)$ *on* $[-1,+1]$.

Problem 2.4 *Let* $f : [-1,+1] \to \mathbb{R}$ *be defined by* $f(x) = x^3 e^{-x^2}$. *Find all local minimizers for* $f(\cdot)$ *on* $[-1,+1]$.

Problem 2.5 *Let* $f : [-2,+2] \to \mathbb{R}$ *be defined by* $f(x) = x^3 e^{-x^2}$. *Find all local minimizers for* $f(\cdot)$ *on* $[-2,+2]$.

Problem 2.6 *Let* $x : [-1,1] \to \mathbb{R}$ *be defined by*

$$x(t) = \begin{cases} |t|, & t \neq 0, \\ 1, & t = 0. \end{cases}$$

Prove that $z(t) = |t|$ *is the only continuous function satisfying* $z(t) = x(t)$ *except at a finite number of points.*

Problem 2.7 *Let* $x : [-1,1] \to \mathbb{R}$ *and* $z : [-1,1] \to \mathbb{R}$ *be defined by* $x(t) = |t|$ *and* $z(t) = t^2$, *respectively. Compute* $d_0(x(\cdot); z(\cdot))$ *and* $d_1(x(\cdot); z(\cdot))$.

Problem 2.8 *Let* $x_N : [-\pi,\pi] \to \mathbb{R}$ *and* $z : [-\pi,\pi] \to \mathbb{R}$ *be defined by* $x_N(t) = \frac{1}{N}\sin(Nt)$ *and* $z(t) = 0$, *respectively. Compute* $d_0(x_N(\cdot); z(\cdot))$ *and* $d_1(x_N(\cdot); z(\cdot))$.

Problem 2.9 *Consider the functional* $J(x(\cdot)) = \int_0^{\pi} \{[\dot{x}(s)]^2 - [x(s)]^2\} ds$ *and let* $\hat{x}(\cdot) = \sin(t)$. *Compute the first and second variation of* $J(\cdot)$ *at* $\hat{x}(\cdot)$ *in the direction of* $\eta(\cdot) \in PWS(0,\pi)$.

Problem 2.10 *Consider the functional* $J(x(\cdot)) = \int_0^{2\pi} [x(s)]^2 ds$ *and let* $\hat{x}(\cdot) = |t-1|$. *Compute the first and second variation of* $J(\cdot)$ *at* $\hat{x}(\cdot)$ *in the direction of* $\eta(\cdot) \in PWC(0,\pi)$.

Problem 2.11 *Let* $J : PWS(0,1) \longrightarrow \mathbb{R}$ *be defined by*

$$J(x(\cdot)) = \int_0^1 x(s)\sqrt{1 + [\dot{x}(s)]^2} ds$$

and let $\hat{x}(s) = \sin(s)$. Show that any $\eta(\cdot) \in PWS(0,1)$ is an admissible direction for $\hat{x}(s) = \sin(s)$. Compute the first and second variation of $J(\cdot)$ at $\hat{x}(\cdot)$ in the direction of $\eta(\cdot)$. You may assume it is allowable to pull the partial derivative with respect to ε inside the integral.

Problem 2.12 *Let $J : PWS(0,1) \longrightarrow \mathbb{R}$ be defined by*

$$J(x(\cdot)) = \int_{t_0}^{t_1} f(s, x(s), \dot{x}(s))ds$$

where $f(t,x,u)$ is a C^2 function. Given any $\hat{x}(\cdot) \in PWS(t_0,t_1)$, show that $\eta(\cdot)$ is an admissible direction for all $\eta(\cdot) \in PWS(0,1)$. Compute the first and second variation of $J(\cdot)$ at $\hat{x}(\cdot)$ in the direction of $\eta(\cdot)$. You may assume it is allowable to pull the partial derivative with respect to ε inside the integral.

Problem 2.13 *Show that for the river crossing problem that the time to cross is given by the integral*

$$J(x(\cdot)) = \int_0^1 \frac{\sqrt{c^2(1 + [\dot{x}\,(s)]^2) - [v(s)]^2} - v(s)\dot{x}\,(s)}{c^2 - [v(s)]^2} ds.$$

Problem 2.14 *Use the Lagrange Multiplier Theorem to find all possible minimizers of*

$$J(x,y,x) = x^2 + y^2 + z^2$$

subject to the constraints

$$g_1(x,y,z) = x^2 + y^2 - 4 = 0,$$
$$g_2(x,y,z) = 6x + 3y + 2z - 6 = 0.$$

Problem 2.15 *Use the Lagrange Multiplier Theorem to find all possible minimizers of*

$$J(x,y,x) = x^2 + y^2 + z^2$$

subject to the constraints

$$g_1(x,y,z) = x^2 + y^2 + z^2 - 1 = 0,$$
$$g_2(x,y,z) = x + y + z - 1 = 0.$$

Advanced Problems

Problem 2.16 *Given any* $x(\cdot) \in PWS(t_0, t_1)$, *let* $\|x(\cdot)\|_0 = \sup\{|x(t)| : t_0 \leq t \leq t_1\}$. *Prove that* $\|\cdot\|_0$ *is a norm on* $PWS(t_0, t_1)$.

Problem 2.17 *Use the finite element method to solve the two point boundary problem*

$$-\ddot{x}(t) + 4x(t) = t, \quad x(0) = x(1) = 0.$$

Chapter 3

The Simplest Problem in the Calculus of Variations

We turn now to the Simplest Problem in the Calculus of Variations (SPCV) with both endpoints fixed and focus on obtaining (first order) necessary conditions. We begin with global minimizers and then return to the problem for local minimizers later.

3.1 The Mathematical Formulation of the SPCV

We shall make the standard assumption that $f = f(t,x,u)$ is a C^2 function of three real variables t, x and u. In particular, $f : \mathcal{D}(f) \subseteq \mathbb{R}^3 \to \mathbb{R}$ is real-valued, continuous and all the partial derivatives

$$f_t(t,x,u), \quad f_x(t,x,u), \quad f_u(t,x,u),$$

$$f_{tx}(t,x,u), \quad f_{tu}(t,x,u), \quad f_{xu}(t,x,u),$$

and

$$f_{tt}(t,x,u), \quad f_{xx}(t,x,u), \quad f_{uu}(t,x,u),$$

exist and are continuous on the domain $\mathcal{D}(f)$. Recall that by Schwarz's Theorem (see page 369 in [15]) this implies all the mixed derivatives of 2 or less are equal. Thus, we have

$$\frac{\partial^2}{\partial t \partial x} f(t,x,u) = \frac{\partial^2}{\partial x \partial t} f(t,x,u), \quad \frac{\partial^2}{\partial t \partial u} f(t,x,u) = \frac{\partial^2}{\partial u \partial t} f(t,x,u)$$

and

$$\frac{\partial^2}{\partial x \partial u} f(t,x,u) = \frac{\partial^2}{\partial u \partial x} f(t,x,u).$$

In most cases the domain $\mathcal{D}(f)$ will have the form $\mathcal{D}(f) = [t_0, t_1] \times \mathbb{R}^2$. Although, there are problems where $f(t,x,u)$ will not be defined for all $[x \ u]^T \in \mathbb{R}^2$ and one should be aware of these cases, the main ideas discussed in this book are not greatly impacted by this detail.

Remark 3.1 *It is important to comment on the "notation" $f(t,x,u)$ used in this book. Using the symbols t, x and u for the independent variables is not standard in many classical books on the calculus of variations. However, the choice of the symbol u as the third independent variable is rather standard in treatments of modern control theory. Consequently, using the same notation for both classical and modern problems avoids having to "switch" notation as the material moves from variational theory to control. Many older texts on the calculus of variations use notation such as $f(t,x,\dot{x})$ or $f(x,y,\dot{y})$ so that one sees terms of the form*

$$\frac{\partial}{\partial \dot{x}} f(t,x,\dot{x}), \quad \frac{\partial}{\partial \dot{y}} f(x,y,\dot{y}) \quad \text{and} \quad \frac{\partial^2}{\partial \dot{y} \partial x} f(x,y,\dot{y}).$$

This notation often leads to a misunderstanding (especially by students new to these subjects) of the difference between the ***independent variable*** *$\dot{x}$ and the* ***derivative of the function*** *$x(\cdot)$. Ewing's book [77] was one of the first texts on the calculus of variations to use the notation $f(x,y,r)$ and Leitmann [120] uses a similar notation $f(t,x,r)$ to avoid this issue. Using t, x and u for the independent variables is mathematically more pleasing, leads to fewer mistakes and is consistent with modern notation.*

Note that if $x : [t_0, t_1] \to \mathbb{R}$ is PWS on $[t_0, t_1]$, then the function $g : [t_0, t_1] \to \mathbb{R}$ defined by

$$g(s) = f(s, x(s), \dot{x}(s))$$

is PWC on $[t_0, t_1]$. Therefore, the integral

$$\int_{t_0}^{t_1} f\left(s, x(s), \dot{x}\left(s\right)\right) ds$$

exists and is finite. Let $X = PWS(t_0, t_1)$ denote the space of all real-valued piecewise smooth functions defined on $[t_0, t_1]$. For each PWS function $x : [t_0, t_1] \to \mathbb{R}$, define the *functional* (a "function of a function") by

$$J(x(\cdot)) = \int_{t_0}^{t_1} f\left(s, x(s), \dot{x}\left(s\right)\right) ds. \tag{3.1}$$

Assume that the points $[t_0\ x_0]^T \in \mathbb{R}^1 \times \mathbb{R}^1$ and $[t_1\ x_1]^T \in \mathbb{R}^1 \times \mathbb{R}^1$ are given and define the set of PWS functions Θ by

$$\Theta = \left\{x(\cdot) \in PWS(t_0, t_1) \colon x\left(t_0\right) = x_0, x\left(t_1\right) = x_1\right\}. \tag{3.2}$$

Observe that $J : X \to \mathbb{R}$ is a real valued function on $X = PWS(t_0, t_1)$.

The Simplest Problem in the Calculus of Variations (SPCV) is the problem of minimizing $J(\cdot)$ on Θ. In particular, the goal is to find $x^*\left(\cdot\right) \in \Theta$ such that

$$\begin{aligned} J(x^*(\cdot)) &= \int_{t_0}^{t_1} f\left(s, x^*(s), \dot{x}^*\left(s\right)\right) ds \leq J(x(\cdot)) \\ &= \int_{t_0}^{t_1} f\left(s, x(s), \dot{x}\left(s\right)\right) ds, \end{aligned} \tag{3.3}$$

for all $x\left(\cdot\right) \in \Theta$.

In the brachistochrone problem,

$$f(t, x, u) = \frac{\sqrt{1 + [u]^2}}{\sqrt{2gx}},$$

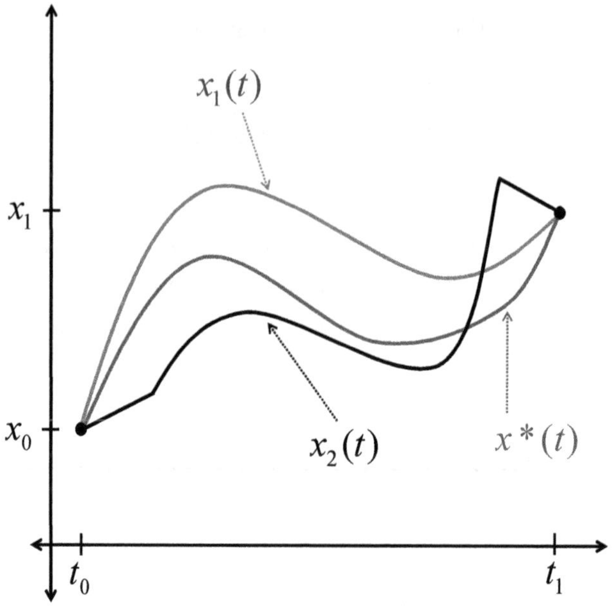

Figure 3.1: The Simplest Problem in the Calculus of Variations

so that

$$J(x(\cdot)) = \int_0^a \frac{\sqrt{1+[\dot{x}(s)]^2}}{\sqrt{2gx(s)}} ds = \int_0^a f(s, x(s), \dot{x}(s))\, ds.$$

For the problem of finding the surface of revolution of minimum area,

$$f(t, x, u) = 2\pi x \sqrt{1+[u]^2},$$

so that

$$J(x(\cdot)) = S = 2\pi \int_{a_0}^{a_1} x(s) \sqrt{1+[\dot{x}(s)]^2} ds = \int_{a_0}^{a_1} f(s, x(s), \dot{x}(s))\, ds.$$

The Simplest Problem in the Calculus of Variations will be the focus of the next few chapters. We will move on to other (more general) problems in later chapters. The basic goal is the development of the classical necessary conditions for (local) minimizers. In the following chapters we shall derive necessary conditions for local minimizers. However, we start with global minimizers.

Before we derive the first necessary condition, we develop a few fundamental lemmas that provide the backbone to much of variational theory. In addition, we review some basic results on differentiation.

3.2 The Fundamental Lemma of the Calculus of Variations

The Fundamental Lemma of the Calculus of Variations (FLCV) is also known as the Du Bois-Reymond Lemma. To set the stage for the lemma, we need some additional notation. Let V_0 denote the set

$$V_0 = PWS_0(t_0,t_1) = \{\eta(\cdot) \in PWS(t_0,t_1) \colon \eta\,(t_0) = \eta\,(t_1) = 0\}\,. \tag{3.4}$$

Observe that $V_0 \subseteq PWS(t_0,t_1)$ is the set of all PWS functions on (t_0,t_1) that vanish at both ends (see Figure 3.2 below). The space V_0 defined by (3.4) is called the space of *admissible variations* for the fixed endpoint problem.

Lemma 3.1 ***(Fundamental Lemma of the Calculus of Variations)***
Part (A): *If $\alpha(\cdot)$ is piecewise continuous on $[t_0,t_1]$ and*

$$\int_{t_0}^{t_1} \alpha\,(s)\,\dot{\eta}\,(s)\,ds = 0 \tag{3.5}$$

for all $\eta(\cdot) \in V_0$, then $\alpha\,(t)$ is constant on $[t_0,t_1]$ except at a finite number of points. In particular, there is a constant c and a finite

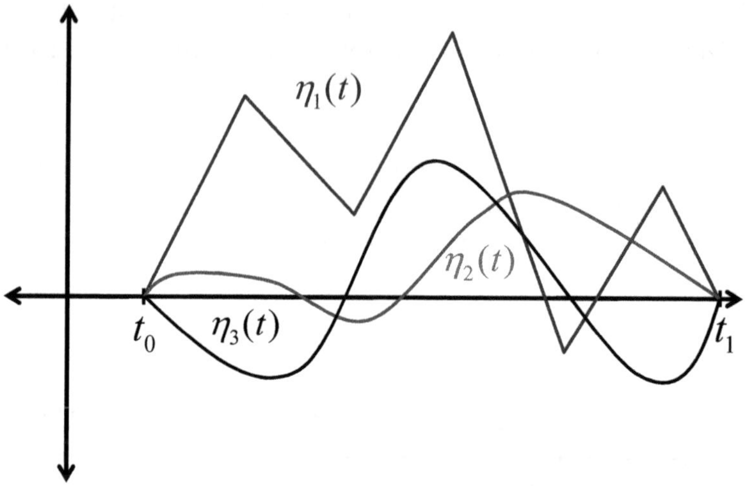

Figure 3.2: Admissible Variations

set of points $t_0 < \hat{t}_1 < \hat{t}_2 < ... < \hat{t}_p < t_1$, in (t_0, t_1) such that for each $t \in (t_0, t_1)$ with $t \neq \hat{t}_i, i = 1, 2, ..., p$

$$\alpha(t) = c.$$

The converse is also true.

Part (B): *If $\alpha(\cdot)$ and $\beta(\cdot)$ are piecewise continuous functions on $[t_0, t_1]$ and*

$$\int_{t_0}^{t_1} [\alpha(s)\eta(s) + \beta(s)\dot{\eta}(s)]\, ds = 0 \tag{3.6}$$

for all $\eta(\cdot) \in V_0$, then there is a constant c such that

$$\beta(t) = c + \int_{t_0}^{t} \alpha(s)\, ds$$

except at a finite number of points. In particular, there is a finite set of points $t_0 < \hat{t}_1 < \hat{t}_2 < ... < \hat{t}_q < t_1$, in (t_0, t_1) such that for *each* $t \in (t_0, t_1)$ *with* $t \neq \hat{t}_i, i = 1, 2, ..., q$, $\dot{\beta}(t)$ *exists and*

$$\dot{\beta}(t) = \alpha(t).$$

The converse is also true.

Proof of Part (A): Assume (3.5) holds for all $\eta(\cdot) \in V_0$. Let

$$c = \frac{1}{t_1 - t_0} \int_{t_0}^{t_1} \alpha(s)\, ds,$$

and define the function $\hat{\eta}(\cdot)$ by

$$\hat{\eta}(t) = \int_{t_0}^{t} [\alpha(s) - c]\, ds.$$

Observe that $\hat{\eta}(\cdot)$ is PWS since it is the integral of a piecewise continuous function. Also,

$$\hat{\eta}(t_0) = \int_{t_0}^{t_0} [\alpha(s) - c]\, ds = 0, \tag{3.7}$$

and

$$\hat{\eta}(t_1) = \int_{t_0}^{t_1} [\alpha(s) - c]\, ds = \int_{t_0}^{t_1} \alpha(s)\, ds - \int_{t_0}^{t_1} c ds$$
$$= \int_{t_0}^{t_1} \alpha(s)\, ds - c[t_1 - t_0] = 0. \tag{3.8}$$

Therefore, it follows from (3.7) and (3.8) that $\hat{\eta}(\cdot) \in V_0$. Hence, the assumption implies that

$$\int_{t_0}^{t_1} \alpha(s) [\frac{d}{ds}\hat{\eta}(s)] ds = 0.$$

However, since c is constant and $\hat{\eta}(t_0) = \hat{\eta}(t_1) = 0$, it follows that

$$\int_{t_0}^{t_1} c[\frac{d}{ds}\hat{\eta}(s)]ds = c[\hat{\eta}(t_1) - \hat{\eta}(t_0)] = 0.$$

Therefore, we have that

$$\int_{t_0}^{t_1} [\alpha(s) - c]\,[\frac{d}{ds}\hat{\eta}(s)]ds = \int_{t_0}^{t_1} \alpha(s)\,[\frac{d}{ds}\hat{\eta}(s)]ds - \int_{t_0}^{t_1} c[\frac{d}{ds}\hat{\eta}(s)]ds = 0$$

and since

$$[\frac{d}{ds}\hat{\eta}(s)] = [\alpha(t) - c],$$

it follows that

$$\int_{t_0}^{t_1} [\alpha(s) - c]^2\, ds = \int_{t_0}^{t_1} [\alpha(s) - c]\,[\frac{d}{ds}\hat{\eta}(s)]ds = 0. \tag{3.9}$$

The assumption that $\alpha(\cdot)$ is piecewise continuous implies that there is a partition of $[t_0, t_1]$, say $t_0 = \hat{t}_0 < \hat{t}_1 < \hat{t}_2 < ... < \hat{t}_p < \hat{t}_{p+1} = t_1$, such that $\alpha(\cdot)$ is continuous on each subinterval $(\hat{t}_i, \hat{t}_{i+1})$. On the other hand (3.9) implies that

$$\int_{\hat{t}_i}^{\hat{t}_{i+1}} [\alpha(s) - c]^2\, ds = 0$$

on these subintervals. Consequently, for $s \in (\hat{t}_i, \hat{t}_{i+1})$ it follows that

$$[\alpha(s) - c]^2 = 0$$

and hence on each subinterval $(\hat{t}_i, \hat{t}_{i+1})$

$$\alpha(s) = c,$$

which establishes the result.

For the converse, simply note that if $\eta(\cdot) \in V_0$ and

$$\alpha(t) = c.$$

for each $t \in (t_0, t_1)$ with $t \neq \hat{t}_i, i = 1, 2, ..., p$, then

$$\int_{t_0}^{t_1} \alpha(s)\,[\frac{d}{ds}\eta(s)]ds = \sum_{i=0}^{p} \int_{\hat{t}_i}^{\hat{t}_{i+1}} \alpha(s)\,[\frac{d}{ds}\eta(s)]ds$$

$$= \sum_{i=0}^{p} \int_{\hat{t}_i}^{\hat{t}_{i+1}} c[\frac{d}{ds}\eta(s)]ds$$

$$= c[\sum_{i=0}^{p} \int_{\hat{t}_i}^{\hat{t}_{i+1}} \frac{d}{ds}\eta(s)\,ds]$$

$$= c\int_{t_0}^{t_1} [\frac{d}{ds}\eta(s)]ds = c\,[\eta(t_1) - \eta(t_0)] = 0,$$

and this proves the converse of Part (A).

Proof of Part (B): Assume (3.6) holds for all for all $\eta(\cdot) \in V_0$. Let $g(\cdot)$ be defined by

$$g(t) = \int_{t_0}^{t} \alpha(s)\,ds,$$

and observe that $g(\cdot)$ is piecewise smooth with

$$\frac{d}{dt}g(t) = \dot{g}(t) = \alpha(t)$$

except at a finite number of points in (t_0, t_1). Integrate (3.6) by parts, to obtain

$$\int_{t_0}^{t_1} \left[\alpha(s)\,\eta(s) + \beta(s)\,[\frac{d}{ds}\eta(s)]\right] ds$$

$$= \int_{t_0}^{t_1} \left[[\frac{d}{ds}g(s)]\eta(s) + \beta(s)\,[\frac{d}{ds}\eta(s)]\right] ds$$

$$= \int_{t_0}^{t_1} \left[[\frac{d}{ds}g(s)]\eta(s)\right] ds + \int_{t_0}^{t_1} \left[\beta(s)\,[\frac{d}{ds}\eta(s)]\right] ds$$

$$= g(t)\,\eta(t)|_{t=t_0}^{t=t_1} - \int_{t_0}^{t_1} \left[g(s)\,[\frac{d}{ds}\eta(s)]\right] ds + \int_{t_0}^{t_1} \left[\beta(s)\,[\frac{d}{ds}\eta(s)]\right] ds$$

$$= \int_{t_0}^{t_1} \left[-g(s)\,[\frac{d}{ds}\eta(s)]\right] ds + \int_{t_0}^{t_1} \left[\beta(s)\,[\frac{d}{ds}\eta(s)]\right] ds$$

$$= \int_{t_0}^{t_1} [-g(s) + \beta(s)]\,[\frac{d}{ds}\eta(s)]ds.$$

Therefore, it follows that

$$\int_{t_0}^{t_1} [-g(s) + \beta(s)]\,\dot{\eta}(s)\,ds = \int_{t_0}^{t_1} [\alpha(s)\,\eta(s) + \beta(s)\,\dot{\eta}(s)]\,ds = 0,$$

for all $\eta(\cdot) \in V_0$. Applying Part (A) yields the existence of a constant c such that

$$-g(t) + \beta(t) = c,$$

except at a finite number of points. Therefore,

$$\beta(t) = c + g(t) = c + \int_{t_0}^{t} \alpha(s)\,ds \tag{3.10}$$

except at a finite number of points. Since $\alpha(\cdot)$ is piecewise continuous, there is a partition of $[t_0, t_1]$, say $t_0 = \hat{t}_0 < \hat{t}_1 < \hat{t}_2 <$

$\ldots < \hat{t}_q < \hat{t}_{q+1} = t_1$ such that on each subinterval $(\hat{t}_i, \hat{t}_{i+1})$, $\beta(t)$ is differentiable and for all $t \neq \hat{t}_i, i = 1, 2, \ldots, q$

$$\dot{\beta}(t) = \alpha(t).$$

Now assume that (3.10) holds. If $\eta(\cdot) \in V_0$, then

$$\begin{aligned}
&\int_{t_0}^{t_1} \left[\alpha(s)\eta(s) + \beta(s)\left[\frac{d}{ds}\eta(s)\right]\right] ds \\
&\quad = \int_{t_0}^{t_1} \left[\left[\frac{d}{ds}\beta(s)\right]\eta(s) + \beta(s)\left[\frac{d}{ds}\eta(s)\right]\right] ds \\
&\quad = \int_{t_0}^{t_1} \frac{d}{ds}\left[\beta(s)\eta(s)\right] ds \\
&\quad = \beta(s)\eta(s)\big|_{s=t_0}^{s=t_1} = 0
\end{aligned}$$

and hence the converse is also true. This completes the proof of the lemma. □

Remark 3.2 ***A Third Important Remark on Notation:*** *Observe that Part (A) of the FLCV implies that $\alpha(t) = c$ except at a finite number of points. If one defines the* ***continuous function*** *$\alpha_c(t) \equiv c$, then it follows that $\alpha(t) = \alpha_c(t) \equiv c$ e.f. and $\alpha_c(t)$ is the only continuous function satisfying $\alpha(t) = \alpha_c(t) \equiv c$ e.f. Moreover, even at points $\hat{t}_i$ where $\alpha(\hat{t}_i) \neq c$, it follows that $\alpha(\hat{t}_i^+) = \alpha(\hat{t}_i^-) = c$. Thus, as noted in Remark 2.10 we identify $\alpha(\cdot)$ with the continuous constant function $\alpha_c(t) \equiv c$.*

Likewise, Part (B) of the FLCV implies that if we define the PWS function $\beta_c(\cdot)$ by

$$\beta_c(t) \equiv c + \int_{t_0}^{t} \alpha(s)\, ds,$$

then it follows that $\beta(\cdot) = \beta_c(\cdot)$ e.f. and $\beta_c(\cdot)$ is the only PWS (continuous) function satisfying $\beta(\cdot) = \beta_c(\cdot)$ e.f. Moreover,

$$\beta(\hat{t}_i^+) = \beta_c(\hat{t}_i^+) = \beta_c(\hat{t}_i^-) = \beta(\hat{t}_i^-).$$

Also, at a corner $\hat{t}$ of $\beta_c(\cdot)$, it follows that

$$\dot{\beta}\left(\hat{t}^{+}\right) = \dot{\beta}_c\left(\hat{t}^{+}\right) = \alpha\left(\hat{t}^{+}\right)$$

and

$$\dot{\beta}\left(\hat{t}^{-}\right) = \dot{\beta}_c\left(\hat{t}^{-}\right) = \alpha\left(\hat{t}^{-}\right).$$

In light of Remarks 2.10 and 2.12, to be consistent ***we shall use the equivalent PWS representation*** *$\beta_c(\cdot)$* ***of*** *$\beta(\cdot)$* ***when it exists and not distinguish between*** *$\beta(\cdot)$* ***and*** *$\beta_c(\cdot)$ except in special circumstances.*

We turn now to the derivation of Euler's Necessary Condition for a global minimizer. Necessary conditions for local minimizers will be treated later.

3.3 The First Necessary Condition for a Global Minimizer

Assume that $x^*(\cdot) \in \Theta$ is a global minimizer for $J(\cdot)$ on Θ. In particular, assume that $x^*(\cdot) \in \Theta$ satisfies

$$\begin{aligned} J(x^*(\cdot)) &= \int_{t_0}^{t_1} f\left(t, x^*(x), \dot{x}^*(t)\right) dt \leq J(x(\cdot)) \\ &= \int_{t_0}^{t_1} f\left(t, x(t), \dot{x}(t)\right) dt, \end{aligned} \tag{3.11}$$

for all $x(\cdot) \in \Theta$.

Let $\eta(\cdot) \in V_0$ and consider the "variation"

$$\varphi(t, \varepsilon) \triangleq x^*(t) + \varepsilon\eta(t). \tag{3.12}$$

Observe that for each $\varepsilon \in \mathbb{R}$, the variation $\varphi(x, \varepsilon)$ satisfies the following conditions:

(i) $\varphi(t_0, \varepsilon) = x^*(t_0) + \varepsilon\eta(t_0) = x_0 + \varepsilon\eta(t_0) = x_0 + \varepsilon 0 = x_0$,

(ii) $\varphi(t_1, \varepsilon) = x^*(t_1) + \varepsilon\eta(t_1) = x_1 + \varepsilon\eta(t_1) = x_1 + \varepsilon 0 = x_1$,

(iii) $\varphi(t, \varepsilon) = x^*(t) + \varepsilon\eta(t) \in PWS(t_0, t_1)$.

It follows that if $\eta(\cdot) \in V_0$, then for all $\varepsilon \in \mathbb{R}$ the variation $\varphi(t, \varepsilon) \triangleq x^*(t) + \varepsilon\eta(t)$ belongs to Θ, i.e. it is admissible. Since $x^*(\cdot) \in \Theta$ minimizes $J(\cdot)$ on Θ, it follows that

$$J(x^*(\cdot)) \leq J(x^*(\cdot) + \varepsilon\eta(\cdot)) \tag{3.13}$$

for all $\varepsilon \in (-\infty, +\infty)$. Define $F : (-\infty, +\infty) \longrightarrow \mathbb{R}$ by

$$F(\varepsilon) = J(x^*(\cdot) + \varepsilon\eta(\cdot)) = \int_{t_0}^{t_1} f(s, x^*(s) + \varepsilon\eta(s), \dot{x}^*(s) + \varepsilon\dot{\eta}(s))ds, \tag{3.14}$$

and note that (3.13) implies that

$$F(0) = J(x^*(\cdot)) \leq J(x^*(\cdot) + \varepsilon\eta(\cdot)) = F(\varepsilon)$$

for all $\varepsilon \in (-\infty, +\infty)$. Therefore, $F(\cdot)$ has a minimum on $(-\infty, +\infty)$ at $\varepsilon^* = 0$ and applying **Theorem** 2.1 it follows that (if the derivative exists)

$$\left.\frac{d}{d\varepsilon}F(\varepsilon)\right|_{\varepsilon=0} = \left.\frac{d}{d\varepsilon}\left[J(x^*(\cdot) + \varepsilon\eta(\cdot))\right]\right|_{\varepsilon=0} = 0. \tag{3.15}$$

Observe that (3.15) must hold for all $\eta(\cdot) \in V_0$. Recalling Definition 2.14, the derivative defined by (3.15) is called the first variation of $J(\cdot)$ at $x^*(\cdot)$ in the direction of $\eta(\cdot)$. In particular, we have established the following necessary condition.

Theorem 3.1 *Assume $x^*(\cdot) \in \Theta$ minimizes $J(\cdot)$ on Θ. If $\eta(\cdot) \in V_0$ and the first variation $\delta J(x^*(\cdot); \eta(\cdot))$ exists, then*

$$\delta J(x^*(\cdot); \eta(\cdot)) = \left.\frac{d}{d\varepsilon}\left[J(x^*(\cdot) + \varepsilon\eta(\cdot))\right]\right|_{\varepsilon=0} = 0. \tag{3.16}$$

In order to apply the previous theorem, we need to know that the first variation exists at $x^*(\cdot)$ for all $\eta(\cdot) \in V_0$, how to compute it and then use it to obtain useful information about the minimizer $x^*(\cdot)$. We first recall Leibniz's formula (see page 245 in [15]).

Lemma 3.2 (Leibniz's Formula) *Suppose that for each $\varepsilon \in [-\delta, \delta]$ the function $g(t, \varepsilon)$ and the partial derivative $\frac{\partial}{\partial \varepsilon} g(t, \varepsilon) = g_\varepsilon(t, \varepsilon)$ are continuous functions of t on the interval $[a, b]$. In addition, assume that the functions $p : [-\delta, \delta] \longrightarrow [a, b]$ and $q : [-\delta, \delta] \longrightarrow [a, b]$ are differentiable. If $F : [-\delta, \delta] \longrightarrow \mathbb{R}$ is defined by*

$$F(\varepsilon) = \int_{p(\varepsilon)}^{q(\varepsilon)} g(s, \varepsilon)\, ds,$$

then $F'(\varepsilon)$ exists and

$$\frac{d}{d\varepsilon} F(\varepsilon) = g(q(\varepsilon), \varepsilon)[\frac{d}{d\varepsilon} q(\varepsilon)] - g(p(\varepsilon), \varepsilon)[\frac{d}{d\varepsilon} p(\varepsilon)] + \int_{p(\varepsilon)}^{q(\varepsilon)} g_\varepsilon(s, \varepsilon)\, ds. \tag{3.17}$$

A special case occurs when $p(\cdot)$ and $q(\cdot)$ are independent of ε. In this case

$$\frac{d}{d\varepsilon} F(\varepsilon) = \int_p^q g_\varepsilon(s, \varepsilon)\, ds.$$

Suppose that $x^*(\cdot)$ and $\eta(\cdot) \in PWS(t_0, t_1)$ and define the function

$$g(t, \varepsilon) = f(t, x^*(t) + \varepsilon\eta(t), \dot{x}^*(t) + \varepsilon\, \dot{\eta}(t)).$$

It follows that

$$F(\varepsilon) = J(x^*(\cdot) + \varepsilon\eta(\cdot)) = \int_{t_0}^{t_1} f(s, x^*(s) + \varepsilon\eta(s), \dot{x}^*(s) + \varepsilon\dot{\eta}(s))ds$$

$$= \int_{t_0}^{t_1} g(s, \varepsilon) ds$$

and the goal is to differentiate $F(\varepsilon)$ at $\varepsilon = 0$. Since $x^*(\cdot)$ and $\eta(\cdot) \in PWS(t_0, t_1)$, it follows that $g(t, \varepsilon)$ is piecewise continuous and there are points $t_0 = \hat{t}_0 < \hat{t}_1 < \hat{t}_2 < ... < \hat{t}_p < \hat{t}_{p+1} = t_1$, such that $g(t, \varepsilon)$ is continuous and bounded on each subinterval $(\hat{t}_i, \hat{t}_{i+1})$. For example, let $\hat{t}_1 < \hat{t}_2 < ... < \hat{t}_p$ be the union of all points where $\dot{x}(\cdot)$ and $\dot{\eta}(\cdot)$ are discontinuous. Observe that

$$\begin{aligned} F(\varepsilon) &= J(x^*(\cdot) + \varepsilon\eta(\cdot)) \\ &= \sum_{i=0}^{p} \int_{\hat{t}_i}^{\hat{t}_{i+1}} f(s, x^*(s) + \varepsilon\eta(s), \dot{x}^*(s) + \varepsilon\dot{\eta}(s))ds \\ &= \sum_{i=0}^{p} \int_{\hat{t}_i}^{\hat{t}_{i+1}} g(s, \varepsilon)ds, \end{aligned}$$

so that in order to use Leibniz's **Lemma** 3.2 above, we must only show that $g(t, \varepsilon) = f(t, x^*(t) + \varepsilon\eta(t), \dot{x}^*(t) + \varepsilon\,\dot{\eta}(t))$ and the partial derivative $g_\varepsilon(t, \varepsilon)$ are continuous on each subinterval $(\hat{t}_i, \hat{t}_{i+1})$. Since $\dot{x}^*(\cdot)$ and $\dot{\eta}(\cdot)$ are continuous on each subinterval $(\hat{t}_i, \hat{t}_{i+1})$ and the integrand $f = f(t, x, u)$ is a smooth function, it follows that $g(t, \varepsilon)$ is continuous and $g_\varepsilon(t, \varepsilon)$ exists and is also continuous on each subinterval $(\hat{t}_i, \hat{t}_{i+1})$. Applying the chain rule one obtains that for $t \in (\hat{t}_i, \hat{t}_{i+1})$

$$\begin{aligned} g_\varepsilon(t, \varepsilon) &= \frac{d}{d\varepsilon}[f(t, x^*(t) + \varepsilon\eta(t), \dot{x}^*(t) + \varepsilon\,\dot{\eta}(t))] \\ &= f_x(t, x^*(t) + \varepsilon\eta(t), \dot{x}^*(t) + \varepsilon\,\dot{\eta}(t))\,\eta(t) \\ &\quad + f_u(t, x^*(t) + \varepsilon\eta(t), \dot{x}^*(t) + \varepsilon\,\dot{\eta}(t))\,\dot{\eta}(t). \end{aligned}$$

Leibniz's **Lemma** 3.2 can now be applied on each subinterval to produce the expression

$$\frac{d}{d\varepsilon}F(\varepsilon) = \frac{d}{d\varepsilon}\int_{t_0}^{t_1} g(s,\varepsilon)ds = \frac{d}{d\varepsilon}\left[\sum_{i=0}^{p}\int_{\hat{t}_i}^{\hat{t}_{i+1}} g(s,\varepsilon)ds\right]$$

$$= \sum_{i=0}^{p}\frac{d}{d\varepsilon}\int_{\hat{t}_i}^{\hat{t}_{i+1}} g(s,\varepsilon)ds = \sum_{i=0}^{p}\int_{\hat{t}_i}^{\hat{t}_{i+1}} g_\varepsilon(s,\varepsilon)ds$$

$$= \int_{t_0}^{t_1} g_\varepsilon(s,\varepsilon)ds,$$

and hence

$$\left.\frac{d}{d\varepsilon}F(\varepsilon)\right|_{\varepsilon=0} = \left.\frac{d}{d\varepsilon}\left[J(x^*(\cdot)+\varepsilon\eta(\cdot))\right]\right|_{\varepsilon=0} = \int_{t_0}^{t_1} g_\varepsilon(s,0)ds. \quad (3.18)$$

Observe that we needed only to compute $g_\varepsilon(t,\varepsilon)$ on each subinterval $(\hat{t}_i,\hat{t}_{i+1})$ where both $\dot{x}^*(\cdot)$ and $\dot{\eta}(\cdot)$ are continuous. Moreover, the chain rule produced

$$g_\varepsilon(s,\varepsilon) = \left[f_x\left(s,x^*(s)+\varepsilon\eta(s),\dot{x}^*(s)+\varepsilon\dot{\eta}(s)\right)\right]\cdot\eta(s)$$
$$+\left[f_u\left(s,x^*(s)+\varepsilon\eta(s),\dot{x}^*(s)+\varepsilon\dot{\eta}(s)\right)\right]\cdot\dot{\eta}(s).$$

Setting $\varepsilon = 0$, it follows that (recall that $f(t,x,u) \in C^2$)

$$g_\varepsilon(s,\varepsilon)|_{\varepsilon=0} = \left[f_x\left(s,x^*(s),\dot{x}^*(s)\right)\right]\eta(s)$$
$$+\left[f_u\left(s,x^*(s),\dot{x}^*(s)\right)\right]\dot{\eta}(s). \quad (3.19)$$

Substituting (3.19) into (3.18) yields the existence and an explicit formula for the first variation $\delta J(x^*(\cdot);\eta(\cdot))$ of $J(\cdot)$ at $x^*(\cdot)$ in the direction of $\eta(\cdot)$. In particular,

$$\delta J(x^*(\cdot);\eta(\cdot)) = \int_{t_0}^{t_1}\{f_x(s,x^*(s),\dot{x}^*(s))\eta(s)$$
$$+ f_u(s,x^*(s),\dot{x}^*(s))\dot{\eta}(s)\}ds. \quad (3.20)$$

In view of **Theorem** 3.1 and the formula (3.20) we have established the following result.

Theorem 3.2 *If $x^*(\cdot) \in \Theta$ minimizes $J(\cdot)$ on Θ, and $\eta(\cdot) \in V_0$, then the first variation $\delta J(x^*(\cdot);\eta(\cdot))$ of $J(\cdot)$ at $x^*(\cdot)$ in the direction of $\eta(\cdot)$ exists and*

$$\int_{t_0}^{t_1} \{[f_x(s,x^*(s),\dot{x}^*(s))]\cdot \eta(s) + [f_u(s,x^*(s),\dot{x}^*(s))]\cdot \dot{\eta}(s)\}ds = 0. \tag{3.21}$$

Although (3.21) is equivalent to "setting the first variation equal to zero", it is not very informative. However, this is where the **Fundamental Lemma of the Calculus of Variations** becomes useful. Applying (3.6) in **Lemma** 3.1 to (3.21) with

$$\alpha(t) = [f_x(t,x^*(t),\dot{x}^*(t))]$$

and

$$\beta(t) = [f_u(t,x^*(t),\dot{x}^*(t))],$$

yields the existence of a constant c such that for all $t \in [t_0, t_1]$

$$[f_u(t,x^*(t),\dot{x}^*(t))] = \beta(t) = c + \int_{t_0}^{t} \alpha(s)\,ds$$

$$= c + \int_{t_0}^{t} [f_x(s,x^*(s),\dot{x}^*(s))]\,ds.$$

Thus, we have derived (proven) the following Euler necessary condition.

Theorem 3.3 ***(Euler Necessary Condition for a Global Minimum)*** *If $x^*(\cdot) \in \Theta$ minimizes $J(\cdot)$ on Θ, then*

(1) there is a constant c such that for all $t \in [t_0, t_1]$,

$$[f_u(t,x^*(t),\dot{x}^*(t))] = c + \int_{t_0}^{t} [f_x(s,x^*(s),\dot{x}^*(s))]\,ds, \tag{3.22}$$

(2) $x^*(t_0) = x_0$,

(3) $x^*(t_1) = x_1$.

(4) Between corners the function $f_u(t, x^*(t), \dot{x}^*(t))$ *is differentiable and*

$$\frac{d}{dt}\left[f_u(t, x^*(t), \dot{x}^*(t))\right] = \left[f_x(t, x^*(t), \dot{x}^*(t))\right]. \tag{3.23}$$

Remark 3.3 *At first cut it may seem strange to include items* (2) $x^*(t_0) = x_0$ *and* (3) $x^*(t_1) = x_1$ *in the statement of the theorem since these conditions are used in defining the constraint set* Θ. *However, later we shall consider problems with "free" end conditions that do not occur in the definition of the corresponding constraint set and the corresponding necessary conditions will yield "natural" boundary conditions (and transversality conditions) to replace* (2) *or* (3) *or both. In such cases it is essential to include these boundary conditions as a fundamental part of the necessary condition. By "repeating" conditions* (2) *and* (3) *for the simplest problem we hope to emphasize the importance of obtaining the correct boundary conditions.*

Equation (3.22) is called *Euler's Integral Equation*, while equation (3.23) is called *Euler's Differential Equation.* Therefore, we have shown that a minimizer $x^*(\cdot)$ of $J(\cdot)$ on Θ must satisfy Euler's Integral Equation and between corners $x^*(\cdot)$ satisfies Euler's Differential Equation. We say that a function $x(\cdot)$ satisfies *Euler's equation* if it is a solution to either Euler's Integral Equation or, where differentiable, Euler's Differential Equation. Euler's equation is one of the most important equations in the calculus of variations. What we have shown is that optimizers for the SPCV must satisfy Euler's equation. However, not all solutions of Euler's equation are minimizers.

Definition 3.1 *Any piecewise smooth function* $x(\cdot)$ *satisfying Euler's Integral Equation*

$$f_u(t, x(t), \dot{x}(t)) = c + \int_{t_0}^{t} f_x(s, x(s), \dot{x}(s))\, ds, \tag{3.24}$$

is called an ***extremal***.

Remark 3.4 *It is very important to note that* ***extremals do not have to satisfy any prescribed boundary conditions***. *In particular,* ***any*** *piecewise smooth function $x(\cdot)$ satisfying Euler's Integral Equation is called an extremal. The Euler Necessary Condition (3.22) implies that any global minimizer of $J(\cdot)$ on Θ is an extremal. As we see later, this is also true for local minimizers.*

Since the right-hand side of the equation

$$f_u(t, x(t), \dot{x}(t)) = c + \int_{t_0}^{t} f_x(s, x(s), \dot{x}(s))\, ds,$$

is a continuous function of t, it follows that if $x(\cdot)$ is an extremal, then the function

$$\psi(t) \triangleq f_u(t, x(t), \dot{x}(t)) = c + \int_{t_0}^{t} f_x(s, x(s), \dot{x}(s))\, ds$$

is continuous. Thus, even when we have a corner at $\hat{t}$, i.e.

$$\dot{x}\left(\hat{t}^+\right) \neq \dot{x}\left(\hat{t}^-\right),$$

the left-hand and right-hand limits of $\psi(t) \triangleq \frac{\partial}{\partial u} f(t, x(t), \dot{x}(t))$ must be equal. In particular,

$$f_u\left(\hat{t}, x(\hat{t}), \dot{x}\left(\hat{t}^+\right)\right) = \psi(\hat{t}^+) = \psi(\hat{t}) = \psi(\hat{t}^-) = f_u(\hat{t}, x(\hat{t}), \dot{x}(\hat{t}^-))$$

for all $\hat{t} \in (t_0, t_1)$. Therefore we have established the following result.

Theorem 3.4 (Weierstrass-Erdmann Corner Condition) *If $x(\cdot) \in PWS(t_0, t_1)$ is an extremal, then*

$$f_u(\hat{t}, x(\hat{t}), \dot{x}(\hat{t}^+)) = f_u(\hat{t}, x(\hat{t}), \dot{x}(\hat{t}^-)). \tag{3.25}$$

for all $\hat{t} \in (t_0, t_1)$.

Definition 3.2 *If $f_{uu}(t,x,u) \neq 0$ for all $(t,x,u) \in \mathcal{D}(f)$, then the integrand $f(t,x,u)$ is said to be* ***non-singular*** *. If $f_{uu}(t,x,u) > 0$ for all $(t,x,u) \in \mathcal{D}(f)$, then the integrand $f(t,x,u)$ is said to be* ***regular*** *and the SPCV is said to be a* ***regular problem.***

If $\hat{t}$ and $\hat{x}$ are fixed, then the function $\rho(u) \triangleq f\left(\hat{t},\hat{x},u\right)$ is called the *figurative* (or *indicatrix)* at $(\hat{t},\hat{x})$. Note that if $f(t,x,u)$ is a regular integrand, then the figurative is a strictly convex function since $\frac{d^2}{du^2}\rho(u) = f_{uu}\left(\hat{t},\hat{x},u\right) > 0$. This implies that $\frac{d}{du}\rho(u) = f_u\left(\hat{t},\hat{x},u\right)$ is a strictly increasing function. This observation leads to the following result.

Theorem 3.5 *If the integrand $f(t,x,u)$ is regular, then all extremals are of class C^2. In particular, extremals for a regular problem cannot have corners.*

Proof: Assume that $f(t,x,u)$ is regular, and suppose that $x(\cdot)$ is an extremal with a corner at $\hat{t}$, i.e.

$$\dot{x}\left(\hat{t}^+\right) \neq \dot{x}\left(\hat{t}^-\right).$$

Without loss of generality we may assume that

$$u_1 \triangleq \dot{x}\left(\hat{t}^+\right) < \dot{x}\left(\hat{t}^-\right) \triangleq u_2.$$

The derivative of the figurative $\rho'(u) \triangleq f_u(\hat{t},x(\hat{t}),u)$ at $(\hat{t},x(\hat{t}))$ is strictly increasing so that

$$f_u(\hat{t},x(\hat{t}),u_1) = \frac{d}{du}\rho(u_1) < \frac{d}{du}\rho(u_2) = f_u(\hat{t},x(\hat{t}),u_2). \tag{3.26}$$

However, the corner condition (3.25) implies that

$$\begin{aligned} f_u(\hat{t},x(\hat{t}),u_1) &= f_u(\hat{t},x(\hat{t}),\dot{x}\left(\hat{t}^+\right)) = f_u(\hat{t},x(\hat{t}),\dot{x}\left(\hat{t}^-\right)) \\ &= f_u(\hat{t},x(\hat{t}),u_2), \end{aligned}$$

which contradicts (3.26). Therefore, $x(\cdot)$ cannot have a corner at $\hat{t}$. Since $\dot{x}(\cdot)$ has no corners, it follows that $\dot{x}(\cdot) \in C^1(t_0,t_1)$ and hence $x(\cdot) \in C^2(t_0,t_1)$ and this completes the proof. □

Actually, Hilbert proved a much stronger result. We state his theorem below.

Theorem 3.6 (Hilbert's Differentiability Theorem) *If $x(\cdot) \in PWS(t_0, t_1)$ is an extremal, $\hat{t}$ is not a corner of $x(\cdot)$, and $f_{uu}(\hat{t}, x(\hat{t}), \dot{x}(\hat{t})) \neq 0$, then there exists a $\delta > 0$ such that $x(\cdot)$ has a continuous second derivative for all $t \in (\hat{t} - \delta, \hat{t} + \delta)$ and*

$$\begin{aligned}[f_{uu}(t, x(t), \dot{x}(t))] \cdot \ddot{x}(t) = &- [f_{ut}(t, x(t), \dot{x}(t))] \\ &- [f_{ux}(t, x(t), \dot{x}(t))] \cdot \dot{x}(t) \\ &+ [f_x(t, x(t), \dot{x}(t))] .\end{aligned} \tag{3.27}$$

If in addition, $f(t, x, u)$ is of class C^p, $p \geq 2$, then any extremal $x(\cdot)$ is also of class C^p on $(\hat{t} - \delta, \hat{t} + \delta)$.

Observe that **Theorem** 3.6 implies that, for regular integrands, all extremals $x(\cdot)$ have continuous second derivatives $\ddot{x}(\cdot)$ since f is assumed to be of class C^2. Therefore, we may differentiate (3.23) by applying the chain rule to obtain

$$\begin{aligned}\frac{d}{dt}[f_u(t, x(t), \dot{x}(t))] = &[f_{ut}(t, x(t), \dot{x}(t))] \\ &+ [f_{ux}(t, x(t), \dot{x}(t))] \cdot \dot{x}(t) + [f_{uu}(t, x(t), \dot{x}(t))] \cdot \ddot{x}(t) .\end{aligned}$$

Hence, the Euler Differential Equation (3.23) becomes the second order differential equation

$$\begin{aligned}[f_{uu}(t, x(t), \dot{x}(t))] \cdot \ddot{x}(t) = &[f_x(t, x(t), \dot{x}(t))] - [f_{ut}(t, x(t), \dot{x}(t))] \\ &- [f_{ux}(t, x(t), \dot{x}(t))] \cdot \dot{x}(t) .\end{aligned} \tag{3.28}$$

Observe that since $f_{uu}(t, x(t), \dot{x}(t)) > 0$, this differential equation may be written as

$$\ddot{x}(t) = \frac{[f_x(t, x(t), \dot{x}(t))] - [f_{ut}(t, x(t), \dot{x}(t))] - [f_{ux}(t, x(t), \dot{x}(t))] \cdot \dot{x}(t)}{[f_{uu}(t, x(t), \dot{x}(t))]} . \tag{3.29}$$

Note that Hilbert's Theorem is valid even if the problem is not regular. The key is that along an extremal $x(\cdot)$, the function

$$\rho(t) = f_{uu}(t, x(t), \dot{x}(t)) \neq 0$$

for all points t where $\dot{x}(t)$ exists. Such extremals have a special name leading to the following definition.

Definition 3.3 *If $x(\cdot) \in PWS(t_0, t_1)$ is an extremal, then $x(\cdot)$ is called a* ***non-singular extremal*** *if $f_{uu}(t, x(t), \dot{x}(t)) \neq 0$ for all $t \in (t_0, t_1)$ where $\dot{x}(t)$ exists. If $x(\cdot) \in PWS(t_0, t_1)$ is an extremal, then $x(\cdot)$ is called a* ***regular extremal*** *if $f_{uu}(t, x(t), \dot{x}(t)) > 0$ for all $t \in (t_0, t_1)$ where $\dot{x}(t)$ exists.*

3.3.1 Examples

We shall go through a few examples to illustrate the application of the necessary condition. It is important to note that at this point we can say very little about the existence of a minimum except in some special cases. The following three examples illustrate that the interval $[t_0, t_1]$ plays an important role in SPCV.

Example 3.1 *Find a PWS function $x^*(\cdot)$ satisfying $x(0) = 0, x(\pi/2) = 1$ and such that $x^*(\cdot)$ minimizes*

$$J(x(\cdot)) = \int_0^{\pi/2} \frac{1}{2}\left([\dot{x}(s)]^2 - [x(s)]^2\right) ds.$$

We note that $t_0 = 0$, $t_1 = \pi/2$, $x_0 = 0$, and $x_1 = 1$. The integrand $f(t, x, u)$ is given by

$$f(t, x, u) = \frac{1}{2}([u]^2 - [x]^2)$$

and hence,

$$\begin{aligned} f_x(t, x, u) &= -x \\ f_u(t, x, u) &= +u \\ f_{uu}(t, x, u) &= +1 > 0. \end{aligned}$$

We see that $f(t, x, u)$ is regular and hence the minimizer cannot have corners. Euler's Equation

$$\frac{d}{dt}\left[f_u\left(t, x^*(t), \dot{x}^*(t)\right)\right] = \left[f_x\left(t, x^*(t), \dot{x}^*(t)\right)\right]$$

becomes

$$\frac{d}{dt}\left[\dot{x}^*(t)\right] = \left[-x^*(t)\right],$$

or equivalently,

$$\ddot{x}^*(t) + x^*(t) = 0.$$

The general solution is

$$x^*(t) = \alpha\cos(t) + \beta\sin(t),$$

and applying the boundary conditions

$$\begin{aligned} 0 &= x^*(0) = \alpha\cos(0) + \beta\sin(0) = \alpha, \\ 1 &= x^*(\pi/2) = \alpha\cos(\pi/2) + \beta\sin(\pi/2) = \beta, \end{aligned}$$

it follows that

$$x^*(t) = \sin(t)$$

is the only solution to the Euler Necessary Condition as given in **Theorem** *3.3. Observe that we do not know if $x^*(t) = \sin(t)$ minimizes $J(\cdot)$. However, if there is a minimizer, then $x^*(t) = \sin(t)$ must be the minimizer since it is the only function satisfying the necessary condition.*

Example 3.2 *Find a PWS function $x^*(\cdot)$ satisfying $x(0) = 0, x(3\pi/2) = 0$ and such that $x^*(\cdot)$ minimizes*

$$J(x(\cdot)) = \int_0^{3\pi/2} \frac{1}{2}\left([\dot{x}(s)]^2 - [x(s)]^2\right) ds.$$

Observe that the integrand $f(t, x, u)$ is the same as in Example 3.1 so that the Euler Equation is the same

$$\ddot{x}^*(t) + x^*(t) = 0$$

and has the general solution

$$x^*(t) = \alpha\cos(t) + \beta\sin(t).$$

The boundary conditions

$$\begin{aligned} 0 &= x^*(0) = \alpha\cos(0) + \beta\sin(0) = \alpha, \\ 0 &= x^*(\pi/2) = \alpha\cos(3\pi/2) + \beta\sin(3\pi/2) = -\beta, \end{aligned}$$

imply that

$$x^* (t) \equiv 0$$

is the only solution to the Euler Necessary Condition as given in **Theorem** *3.3. Again, at this point we do not know if* $x^*(t) \equiv 0$ *minimizes* $J(\cdot)$. *However, if there is a minimizer, then* $x^*(t) \equiv 0$ *must be the minimizer since it is the only function satisfying the necessary condition.*

Example 3.3 *Find a PWS function* $x^*(\cdot)$ *satisfying* $x(0) = 0, x(2\pi) = 0$ *and such that* $x^*(\cdot)$ *minimizes*

$$J(x(\cdot)) = \int_0^{2\pi} \frac{1}{2}\left([\dot{x}(s)]^2 - [x(s)]^2\right) ds.$$

Again, the integrand $f(t, x, u)$ *is the same as in Example 3.1 so that the Euler Equation is*

$$\ddot{x}^* (t) + x^* (t) = 0,$$

and has the general solution

$$x^* (t) = \alpha \cos(t) + \beta \sin(t).$$

However, the boundary conditions

$$0 = x^* (0) = \alpha \cos(0) + \beta \sin(0) = \alpha,$$
$$0 = x^* (2\pi) = \alpha \cos(2\pi) + \beta \sin(2\pi) = \alpha,$$

only imply that

$$x^* (t) = \beta \sin(t).$$

Therefore, there are infinitely many solutions to the Euler Necessary Condition as given in **Theorem** *3.3, and we do not know if any of these functions* $x^*(t) = \beta \sin(t)$ *actually minimizes* $J(\cdot)$.

Example 3.4 *Find a PWS function* $x^*(\cdot)$ *satisfying* $x(-1) = 0, x(1) = 1$ *and such that* $x^*(\cdot)$ *minimizes*

$$J(x(\cdot)) = \int_{-1}^{1} [x(s)]^2 [\dot{x}(s) - 1]^2 ds.$$

The integrand $f(t,x,u)$ *is given by*

$$f(t,x,u) = [x]^2 [u-1]^2$$

and hence,

$$\begin{aligned} f_x(t,x,u) &= 2x[u-1]^2 \\ f_u(t,x,u) &= 2[x]^2[u-1] \\ f_{uu}(t,x,u) &= 2[x]^2 \geq 0. \end{aligned}$$

Note that the integrand is not regular since $f_{uu}(t,0,u) = 0$*. The Euler equation is*

$$[f_u(t,x^*(t),\dot{x}^*(t))] = c + \int_{-1}^{t} [f_x(s,x^*(s),\dot{x}^*(s))]\, ds,$$

or equivalently,

$$2[x^*(t)]^2[\dot{x}^*(t)-1] = c + \int_{-1}^{t} \left[2x^*(s)[\dot{x}^*(s)-1]^2\right] ds.$$

This equation is not as simple as in the previous examples. However, it is possible to find the solution to this problem by "inspection" of the cost function. Observe that

$$J(x(\cdot)) = \int_{-1}^{1} [x(s)]^2 [\dot{x}(s)-1]^2\, ds \geq 0$$

for all functions $x(\cdot)$*, and*

$$J(x(\cdot)) = \int_{-1}^{1} [x(s)]^2 [\dot{x}(s)-1]^2\, ds = 0$$

if $x(s) = 0$ *or* $\dot{x}(s) - 1 = 0$*. Consider the function defined by*

$$x^*(t) = \begin{cases} 0, & -1 \leq t \leq 0, \\ t, & 0 \leq t \leq 1. \end{cases}$$

Note that

$$J(x^*(\cdot)) = \int_{-1}^{1} [x^*(s)]^2 [\dot{x}^*(s) - 1]^2 ds$$

$$= \int_{-1}^{0} [x^*(s)]^2 [\dot{x}^*(s) - 1]^2 ds + \int_{0}^{1} [x^*(s)]^2 [\dot{x}^*(s) - 1]^2 ds$$

$$= \int_{-1}^{0} [0]^2 [0 - 1]^2 ds + \int_{0}^{1} [s]^2 [1 - 1]^2 ds = 0,$$

and hence

$$J(x^*(\cdot)) = 0 \leq \int_{-1}^{1} [x(s)]^2 [\dot{x}(s) - 1]^2 ds = J(x(\cdot)),$$

for all $x(\cdot)$. *Hence,* $x^*(\cdot)$ *is a global minimizer for* $J(\cdot)$ *on*

$$\Theta = \{x(\cdot) \in PWS(-1,1) \colon x(-1) = 0, x(1) = 1\}.$$

Remark 3.5 *This example illustrates one important point about "solving" problems.* ***Always think about the problem before you start to "turn the crank" and compute.***

Example 3.5 *Minimize the functional* $J(x(\cdot)) = \int_{0}^{1}[\dot{x}(s)]^3 ds$, *subject to the endpoint conditions* $x(0) = 0$ *and* $x(1) = 1$. *Here,* $f(t,x,u) = u^3$, $f_u(t,x,u) = 3u^2$, *and* $f_x(t,x,u) = 0$. *Euler's equation becomes*

$$3[\dot{x}^*(t)]^2 = f_u(t, x^*(t), \dot{x}^*(t)) = c + \int_{0}^{t} f_x(s, x^*(s), \dot{x}^*(s))ds$$

$$= c + \int_{0}^{t} 0 ds = c,$$

or equivalently,

$$3[\dot{x}^*(t)]^2 = c.$$

Therefore,

$$\dot{x}^*(t) = \pm\sqrt{c/3} = \pm k$$

and all we know is that $x^*(t)$ *is piecewise linear. Since* $x^*(0) = 0$ *and* $x^*(1) = 1$*, a possible candidate is*

$$x^*(t) = t.$$

Although we have derived a first order necessary condition for the simplest problem, the basic idea can be extended to very general problems. In particular, we shall see that the simplest problem is a special case of a class of "infinite dimensional" optimization problems. We shall discuss this framework in later sections and apply this to the problem of finding local minimizers for the simplest problem. However, we first discuss some other applications of the FLCV.

3.4 Implications and Applications of the FLCV

In the previous section we applied the FLCV to develop the Euler Necessary Condition **Theorem** 3.3 for a global minimum. However, the FLCV also plays a key role in the development of many ideas that provide the basis for the modern theories of distributions and partial differential equations. Although these ideas are important and interesting, a full development of the material lies outside the scope of these notes. However, we present two simple examples to illustrate other applications of the FLCV and to provide some historical perspective on the role the calculus of variations has played in the development of modern mathematics.

3.4.1 Weak and Generalized Derivatives

Recall that Part (B) of the FLCL **Lemma** 3.1 states that if $\alpha(\cdot)$ and $\beta(\cdot)$ are piecewise continuous on $[t_0, t_1]$ and

$$\int_{t_0}^{t_1} [\alpha(s)\eta(s) + \beta(s)\dot{\eta}(s)]\, ds = 0 \tag{3.30}$$

for all $\eta(\cdot) \in V_0$, then there is a constant c such that

$$\beta(t) = c + \int_{t_0}^{t} \alpha(s)\, ds$$

except at a finite number of points. The converse is also true. In particular, $\beta(\cdot) = \beta_c(\cdot)$ *e.f.* where $\beta_c(\cdot)$ is the PWS function defined by

$$\beta_c(t) \equiv c + \int_{t_0}^{t} \alpha(s)\, ds$$

and at points t where $\alpha(\cdot)$ is continuous

$$\dot{\beta}_c(t) = \alpha(t).$$

If we rewrite (3.30) as

$$\int_{t_0}^{t_1} \beta(s)\dot{\eta}(s)\, ds = -1 \int_{t_0}^{t_1} \alpha(s)\eta(s)\, ds, \tag{3.31}$$

then the expression (3.31) can be used to define the "weak derivative" of a piecewise continuous function $\beta(\cdot)$.

Definition 3.4 *Let $\beta(\cdot) \in PWC(t_0, t_1)$. We say that $\beta(\cdot)$ has a* ***weak derivative on*** $[t_0, t_1]$*, if there is a PWC function $\alpha(\cdot) \in PWC(t_0, t_1)$ such that*

$$\int_{t_0}^{t_1} \beta(s)\dot{\eta}(s)\, ds = (-1)^1 \int_{t_0}^{t_1} \alpha(s)\eta(s)\, ds$$

for all $\eta(\cdot) \in V_0$. The function $\alpha(\cdot) \in PWC(t_0, t_1)$ is called the weak derivative of $\beta(\cdot)$ on $[t_0, t_1]$.

Remark 3.6 *It is important to note that the concept of a weak derivative* ***as defined here*** *is dependent on the specific interval $[t_0, t_1]$. In particular, it is possible that a function can have a weak derivative on the interval $[-1, 0]$, and a weak derivative on $[0, +1]$, but not have a weak derivative on the interval $[-1, +1]$.*

Observe that Part (B) of the FLCV implies that if $\beta(\cdot) \in PWC(t_0, t_1)$ has a weak derivative on $[t_0, t_1]$, say $\alpha(\cdot) \in PWC(t_0, t_1)$, then $\beta(\cdot)$ has a ordinary derivative except at a finite number of points in $[t_0, t_1]$ and at points where $\alpha(\cdot)$ is continuous

$$\dot{\beta}(t) = \alpha(t).$$

Therefore, if the weak derivative of $\beta(\cdot)$ exist, then the ordinary (strong) derivative of $\beta(\cdot)$ exists except at a finite number of points and is given by $\dot{\beta}(t) = \alpha(t)$. The FLCV also implies the converse is true. Moreover, as noted in Remark 3.2 above there is a unique PWS function $\beta_c(\cdot)$ such that $\beta(\cdot) = \beta_c(\cdot)$ $e.f.$ and we can identify $\beta(\cdot)$ with its "equivalent" PWS representation $\beta_c(\cdot)$. Thus, with this convention one can say that if $\beta(\cdot) \in PWC(t_0, t_1)$ has a weak derivative on $[t_0, t_1]$, then $\beta(\cdot) \in PWS(t_0, t_1)$.

It may appear that the notion of a *weak derivative* does not bring anything very new to the table and in one dimension this is partially true because of the FLCV. However, consider how one might extend the notion to higher order derivatives. A natural extension to higher order derivatives would be to define a function $\alpha(\cdot) \in PWC(t_0, t_1)$ to be a *weak second derivative* of $\beta(\cdot) \in PWC(t_0, t_1)$, if

$$\int_{t_0}^{t_1} \beta(s)\,\ddot{\eta}(s)\,ds = (-1)^2 \int_{t_0}^{t_1} \alpha(s)\,\eta(s)\,ds \tag{3.32}$$

for all $\eta(\cdot) \in V_0$ with $\dot{\eta}(\cdot) \in V_0$. Later we shall see that extensions of the FLCV can be used to show that (3.31) implies that $\dot{\beta}(\cdot) \in PWS(t_0, t_1)$ and $\ddot{\beta}(\cdot)$ exists except at a finite number of points. It is easy to see, the key idea is to use integration by parts (like in the proof of Part (B) of the FLCV) to "move" the derivatives from $\beta(\cdot) \in PWC(t_0, t_1)$ to the functions $\eta(\cdot)$. Again, it may appear that the notion of a *weak* 2nd *derivative* does not bring anything new to the table. However, the real power of of this idea comes when it is applied to functions of several variables.

Example 3.6 *Consider the PWC function* $\beta(\cdot)$ *defined on* $[-1,+1]$ *by*

$$\beta(t) = |t|$$

and $\alpha(\cdot)$ *defined by*

$$\alpha(t) = \begin{cases} -1, & -1 \le t \le 0, \\ +1, & 0 < t \le +1. \end{cases}$$

If $\eta(\cdot) \in V_0$, *then*

$$\int_{-1}^{+1} \beta(s)\dot{\eta}(s)\,ds = \int_{-1}^{0} |s|\,\dot{\eta}(s)\,ds + \int_{0}^{+1} |s|\,\dot{\eta}(s)\,ds$$

$$= -\int_{-1}^{0} s\dot{\eta}(s)\,ds + \int_{0}^{+1} s\dot{\eta}(s)\,ds$$

$$= -\left[s\eta(s)\right]|_{s=-1}^{s=0} + \int_{-1}^{0} \eta(s)\,ds + \left[s\eta(s)\right]|_{s=0}^{s=+1}$$

$$-\int_{0}^{+1} \eta(s)\,ds$$

$$= -0 + \int_{-1}^{0} \eta(s)\,ds + 0 - \int_{0}^{+1} \eta(s)\,ds$$

$$= \int_{-1}^{0} \eta(s)\,ds - \int_{0}^{+1} \eta(s)\,ds.$$

On the other hand

$$\int_{-1}^{+1} \alpha(s)\,\eta(s)\,ds = \int_{-1}^{0} \alpha(s)\,\eta(s)\,ds + \int_{0}^{+1} \alpha(s)\,\eta(s)\,ds$$
$$= -\int_{-1}^{0} \eta(s)\,ds + \int_{0}^{+1} \eta(s)\,ds,$$

so that

$$\int_{-1}^{+1} \beta(s)\,\dot{\eta}(s)\,ds = \int_{-1}^{0} \eta(s)\,ds - \int_{0}^{+1} \eta(s)\,ds$$
$$= -\left[-\int_{-1}^{0} \eta(s)\,ds + \int_{0}^{+1} \eta(s)\,ds\right]$$
$$= -\int_{-1}^{+1} \alpha(s)\,\eta(s)\,ds.$$

Hence, $\beta(t) = |t|$ *has a PWC weak derivative on* $[-1,+1]$ *and the weak derivative is* $\alpha(\cdot)$.

Example 3.7 *Consider the PWC function* $\beta(\cdot)$ *defined on* $[-1,+1]$ *by*

$$\beta(t) = \begin{cases} -1/2, & -1 \le t \le 0, \\ +1/2, & 0 \le t \le +1. \end{cases}$$

We show that $\beta(\cdot)$ ***does not have a weak derivative*** *on* $[-1,+1]$. *Assume the contrary, that there is a function* $\alpha(\cdot) \in PWS(-1,+1)$ *such that*

$$\int_{-1}^{+1} \beta(s)\,\dot{\eta}(s)\,ds = -\int_{-1}^{+1} \alpha(s)\,\eta(s)\,ds$$

for all $\eta(\cdot) \in V_0(-1,+1)$. *Note that*

$$\int_{-1}^{+1} \beta(s)\dot{\eta}(s)\,ds = \int_{-1}^{0} (-1/2)\dot{\eta}(s)\,ds + \int_{0}^{+1} (+1/2)\dot{\eta}(s)\,ds$$
$$= \frac{1}{2}\left[-\int_{-1}^{0} \dot{\eta}(s)\,ds + \int_{0}^{+1} \dot{\eta}(s)\,ds\right]$$
$$= \frac{1}{2}\left[-\eta(0^-) - \eta(0^+)\right] = -\frac{1}{2}\left[\eta(0^-) + \eta(0^+)\right]$$
$$= -\eta(0),$$

since $\eta(\cdot) \in V_0(-1,+1)$ *is continuous. Thus, if*

$$\int_{-1}^{+1} \beta(s)\dot{\eta}(s)\,ds = -\int_{-1}^{+1} \alpha(s)\eta(s)\,ds$$

for all $\eta(\cdot) \in V_0(-1,+1)$, *then*

$$\int_{-1}^{+1} \alpha(s)\eta(s)\,ds = \eta(0) \tag{3.33}$$

for all $\eta(\cdot) \in V_0(-1,+1)$. *To see that there is no function* $\alpha(\cdot) \in PWS(-1,+1)$ *satisfying (3.33) for all* $\eta(\cdot) \in V_0(-1,+1)$, *assume such a function* $\alpha(\cdot) \in PWS(-1,+1)$ *exists and let* $\hat{\alpha}$ *be such that*

$$|\alpha(s)| \leq \hat{\alpha}, \quad -1 \leq s \leq +1.$$

For $m = 1, 2, 3, \ldots$ *let* $\eta^m(\cdot) \in V_0(-1,+1)$ *be given by*

$$\eta^m(t) = \begin{cases} +m(t - 1/m), & -1/m \leq t \leq 0, \\ -m(t + 1/m), & 0 \leq t \leq 1/m, \\ 0, & \textit{elsewhere}, \end{cases}$$

and note that $\eta^m(0) = 1$ *while*

$$\int_{-1}^{+1} \eta^m(s)\,ds = \int_{-1}^{+1} |\eta^m(s)|\,ds = \frac{1}{m}.$$

Select $M > 1$ so that

$$\frac{\hat{\alpha}}{M} < 1/2$$

and observe that

$$\left| \int_{-1}^{+1} \alpha(s)\,\eta^M(s)\,ds \right| \le \int_{-1}^{+1} \left|\alpha(s)\,\eta^M(s)\right| ds \le |\hat{\alpha}| \int_{-1}^{+1} \left|\eta^M(s)\right| ds$$
$$= \frac{\hat{\alpha}}{M} < 1/2.$$

However, $\eta^M(\cdot) \in V_0(-1,+1)$ and $\eta^M(0) = 1$, but

$$\int_{-1}^{+1} \alpha(s)\,\eta^M(s)\,ds \le 1/2 < 1 = \eta^M(0)$$

and hence (3.33) does not hold for $\eta^M(\cdot) \in V_0(-1,+1)$. Consequently,

$$\beta(t) = \begin{cases} -1/2, & -1 \le t \le 0, \\ +1/2, & 0 \le t \le +1, \end{cases}$$

does not have a weak derivative on $[-1,+1]$. Observe that $\beta(\cdot)$ does have a weak derivative on $[-1,0]$ and a weak derivative on $[0,+1]$ and on each of these intervals the weak derivative is zero.

Remark 3.7 *The extension of weak derivatives to a more general setting requires the development of "distribution theory" and will not be discussed in this book. However, this extension leads to the modern definition of a generalized derivative (or distribution) that covers the example above. In particular, for $\beta(\cdot)$ defined on $(-\infty,+\infty)$ by*

$$\beta(t) = \begin{cases} 0, & t \le 0, \\ +1, & 0 < t. \end{cases}$$

the generalized derivative of $\beta(\cdot)$ (on $\mathbb{R}$) is the Dirac "delta function", denoted by $\delta(\cdot)$ and is not a PWC function. In fact, $\delta(\cdot)$ is not a function in the usual sense and hence the generalized derivative of $\beta(\cdot)$ is a distribution (see the references [2], [88] and [161]). Modern theories of partial differential equations make extensive *use of weak and generalized derivatives. These derivatives are used to define weak solutions that are key to understanding both theoretical and computational issues in this field (see [55], [56], [76], [96], [126] and [179]).*

3.4.2 Weak Solutions to Differential Equations

In order to set up the finite element method in Section 2.4.5 to solve the two-point boundary value problem

$$-\ddot{x}(t) + x(t) = f(t), \quad 0 < t < 1, \tag{3.34}$$

subject to the Dirichlet boundary conditions

$$x(0) = 0, \quad x(1) = 0, \tag{3.35}$$

we discussed strong and weak solutions to (3.34) - (3.35). If $x(t)$ is a solution of (3.34) - (3.35) in the classical sense, then multiplying both sides of (3.34) by $\eta(\cdot) \in V_0 = PWS_0(0,1)$ and integration by parts produced the variational equation

$$\int_0^1 \dot{x}(t)\dot{\eta}(t)dt + \int_0^1 x(t)\eta(t)dt = \int_0^1 f(t)\eta(t)dt, \tag{3.36}$$

which must hold for all $\eta(\cdot) \in PWS(0,1)$ satisfying

$$\eta(0) = 0, \quad \eta(1) = 0. \tag{3.37}$$

Thus,

$$\begin{aligned}\int_0^1 \dot{x}(t)\dot{\eta}(t)dt &= -\int_0^1 x(t)\eta(t)dt + \int_0^1 f(t)\eta(t)dt \\ &= (-1)^1 \int_0^1 [x(t) - f(t)]\eta(t)dt\end{aligned}$$

for all $\eta(\cdot) \in V_0$ and hence $\beta(t) = \dot{x}(t)$ satisfies

$$\int_0^1 \beta(t)\dot{\eta}(t)dt = (-1)^1 \int_0^1 \alpha(t)\eta(t)dt$$

for all $\eta(\cdot) \in V_0$, where $\alpha(t) = [x(t) - f(t)]$. Consequently, $\beta(t) = \dot{x}(t)$ has a weak derivative on $[0, 1]$ given by

$$[x(t) - f(t)]$$

and it follows that if $x(\cdot) \in PWS_0(0, 1)$, then is a weak solution of the two-point boundary value problem (3.34) - (3.35).

Weak and generalized (distributional) derivatives and the notion of weak solutions to differential equations are key concepts in modern analysis. In multi-dimensional settings where one is interested in partial differential equations, the mathematical background required to properly address the theory of partial differential equations is more complex. However, the basic ideas have their roots in the classical problems discussed above. For more advanced readers we suggest the references [64], [65], [89] and [186].

3.5 Problem Set for Chapter 3

Consider the **Simplest Problem in the Calculus of Variations (SPCV)**: Find $x^*(\cdot)$ to minimize the cost function

$$J(x(\cdot)) = \int_{t_0}^{t_1} f(s, x(s), \dot{x}(s))ds,$$

subject to

$$x(t_0) = x_0, \quad x(t_1) = x_1.$$

For each of the following problems:

(A) Write out the integrand $f(t, x, u)$.

(B) Determine the endpoints t_0 and t_1.

(C) Determine the endpoints x_0 and x_1.

(D) Compute all the partial derivatives $f_t(t, x, u)$, $f_x(t, x, u)$, $f_u(t, x, u)$, $f_{tt}(t, x, u)$, $f_{xx}(t, x, u)$, $f_{uu}(t, x, u)$, $f_{tx}(t, x, u)$, $f_{tu}(t, x, u)$ and $f_{xu}(t, x, u)$.

(E) What can you say about possible minimizing functions $x^*(\cdot)$ for these problems? Write a short summary of what you know and don't know about each problem.

Here $\dot{x}(t) = \frac{dx(t)}{dt}$ is the derivative.

Problem 3.1 *Minimize the functional* $J(x(\cdot)) = \int_0^1 \dot{x}(s)\, ds$, *subject to the endpoint conditions* $x(0) = 0$ *and* $x(1) = 1$.

Problem 3.2 *Minimize the functional* $J(x(\cdot)) = \int_0^1 x(s)\, \dot{x}(s)\, ds$, *subject to the endpoint conditions* $x(0) = 0$ *and* $x(1) = 1$.

Problem 3.3 *Minimize the functional* $J(x(\cdot)) = \int_0^1 s x(s)\, \dot{x}(s)\, ds$, *subject to the endpoint conditions* $x(0) = 0$ *and* $x(1) = 1$.

Problem 3.4 *Minimize the functional* $J(x(\cdot)) = \int_0^b [\dot{x}(s)]^3 ds$, *subject to the endpoint conditions* $x(0) = 0$ *and* $x(b) = x_1$.

Problem 3.5 *Minimize the functional*

$$[J(x(\cdot)) = \int_0^1 \{[\dot{x}(s)]^2 + [x(s)]^2 + 2e^s x(s)\} ds,$$

subject to the endpoint conditions $x(0) = 0$ *and* $x(1) = e/2$.

Problem 3.6 *Minimize the functional* $J(x(\cdot)) = \int_1^2 s^{-3} [\dot{x}(s)]^2 ds$, *subject to the endpoint conditions* $x(1) = 1$ *and* $x(2) = 16$.

Problem 3.7 *Minimize the functional*

$$J(x(\cdot)) = \int_0^4 [\dot{x}(s) - 1]^2 [\dot{x}(s) + 1]^2\, ds,$$

subject to the endpoint conditions $x(0) = 0$ *and* $x(4) = 2$.

Problem 3.8 *Minimize the functional*

$$J(x(\cdot)) = \int_0^{\pi/2} \{[\dot{x}(s)]^2 - [x(s)]^2\} ds,$$

subject to the endpoint conditions $x(0) = 0$ *and* $x(\pi/2) = 0$.

Problem 3.9 *Minimize the functional*

$$J(x(\cdot)) = \int_0^{\pi} \{[\dot{x}(s)]^2 - [x(s)]^2\} ds,$$

subject to the endpoint conditions $x(0) = 0$ *and* $x(\pi) = 0$.

Problem 3.10 *Minimize the functional*

$$J(x(\cdot)) = \int_0^{3\pi/2} \{[\dot{x}(s)]^2 - [x(s)]^2\} ds,$$

subject to the endpoint conditions $x(0) = 0$ *and* $x(3\pi/2) = 0$.

Problem 3.11 *Minimize the functional*

$$J(x(\cdot)) = \int_0^{b} x(s)\sqrt{1 + [\dot{x}(s)]^2} ds,$$

subject to the endpoint conditions $x(0) = 1$ *and* $x(b) = 2$.

Problem 3.12 *Minimize the functional*

$$J(x(\cdot)) = \int_0^{b} \sqrt{\frac{1 + [\dot{x}(s)]^2}{2gx(s)}} ds,$$

subject to the endpoint conditions $x(0) = 1$ *and* $x(b) = 0$.

Problem 3.13 *Minimize the functional*

$$J(x(\cdot)) = \int_1^2 \{[\dot{x}(s)]^2 - 2sx(s)\}ds,$$

subject to the endpoint conditions $x(1) = 0$ *and* $x(2) = -1$.

Problem 3.14 *Minimize the functional*

$$J(x(\cdot)) = \int_0^\pi \{[x(s)]^2[1 - [\dot{x}(s)]^2\}ds,$$

subject to the endpoint conditions $x(0) = 0$ *and* $x(\pi) = 0$.

Problem 3.15 *Minimize the functional*

$$J(x(\cdot)) = \int_1^3 \{[3s - x(s)]x(s)\}ds,$$

subject to the endpoint conditions $x(1) = 1$ *and* $x(3) = 9/2$.

Problem 3.16 *Minimize the functional*

$$J(x(\cdot)) = 4\pi\rho v^2 \int_0^L \{[\dot{x}(s)]^3 x(s)\}ds,$$

subject to the endpoint conditions $x(0) = 1$ *and* $x(L) = R$. *Here,* ρ, v^2, $L > 0$ *and* $R > 0$ *are all constants.*

Problem 3.17 *Minimize the functional*

$$J(x(\cdot)) = \int_1^2 \{\dot{x}(s)[1 + s^2\dot{x}(s)]\}ds,$$

subject to the endpoint conditions $x(1) = 3$ *and* $x(2) = 5$.

Advanced Problems

Problem 3.18 *Show that if* $x^*(\cdot)$ *minimize the functional*

$$J(x(\cdot)) = \int_0^1 \frac{1}{2}\left\{[\dot{x}(s)]^2 + [x(s)]^2 - 2e^s x(s)\right\} ds,$$

then $x^*(\cdot)$ *satisfies the two point boundary value problem*

$$-\ddot{x}(t) + x(t) = e^t, \quad x(0) = x(1) = 0.$$

Problem 3.19 *Use the finite element method to solve the two point boundary problem*

$$-\ddot{x}(t) + x(t) = e^t, \quad x(0) = x(1) = 0.$$

Chapter 4

Necessary Conditions for Local Minima

We turn now to the problem of obtaining necessary conditions for local minimizers. As in the previous chapters, let $X = PWS(t_0, t_1)$ denote the space of all real-valued piecewise smooth functions defined on $[t_0, t_1]$. For each PWS function $x : [t_0, t_1] \to \mathbb{R}$, define the *functional* $J : X \to \mathbb{R}$ (a "function of a function") by

$$J(x(\cdot)) = \int_{t_0}^{t_1} f\left(s, x(s), \dot{x}\left(s\right)\right) ds. \tag{4.1}$$

Assume that the points $[t_0 \;\; x_0]^T$ and $[t_1 \;\; x_1]^T$ are given and define the subset Θ of $PWS(t_0, t_1)$ by

$$\Theta = \left\{x(\cdot) \in PWS(t_0, t_1) : x\left(t_0\right) = x_0, \;\; x\left(t_1\right) = x_1\right\}. \tag{4.2}$$

Observe that $J : X \to \mathbb{R}$ is a real valued function on X.

The Simplest Problem in the Calculus of Variations (the fixed endpoint problem) is the problem of minimizing $J(\cdot)$ on

Θ. In particular, the goal is to find $x^*(\cdot) \in \Theta$ such that

$$J(x^*(\cdot)) = \int_{t_0}^{t_1} f(s, x^*(s), \dot{x}^*(s))ds \le J(x(\cdot))$$
$$= \int_{t_0}^{t_1} f(s, x(s), \dot{x}(s))ds,$$

for all $x(\cdot) \in \Theta$.

The basic goal in this chapter is the development of the classical necessary conditions for ***local*** minimizers. We begin with a review of the basic definitions.

4.1 Weak and Strong Local Minimizers

In order to define local minimizers for the SPCV, we must have a measure of distance between functions in $PWS(t_0, t_1)$. Given two functions $x(\cdot)$ and $z(\cdot) \in PWS(t_0, t_1)$ there are many choices for a distance function, but we will focus on the weak and strong metrics defined in section 2.3.1. Recall that the d_0 **distance between** $x(\cdot)$ and $z(\cdot)$ is defined by

$$d_0(x(\cdot), z(\cdot)) \triangleq \sup_{t_0 \le t \le t_1} \{|x(t) - z(t)|\}. \tag{4.3}$$

In this case we can define a norm on $PWS(t_0, t_1)$ by

$$\|x(\cdot)\|_0 = \sup_{t_0 \le t \le t_1} \{|x(t)|\} \tag{4.4}$$

and note that

$$d_0(x(\cdot), z(\cdot)) = \|x(\cdot) - z(\cdot)\|_0 .$$

Given $\hat{x}(\cdot) \in PWS(t_0, t_1)$ and $\delta > 0$, the $U_0(\hat{x}(\cdot), \delta)$**-neighborhood (or Strong Neighborhood) of** $\hat{x}(\cdot)$ is defined to be the open ball

$$U_0(\hat{x}(\cdot),\delta) = \{x(\cdot) \in PWS(t_0,t_1) \colon d_0(\hat{x}(\cdot),x(\cdot)) < \delta\}.$$

Likewise, the d_1 **distance between** $x(\cdot)$ and $z(\cdot)$ is defined by

$$\begin{aligned} d_1(x(\cdot),z(\cdot)) &= \sup\{|x(t)-z(t)| : t_0 \le t \le t_1\} \\ &\quad + \sup\{|\dot{x}(t)-\dot{z}(t)| : t_0 \le t \le t_1, \ \ t \ne \hat{t}_i\} \qquad (4.5) \\ &= d_0(x(\cdot),z(\cdot)) + \sup\{|\dot{x}(t)-\dot{z}(t)| : t_0 \le t \le t_1, \\ &\qquad t \ne \hat{t}_i\}. \end{aligned}$$

In this case the 1-norm is defined on $PWS(t_0,t_1)$ by

$$\|x(\cdot)\|_1 = d_1(x(\cdot),0(\cdot)), \tag{4.6}$$

where $0(\cdot)$ is the zero function and as before,

$$d_1(x(\cdot),z(\cdot)) = \|x(\cdot)-z(\cdot)\|_1. \tag{4.7}$$

If $\hat{x}(\cdot) \in PWS(t_0,t_1)$ and $\delta > 0$, the $U_1(\hat{x}(\cdot),\delta)$-**neighborhood (or Weak Neighborhood) of** $\hat{x}(\cdot)$ is defined to be the open ball

$$U_1(\hat{x}(\cdot),\delta) = \{x(\cdot) \in PWS(t_0,t_1) \colon d_1(\hat{x}(\cdot),x(\cdot)) < \delta\}.$$

Remark 4.1 *Recall that if $d_1(x(\cdot),z(\cdot)) = 0$, then $x(t) = z(t)$ for all $t \in [t_0,t_1]$ and $\dot{x}(t) = \dot{z}(t)$ e.f. Also,*

$$d_0(x(\cdot),z(\cdot)) \le d_1(x(\cdot),z(\cdot))$$

and it follows that if $d_1(x(\cdot),z(\cdot)) < \delta$, then $d_0(x(\cdot),z(\cdot)) \le \delta$. This is an important inequality since it implies that

$$U_1(\hat{x}(\cdot),\delta) \subset U_0(\hat{x}(\cdot),\delta) \subset PWS(t_0,t_1), \tag{4.8}$$

so that the $U_1(\hat{x}(\cdot),\delta)$-neighborhood $U_1(\hat{x}(\cdot),\delta)$ is smaller than the $U_0(\hat{x}(\cdot),\delta)$-neighborhood $U_0(\hat{x}(\cdot),\delta)$.

In addition to defining global minimizers, the metrics d_0 and d_1 defined on $PWS(t_0,t_1)$ allows us to define two types of local minimizers for the SPCV.

Definition 4.1 *The function $x^*(\cdot) \in \Theta$, provides a* ***global minimizer*** *for $J(\cdot)$ on Θ if*

$$J(x^*(\cdot)) \leq J(x(\cdot))$$

for all $x(\cdot) \in \Theta$.

Definition 4.2 *The function $x^*(\cdot) \in \Theta$, provides a* ***strong local minimizer*** *for $J(\cdot)$ on Θ if there is a $\delta > 0$ such that*

$$J(x^*(\cdot)) \leq J(x(\cdot))$$

for all $x(\cdot) \in U_0(x^*(\cdot), \delta) \cap \Theta$.

Definition 4.3 *The function $x^*(\cdot) \in \Theta$, provides a* ***weak local minimizer*** *for $J(\cdot)$ on Θ if there is a $\delta > 0$ such that*

$$J(x^*(\cdot)) \leq J(x(\cdot))$$

for all $x(\cdot) \in U_1(x^*(\cdot), \delta) \cap \Theta$.

Remark 4.2 *Recall that $U_1(x^*(\cdot), \delta) \subset U_0(x^*(\cdot), \delta) \subset PWS(t_0, t_1)$. Therefore, it follows that a global minimizer is a strong local minimizer, and a strong local minimizer is a weak local minimizer. It is important to note that a necessary condition for weak local minimum is also a necessary condition for a strong local minimum, and a necessary condition for strong local minimum is also a necessary condition for a global minimum. In particular, any necessary condition for a weak local minimum applies to strong and global minima. However, a necessary condition obtained by assuming that $x^*(\cdot)$ is a global minimum may not apply to a local minimum.* ***The important point is that if one can derive a necessary condition assuming only that $x^*(\cdot)$ is a weak local minimizer, then it is more powerful (i.e. applies to more problems) than a necessary condition obtained by assuming that $x^*(\cdot)$ is a strong or global minimizer.***

Remark 4.3 *In the following sections we derive four necessary conditions for weak and strong local minimizers for the SPCV.*

These necessary conditions are numbered I, II, III and IV. This numbering system is used to match what appears in the classical work of Gilbert Bliss [29] and follows the convention used by Ewing [77]. The numbers do not reflect the historical development of the conditions and should not be thought of as an order for solution of practical problems.

4.2 The Euler Necessary Condition - (I)

In this section we extend the Euler Necessary Condition from the previous chapter to weak local minimizers. Assume that $x^*(\cdot) \in \Theta$ is a weak local minimizer for $J(\cdot)$ on Θ. In particular, there is a $\delta > 0$ such that

$$\begin{aligned} J(x^*(\cdot)) &= \int_{t_0}^{t_1} f(s, x^*(s), \dot{x}^*(s))ds \leq J(x(\cdot)) \\ &= \int_{t_0}^{t_1} f(s, x(s), \dot{x}(s))ds, \end{aligned} \tag{4.9}$$

for all $x(\cdot) \in U_1(x^*(\cdot), \delta) \cap \Theta$.

Let $\eta(\cdot) \in V_0$ and consider the variation

$$\varphi(t, \varepsilon) \triangleq x^*(t) + \varepsilon\eta(t). \tag{4.10}$$

Recall that for each $\varepsilon \in \mathbb{R}$ the variation $\varphi(t, \varepsilon)$ satisfies the following conditions:

(i) $\varphi(t_0, \varepsilon) = x^*(t_0) + \varepsilon\eta(t_0) = x_0 + \varepsilon\eta(t_0) = x_0 + \varepsilon 0 = x_0$,

(ii) $\varphi(t_1, \varepsilon) = x^*(t_1) + \varepsilon\eta(t_1) = x_1 + \varepsilon\eta(t_1) = x_1 + \varepsilon 0 = x_1$,

(iii) $\varphi(t, \varepsilon) = x^*(t) + \varepsilon\eta(t) \in PWS(t_0, t_1)$.

It follows that if $\eta(\cdot) \in V_0$, then for all $\varepsilon \in \mathbb{R}$, $\varphi(\cdot,\varepsilon) = x^*(\cdot) + \varepsilon\eta(\cdot) \in \Theta$. However, it is not always true that $\varphi(\cdot,\varepsilon) = x^*(\cdot) + \varepsilon\eta(\cdot) \in U_1(x^*(\cdot),\delta)$ unless ε is small. Let

$$\begin{aligned}\|\eta(\cdot)\|_1 &= \sup\left\{|\eta(t)| : t_0 \le t \le t_1\right\} \\ &\quad + \sup\left\{|\dot{\eta}(t)| : t_0 \le t \le t_1, t \ne \hat{t}_i\right\}\end{aligned}$$

and note that $\|\eta(\cdot)\|_1 = 0$ if and only if $\eta(t) = 0$ for all $t \in [t_0, t_1]$. The case $\|\eta(\cdot)\|_1 = 0$ is trivial, so assume that $\|\eta(\cdot)\|_1 \ne 0$ and select ε such that

$$\frac{-\delta}{\|\eta(\cdot)\|_1} < \varepsilon < \frac{\delta}{\|\eta(\cdot)\|_1}.$$

If $\varepsilon \in (\frac{-\delta}{\|\eta(\cdot)\|_1}, \frac{\delta}{\|\eta(\cdot)\|_1})$, then the distance between $\varphi(\cdot,\varepsilon)$ and $x^*(\cdot)$ is given by

$$\begin{aligned}
&d_1(x^*(\cdot),\varphi(\cdot,\varepsilon)) \\
&= \sup_{t_0\le t\le t_1}\{|x^*(t)-\varphi(t,\varepsilon)|\} + \sup_{t_0\le t\le t_1,\ t\ne\hat{t}_i}\{|\dot{x}^*(t) - \frac{\partial\varphi(t,\varepsilon)}{\partial t}|\} \\
&= \sup_{t_0\le t\le t_1}\{|x^*(t) - [x^*(t)+\varepsilon\eta(t)]|\} \\
&\quad + \sup_{t_0\le t\le t_1,\ t\ne\hat{t}_i}\{|\dot{x}^*(t) - [\dot{x}^*(t)+\varepsilon\dot{\eta}(t)]|\} \\
&= \sup_{t_0\le t\le t1}\{|\varepsilon\eta(t)|\} + \sup_{t_0\le t\le t_1,\ t\ne\hat{t}_i}\{|\varepsilon\dot{\eta}(t)|\} \\
&= |\varepsilon|[\sup_{t_0\le t\le t_1}\{|\eta(t)|\} + \sup_{t_0\le t\le t_1,\ t\ne t_i}\{|\dot{\eta}(t)|\}] \\
&= |\varepsilon|\,\|\eta(\cdot)\|_1 < \frac{\delta}{\|\eta(\cdot)\|_1}\|\eta(\cdot)\|_1 = \delta.
\end{aligned}$$

Therefore, if $\varepsilon \in (\frac{-\delta}{\|\eta(\cdot)\|_1}, \frac{\delta}{\|\eta(\cdot)\|_1})$, then $\varphi(\cdot,\varepsilon) = x^*(\cdot) + \varepsilon\eta(\cdot) \in U_1(x^*(\cdot),\delta)\cap\Theta$ and is admissible. Since $x^*(\cdot) \in \Theta$ minimizes $J(\cdot)$ on $U_1(x^*(\cdot),\delta)\cap\Theta$, it follows that

$$J(x^*(\cdot)) \le J(x^*(\cdot) + \varepsilon\eta(\cdot)) \tag{4.11}$$

for all $\varepsilon \in (\frac{-\delta}{\|\eta(\cdot)\|_1}, \frac{\delta}{\|\eta(\cdot)\|_1})$. Define $F : (\frac{-\delta}{\|\eta(\cdot)\|_1}, \frac{\delta}{\|\eta(\cdot)\|_1}) \longrightarrow \mathbb{R}$ by

$$F(\varepsilon) = J(x^*(\cdot) + \varepsilon\eta(\cdot)) = \int_{t_0}^{t_1} f(s, x^*(s) + \varepsilon\eta(s), \dot{x}^*(s) + \varepsilon\dot{\eta}(s))ds, \tag{4.12}$$

and note that the equation (4.11) implies that

$$F(0) = J(x^*(\cdot)) \leq J(x^*(\cdot) + \varepsilon\eta(\cdot)) = F(\varepsilon)$$

for all $\varepsilon \in (\frac{-\delta}{\|\eta(\cdot)\|_1}, \frac{\delta}{\|\eta(\cdot)\|_1})$. Therefore, $F(\cdot)$ has a minimum on the open interval $(\frac{-\delta}{\|\eta(\cdot)\|_1}, \frac{\delta}{\|\eta(\cdot)\|_1})$ at $\varepsilon^* = 0$. Applying the simple first order necessary condition **Theorem** 2.1, it follows that

$$\frac{d}{d\varepsilon}F(\varepsilon)\bigg|_{\varepsilon=0} = \frac{d}{d\varepsilon}\left[J(x^*(\cdot) + \varepsilon\eta(\cdot))\right]\bigg|_{\varepsilon=0} = 0. \tag{4.13}$$

Observe that (4.13) holds for all $\eta(\cdot) \in V_0$ and we have established the following necessary condition.

Theorem 4.1 *If $x^*(\cdot) \in \Theta$ provides a weak local minimum for $J(\cdot)$ on Θ and the first variation $\delta J(x^*(\cdot);\eta(\cdot))$ exists, then*

$$\delta J(x^*(\cdot);\eta(\cdot)) = \frac{d}{d\varepsilon}[J(x^*(\cdot) + \varepsilon\eta(\cdot))]\bigg|_{\varepsilon=0} = 0, \tag{4.14}$$

for all $\eta(\cdot) \in V_0$.

This result is identical to **Theorem** 3.1, which was established for a global minimizer in Section 3.3. Thus, **Theorem** 3.1 is valid for weak and strong local minimizers. We know that

$$\delta J(x^*(\cdot);\eta(\cdot)) = \int_{t_0}^{t_1} \{f_x(s, x^*(s), \dot{x}^*(s))\,\eta(s) + f_u(s, x^*(s), \dot{x}^*(s))\,\dot{\eta}(s)\}ds. \tag{4.15}$$

In view of **Theorem** 4.1 and the formula (4.15) we have that if $x^*(\cdot) \in \Theta$ provides a weak local minimum for $J(\cdot)$ on Θ, then

$$\int_{t_0}^{t_1} \{f_x(s, x^*(s), \dot{x}^*(s)) \cdot \eta(s) + f_u(s, x^*(s), \dot{x}^*(s)) \cdot \dot{\eta}(s)\} \, ds = 0, \tag{4.16}$$

for all $\eta(\cdot) \in V_0$. The Fundamental Lemma of the Calculus of Variations (**Lemma** 3.1) yields the existence of a constant c such that for all $t \in [t_0, t_1]$

$$f_u(t, x^*(t), \dot{x}^*(t)) = c + \int_{t_0}^{t} f_x(s, x^*(s), \dot{x}^*(s)) \, ds.$$

Thus, we have established the following result.

Theorem 4.2 (Euler Necessary Condition - (I)) *If $x^*(\cdot) \in \Theta$ provides a weak local minimum for $J(\cdot)$ on Θ, then*

(E-1) there is a constant c such that for all $t \in [t_0, t_1]$,

$$f_u(t, x^*(t), \dot{x}^*(t)) = c + \int_{t_0}^{t} f_x(s, x^*(s), \dot{x}^*(s)) \, ds, \tag{4.17}$$

(E-2) $x^*(t_0) = x_0$,

(E-3) $x^*(t_1) = x_1$.

(E-4) Between corners of $x^*(\cdot)$ the function $f_u(t, x^*(t), \dot{x}^*(t))$ is differentiable and if t is not a corner of $x^*(\cdot)$, then

$$\frac{d}{dt} f_u(t, x^*(t), \dot{x}^*(t)) = f_x(t, x^*(t), \dot{x}^*(t)). \tag{4.18}$$

Recall that equation (4.17) is called *Euler's Integral Equation*, while equation (4.18) is called *Euler's Differential Equation*. **Euler Necessary Condition - (I)** (**Theorem** 4.2) implies that any local minimizer must be an extremal (i.e. a PWS function satisfying (4.17)). Thus, if $x^*(\cdot) \in \Theta$ is a weak local minimizer for $J(\cdot)$ on Θ, then $x^*(\cdot)$ satisfies the **Weierstrass-Erdmann Corner Condition**

$$f_u(\hat{t}, x^*(\hat{t}), \dot{x}^*(\hat{t}^+)) = f_u(\hat{t}, x^*(\hat{t}), \dot{x}^*(\hat{t}^-)) \tag{4.19}$$

for all $\hat{t} \in (t_0, t_1)$.

If f is regular, then extremals cannot have corners. In addition, if $f(t,x,u)$ is of class C^p, $p \geq 2$, then **Hilbert's Differentiability Theorem** 3.6 implies that $x^*(\cdot)$ is also of class C^p and satisfies

$$\begin{aligned} \ddot{x}^*(t) = {} & \frac{[f_t(t, x^*(t), \dot{x}^*(t))] - [f_{ut}(t, x^*(t), \dot{x}^*(t))]}{[f_{uu}(t, x^*(t), \dot{x}^*(t))]} \\ & - \frac{[f_{ux}(t, x^*(t), \dot{x}^*(t))] \cdot \dot{x}^*(t)}{[f_{uu}(t, x^*(t), \dot{x}^*(t))]}. \end{aligned} \tag{4.20}$$

It is important to emphasize that the **Euler Necessary Condition - (I)** has four parts. Part *(E-2)* $x^*(t_0) = x_0$ and Part *(E-3)* $x^*(t_1) = x_1$, are also covered by the fact that $x^*(\cdot) \in \Theta$. However, for more general problems to be considered later, these boundary conditions will change and become more significant.

4.3 The Legendre Necessary Condition - (III)

The Euler and Weierstrass necessary conditions are first order conditions. We turn now to second order conditions. We begin just as we did for the Euler Necessary Condition. Assume that $x^*(\cdot) \in \Theta$ is a weak local minimizer for $J(\cdot)$ on Θ. In particular, there is a $\delta > 0$ such that

$$J(x^*(\cdot)) = \int_{t_0}^{t_1} f(s, x^*(s), \dot{x}^*(s))ds \leq J(x(\cdot)) = \int_{t_0}^{t_1} f(s, x(s), \dot{x}(s))ds,$$

for all $x(\cdot) \in U_1(x^*(\cdot), \delta) \cap \Theta$.

Let $\eta(\cdot) \in V_0$ and again we consider the classical variation

$$\varphi(x,\varepsilon) \triangleq x^*(t) + \varepsilon\eta(t).$$

As before, if $\varepsilon \in (\frac{-\delta}{\|\eta(\cdot)\|_1}, \frac{\delta}{\|\eta(\cdot)\|_1})$, then $d_1(x^*(\cdot), \varphi(\cdot,\varepsilon)) < \delta$ and $\varphi(\cdot,\varepsilon) = x^*(\cdot) + \varepsilon\eta(\cdot) \in U_1(x^*(\cdot),\delta) \cap \Theta$ and is admissible. Since $x^*(\cdot) \in \Theta$ minimizes $J(\cdot)$ on $U_1(x^*(\cdot),\delta) \cap \Theta$, it follows that

$$J(x^*(\cdot)) \leq J(x^*(\cdot) + \varepsilon\eta(\cdot))$$

for all $\varepsilon \in (\frac{-\delta}{\|\eta(\cdot)\|_1}, \frac{\delta}{\|\eta(\cdot)\|_1})$. Define $F : (\frac{-\delta}{\|\eta(\cdot)\|_1}, \frac{\delta}{\|\eta(\cdot)\|_1}) \longrightarrow \mathbb{R}$ by

$$F(\varepsilon) = J(x^*(\cdot) + \varepsilon\eta(\cdot)) = \int_{t_0}^{t_1} f(s, x^*(s) + \varepsilon\eta(s), \dot{x}^*(s) + \varepsilon\dot{\eta}(s))ds,$$

and note that

$$F(0) = J(x^*(\cdot)) \leq J(x^*(\cdot) + \varepsilon\eta(\cdot)) = F(\varepsilon)$$

for all $\varepsilon \in (\frac{-\delta}{\|\eta(\cdot)\|_1}, \frac{\delta}{\|\eta(\cdot)\|_1})$. Therefore, $F(\cdot)$ has a minimum on $(\frac{-\delta}{\|\eta(\cdot)\|_1}, \frac{\delta}{\|\eta(\cdot)\|_1})$ at $\varepsilon^* = 0$. This time we apply the second order condition as stated in **Theorem** 2.2 from Chapter 2. In particular, if $\frac{d^2}{d\varepsilon^2}F(\varepsilon)\Big|_{\varepsilon=0}$ exists then

$$\frac{d^2}{d\varepsilon^2}F(\varepsilon)\Big|_{\varepsilon=0} \triangleq \delta^2 J(x^*(\cdot);\eta(\cdot)) = \frac{d^2}{d\varepsilon^2}\left[J(x^*(\cdot) + \varepsilon\eta(\cdot))\right]\Big|_{\varepsilon=0} \geq 0. \tag{4.21}$$

Observe that (4.21) holds for all $\eta(\cdot) \in V_0$.

To use (4.21) we must compute the second variation $\delta^2 J(x^*(\cdot);\eta(\cdot))$. The first variation of

$$F(\varepsilon) = \int_{t_0}^{t_1} f(s, x^*(s) + \varepsilon\eta(s), \dot{x}^*(s) + \varepsilon\dot{\eta}(s))ds$$

is given by

$$\frac{d}{d\varepsilon}F(\varepsilon) = \frac{d}{d\varepsilon}\int_{t_0}^{t_1} f(s, x^*(s) + \varepsilon\eta(s), \dot{x}^*(s) + \varepsilon\dot{\eta}(s))ds$$

$$= \int_{t_0}^{t_1} [f_x(s, x^*(s) + \varepsilon\eta(s), \dot{x}^*(s) + \varepsilon\dot{\eta}(s))] \cdot \eta(s)ds$$

$$+ \int_{t_0}^{t_1} [f_u(s, x^*(s) + \varepsilon\eta(s), \dot{x}^*(s) + \varepsilon\dot{\eta}(s))] \cdot \dot{\eta}(s)\, ds.$$

Differentiating once again yields

$$\frac{d^2}{d\varepsilon^2}F(\varepsilon) = \frac{d}{d\varepsilon}\int_{t_0}^{t_1} [f_x(s, x^*(s) + \varepsilon\eta(s), \dot{x}^*(s) + \varepsilon\dot{\eta}(s))] \cdot \eta(s)ds$$

$$+ \frac{d}{d\varepsilon}\int_{t_0}^{t_1} [f_u(s, x^*(s) + \varepsilon\eta(s), \dot{x}^*(s) + \varepsilon\dot{\eta}(s))] \cdot \dot{\eta}(s)\, ds$$

$$= \int_{t_0}^{t_1} [f_{xx}(s, x^*(s) + \varepsilon\eta(s), \dot{x}^*(s) + \varepsilon\dot{\eta}(s))] \cdot [\eta(s)]^2 ds$$

$$+ \int_{t_0}^{t_1} [f_{xu}(s, x^*(s) + \varepsilon\eta(s), \dot{x}^*(s) + \varepsilon\dot{\eta}(s))] \cdot \eta(s) \cdot \dot{\eta}(s)ds$$

$$+ \int_{t_0}^{t_1} [f_{ux}(s, x^*(s) + \varepsilon\eta(s), \dot{x}^*(s) + \varepsilon\dot{\eta}(s))] \cdot \dot{\eta}(s) \cdot \eta(s)ds$$

$$+ \int_{t_0}^{t_1} [f_{uu}(s, x^*(s) + \varepsilon\eta(s), \dot{x}^*(s) + \varepsilon\dot{\eta}(s))] \cdot [\dot{\eta}(s)]^2 ds,$$

and setting $\varepsilon = 0$ produces

$$\left.\frac{d^2}{d\varepsilon^2}F(\varepsilon)\right|_{\varepsilon=0} = \int_{t_0}^{t_1} [f_{xx}(s,x^*(s),\dot{x}^*(s))]\cdot[\eta(s)]^2 ds + \int_{t_0}^{t_1} [f_{xu}(s,x^*(s),\dot{x}^*(s))]\cdot\eta(s)\cdot\dot{\eta}(s)\, ds + \int_{t_0}^{t_1} [f_{ux}(s,x^*(s),\dot{x}^*(s))]\cdot\dot{\eta}(s)\cdot\eta(s) ds + \int_{t_0}^{t_1} [f_{uu}(s,x^*(s),\dot{x}^*(s))]\cdot[\dot{\eta}(s)]^2 ds.$$

Since, $f_{ux}(s,x^*(s),\dot{x}^*(s)) = f_{xu}(s,x^*(s),\dot{x}^*(s))$, it follows that

$$\left.\frac{d^2}{d\varepsilon^2}F(\varepsilon)\right|_{\varepsilon=0} = \int_{t_0}^{t_1} [f_{xx}(s,x^*(s),\dot{x}^*(s))]\cdot[\eta(s)]^2 ds + \int_{t_0}^{t_1} [2f_{xu}(s,x^*(s),\dot{x}^*(s))]\cdot\eta(s)\cdot\dot{\eta}(s) ds + \int_{t_0}^{t_1} [f_{uu}(s,x^*(s),\dot{x}^*(s))]\cdot[\dot{\eta}(s)]^2 ds. \tag{4.22}$$

In order to simplify notation, we set

$$f^*_{xx}(t) = f_{xx}(t,x^*(t),\dot{x}^*(t)), \tag{4.23}$$

$$f^*_{xu}(t) = f_{xu}(t,x^*(t),\dot{x}^*(t)), \tag{4.24}$$

and

$$f^*_{uu}(t) = f_{uu}(t,x^*(t),\dot{x}^*(t)). \tag{4.25}$$

Therefore, we have established that the second variation is given by

$$\delta^2 J(x^*(\cdot);\eta(\cdot)) = \int_{t_0}^{t_1} \{f_{xx}^*(s)[\eta(s)]^2 + 2f_{xu}^*(s)[\eta(s)\dot{\eta}(s)] + f_{uu}^*(s)[\dot{\eta}(s)]^2\}ds. \tag{4.26}$$

Consequently, (4.21) is equivalent to the condition that

$$\int_{t_0}^{t_1} \{f_{xx}^*(s)[\eta(s)]^2 + 2f_{xu}^*(s)[\eta(s)\dot{\eta}(s)] + f_{uu}^*(s)[\dot{\eta}(s)]^2\}ds \geq 0$$

holds for all $\eta(\cdot) \in V_0$. Therefore, we have established the following necessary condition.

Theorem 4.3 *If $x^*(\cdot) \in \Theta$ provides a weak local minimum for $J(\cdot)$ on Θ, then*

$$\int_{t_0}^{t_1} \{f_{xx}^*(s)[\eta(s)]^2 + 2f_{xu}^*(s)[\eta(s)\dot{\eta}(s)] + f_{uu}^*(s)[\dot{\eta}(s)]^2\}ds \geq 0, \tag{4.27}$$

for all $\eta(\cdot) \in V_0$.

Theorem 4.3 above is not very useful as stated. We need to extract useful information about $x^*(\cdot)$ from this inequality. The first result along this line is the Legendre Necessary Condition.

Theorem 4.4 (Legendre Necessary Condition - (III)) *If $x^*(\cdot) \in \Theta$ provides a weak local minimum for $J(\cdot)$ on Θ, then,*

(L-1) $f_{uu}^*(t) = f_{uu}(t, x^*(t), \dot{x}^*(t)) \geq 0$, for all $t_0 \leq t \leq t_1$,

(L-2) $x^*(t_0) = x_0$,

(L-3) $x^*(t_1) = x_1$.

Remark 4.4 *It is important to note that condition (L-1) holds at corners. In particular, since*

$$f_{uu}(t, x^*(t), \dot{x}^*(t)) \geq 0 \tag{4.28}$$

it follows that

$$f_{uu}\left(t, x^*(t), \dot{x}^*\left(t^+\right)\right) \geq 0$$

for all $t \in [t_0, t_1)$, *and*

$$f_{uu}\left(t, x^*(t), \dot{x}^*\left(t^-\right)\right) \geq 0$$

for all $t \in (t_0, t_1]$.

Remark 4.5 *The condition that*

$$f_{uu}\left(t, x^*(t), \dot{x}^*(t)\right) \geq 0 \tag{4.29}$$

for all $t \in [t_0, t_1]$ *is called the* ***Legendre Condition***. *Note that the Legendre Necessary Condition is often easy to check. In the case where* $f_{uu}(t, x, u) \geq 0$ *for all* (t, x, u), *it is not very helpful. However, the* ***Strengthened Legendre Condition***

$$f_{uu}\left(t, x^*(t), \dot{x}^*(t)\right) > 0 \tag{4.30}$$

for all $t \in [t_0, t_1]$ *will be very useful.*

Recall that an extremal $x(\cdot) \in PWS(t_0, t_1)$ is called a regular extremal if

$$f_{uu}(t, x(t), \dot{x}(t)) > 0$$

for all $t \in [t_0, t_1]$ such that $\dot{x}(t)$ exists. Also, if $f(t, x, u)$ is a regular integrand, then all extremals are regular.

Example 4.1 *Consider the functional* $J(x(\cdot)) = \int_0^1 [\dot{x}(s)]^3 ds$. *Here,* $f(t, x, u) = u^3$, $f_u(t, x, u) = 3u^2$, $f_{uu}(t, x, u) = 6t$, *and* $f_x(t, x, u) = 0$. *As noted in* ***Example 3.5*** *all extremals are piecewise linear functions. In particular, Euler's Integral Equation is given by*

$$3[\dot{x}^*(t)]^2 = f_u(t, x^*(t), \dot{x}^*(t)) = c + \int_0^t f_x(s, x^*(s), \dot{x}^*(s)) ds$$

$$= c + \int_0^x 0 ds = c,$$

or equivalently,

$$3[\dot{x}^*(t)]^2 = c.$$

Therefore,

$$\dot{x}^*(t) = \pm\sqrt{c/3} = \pm k$$

and hence it follows that $x^*(\cdot)$ is piecewise linear with slope restricted to either $\pm k$. Hence,

$$f_{uu}(t, x^*(t), \dot{x}^*(t)) = 6\dot{x}^*(t) = \pm 6k$$

and as long as $k \neq 0$ the extremal is non-singular. On the other hand, the only regular extremals are those satisfying

$$f_{uu}(t, x(t), \dot{x}^*(t)) = 6\dot{x}^*(t) = \pm 6k > 0,$$

which means that the derivative must always be positive. In particular,

$$x^*(t) = mt + r$$

with $m > 0$.

Example 4.2 *Consider the problem of minimizing the functional*

$$J(x(\cdot)) = \int_0^1 [\dot{x}(s)]^3 ds$$

subject to the endpoint conditions $x(0) = 0$ and $x(1) = b$. We know that if $\dot{x}^*(\cdot)$ is a weak local minimizer it is an extremal so that

$$\dot{x}^*(t) = \pm 6k$$

for some k. Applying the Legendre Necessary Condition, it must be the case that

$$f_{uu}(t, x^*(t), \dot{x}^*(t)) = 6\dot{x}^*(t) = \pm 6k \triangleq m \geq 0, \tag{4.31}$$

and the derivative cannot change sign. Thus, $x^*(t) = mt + r$ for all $t \in [0, 1]$ where $m \geq 0$. The endpoint conditions $x(0) = 0$ and $x(1) = b$ imply that $m = b$ and $x^*(t) = bt$ is the only possible minimizer. If $b < 0$, then $x^*(t) = bt$ fails to satisfy the Legendre Condition (4.31) and there is no local minimizer. If $b \geq 0$, then $x^*(t) = bt$ will satisfy the Legendre Condition (4.31) and perhaps can be a minimizer.

4.4 Jacobi Necessary Condition - (IV)

The general second order necessary condition **Theorem** 4.3 can also be used to obtain another necessary condition due to Karl Gustav Jacob Jacobi. In 1837 Jacobi used some of Legendre's basic ideas on the second variation and constructed what is known as Jacobi's (second order) Necessary Condition. In order to state the result we need to introduce some additional terms and definitions.

Recall that if $x^*(\cdot) \in \Theta$ provides a weak local minimum for $J(\cdot)$ on Θ, then the general result **Theorem** 4.3 implies that

$$\delta^2 J(x^*(\cdot);\eta(\cdot)) = \int_{t_0}^{t_1} \{f^*_{xx}(s)[\eta(s)]^2 + 2f^*_{xu}(s)[\eta(s)\dot{\eta}(s)] + f^*_{uu}(s)[\dot{\eta}(s)]^2\}ds \geq 0, \tag{4.32}$$

for all $\eta(\cdot) \in V_0$. Jacobi noted that if $\eta_o(\cdot) \in V_0$ is defined to be the zero function, $\eta_o(t) \equiv 0$ for all $t \in [t_0, t_1]$, then

$$\delta^2 J(x^*(\cdot);\eta_o(\cdot)) = \int_{t_0}^{t_1} \{f^*_{xx}(s)[0]^2 + 2f^*_{xu}(s)[0\dot{0}] + f^*_{uu}(s)[0]^2\}ds = 0. \tag{4.33}$$

Again, remember that the functions $f^*_{xx}(\cdot)$, $f^*_{xu}(\cdot)$, and $f^*_{uu}(\cdot)$ are fixed functions of t given by

$$f^*_{xx}(t) = f_{xx}(t, x^*(t), \dot{x}^*(t)),$$

$$f^*_{xu}(t) = f_{xu}(t, x^*(t), \dot{x}^*(t)),$$

and

$$f^*_{uu}(t) = f_{uu}(t, x^*(t), \dot{x}^*(t)),$$

respectively.

Using this notation we define the function $\mathcal{F}(t, \eta, \xi)$ by

$$\mathcal{F}(t,\eta,\xi) = \frac{1}{2}[f^*_{xx}(t)\eta^2 + 2f^*_{xu}(t)\eta\xi + f^*_{uu}(t)\xi^2] \tag{4.34}$$

and consider the functional $\mathcal{J} : PWS(t_0, t_1) \longrightarrow \mathbb{R}$ given by

$$\mathcal{J}(\eta(\cdot)) = \int_{t_0}^{t_1} \mathcal{F}(s, \eta(s), \dot{\eta}(s))ds. \tag{4.35}$$

Let $\Theta_S \subset PWS(t_0, t_1)$ be defined by

$$\Theta_S = \{\eta(\cdot) \in PWS(t_0, t_1) \colon \eta\,(t_0) = 0, \eta\,(t_1) = 0\} = V_0, \tag{4.36}$$

and consider the so called Accessory (Secondary) Minimum Problem.

The Accessory (Secondary) Minimum Problem: Find $\eta^*(\cdot) \in \Theta_S$, such that

$$\mathcal{J}(\eta^*(\cdot)) = \int_{t_0}^{t_1} \mathcal{F}(s, \eta^*(s), \dot{\eta}^*(s))ds \leq \mathcal{J}(\eta(\cdot))$$
$$= \int_{t_0}^{t_1} \mathcal{F}(s, \eta(s), \dot{\eta}(s))ds,$$

for all $\eta\,(\cdot) \in \Theta_S$.

There are two key observations that make the Accessory Minimum Problem important and useful.

(1) The answer to the Accessory Minimum Problem is known. In view of (4.32) and (4.33), we know that if $x^*(\cdot) \in \Theta$ provides a weak local minimum for $J(\cdot)$ on Θ, then the zero function $\eta_o\,(t) \equiv 0$, satisfies

$$\mathcal{J}(\eta_o(\cdot)) = \int_{t_0}^{t_1} \mathcal{F}(s, \eta_o(s), \dot{\eta}_o(s))ds = 0 \leq \mathcal{J}(\eta(\cdot))$$
$$= \int_{t_0}^{t_1} \mathcal{F}(s, \eta(s), \dot{\eta}(s))ds,$$

for all $\eta\,(\cdot) \in V_0 = \Theta_S$. In particular, $\eta^*(x) = \eta_o\,(x) \equiv 0$ is a **global minimizer for $\mathcal{J}$ on Θ_S**.

(2) The Accessory Minimum Problem is a special case of the Simplest Problem in the Calculus of Variations with the change of variables

$$\begin{aligned} (t, x, u) &\longleftrightarrow (t, \eta, \xi), \\ f(t, x, u) &\longleftrightarrow \mathcal{F}(t, \eta, \xi), \\ J(x(\cdot)) &\longleftrightarrow \mathcal{J}(\eta(\cdot)), \\ (t_0, x_0) &\longleftrightarrow (t_0, 0), \\ (t_1, x_1) &\longleftrightarrow (t_1, 0), \end{aligned}$$

and

$$\Theta \longleftrightarrow \Theta_S.$$

Therefore, we can apply the Euler Necessary Condition to the Accessory Problem. In particular, if $\eta^*(\cdot) \in \Theta_S$, is any minimizer of

$$\mathcal{J}(\eta(\cdot)) = \int_{t_0}^{t_1} \mathcal{F}(s, \eta(s), \dot{\eta}(s))ds$$

on Θ_S, then there is a constant c such that $\eta^*(\cdot)$ satisfies the Euler's Integral Equation

$$[\mathcal{F}_\xi(t, \eta^*(s), \dot{\eta}^*(s))] = c + \int_{t_0}^{t} [\mathcal{F}_\eta(s, \eta^*(s), \dot{\eta}^*(s))]\, ds. \tag{4.37}$$

In addition, between corners the function $\mathcal{F}_\xi(t, \eta^*(t), \dot{\eta}^*(t))$ is differentiable and

$$\frac{d}{dt}[\mathcal{F}_\xi(t, \eta^*(t), \dot{\eta}^*(t))] = [\mathcal{F}_\eta(t, \eta^*(t), \dot{\eta}^*(t))]. \tag{4.38}$$

The equation

$$[\mathcal{F}_\xi(t, \eta(t), \dot{\eta}(t))] = c + \int_{t_0}^{t} [\mathcal{F}_\eta(s, \eta(s), \dot{\eta}(s))]\, ds \tag{4.39}$$

is called Jacobi's Integral Equation and

$$\frac{d}{dt}\left[\mathcal{F}_{\xi}\left(t,\eta(t),\dot{\eta}\left(t\right)\right)\right]=\left[\mathcal{F}_{\eta}\left(t,\eta(t),\dot{\eta}\left(t\right)\right)\right] \tag{4.40}$$

is called Jacobi's Differential Equation. Observe that Jacobi's Equation is Euler's Equation for the case where $f(t,x,u)$ is replaced by $\mathcal{F}(t,\eta,\xi)$.

Definition 4.4 *A PWS function $\eta(\cdot)$ satisfying Jacobi's Integral Equation (4.39) (or (4.40)) is called a* ***secondary extremal.***

We are interested in secondary extremals and what they can tell us about the minimizer $x^*(\cdot)$ to the original SPCV. Thus, it is important to look at the specific form of the Jacobi equations. Since

$$\mathcal{F}(t,\eta,\xi)=\frac{1}{2}[f^*_{xx}(t)\eta^2+2f^*_{xu}(t)\eta\xi+f^*_{uu}(t)\xi^2] \tag{4.41}$$

it is obvious that

$$\mathcal{F}_{\eta}(t,\eta,\xi)=\frac{\partial}{\partial\eta}\mathcal{F}(t,\eta,\xi)=[f^*_{xx}(t)\eta+f^*_{xu}(t)\xi], \tag{4.42}$$

$$\mathcal{F}_{\xi}(x,\eta,\xi)=\frac{\partial}{\partial\xi}\mathcal{F}(t,\eta,\xi)=[f^*_{xu}(t)\eta+f^*_{uu}(t)\xi] \tag{4.43}$$

and

$$\mathcal{F}_{\xi\xi}(t,\eta,\xi)=\frac{\partial^2}{\partial\xi^2}\mathcal{F}(t,\eta,\xi)=[f^*_{uu}(t)]=f_{uu}(t,x^*(t),\dot{x}^*(t)).$$

Legendre's Necessary Condition applied to the Accessory Minimum Problem implies that

$$\mathcal{F}_{\xi\xi}(t,\eta,\xi)=f_{uu}(t,x^*(t),\dot{x}^*(t))\geq 0.$$

However, in order to go further, we must assume that $x^*(\cdot)$ is a non-singular extremal. In this case

$$\mathcal{F}_{\xi\xi}(t,\eta,\xi)=f_{uu}(t,x^*(t),\dot{x}^*(t))>0, \tag{4.44}$$

which in turn implies that the corresponding Accessory Minimum Problem is regular. In particular, we know that all secondary extremals $\eta(\cdot)$ are smooth and, in this case, we need only consider Jacobi's Differential Equation (4.40). In view of (4.41) Jacobi's Differential Equation has the form

$$\frac{d}{dt}[f_{xu}^*(t)\eta(t) + f_{uu}^*(t)\dot{\eta}(t)] = [f_{xx}^*(t)\eta(t) + f_{xu}^*(t)\dot{\eta}(t)]. \qquad (4.45)$$

Remark 4.6 *We will focus on Jacobi's Differential Equation (4.45). Recall that solutions of Jacobi's Differential Equation are secondary extremals. It is important to note that Jacobi's Differential Equation is a second order linear differential equation in* $\eta(\cdot)$*. Consequently, Jacobi's Differential Equation with initial conditions of the form* $\eta(\hat{t}) = \hat{p}$ *and* $\dot{\eta}(\hat{t}) = \hat{v}$ *has a unique solution. This point is important in the proof of Jacobi's Necessary Condition.*

Example 4.3 *Minimize the functional*

$$J(x(\cdot)) = \int_0^{\pi/2} \left\{[\dot{x}(s)]^2 - [x(s)]^2\right\} ds,$$

subject to the endpoint conditions $x(0) = 0$ *and* $x(\pi/2) = 0$*. Here* $f(t,x,u) = u^2 - x^2$, $f_x(t,x,u) = -2x$, $f_{xx}(t,x,u) = -2$, $f_u(t,x,u) = 2u$, $f_{uu}(t,x,u) = 2$, *and* $f_{xu}(t,x,u) = f_{xu}(t,x,u) = 0$*. Thus, if* $x^*(\cdot)$ *is any minimizer of* $J(\cdot)$ *on* Θ,

$$f_{ux}^*(t) = f_{ux}(t, x^*(t), \dot{x}^*(t)) = 0,$$
$$f_{xu}^*(t) = f_{xu}(t, x^*(t), \dot{x}^*(t)) = 0,$$
$$f_{xx}^*(t) = f_{xx}(t, x^*(t), \dot{x}^*(t)) = -2,$$

and

$$f_{uu}^*(t) = f_{uu}(t, x^*(t), \dot{x}^*(t)) = 2 > 0.$$

Therefore, Jacobi's Equation (4.45)

$$\frac{d}{dt}[f_{xu}^*(t)\eta(t) + f_{uu}^*(t)\dot{\eta}(t)] = [f_{xx}^*(t)\eta(x) + f_{xu}^*(t)\dot{\eta}(t)],$$

reduces to

$$\frac{d}{dt}[0 \cdot \eta(t) + 2\dot{\eta}(t)] = [-2\eta(t) + 0 \cdot \dot{\eta}(t)],$$

or equivalently,

$$\ddot{\eta}(t) = -\eta(t).$$

This implies that all secondary extremals have the form

$$\eta(t) = c_1 \cos(t) + c_2 \sin(t).$$

Definition 4.5 *A value $\hat{t}_c$ is said to be a* ***conjugate value to*** t_0*, if $t_0 < \hat{t}_c$, and there is a solution $\eta_c(\cdot)$ to Jacobi's Equation (4.45) satisfying (i) $\eta_c(t_0) = \eta_c(\hat{t}_c) = 0$ and $\eta_c(t) \neq 0$, for some $t \in (t_0, \hat{t}_c)$. In particular, $\eta_c(\cdot)$ does not vanish on $(t_0, \hat{t}_c)$. The point $[\hat{t}_c \;\; x^*(\hat{t}_c)]^T \in \mathbb{R}^2$ on the graph of $x^*(\cdot)$ is said to be a* ***conjugate point to the initial point*** $[t_0 \;\; x^*(t_0)]^T \in \mathbb{R}^2$.

Figure 4.1 illustrates the definition. We now state the Jacobi Necessary Condition.

Theorem 4.5 (Jacobi's Necessary Condition - (IV)) *Assume that $x^*(\cdot) \in \Theta$ provides a weak local minimum for $J(\cdot)$ on Θ. If $x^*(\cdot)$ is smooth and regular, then*

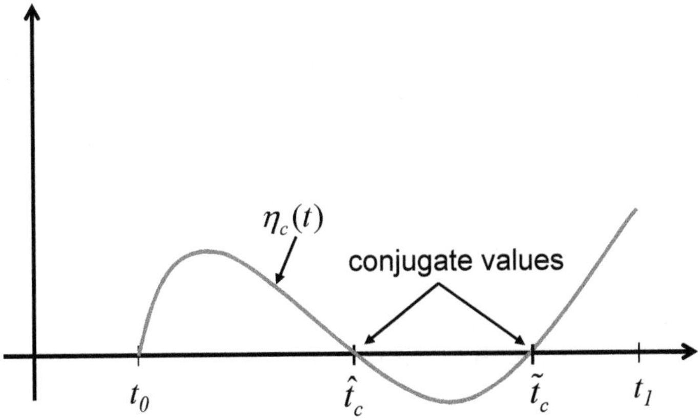

Figure 4.1: Definition of a Conjugate Value

(J-1) there cannot be a value $\hat{t}_c$ *conjugate to* t_0 *with*

$$\hat{t}_c < t_1, \tag{4.46}$$

(J-2) $x^*(t_0) = x_0$,

(J-3) $x^*(t_1) = x_1$.

4.4.1 Proof of the Jacobi Necessary Condition

Assume that $x^*(\cdot) \in \Theta$ provides a weak local minimum for $J(\cdot)$ on Θ, $x^*(\cdot)$ is smooth and $f_{uu}(t, x^*(t), \dot{x}^*(t)) > 0$. The proof is by contradiction. Assume that there is a value $\hat{t}_c$ conjugate to t_0 with

$$\hat{t}_c < t_1.$$

Without loss in generality we may assume that $\hat{t}_c$ is the "first" conjugate value so that (see Figure 4.1), there is a secondary extremal $\eta_c(\cdot) \in PWS(t_0, t_1)$ such that $\eta_c(\cdot)$ satisfies the Jacobi Equation

$$\frac{d}{dt}[f_{xu}^*(t)\eta_c(t) + f_{uu}^*(t)\dot{\eta}_c(t)] = [f_{xx}^*(t)\eta_c(t) + f_{xu}^*(t)\dot{\eta}_c(t)],$$

with $\eta_c(t_0) = 0$, $\eta_c(\hat{t}_c) = 0$ and

$$\eta_c(t) \neq 0, \quad t_0 < t < \hat{t}_c. \tag{4.47}$$

Since $\mathcal{F}_{\xi\xi}(t, \eta, \xi) = f_{uu}(t, x^*(t), \dot{x}^*(t)) > 0$, the accessory problem is regular and Hilbert's Differentiability Theorem implies that all secondary extremals are smooth. Thus, $\eta_c(\cdot)$ cannot have a corner. Let $\hat{\eta}(t)$ be the piecewise smooth function defined by

$$\hat{\eta}(t) = \begin{cases} \eta_c(t), & t_0 \le t \le \hat{t}_c \\ 0, & \hat{t}_c \le t \le t_1 \end{cases}$$

and note that $\hat{\eta}(t) \neq 0$ for $t_0 < t < \hat{t}_c$ (see Figure 4.2). We shall show that $\hat{\eta}(\cdot) \in V_0 = \Theta_S$ minimizes

$$\mathcal{J}(\eta(\cdot)) = \int_{t_0}^{t_1} \mathcal{F}(s, \eta(s), \dot{\eta}(s))ds,$$

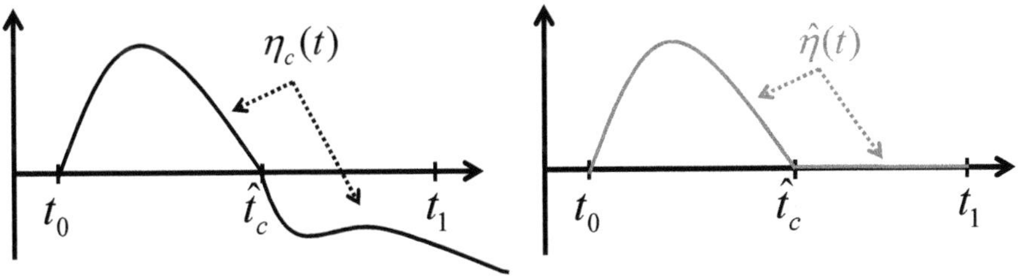

Figure 4.2: Definition of $\hat{\eta}(\cdot)$

and hence is also a secondary extremal. Observe that for each $t_0 < t < \hat{t}_c$,

$$\begin{aligned}
&\frac{d}{dt}\left[\hat{\eta}(t)\mathcal{F}_\xi(t, \hat{\eta}(t), \frac{d}{dt}\hat{\eta}(t))\right] \\
&\quad = \hat{\eta}(t)\left[\frac{d}{dt}\mathcal{F}_\xi(t, \hat{\eta}(t), \frac{d}{dt}\hat{\eta}(t))\right] + \frac{d}{dt}\hat{\eta}(t)\left[\mathcal{F}_\xi(t, \hat{\eta}(t), \frac{d}{dt}\hat{\eta}(t))\right] \\
&\quad = \eta_c(t)\left[\frac{d}{dt}\mathcal{F}_\xi(t, \eta_c(t), \dot{\eta}_c(t))\right] + \dot{\eta}_c(t)\left[\mathcal{F}_\xi(t, \eta_c(t), \dot{\eta}_c(t))\right] \\
&\quad = \eta_c(t)\left[\frac{d}{dt}\left[f_{xu}^*(t)\eta_c(t) + \hat{f}_{uu}(t)\dot{\eta}_c(t)\right]\right] \\
&\qquad + \dot{\eta}_c(t)\left[f_{xu}^*(t)\eta_c(t) + f_{uu}^*(t)\dot{\eta}_c(t)\right] \\
&\quad = \eta_c(t)\left[f_{xx}^*(t)\eta_c(t) + f_{xu}^*(t)\dot{\eta}_c(t)\right] \\
&\qquad + \dot{\eta}_c(t)\left[f_{xu}^*(t)\eta_c(t) + f_{uu}^*(t)\dot{\eta}_c(t)\right] \\
&\quad = \left[f_{xx}^*(t)[\eta_c(t)]^2 + f_{xu}^*(t)\eta_c(t)\dot{\eta}_c(t)\right] \\
&\qquad + \left[f_{ux}^*(t)\eta_c(t)\dot{\eta}_c(t) + f_{uu}^*(t)[\dot{\eta}_c(t)]^2\right] \\
&\quad = 2\mathcal{F}(t, \eta_c(t), \dot{\eta}_c(t)).
\end{aligned}$$

Hence, it follows that

$$2\mathcal{J}(\hat{\eta}(\cdot)) = \int_{t_0}^{t_1} 2\mathcal{F}(s, \hat{\eta}(s), \frac{d}{ds}\hat{\eta}(s))ds$$

$$= \int_{t_0}^{\hat{t}_c} 2\mathcal{F}(s, \eta_c(s), \dot{\eta}_c(s))ds + \int_{\hat{t}_c}^{t_1} 2\mathcal{F}(s, 0, 0)ds$$

$$= \int_{t_0}^{\hat{t}_c} 2\mathcal{F}(s, \eta_c(s), \dot{\eta}_c(s))ds$$

$$= \int_{t_0}^{\hat{t}_c} \frac{d}{ds}\left[\hat{\eta}(s) \cdot \mathcal{F}_{\xi}(s, \hat{\eta}(s), \frac{d}{ds}\hat{\eta}(s))\right] ds$$

$$= \left[\hat{\eta}(t) \cdot \mathcal{F}_{\xi}(t, \hat{\eta}(t), \frac{d}{ds}\hat{\eta}(t))\right]\Bigg|_{t=t_0}^{t=\hat{t}_c} = 0.$$

Therefore,

$$2\mathcal{J}(\hat{\eta}(\cdot)) = \int_{t_0}^{t_1} 2\mathcal{F}(s, \hat{\eta}(s), \frac{d}{ds}(s)\hat{\eta})ds = 0 \leq 2\mathcal{J}(\eta(\cdot))$$

for all $\eta(\cdot) \in V_0 = \Theta_S$, and hence $\hat{\eta}(\cdot)$ minimizes $\mathcal{J}(\eta(\cdot))$ on Θ_S. However, this means that $\hat{\eta}(\cdot)$ satisfies Jacobi's Equation (4.40) and is a secondary extremal.

Since secondary extremals cannot have corners, it follows that

$$\frac{d}{dt}\hat{\eta}(\hat{t}_c) = \frac{d}{dt}\hat{\eta}(\hat{t}_c^-) = \frac{d}{dt}\hat{\eta}(\hat{t}_c^+) = 0,$$

and, by Hilbert's Differentiability Theorem, $\hat{\eta}(\cdot) \in C^2$. Thus, $\hat{\eta}(\cdot)$ satisfies the linear second order initial value problem

$$\frac{d}{dt}[f_{xu}^*(t)\eta(t) + f_{uu}^*(t)\dot{\eta}(t)] = [f_{xx}^*(t)\eta(t) + f_{xu}^*(t)\dot{\eta}(t)],$$

with initial condition

$$\eta(\hat{t}_c) = 0, \quad \dot{\eta}(\hat{t}_c) = 0.$$

However, the only solution to a linear second order initial value problem, with zero initial data, is $\hat{\eta}(t) \equiv 0$. It follows that for $t_0 < t < \hat{t}_c$,

$$\eta_c(t) = \hat{\eta}(t) = 0,$$

which contradicts (4.47) above. Therefore, there cannot be a value $\hat{t}_c$ with $t_0 < \hat{t}_c < t_1$ conjugate to t_0 and this completes the proof. □

4.5 Weierstrass Necessary Condition - (II)

The three previous necessary conditions are valid for weak local minimizers. In this section we assume we have a strong local minimizer and derive the Weierstrass' Necessary Condition for a **strong local minimum** (Karl Theodor Wilhelm Weierstrass - 1879). Although the techniques are similar to the idea used for Euler's Necessary condition, the "variations" are different. Weierstrass' Necessary Condition is much closer to the Maximum Principle we will study in Optimal Control. In particular, for the SPCV the Weierstrass Necessary Condition can be stated as a Maximum Principle which is equivalent to the Pontryagin Maximum principle in optimal control. In order to formulate the Weierstrass Necessary Condition, we need to define the Excess Function.

Definition 4.6 *The **Weierstrass Excess Function** $E : [t_0, t_1] \times \mathbb{R}^3 \longrightarrow \mathbb{R}$ is defined by*

$$E(t, x, u, v) = [f(t, x, v) - f(t, x, u)] - [v - u] f_u(t, x, u) \quad (4.48)$$

for all $[t_0, t_1] \times \mathbb{R}^3$.

Theorem 4.6 (Weierstrass Necessary Condition - (II)) *If $x^*(\cdot) \in \Theta$ provides a strong local minimum for $J(\cdot)$ on Θ, then,*

(W-1) $E(t, x^*(t), \dot{x}^*(t), v) \geq 0$ *for all* $t \in [t_0, t_1]$ *and* $v \in \mathbb{R}$,

(W-2) $x^*(t_0) = x_0$,

(W-3) $x^*(t_1) = x_1$.

Condition (*W-1*)

$$E(t, x^*(t), \dot{x}^*(t), v) \geq 0, \quad (4.49)$$

is the essential new information in Weierstrass' Necessary Condition. Moreover, (4.49) holds at all $t \in [t_0, t_1]$, including corners. In particular, for all $v \in \mathbb{R}$

$$E(t, x^*(t), \dot{x}^*(t^+), v) \geq 0, \tag{4.50}$$

and

$$E(t, x^*(t), \dot{x}^*(t^-), v) \geq 0. \tag{4.51}$$

Before proving the Weierstrass Necessary Condition **Theorem** 4.6 we note some results that follow from this theorem. First we restate the Weierstrass Necessary Condition as a Maximum Principle by defining a new function. Given $x^*(\cdot)$, let

$$H(t, v) \triangleq -E(t, x^*(t), \dot{x}^*(t), v) \tag{4.52}$$

and observe that Weierstrass' Necessary Condition may be written as

$$H(t, v) = -E(t, x^*(t), \dot{x}^*(t), v) \leq 0$$

for all $v \in \mathbb{R}$. However, if $v = u^*(t) = \dot{x}^*(t)$, then using the definition of the excess function, one has that

$$\begin{aligned}
H(t, \dot{x}^*(t)) &= H(t, u^*(t)) \\
&= -E(t, x^*(t), \dot{x}^*(t), u^*(t)) \\
&= -\{[f(t, x^*(t), u^*(t)) - f(t, x^*(t), \dot{x}^*(t))] \\
&\quad - [u^*(t) - \dot{x}^*(t)] f_u(t, x^*(t), \dot{x}^*(t)\} \\
&= -\{[f(t, x^*(t), \dot{x}^*(t)) - f(t, x^*(t), \dot{x}^*(t))] \\
&\quad - [\dot{x}^*(t) - \dot{x}^*(t)] f_u(t, x^*(t), \dot{x}^*(t))\} \\
&= 0.
\end{aligned}$$

Consequently,

$$H(t, v) \leq 0 = H(t, \dot{x}^*(t)) = H(t, u^*(t)),$$

for all $v \in \mathbb{R}$, and we have the following equivalent version of Weierstrass' Necessary Condition.

Theorem 4.7 (Weierstrass Maximum Principle) *If* $x^*(\cdot) \in \Theta$ *provides a strong local minimum for* $J(\cdot)$ *on* Θ *and* $H(t,v) = -E(t, x^*(t), \dot{x}^*(t), v)$, *then,*

(WMP-1) $v = u^*(t) = \dot{x}^*(t)$ *maximizes* $H(t,v)$,
i.e. for all $t \in [t_0, t_1]$,

$$H(t, u^*(t)) = H(t, \dot{x}^*(t)) = \max_{v\in\mathbb{R}} H(t,v) = 0, \tag{4.53}$$

(WMP-2) $x^*(t_0) = x_0$,

(WMP-3) $x^*(t_1) = x_1$.

In addition to the above necessary conditions, one can add stronger corner conditions when $x^*(\cdot) \in \Theta$ provides a strong local minimum for $J(\cdot)$ on Θ. In particular, observe that (4.50) and (4.51) imply that if $\hat{t}$ is a corner of $x^*(\cdot)$, then for any $v \in \mathbb{R}$

$$E(\hat{t}, x^*(\hat{t}), \dot{x}^*(\hat{t}^+), v) \geq 0 \quad \text{and} \quad E(\hat{t}, x^*(\hat{t}), \dot{x}^*(\hat{t}^-), v) \geq 0.$$

Therefore,

$$\begin{aligned}\{[f(\hat{t}, x^*(\hat{t}), v) - f(\hat{t}, x^*(\hat{t}), \dot{x}^*(\hat{t}^+))] \\ - [v - \dot{x}^*(\hat{t}^+)] f_u(\hat{t}, x^*(\hat{t}), \dot{x}^*(\hat{t}^+))\} \geq 0\end{aligned} \tag{4.54}$$

and

$$\begin{aligned}\{[f(\hat{t}, x^*(\hat{t}), v) - f(\hat{t}, x^*(\hat{t}), \dot{x}^*(\hat{t}^-))] \\ - [v - \dot{x}^*(\hat{t}^-)] f_u(\hat{t}, x^*(\hat{t}), \dot{x}^*(\hat{t}^-))\} \geq 0\end{aligned} \tag{4.55}$$

both hold for any $v \in \mathbb{R}$. Setting $v = \dot{x}^*(\hat{t}^-)$ in equation (4.54) and set $v = \dot{x}^*(\hat{t}^+)$ in equation (4.55) yields

$$\begin{aligned}E^+ \triangleq {} & [f(\hat{t}, x^*(\hat{t}), \dot{x}^*(\hat{t}^-)) - f(\hat{t}, x^*(\hat{t}), \dot{x}^*(\hat{t}^+))] \\ & -[\dot{x}^*(\hat{t}^-) - \dot{x}^*(\hat{t}^+)] f_u(\hat{t}, x^*(\hat{t}), \dot{x}^*(\hat{t}^+)) \geq 0\end{aligned} \tag{4.56}$$

and

$$\begin{aligned}E^- \triangleq {} & [f(\hat{t}, x^*(\hat{t}), \dot{x}^*(\hat{t}^+)) - f(\hat{t}, x^*(\hat{t}), \dot{x}^*(\hat{t}^-))] \\ & -[\dot{x}^*(\hat{t}^+) - \dot{x}^*(\hat{t}^-)] f_u(\hat{t}, x^*(\hat{t}), \dot{x}^*(\hat{t}^-)) \geq 0,\end{aligned}$$

respectively. Note that $E^+ = -E^-$ since

$$\begin{aligned}
&-\{[f(\hat{t}, x^*(\hat{t}), \dot{x}^*(\hat{t}^+)) - f(\hat{t}, x^*(\hat{t}), \dot{x}^*(\hat{t}^-))] \\
&\quad - [\dot{x}^*(\hat{t}^+) - \dot{x}^*(\hat{t}^-)] f_u(\hat{t}, x^*(\hat{t}), \dot{x}^*(\hat{t}^-))\} \\
&= \{[f(\hat{t}, x^*(\hat{t}), \dot{x}^*(\hat{t}^-)) - f(\hat{t}, x^*(\hat{t}), \dot{x}^*(\hat{t}^+))] \\
&\quad - [\dot{x}^*(\hat{t}^-) - \dot{x}^*(\hat{t}^+)] f_u(\hat{t}, x^*(\hat{t}), \dot{x}^*(\hat{t}^-))\} \\
&= \{[f(\hat{t}, x^*(\hat{t}), \dot{x}^*(\hat{t}^-)) - f(\hat{t}, x^*(\hat{t}), \dot{x}^*(\hat{t}^+))] \\
&\quad - [\dot{x}^*(\hat{t}^-) - \dot{x}^*(\hat{t}^+)] f_u(\hat{t}, x^*(\hat{t}), \dot{x}^*(\hat{t}^+))\},
\end{aligned}$$

where the last equality follows by replacing the term $f_u(\hat{t}, x^*(\hat{t}), \dot{x}^*(\hat{t}^-))$ by $f_u(\hat{t}, x^*(\hat{t}), \dot{x}^*(\hat{t}^+))$. This step is valid because of the Weierstrass-Erdmann Corner Condition **Theorem** 3.4. In particular, we have shown that $E^+ = -E^- \leq 0$ and from inequality (4.56) $E^+ \geq 0$ so that $E^+ = 0 = -E^- = 0 = E^-$.

In particular,

$$\begin{aligned}
&[f(\hat{t}, x^*(\hat{t}), \dot{x}^*(\hat{t}^-)) - f(\hat{t}, x^*(\hat{t}), \dot{x}^*(\hat{t}^+))] \\
&\quad - [\dot{x}^*(\hat{t}^-) - \dot{x}^*(\hat{t}^+)] f_u(\hat{t}, x^*(\hat{t}), \dot{x}^*(\hat{t}^+)) \\
&= [f(\hat{t}, x^*(\hat{t}), \dot{x}^*(\hat{t}^+)) - f(\hat{t}, x^*(\hat{t}), \dot{x}^*(\hat{t}^-))] \\
&\quad - [f(\hat{t}, x^*(\hat{t}), \dot{x}^*(\hat{t}^+)) - f(\hat{t}, x^*(\hat{t}), \dot{x}^*(\hat{t}^-))].
\end{aligned}$$

Rearranging terms and again using the fact that

$$f_u(\hat{t}, x^*(\hat{t}), \dot{x}^*(\hat{t}^-)) = f_u(\hat{t}, x^*(\hat{t}), \dot{x}^*(\hat{t}^+))$$

yields

$$\begin{aligned}
&2f(\hat{t}, x^*(\hat{t}), \dot{x}^*(\hat{t}^-)) - 2\dot{x}_*(\hat{t}^-) f_u(\hat{t}, x^*(\hat{t}), \dot{x}^*(\hat{t}^-)) \\
&= 2f(\hat{t}, x^*(\hat{t}), \dot{x}^*(\hat{t}^+)) - 2\dot{x}_*(\hat{t}^+) f_u(\hat{t}, x^*(\hat{t}), \dot{x}^*(\hat{t}^+)).
\end{aligned}$$

Dividing both sides of this expression by 2 provides a proof of the following result.

Theorem 4.8 (Second Corner Condition of Erdmann) *If $x^*(\cdot) \in \Theta$ provides a strong local minimum for $J(\cdot)$ on Θ and $\hat{t}$ is a corner, then*

$$\begin{aligned}
&f(\hat{t}, x^*(\hat{t}), \dot{x}^*(\hat{t}^+)) - \dot{x}^*(\hat{t}^+) \cdot f_u(\hat{t}, x^*(\hat{t}), \dot{x}^*(\hat{t}^+)) \\
&= f(\hat{t}, x^*(\hat{t}), \dot{x}^*(\hat{t}^-)) - \dot{x}^*(\hat{t}^-) \cdot f_u(\hat{t}, x^*(\hat{t}), \dot{x}^*(\hat{t}^-)).
\end{aligned}$$

The remainder of this section is devoted to proving the Weierstrass Necessary Condition **Theorem** 4.6.

4.5.1 Proof of the Weierstrass Necessary Condition

Assume that $x^*(\cdot) \in \Theta$ provides a strong local minimum for $J(\cdot)$ on Θ. Therefore, there is a $\delta > 0$ such that

$$J(x^*(\cdot)) = \int_{t_0}^{t_1} f(s, x^*(s), \dot{x}^*(s))ds \leq J(x(\cdot)) = \int_{t_0}^{t_1} f(s, x(s), \dot{x}(s))ds, \quad (4.57)$$

for all $x(\cdot) \in U_0(x^*(\cdot), \delta) \cap \Theta$. Recall that strong local minimizers are also weak local minimizers so that Euler's Necessary Condition must hold. In particular, there is a constant c such that

$$[f_u(t, x^*(t), \dot{x}^*(t))] = c + \int_{t_0}^{t} [f_x(s, x^*(s), \dot{x}^*(s))]\, ds,$$

and, between corners,

$$\frac{d}{dt}[f_u(t, x^*(t), \dot{x}^*(t))] = [f_x(t, x^*(t), \dot{x}^*(t))].$$

Let $t_0 < \hat{t}_1 < \hat{t}_2 < \cdots < \hat{t}_p < t_1$ be the corners of $x^*(\cdot)$. On each subinterval $(\hat{t}_i, \hat{t}_{i+1})$ the minimizer $x^*(\cdot)$ and $\dot{x}^*(\cdot)$ are continuous (see Figure 4.3).

Select a subinterval $(\hat{t}_j, \hat{t}_{j+1})$ and let z be any point in $(\hat{t}_j, \hat{t}_{j+1})$. In particular, z is not a corner and there is a $\rho > 0$ such that

$$\hat{t}_j < z < z + \rho < \hat{t}_{j+1}.$$

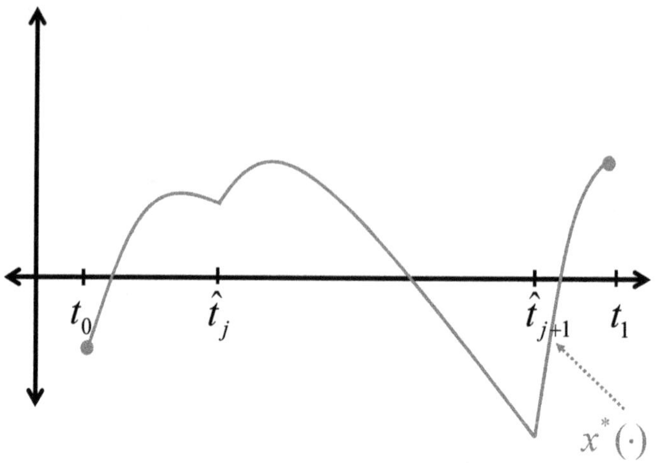

Figure 4.3: The Strong Local Minimizer

Note that if α is any number satisfying $z < \alpha \leq z + \rho$, then $[z, \alpha] \subset (\hat{t}_j, \hat{t}_{j+1})$, so that $x^*(\cdot)$ is smooth on $[z, \alpha]$. In particular, the derivative $\dot{x}^*(t)$ is continuous on $[z, z + \rho]$ so there exists a bound $M = M(\rho) \geq 0$ such that if $z \leq t \leq z + \rho$, then

$$|\dot{x}^*(t)| \leq |M(\rho)| .$$

Let $\hat{M} = \hat{M}(v, \rho) = |v| + M(\rho)$ and let $\alpha = \alpha(\delta, v)$ be a number satisfying $z < \alpha \leq z + \rho$ and

$$\hat{M} \cdot |\alpha - z| < \delta. \tag{4.58}$$

Observe that if $z \leq \varepsilon \leq \alpha$, then

$$|v| \cdot |\varepsilon - z| < \hat{M} \cdot |\alpha - z| < \delta. \tag{4.59}$$

The Mean Value Theorem implies that if $z \leq t \leq z + \rho$, then there is a $\hat{t} \in (z, t)$ such that $x^*(t) - x^*(z) = \dot{x}^*(\hat{t})(t - z)$ and hence

$$|x^*(t) - x^*(z)| \leq \left|\dot{x}^*(\hat{t})\right| \cdot |t - z| < \hat{M} \cdot |\alpha - z| < \delta. \tag{4.60}$$

We shall use these inequalities to prove the results.

For $v \in \mathbb{R}$ and α as above with $z \leq \varepsilon \leq \alpha$, define the functions

$$x_v(t) = x^*(z) + v(t - z),$$

and

$$\theta(t, \varepsilon) = x^*(t) + [x_v(\varepsilon) - x^*(\varepsilon)]\frac{(\alpha - t)}{(\alpha - \varepsilon)},$$

respectively.

Using these functions we construct a "variation" $\varphi(\cdot, \varepsilon)$ of $x^*(t)$ by

$$\varphi(t, \varepsilon) = \begin{cases} x^*(t), & t_0 \leq t < z, \\ x_v(t), & z \leq t \leq \varepsilon, \\ \theta(t, \varepsilon), & \varepsilon \leq t \leq \alpha, \\ x^*(t), & \alpha \leq t \leq t_1. \end{cases}.$$

Figure 4.4 provides a plot of $\varphi(\cdot, \varepsilon)$ in terms of its pieces. It is clear that for each $z \leq \varepsilon \leq \alpha$, the variation $\varphi(\cdot, \varepsilon)$ is piecewise smooth, $\varphi(t_0, \varepsilon) = x_0$, and $\varphi(t_1, \varepsilon) = x_1$. Thus, $\varphi(\cdot, \varepsilon) \in \Theta$. Also,

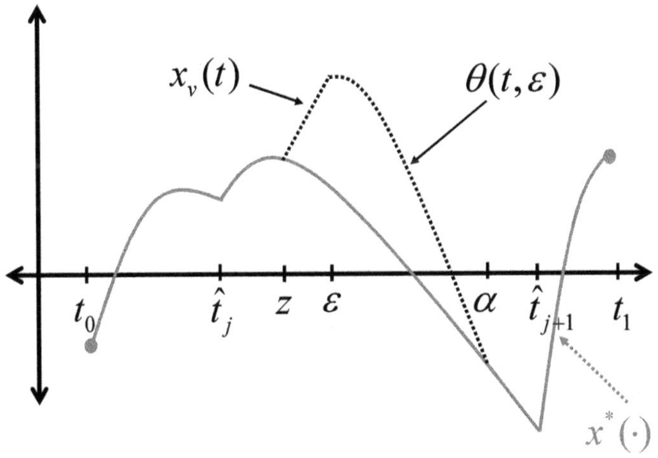

Figure 4.4: Defining the Variation: Step One

observe that

$$|x^*(t) - \varphi(t,\varepsilon)| = \begin{cases} 0, & t_0 \le t < z, \\ |x^*(t) - x_v(t)|, & z \le t \le \varepsilon \\ |x^*(t) - \theta\,(t,\varepsilon)\,|, & \varepsilon < t \le \alpha \\ 0, & \alpha \le t \le t_1, \end{cases}$$

$$= \begin{cases} 0, & t_0 \le t < z, \\ |x^*(t) - [x^*(z) + v(t-z)]|, & z \le t \le \varepsilon \\ |x^*(t) - \{x^*(t) + [x_v(\varepsilon) - x^*(\varepsilon)]\frac{(\alpha-t)}{(\alpha-\varepsilon)}\}|, & \\ \quad \varepsilon < t \le \alpha & \\ 0, & \alpha \le t \le t_1, \end{cases}$$

so that for $z \le t \le \varepsilon < \alpha$

$$\begin{aligned} |x^*(t) - [x^*(z) + v(t-z)]| &\le |x^*(t) - x^*(z)| + |v(t-z)| \\ &\le \left|\dot{x}^*(\hat{t})\right| \cdot |t-z| + |v| \cdot |t-z| \\ &\le M(\rho) \cdot |\varepsilon - z| + |v| \cdot |\varepsilon - z| \\ &< (M(\rho) + |v|) \cdot |\varepsilon - z| \\ &< \hat{M} \cdot |\varepsilon - z| < \hat{M} \cdot |\alpha - z| < \delta. \end{aligned}$$

Therefore, for $z \le t \le \varepsilon < \alpha$ it follows that

$$|x^*(t) - \varphi(t,\varepsilon)| = |x^*(t) - x_v(t)| < M \cdot |\varepsilon - z| < \delta. \tag{4.61}$$

Now consider the case where $z < \varepsilon < t \le \alpha$. For this case it follows that

$$\begin{aligned} |x^*(t) - \varphi(t,\varepsilon)| &= |x^*(t) - \{x^*(t) + [x_v(\varepsilon) - x^*(\varepsilon)]\frac{(\alpha-t)}{(\alpha-\varepsilon)}\}| \\ &= |[x_v(\varepsilon) - x^*(\varepsilon)]\frac{(\alpha-t)}{(\alpha-\varepsilon)}| \\ &= |[x^*(z) + v(\varepsilon - z) - x^*(\varepsilon)]\frac{(\alpha-t)}{(\alpha-\varepsilon)}| \\ &= |[x^*(z) - x^*(\varepsilon) + v(\varepsilon - z)]\frac{(\alpha-t)}{(\alpha-\varepsilon)}| \\ &\le [|x^*(z) - x^*(\varepsilon)| + |v(\varepsilon - z)|] \cdot |\frac{(\alpha-t)}{(\alpha-\varepsilon)}|. \end{aligned}$$

Applying the Mean Value Theorem to $x^*(\cdot)$ on the interval $[z, \varepsilon]$ yields a $\widehat{\epsilon} \in (z, \varepsilon)$ such that

$$x^*(z) - x^*(\varepsilon) = \dot{x}^*(\widehat{\epsilon})[\varepsilon - z],$$

and hence we have

$$\begin{aligned}
|x^*(t) - \varphi(t,\varepsilon)| &\leq [|x^*(z) - x^*(\varepsilon)| + |v(\varepsilon - z)|] \cdot |\frac{(\alpha - t)}{(\alpha - \varepsilon)}| \\
&\leq [|\dot{x}^*(\widehat{\epsilon})|\,|\varepsilon - z| + |v| \cdot |\varepsilon - z|] \left|\frac{(\alpha - t)}{(\alpha - \varepsilon)}\right| \\
&\leq \hat{M} \cdot |\varepsilon - z| \cdot \left|\frac{(\alpha - \varepsilon)}{(\alpha - \varepsilon)}\right| \\
&= \hat{M} \cdot |\varepsilon - z| < \delta.
\end{aligned}$$

Consequently, for $z < \varepsilon < t \leq \alpha$,

$$\begin{aligned}
|x^*(t) - \varphi(t,\varepsilon)| &= |x^*(t) - \{x^*(t) + [x_v(\varepsilon) - x^*(\varepsilon)]\frac{(\alpha - t)}{(\alpha - \varepsilon)}\}| \\
&< \hat{M} \cdot |\varepsilon - z| < \delta. \qquad (4.62)
\end{aligned}$$

Therefore, it follows from (4.61) and (4.62) that for each $t \in [t_0, t_1]$ and $z \leq \varepsilon \leq \alpha$

$$|x^*(t) - \varphi(t,\varepsilon)| < \hat{M} \cdot |\varepsilon - z| < \delta.$$

This last inequality implies that

$$d_0(x^*(\cdot), \varphi(\cdot,\varepsilon)) = \sup_{t_0 \leq t \leq t_1} |x^*(t) - \varphi(t,\varepsilon)| < \hat{M} \cdot |\varepsilon - z| < \delta \quad (4.63)$$

for all $z \leq \varepsilon \leq \alpha$ and hence the variation $\varphi(\cdot,\varepsilon) \in U_0(x^*(\cdot),\delta) \cap \Theta$ for all ε satisfying $z \leq \varepsilon \leq \alpha$. Also note that when $\varepsilon = z$, then $\varphi(t,\varepsilon)|_{\varepsilon=z} = \varphi(t,z)$ is given by

$$\begin{aligned}
\varphi(t,z) &= \begin{cases} x^*(t), & t_0 \leq t \leq z, \\ \theta(t,z), & z < t \leq \alpha, \\ x^*(t), & \alpha \leq t \leq t_1, \end{cases} \\
&= \begin{cases} x^*(t), & t_0 \leq t \leq z, \\ x^*(t) + [x_v(z) - x^*(z)]\frac{(\alpha - t)}{(\alpha - z)}, & z < t \leq \alpha, \\ x^*(t), & \alpha \leq t \leq t_1. \end{cases}
\end{aligned}$$

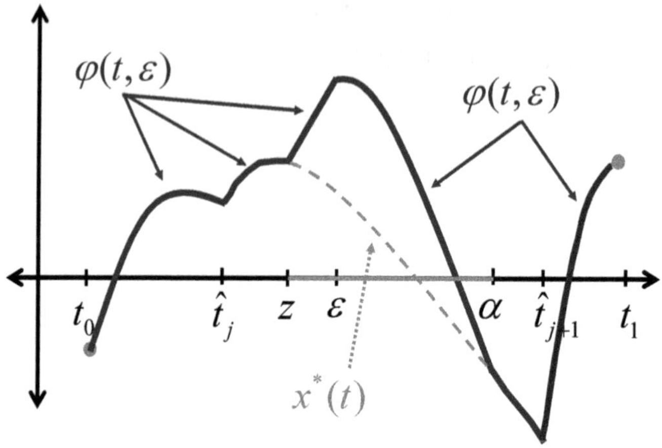

Figure 4.5: Defining the Variation: Step Two

Hence, it follows that

$$\varphi(t,z) = \varphi(t,\varepsilon)\big|_{\varepsilon=z} = \begin{cases} x^*(t), & t_0 \le t \le z, \\ x^*(t), & z < t \le \alpha, \\ x^*(t), & \alpha \le t \le t_1, \end{cases} = x^*(t), \;\; t_0 \le t < t_1. \tag{4.64}$$

Define the function $F : [z,\alpha] \longrightarrow \mathbb{R}$ by

$$\begin{aligned} F(\varepsilon) &= J(\varphi(\cdot,\varepsilon)) - J(x^*(\cdot)) \\ &= \int_{t_0}^{t_1} f(s,\varphi(s,\varepsilon),\frac{\partial\varphi(s,\varepsilon)}{\partial s})ds - \int_{t_0}^{t_1} f(s,x^*(s),\dot{x}^*(s))ds, \end{aligned} \tag{4.65}$$

and observe that (4.64) implies

$$F(z) = J(\varphi(\cdot,z)) - J(x^*(\cdot)) = J(x^*(\cdot)) - J(x^*(\cdot)) = 0.$$

The error bound (4.63) implies that $\varphi(\cdot,\varepsilon) \in U_0(x^*(\cdot),\delta) \cap \Theta$ for all ε satisfying $z \le \varepsilon \le \alpha$ so that

$$J(x^*(\cdot)) \le J(\varphi(\cdot,\varepsilon)),$$

and hence

$$F(z) = 0 \le J(\varphi(\cdot, \varepsilon)) - J(x^*(\cdot)) = F(\varepsilon).$$

Therefore, $F : [z, \alpha] \longrightarrow \mathbb{R}$ has a minimum at the left endpoint $\varepsilon = z$. We now apply the one dimensional result **Theorem** 2.1 from Chapter 1. In particular, it follows that

$$\left.\frac{d}{d\varepsilon}F(\varepsilon)\right|_{\varepsilon=z} \ge 0, \tag{4.66}$$

and to complete the proof we need to compute this derivative.

The computation of $\frac{d}{d\varepsilon}F(\varepsilon)$ requires some preparation. Observe that

$$x_v(z) = x^*(z) + v(z - z) = x^*(z), \tag{4.67}$$

and

$$\dot{x}_v(z) = \left.\frac{d}{dt}[x_v(t)]\right|_{t=z} = \left.\frac{d}{dt}[x^*(z) + v(t - z)]\right|_{t=z} = v|_{t=z} = v. \tag{4.68}$$

From the definition of $\theta(t, \varepsilon)$,

$$\theta(t, \varepsilon) = x^*(t) + [x_v(\varepsilon) - x^*(\varepsilon)]\frac{(\alpha - t)}{(\alpha - \varepsilon)},$$

and (4.67) it follows that

$$\theta(t, z) = x^*(t) + [x_v(z) - x^*(z)]\frac{(\alpha - t)}{(\alpha - z)} = x^*(t), \tag{4.69}$$

and

$$\theta_t(t, \varepsilon) = \frac{\partial}{\partial t}\theta(t, \varepsilon) = \frac{\partial}{\partial t}\left\{x^*(t) + [x_v(\varepsilon) - x^*(\varepsilon)]\frac{(\alpha - t)}{(\alpha - \varepsilon)}\right\} \tag{4.70}$$

$$= \dot{x}^*(t) - \frac{[x_v(\varepsilon) - x^*(\varepsilon)]}{(\alpha - \varepsilon)}.$$

Also, it follows from (4.67) and (4.70) that

$$\theta_t(t, z) = \theta_t(t, \varepsilon)|_{\varepsilon=z} = \dot{x}^*(t) - \frac{[x_v(z) - x^*(z)]}{(\alpha - z)} = \dot{x}^*(t),$$

so that at $t = z$

$$\theta_t(z,z) = \dot{x}^*(z). \tag{4.71}$$

Differentiating

$$\theta(t,\varepsilon) = x^*(t) + [x_v(\varepsilon) - x^*(\varepsilon)]\frac{(\alpha - t)}{(\alpha - \varepsilon)}$$

with respect to ε yields

$$\begin{aligned}\theta_\varepsilon(t,\varepsilon) = \frac{\partial}{\partial \epsilon}\theta(t,\varepsilon) &= \frac{\partial}{\partial \varepsilon}\left\{x^*(t) + [x_v(\varepsilon) - x^*(\varepsilon)]\frac{(\alpha - t)}{(\alpha - \varepsilon)}\right\} \qquad (4.72)\\ &= [\dot{x}_v(\varepsilon) - \dot{x}^*(\varepsilon)]\frac{(\alpha - t)}{(\alpha - \varepsilon)} + [x_v(\varepsilon) - x^*(\varepsilon)]\frac{(\alpha - t)}{(\alpha - \varepsilon)^2},\end{aligned}$$

and hence

$$\begin{aligned}\theta_\varepsilon(z,z) = \theta_\varepsilon(t,\varepsilon)|_{t=\varepsilon=z} &= [\dot{x}_v(z) - \dot{x}^*(z)]\frac{(\alpha - z)}{(\alpha - z)}\\ &+ [x_v(z) - x^*(z)]\frac{(\alpha - z)}{(\alpha - z)^2}\\ &= [\dot{x}_v(z) - \dot{x}^*(z)] + [x_v(z) - x^*(z)]\frac{1}{(\alpha - z)}.\end{aligned}$$

It follows from equations (4.67) and (4.68) that $x_v(z) - x^*(z) = 0$ and $\dot{x}_v(z) = v$, respectively. Thus,

$$\theta_\varepsilon(z,z) = \theta_\varepsilon(t,\varepsilon)|_{t=\varepsilon=z} = v - \dot{x}^*(z). \tag{4.73}$$

Likewise,

$$\begin{aligned}\theta_\varepsilon(\alpha,z) &= [\dot{x}_v(z) - \dot{x}^*(z)]\frac{(\alpha - \alpha)}{(\alpha - z)}\\ &\quad + [x_v(z) - x^*(z)]\frac{(\alpha - \alpha)}{(\alpha - z)^2} = 0.\end{aligned} \tag{4.74}$$

We are now ready to compute $\frac{d}{d\varepsilon}F(\varepsilon)\Big|_{\varepsilon=z}$. First note that

$$\begin{aligned}
F(\varepsilon) &= J(\varphi(\cdot,\varepsilon)) - J(x^*(\cdot)) \\
&= \int_{t_0}^{t_1} [f(s,\varphi(s,\varepsilon),\varphi_s(s,\varepsilon)) - f(s,x^*(s),\dot{x}^*(s))]ds \\
&= \int_{t_0}^{z} [f(s,\varphi(s,\varepsilon),\varphi_s(s,\varepsilon)) - f(s,x^*(s),\dot{x}^*(s))]ds \\
&\quad + \int_{z}^{\varepsilon} [f(s,\varphi(s,\varepsilon),\varphi_s(s,\varepsilon)) - f(s,x^*(s),\dot{x}^*(s))]ds \\
&\quad + \int_{\varepsilon}^{\alpha} [f(s,\varphi(s,\varepsilon),\varphi_s(s,\varepsilon)) - f(s,x^*(s),\dot{x}^*(s))]ds \\
&\quad + \int_{\alpha}^{t_1} [f(s,\varphi(s,\varepsilon),\varphi_s(s,\varepsilon)) - f(s,x^*(s),\dot{x}^*(s))]ds.
\end{aligned}$$

Since $\varphi(t,\varepsilon) = x^*(t)$ on the intervals $[t_0, z]$ and $[\alpha, t_1]$, it follows that

$$\begin{aligned}
F(\varepsilon) &= \int_{z}^{\varepsilon} [f(s,\varphi(s,\varepsilon),\varphi_s(s,\varepsilon)) - f(s,x^*(s),\dot{x}^*(s))]ds \\
&\quad + \int_{\varepsilon}^{\alpha} [f(s,\varphi(s,\varepsilon),\varphi_s(s,\varepsilon)) - f(s,x^*(s),\dot{x}^*(s))]ds,
\end{aligned}$$

or equivalently,

$$\begin{aligned}
F(\varepsilon) &= \int_{z}^{\varepsilon} f(s,\varphi(s,\varepsilon),\varphi_s(s,\varepsilon))ds + \int_{\varepsilon}^{\alpha} f(s,\varphi(s,\varepsilon),\varphi_s(s,\varepsilon))ds \\
&\quad - \int_{z}^{\alpha} f(s,x^*(s),\dot{x}^*(s))ds.
\end{aligned}$$

Therefore,

$$\begin{aligned}\frac{d}{d\varepsilon}F(\varepsilon) &= \frac{d}{d\varepsilon}\left\{\int_z^{\varepsilon} f(s,\varphi(s,\varepsilon),\varphi_s(s,\varepsilon))ds\right. \\ &\quad \left. + \int_{\varepsilon}^{\alpha} f(s,\varphi(s,\varepsilon),\varphi_s(s,\varepsilon))ds\right\} \\ &\quad - \frac{d}{d\varepsilon}\left\{\int_z^{\alpha} f(s,x^*(s),\dot{x}^*(s))ds\right\} \\ &= \frac{d}{d\varepsilon}\left\{\int_z^{\varepsilon} f(s,\varphi(s,\varepsilon),\varphi_s(s,\varepsilon))ds\right. \\ &\quad \left. + \int_{\varepsilon}^{\alpha} f(s,\varphi(s,\varepsilon),\varphi_s(s,\varepsilon))ds\right\} \\ &= \frac{d}{d\varepsilon}\left\{\int_z^{\varepsilon} f(s,\varphi(s,\varepsilon),\varphi_s(s,\varepsilon))ds\right. \\ &\quad \left. - \int_{\alpha}^{\varepsilon} f(s,\varphi(s,\varepsilon),\varphi_s(s,\varepsilon))ds\right\}\end{aligned}$$

and the definition of $\varphi(\cdot,\varepsilon)$ implies that

$$\begin{aligned}\frac{d}{d\varepsilon}F(\varepsilon) &= \frac{d}{d\varepsilon}\left\{\int_z^{\varepsilon} f(s,x_v(s),\dot{x}_v(s))ds\right\} \\ &\quad - \frac{d}{d\varepsilon}\left\{\int_{\alpha}^{\varepsilon} f(s,\theta(s,\varepsilon),\theta_s(s,\varepsilon))ds\right\}. \end{aligned} \tag{4.75}$$

Applying Leibniz's Formula to the first integral yields

$$\frac{d}{d\varepsilon}\left\{\int_z^{\varepsilon} f(s,x_v(s),\dot{x}_v(s))ds\right\} = f(\varepsilon,x_v(\varepsilon),\dot{x}_v(\varepsilon)),$$

and differentiating the second integral produces

$$\frac{d}{d\varepsilon}\left\{\int_{\alpha}^{\varepsilon} f(s,\theta(s,\varepsilon),\theta_s(s,\varepsilon))ds\right\} = f(\varepsilon,\theta(\varepsilon,\varepsilon),\theta_t(\varepsilon,\varepsilon))$$
$$+\int_{\alpha}^{\varepsilon}[f_x(s,\theta(s,\varepsilon),\theta_s(s,\varepsilon))]\theta_{\varepsilon}(s,\varepsilon)ds$$
$$+\int_{\alpha}^{\varepsilon}[f_u(s,\theta(s,\varepsilon),\theta_s(s,\varepsilon))]\theta_{s\varepsilon}(s,\varepsilon)ds.$$

Setting $\varepsilon = z$ we find that

$$\begin{aligned}\left.\frac{d}{d\varepsilon}F(\varepsilon)\right|_{\varepsilon=z} &= f(z,x_v(z),\dot{x}_v(z)) - f(z,\theta(z,z),\theta_t(z,z))\\ &\quad -\{\int_{\alpha}^{z}[f_x(s,\theta(s,z),\theta_s(s,z))]\theta_{\varepsilon}(s,z)ds\\ &\quad +\int_{\alpha}^{z}[f_u(s,\theta(s,z),\theta_s(s,z))]\theta_{s\varepsilon}(s,z)ds\}\\ &= f(z,x_v(z),\dot{x}_v(z)) - f(z,\theta(z,z),\theta_t(z,z))\\ &\quad -\int_{\alpha}^{z}\{[f_x(s,\theta(s,z),\theta_s(s,z))]\theta_{\varepsilon}(s,z)\\ &\quad +[f_u(s,\theta(s,z),\theta_s(s,z))]\theta_{s\varepsilon}(s,z)\}ds.\end{aligned}$$

Equations (4.67), (4.68), (4.69) and (4.71) yield that $x_v(z) = x^*(z)$, $\dot{x}_v(z) = v$, $\theta(t,z) = x^*(t)$, and $\theta_t(t,z) = \dot{x}^*(t)$, respectively. Therefore,

$$\begin{aligned}\left.\frac{d}{d\varepsilon}F(\varepsilon)\right|_{\varepsilon=z} &= f(z,x^*(z),v) - f(z,x^*(z),\dot{x}^*(z))\\ &\quad -\int_{\alpha}^{z}\{[f_x(s,x^*(s),\dot{x}^*(s))]\theta_{\varepsilon}(t,z)\\ &\quad +[f_u(s,x^*(s),\dot{x}^*(s))]\theta_{t\varepsilon}(s,z)\}ds,\end{aligned}$$

and now we use the fact that $x^*(t)$ satisfies Euler's equation (i.e. $\frac{d}{ds}[f_u(s,x^*(s),\dot{x}^*(s))] = [f_x(s,x^*(s),\dot{x}^*(s))]$), to obtain

$$\begin{aligned}
\left.\frac{d}{d\varepsilon}F(\varepsilon)\right|_{\varepsilon=z} &= f(z,x^*(z),v) - f(z,x^*(z),\dot{x}^*(z)) \\
&\quad - \int_{\alpha}^{z} \left\{ \frac{d}{ds}[f_u(s,x^*(s),\dot{x}^*(s))]\theta_\varepsilon(s,z) \right. \\
&\quad \left. + [f_u(s,x^*(s),\dot{x}^*(s))]\theta_{s\varepsilon}(s,z) \right\} ds \\
&= f(z,x^*(z),v) - f(z,x^*(z),\dot{x}^*(z)) \\
&\quad - \int_{\alpha}^{z} \frac{d}{ds}\left\{[f_u(s,x^*(s),\dot{x}^*(s))]\theta_\varepsilon(s,z)\right\} ds \\
&= f(z,x^*(z),v) - f(z,x^*(z),\dot{x}^*(z)) \\
&\quad - \left\{[f_u(s,x^*(s),\dot{x}^*(s))]\theta_\varepsilon(s,z)\right\}\Big|_{s=\alpha}^{s=z}.
\end{aligned}$$

Finally, we have

$$\begin{aligned}
\left.\frac{d}{d\varepsilon}F(\varepsilon)\right|_{\varepsilon=z} &= f(z,x^*(z),v) - f(z,x^*(z),\dot{x}^*(z)) \\
&\quad - \{[f_u(z,x^*(z),\dot{x}^*(z))]\theta_\varepsilon(z,z) \\
&\quad - [f_u(\alpha,x^*(\alpha),\dot{x}^*(\alpha))]\theta_\varepsilon(\alpha,z)\}.
\end{aligned}$$

Substituting $\theta_\varepsilon(z,z) = v - \dot{x}^*(z)$ (from equation (4.73)) and $\theta_\varepsilon(\alpha,z) = 0$ (from equation (4.74)) into the above equation yields

$$\begin{aligned}
\left.\frac{d}{d\varepsilon}F(\varepsilon)\right|_{\varepsilon=z} &= f(z,x^*(z),v) - f(z,x^*(z),\dot{x}^*(z)) \\
&\quad - \{[f_u(z,x^*(z),\dot{x}^*(z))][v - \dot{x}^*(z)] \\
&\quad - [f_u(\alpha,x^*(\alpha),\dot{x}^*(\alpha))]\cdot 0\}.
\end{aligned}$$

Hence, we have established that $\frac{d}{d\varepsilon}F(z) = \left.\frac{d}{d\varepsilon}F(\varepsilon)\right|_{\varepsilon=z}$ is given by

$$\begin{aligned}
\left.\frac{d}{d\varepsilon}F(\varepsilon)\right|_{\varepsilon=z} &= f(z,x^*(z),v) - f(z,x^*(z),\dot{x}^*(z)) \\
&\quad - [v - \dot{x}^*(z)][f_u(z,x^*(z),\dot{x}^*(z))],
\end{aligned}$$

or equivalently,

$$\frac{d}{d\varepsilon}F(z) = \frac{d}{d\varepsilon}F(\varepsilon)\bigg|_{\varepsilon=z} = E(z, x^*(z), \dot{x}^*(z), v). \tag{4.76}$$

Returning to (4.66), we have now shown that at the point z

$$E(z, x^*(z), \dot{x}^*(z), v) = \frac{d}{d\varepsilon}F(\varepsilon)\bigg|_{\varepsilon=z} \geq 0,$$

which establishes condition (W-1), for all $v \in \mathbb{R}$ and any $z \in [t_0, t_1]$ where $z \neq \hat{t}_j$. If $z = \hat{t}_j$, the same argument shows that

$$E(z, x^*(z), \dot{x}^*(z^+), v) = \frac{d}{d\varepsilon}F(\varepsilon)\bigg|_{\varepsilon=z} \geq 0,$$

and by using the interval $\alpha \leq \varepsilon \leq z$ to the left of z, a similar argument yields

$$E(z, x^*(z), \dot{x}^*(z^-), v) = \frac{d}{d\varepsilon}F(\varepsilon)\bigg|_{\varepsilon=z} \geq 0.$$

This completes the proof when z is not a corner of $x^*(\cdot)$. However, note that the argument holds to the right (or left) if z is a corner of $x^*(\cdot)$. Thus, the proof above can be applied even at corner points and hence $E(z, x^*(z), \dot{x}^*(z^-), v) = \frac{d}{d\varepsilon}F(\varepsilon)\big|_{\varepsilon=z} \geq 0$ is valid for any $z \in [t_0, t_1]$. Since $z \in [t_0, t_1]$ is arbitrary, this completes the proof of the theorem. □

4.5.2 Weierstrass Necessary Condition for a Weak Local Minimum

To motivate this section, we return to **Example** 4.2 with $b = 1$ and see how **Theorem** 4.7 applies.

Example 4.4 *Minimize the functional* $J(x(\cdot)) = \int_0^1 [\dot{x}(s)]^3 ds$*, subject to the endpoint conditions* $x(0) = 0$ *and* $x(1) = 1$*. Here,*

$f(t,x,u) = u^3$, $f_u(t,x,u) = 3u^2$, and $f_x(t,x,u) = 0$. Euler's equation becomes

$$3[\dot{x}^*(t)]^2 = f_u(t,x^*(t),\dot{x}^*(t)) = c + \int_0^t f_x(s,x^*(s),\dot{x}^*(s))ds$$

$$= c + \int_0^t 0 ds = c,$$

or equivalently,

$$3[\dot{x}^*(t)]^2 = c.$$

Therefore,

$$\dot{x}^*(t) = \pm\sqrt{c/3} = \pm k$$

and all we know is that $x^*(\cdot)$ is piecewise linear. Since $x^*(0) = 0$ and $x^*(1) = 1$, a possible candidate is

$$x^*(t) = t.$$

We check Weierstrass' Necessary Condition. The excess function is given by

$$\begin{aligned} E(t,x,u,v) &= [f(t,x,v) - f(t,x,u)] - [v-u]\cdot f_u(t,x,u) \\ &= [v^3 - u^3] - [v-u]\cdot 3u^2 \end{aligned}$$

so that when $x^*(t) = t$, $\dot{x}^*(t) = 1$ and

$$E(t,x^*(t),\dot{x}^*(t),v) = [v^3 - 1] - [v-1]\cdot 3.$$

*If $x^*(t) = t$ was a **strong** local minimizer, then*

$$[v^3 - 1] - [v-1]\cdot 3 = E(t,x^*(t),\dot{x}^*(t),v) \geq 0$$

must hold for all $v \in \mathbb{R}$. However, if $v = -3$, then

$$\begin{aligned} [v^3 - 1] - [v-1]\cdot 3 &= [-27-1] - [-3-1]\cdot 3 \\ &= -28 + 4\cdot 3 = -16 < 0 \end{aligned}$$

and hence

$$E(t, x^*(t), \dot{x}^*(t), -3) \ngeq 0.$$

This shows that $x^*(t) = t$ cannot be a ***strong*** *local minimizer! Since $x^*(t) = t$ is not a strong local minimum it can not be a global minimizer. However, it is still possible that $x^*(t) = t$ is a weak local minimizer.*

This example illustrates that we need more ("better") conditions to help with such problems. For example, the excess function in **Example 4.4** above is given by

$$E(t, x^*(t), \dot{x}^*(t), v) = E(t, t, 1, v) = [v^3 - 1] - [v - 1]3.$$

Observe that (see Figure 4.6) if v is close to $\dot{x}^*(v) = 1$, then

$$E(t, x^*(t), \dot{x}^*(t), v) = v^3 - 3v + 2 \geq 0$$

and the excess function is non-negative in this restricted regime. Thus, if we restrict v to be near $\dot{x}^*(t)$, then we can improve on the basic Weierstrass Necessary Condition. In particular, we can obtain a necessary condition for a weak local minimum.

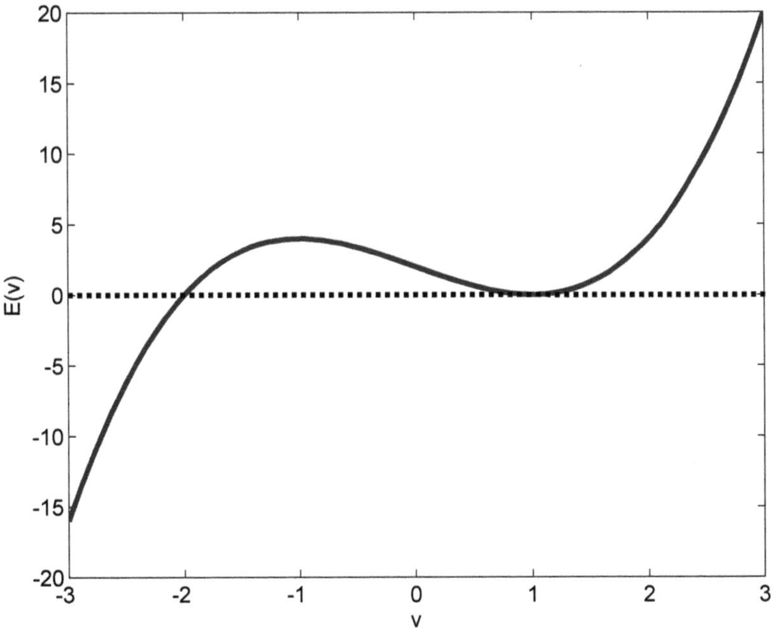

Figure 4.6: Plot of the Excess Function

Assume that $x^*(\cdot) \in \Theta$ is a weak local minimizer for $J(\cdot)$ on Θ. In particular, there is a $\delta > 0$ such that

$$J(x^*(\cdot)) = \int_{t_0}^{t_1} f(s, x^*(s), \dot{x}^*(s))ds \leq J(x(\cdot))$$
$$= \int_{t_0}^{t_1} f(s, x(s), \dot{x}(s))ds,$$

for all $x(\cdot) \in U_1(x^*(\cdot), \delta) \cap \Theta$. Looking back at the proof of **Theorem** 4.6, first select $\alpha_0 > z$ such that $|\dot{x}^*(t) - \dot{x}^*(z)| < \delta/4$ for all $z \leq t \leq \alpha_0$, and assume that

$$|\frac{d}{dt}x^*(z) - v| < \delta/4.$$

Since

$$d_0(x^*(\cdot), \varphi(\cdot, \varepsilon)) = \sup_{t_0 \leq t \leq t_1} |x^*(t) - \varphi(t, \varepsilon)| \leq |v| \cdot |\alpha - z|,$$

we can select $\alpha > z$ such that $z < \alpha \leq \alpha_0$ and

$$d_0(x^*(\cdot), \varphi(\cdot, \varepsilon)) \leq |v| \cdot |\alpha - z| < \delta/2.$$

Consider the error in the derivatives given by

$$|\dot{x}^*(t) - \frac{\partial \varphi(t, \varepsilon)}{\partial t}| = \begin{cases} 0, & t_0 \leq t < z, \\ |\frac{d}{dt}x^*(t) - \frac{d}{dt}x_v(t)|, & z \leq t \leq \varepsilon \\ |\frac{d}{dt}x^*(t) - \frac{d}{dt}\theta(t, \varepsilon)|, & \varepsilon < t \leq \alpha \\ 0, & \alpha \leq t \leq t_1, \end{cases}$$

$$= \begin{cases} 0, & t_0 \leq t < z, \\ |\frac{d}{dt}x^*(t) - \frac{d}{dt}[x^*(z) + v(t - z)]|, & z \leq t \leq \varepsilon \\ |\frac{d}{dt}x^*(t) - \frac{d}{dt}\{x^*(t) + [x_v(\varepsilon) - x^*(\varepsilon)]\frac{(\alpha - t)}{(\alpha - \varepsilon)}\}|, & \\ & \varepsilon < t \leq \alpha \\ 0, & \alpha \leq t \leq t_1, \end{cases}$$

$$= \begin{cases} 0, & t_0 \leq t < z, \\ |\frac{d}{dt}x^*(t) - v|, & z \leq t \leq \varepsilon \\ |\frac{d}{dt}[x_v(\varepsilon) - x^*(\varepsilon)]\frac{(\alpha - t)}{(\alpha - \varepsilon)}|, & \varepsilon < t \leq \alpha \\ 0, & \alpha \leq t \leq t_1, \end{cases} .$$

Let $\alpha_1 = [z+\alpha]/2$ and assume that $\varepsilon < \alpha_1 < \alpha$. Select $z \le \hat{\varepsilon} < \alpha_1$ such that

$$|v||(\varepsilon - z)|\frac{1}{|(\alpha-\varepsilon)|} < \frac{|v||(\hat{\varepsilon}-z)|}{|(\alpha-\alpha_1)|} < \frac{\delta}{2},$$

and note that for $\varepsilon < t \le \alpha$

$$\begin{aligned} |\dot{x}^*(t) - \frac{\partial\varphi(t,\varepsilon)}{\partial t}| &= |[x_v(\varepsilon) - x^*(\varepsilon)]\frac{1}{(\alpha-\varepsilon)}| \\ &= |[x^*(\varepsilon) + v(\varepsilon - z) - x^*(\varepsilon)]\frac{1}{(\alpha-\varepsilon)}| \\ &= \frac{|v(\varepsilon-z)|}{|(\alpha-\varepsilon)|}. \end{aligned}$$

If $z \le \varepsilon < \hat{\varepsilon} < \alpha_1 < \alpha \le \alpha_0$, then

$$\begin{aligned} |\dot{x}^*(t) - \frac{\partial\varphi(t,\varepsilon)}{\partial t}| &= |\dot{x}^*(t) - v| \le |\dot{x}^*(t) - \dot{x}^*(z)| + |\dot{x}^*(z) - v| \\ &< \frac{\delta}{4} + \frac{\delta}{4} \le \frac{\delta}{2}, \end{aligned}$$

for $z \le t \le \varepsilon$, and

$$|\dot{x}^*(t) - \frac{\partial\varphi(t,\varepsilon)}{\partial t}| = \frac{|v(\varepsilon-z)|}{|(\alpha-\varepsilon)|} \le \frac{|v||(\hat{\varepsilon}-z)|}{|(\alpha-\varepsilon)|} \le \frac{|v||(\hat{\varepsilon}-z)|}{|(\alpha-\alpha_1)|} < \frac{\delta}{2},$$

for $\varepsilon < t \le \alpha$. Therefore, we have shown that if $z \le \varepsilon < \hat{\varepsilon}$, and

$$|\dot{x}^*(z) - v| < \delta/4,$$

then

$$\begin{aligned} d_1(x^*(\cdot), \varphi(\cdot,\varepsilon)) &= d_0(x^*(\cdot), \varphi(\cdot,\varepsilon)) + \sup_{t_0 \le t \le t_1,\ t \ne \hat{t}_i} |\dot{x}^*(t) - \frac{\partial\varphi(t,\varepsilon)}{\partial t}| \\ &< \frac{\delta}{2} + \frac{\delta}{2} = \delta. \end{aligned}$$

This inequality shows that $\varphi(\cdot,\varepsilon) \in U_1(x^*(\cdot),\delta) \cap \Theta$ for all $\varepsilon \in [z, \hat{\varepsilon})$. Now we repeat the proof of **Theorem** 4.6 with F :

$[z, \hat{\varepsilon}) \longrightarrow \mathbb{R}$ defined by

$$F(\varepsilon) = J(\varphi(\cdot, \varepsilon)) - J(x^*(\cdot))$$
$$= \int_{t_0}^{t_1} f(s, \varphi(s, \varepsilon), \frac{\partial \varphi(s, \varepsilon)}{\partial s})ds - \int_{t_0}^{t_1} f(s, x^*(s), \dot{x}^*(s))ds$$

on the sub-interval $[z, \hat{\varepsilon}) \subset [z, \alpha)$. This establishes the following Weierstrass Necessary Condition for a weak local minimizer. Note that this condition is valid only for v close to $\dot{x}^*(t)$.

Theorem 4.9 (Restricted Weierstrass Necessary Condition) *If $x^*(\cdot) \in \Theta$ provides a weak local minimum for $J(\cdot)$ on Θ, then, there is a $\rho > 0$ such that*

(RW-1) $E(t, x^*(t), \dot{x}^*(t), v) \geq 0$ for all $t \in [t_0, t_1]$ and v satisfying

$$|\dot{x}^*(t) - v| < \rho, \tag{4.77}$$

and

(RW-2) $x^*(t_0) = x_0$,

(RW-3) $x^*(t_1) = x_1$.

Using this restricted form of the Weierstrass Necessary Condition, one can prove the Legendre Necessary Condition without using second order variations. For completeness we provide this proof below.

4.5.3 A Proof of Legendre's Necessary Condition

Assume $x^*(\cdot) \in \Theta$ provides a weak local minimum for $J(\cdot)$ on Θ. If (4.28) does not hold, then there is a fixed $\hat{z} \in [t_0, t_1]$ such that

$$f_{uu}(\hat{z}, x^*(\hat{z}), \dot{x}^*(\hat{z})) < 0.$$

We consider the case where $\hat{z} \in (t_0, t_1)$ is not a corner. The case at a corner is treated the same except one works to the left or to the right. Since the function

$$\Psi(u) \triangleq f_{uu}(\hat{z}, x^*(\hat{z}), \dot{x}^*(\hat{z}) + u)$$

is continuous, there is a $\rho_1 > 0$ such that

$$\Psi(u) \triangleq f_{uu}(\hat{z}, x^*(\hat{z}), \dot{x}^*(\hat{z}) + u) < 0$$

for all $u \in [-\rho_1, \rho_1]$. However, applying **Theorem** 4.9, there is a $\rho > 0$ such that if $|v - \dot{x}^*(\hat{z})| < \rho$, then

$$E(\hat{z}, x^*(\hat{z}), \dot{x}^*(\hat{z}), v) \geq 0.$$

Let $\gamma(v)$ be the function

$$\gamma(v) = f(\hat{z}, x^*(\hat{z}), v),$$

and select $\hat{v}$ such that $0 < |\hat{v} - \dot{x}^*(\hat{z})| < \min\{\rho, \rho_1\}$. Note that $\gamma(\cdot)$ is twice continuously differentiable and hence by Taylor's Theorem there is a $\hat{\lambda}$, $0 < \hat{\lambda} < 1$, such that

$$\begin{aligned}\gamma(\hat{v}) &= \gamma(\dot{x}^*(\hat{z})) + (\hat{v} - \dot{x}^*(\hat{z}))[\frac{d}{dv}\gamma(\dot{x}^*(\hat{z}))] \\ &+ \frac{1}{2}(\hat{v} - \dot{x}^*(\hat{z}))^2[\frac{d^2}{dv^2}\gamma(\dot{x}^*(\hat{z}) + \hat{\lambda}(\hat{v} - \dot{x}^*(\hat{z}))].\end{aligned} \tag{4.78}$$

However,

$$\begin{aligned}&\gamma(\hat{v}) - \gamma(\dot{x}^*(\hat{z})) - (\hat{v} - \dot{x}^*(\hat{z}))[\frac{d}{dv}\gamma(\dot{x}^*(\hat{z}))] \\ &\quad = f(\hat{z}, x^*(\hat{z}), \hat{v}) - f(\hat{z}, x^*(\hat{z}), \dot{x}^*(\hat{z})) \\ &\quad\quad - (\hat{v} - \dot{x}^*(\hat{z}))[f_u(\hat{z}, x^*(\hat{z}), \dot{x}^*(\hat{z}))] \\ &\quad = E(\hat{z}, x^*(\hat{z}), \dot{x}^*(\hat{z}), \hat{v}),\end{aligned}$$

and (4.78) implies

$$\begin{aligned}E(\hat{z}, x^*(\hat{z}), \dot{x}^*(\hat{z}), \hat{v}) &= \gamma(\hat{v}) - \gamma(\dot{x}^*(\hat{z})) - (\hat{v} - \dot{x}^*(\hat{z}))[\frac{d}{dv}\gamma(\dot{x}^*(\hat{z}))] \\ &= \frac{1}{2}(\hat{v} - \dot{x}^*(\hat{z}))^2[\frac{d^2}{dv^2}\gamma\dot{x}^*(\hat{z}) + \hat{\lambda}(\hat{v} - \dot{x}^*(\hat{z}))] \\ &= \frac{1}{2}(\hat{v} - \dot{x}^*(\hat{z}))^2[f_{uu}(\hat{z}, x^*(\hat{z}), \dot{x}^*(\hat{z}) \\ &\quad + \hat{\lambda}(\hat{v} - \dot{x}^*(\hat{z})))].\end{aligned}$$

Therefore,

$$E(\hat{z}, x^*(\hat{z}), \dot{x}^*(\hat{z}), \hat{v}) = \frac{1}{2}(\hat{v} - \dot{x}^*(\hat{z}))^2[f_{uu}(\hat{z}, x^*(\hat{z}), \dot{x}^*(\hat{z}) + \hat{u})] \tag{4.79}$$

where $\hat{u} \triangleq \hat{\lambda}(\hat{v} - \dot{x}^*(\hat{z}))$, satisfies

$$0 < |\hat{u}| = |\hat{\lambda}||(\hat{v} - \dot{x}^*(\hat{z}))| < |(\hat{v} - \dot{x}^*(\hat{z}))| < \min\{\rho, \rho_1\}.$$

Since $|(\hat{v} - \dot{x}^*(\hat{z}))| < \rho$ it follows that

$$E(\hat{z}, x^*(\hat{z}), \dot{x}^*(\hat{z}), \hat{u}) \geq 0,$$

but, on the other hand, since $\hat{u} \in [-\rho_1, \rho_1]$ it follows that

$$\Psi(\hat{u}) \triangleq f_{uu}(\hat{z}, x^*(\hat{z}), \dot{x}^*\,(\hat{z}) + \hat{u}) < 0.$$

In view of (4.79) we have

$$\begin{aligned} 0 &\leq E(\hat{z}, x^*(\hat{z}), \dot{x}^*(\hat{z}), \hat{u}) \\ &= \frac{1}{2}(\hat{v} - \dot{x}^*(\hat{z}))^2[f_{uu}(\hat{z}, x^*(\hat{z}), \dot{x}^*(\hat{z}) + \hat{u})] < 0, \end{aligned}$$

which is impossible. Hence the assumption that there is a fixed $\hat{z} \in [t_0, t_1]$ such that

$$f_{uu}\,(\hat{z}, x^*(\hat{z}), \dot{x}^*(\hat{z})) < 0$$

must be false. This proves the theorem. □

4.6 Applying the Four Necessary Conditions

At this point we have four necessary conditions. Three conditions (Euler Necessary Condition - (I), Legendre Necessary Condition - (III), and Jacobi Necessary Condition - (IV)) hold for weak local minimum, while the Weierstrass Necessary Condition - (II) holds only for a strong local minimum. In terms of how to proceed, it is suggested that one starts with the Euler NC - (I) and then check the Legendre NC - (III). After this step, when appropriate, check the Jacobi NC - (IV) before turning to the Weierstrass NC - (II). We illustrate this by revisiting **Example 4.4**.

Example 4.5 *Consider the problem of minimizing the functional* $J(x(\cdot)) = \int_0^1 [\dot{x}(s)]^3 ds$ *subject to the endpoint conditions* $x(0) = 0$ *and* $x(1) = 1$. *We already know that if* $x^*(\cdot)$ *is a weak local minimizer, then it is an extremal so that*

$$\dot{x}^*(t) = \pm k$$

for some k. *Applying the Legendre Necessary Condition, it must be the case that*

$$f_{uu}(t, x^*(t), \dot{x}^*(t)) = 6\dot{x}^*(t) = \pm 6k \triangleq m \geq 0,$$

and hence the derivative cannot change sign. Thus, $x^*(t) = mt + r$ *for all* $t \in [0,1]$ *where* $m \geq 0$. *However, the endpoint conditions* $x(0) = 0$ *and* $x(1) = 1$ *imply that* $m = 1$ *and* $x^*(t) = t$ *is the only possible minimizer. Note that* $x^*(t) = t$ *is smooth and*

$$f_{uu}(t, x^*(t), \dot{x}^*(t)) = 6\dot{x}^*(t) = 6 > 0,$$

so we can apply the Jacobi Necessary Condition - (IV). Since

$$f(t, x, u) = u^3,$$

it follows that

$$f_{xu}(t, x^*(t), \dot{x}^*(t)) = 0, \quad f_{xx}(t, x^*(t), \dot{x}^*(t)) = 0$$

and

$$f_{uu}(t, x^*(t), \dot{x}^*(t)) = 6\dot{x}^*(t) = 6.$$

Hence, Jacobi's Equation is given by

$$\frac{d}{dt}[0 \cdot \eta(t) + 6\dot{\eta}(t)] = [0 \cdot \eta(t) + 0 \cdot \dot{\eta}(t)] = 0,$$

or equivalently,

$$\ddot{\eta}(t) = 0.$$

Thus, all secondary extremals have the form

$$\eta(t) = pt + q,$$

and if there is a value $0 < \hat{t}_c < 1$ *such that*

$$\eta(0) = q = 0$$

and

$$\eta(\hat{t}_c) = p\hat{t}_c + q = 0,$$

then

$$\eta(t) \equiv 0$$

for all $t \in \mathbb{R}$. *Hence there are no values conjugate to* $t_0 = 0$ *and Jacobi's Necessary Condition - (IV) is satisfied.*

Summarizing, we know that $x^*(t) = t$ *satisfies all the necessary conditions for a weak local minimum, but fails to satisfy Weierstrass' Necessary Condition. All we can say at this point is that* $x^*(t) = t$ *is not a strong local minimizer.*

4.7 Problem Set for Chapter 4

Consider the **Simplest Problem in the Calculus of Variations (SPCV)**: Find $x^*(\cdot)$ to minimize the cost function

$$J(x(\cdot)) = \int_{t_0}^{t_1} f(s, x(s), \dot{x}(s))ds,$$

subject to

$$x(t_0) = x_0, \quad x(t_1) = x_1.$$

Use the four necessary conditions to completely analyze the problems below. State exactly what you can say about each problem at this point. Be sure to distinguish between weak, strong and global minimizers when possible.

Problem 4.1 *Minimize the functional* $J(x(\cdot)) = \int_0^1 \dot{x}(s)\, ds$, *subject to the endpoint conditions* $x(0) = 0$ *and* $x(1) = 1$.

Problem 4.2 *Minimize the functional* $J(x(\cdot)) = \int_0^1 x(s)\, \dot{x}(s)\, ds$, *subject to the endpoint conditions* $x(0) = 0$ *and* $x(1) = 1$.

Problem 4.3 *Minimize the functional* $J(x(\cdot)) = \int_0^1 sx\,(s)\,\dot{x}\,(s)\,ds$, *subject to the endpoint conditions* $x\,(0) = 0$ *and* $x\,(1) = 1$.

Problem 4.4 *Minimize the functional* $J(x(\cdot)) = \int_0^b [\dot{x}\,(s)]^3 ds$, *subject to the endpoint conditions* $x\,(0) = 0$ *and* $x\,(b) = x_1$.

Problem 4.5 *Minimize the functional*

$$J(x(\cdot)) = \int_0^1 \{[\dot{x}\,(s)]^2 + [x\,(s)]^2 + 2e^s x\,(s)\}ds,$$

subject to the endpoint conditions $x\,(0) = 0$ *and* $x\,(1) = e/2$.

Problem 4.6 *Minimize the functional* $J(x(\cdot)) = \int_1^2 s^{-3}[\dot{x}\,(s)]^2 ds$, *subject to the endpoint conditions* $x\,(1) = 1$ *and* $x\,(2) = 16$.

Problem 4.7 *Minimize the functional*

$$J(x(\cdot)) = \int_0^4 [\dot{x}\,(s) - 1]^2 \times [\dot{x}\,(s) + 1]^2\, ds,$$

subject to the endpoint conditions $x\,(0) = 0$ *and* $x\,(4) = 2$.

Problem 4.8 *Minimize the functional*

$$J(x(\cdot)) = \int_0^{\pi/2} \{[\dot{x}\,(s)]^2 - [x\,(s)]^2\}ds,$$

subject to the endpoint conditions $x\,(0) = 0$ *and* $x\,(\pi/2) = 0$.

Problem 4.9 *Minimize the functional*

$$J(x(\cdot)) = \int_0^{\pi} \{[\dot{x}\,(s)]^2 - [x\,(s)]^2\}ds,$$

subject to the endpoint conditions $x\,(0) = 0$ *and* $x\,(\pi) = 0$.

Problem 4.10 *Minimize the functional*

$$J(x(\cdot)) = \int_0^{3\pi/2} \{[\dot{x}(s)]^2 - [x(s)]^2\}ds,$$

subject to the endpoint conditions $x(0) = 0$ *and* $x(3\pi/2) = 0$.

Problem 4.11 *Minimize the functional*

$$J(x(\cdot)) = \int_0^{b} x(s)\sqrt{1 + [\dot{x}(s)]^2}ds,$$

subject to the endpoint conditions $x(0) = 1$ *and* $x(b) = 2$.

Problem 4.12 *Minimize the functional* $J(x(\cdot)) = \int_0^b \sqrt{\frac{1+[\dot{x}(s)]^2}{2gx(s)}}ds$, *subject to the endpoint conditions* $x(0) = 1$ *and* $x(b) = 0$.

Problem 4.13 *Minimize the functional*

$$J(x(\cdot)) = \int_1^{2} \{[\dot{x}(s)]^2 - 2sx(s)\}ds,$$

subject to the endpoint conditions $x(1) = 0$ *and* $x(2) = -1$.

Problem 4.14 *Minimize the functional*

$$J(x(\cdot)) = \int_0^{\pi} \{[x(s)]^2(1 - [\dot{x}(s)]^2)\}ds,$$

subject to the endpoint conditions $x(0) = 0$ *and* $x(\pi) = 0$.

Problem 4.15 *Minimize the functional*

$$J(x(\cdot)) = \int_1^{3} \{[3s - x(s)]x(s)\}ds,$$

subject to the endpoint conditions $x(1) = 1$ *and* $x(3) = 9/2$.

Problem 4.16 *Minimize the functional*

$$J(x(\cdot)) = 4\pi\rho v^2 \int_0^L \{[\dot{x}(s)]^3 x(s)\}ds,$$

subject to the endpoint conditions $x(0) = 1$ *and* $x(L) = R$*. Here,* ρ*,* v^2*,* $L > 0$ *and* $R > 0$ *are all constants.*

Problem 4.17 *Minimize the functional*

$$J(x(\cdot)) = \int_1^2 \{\dot{x}(s)[1 + s^2\dot{x}(s)]\}ds,$$

subject to the endpoint conditions $x(1) = 3$ *and* $x(2) = 5$*.*

Advanced Problems

Problem 4.18 *Consider the problem of minimizing the functional*

$$J(x(\cdot)) = \int_{-1}^{1} \left\{[\dot{x}(s) + 1]^2[\dot{x}(s) - 1]^2\right\} ds \geq 0,$$

with $x(-1) = x(1) = 0$*. Show that there is a sequence of functions* $x_N(\cdot)$ *such that*

$$J(x_N(\cdot)) \longrightarrow 0$$

and $d_0(x_N(\cdot), 0) \to 0$*, but* $x_0(t) = 0$ *is not a minimizer of* $J(\cdot)$*.*

Problem 4.19 *Consider the problem of minimizing the functional*

$$J(x(\cdot)) = \int_0^1 \left\{[x(s)]^2 + [\dot{x}(s)]^2(1 - [\dot{x}(s)]^2)\right\} ds,$$

with $x(0) = x(1) = 0$*. Show that* $x_0(t) = 0$ *is a weak local minimizer, but not a strong local minimizer.*

Problem 4.16 Minimize the functional

Chapter 5

Sufficient Conditions for the Simplest Problem

Although it is difficult to derive "useful" sufficient conditions, there are some results that combine the necessary conditions. We shall not spend much time on these conditions. However, there is merit in looking at some of the techniques used to derive such conditions. First recall Hilbert's form of Euler's Equations.

Observe that **Hilbert's Differentiability Theorem 3.6** implies that smooth extremals $x(\cdot)$ have continuous second derivatives. Therefore, we may again differentiate Euler's Differential Equation

$$\frac{d}{dt}\left[f_u\left(t, x(t), \dot{x}\left(t\right)\right)\right] = \left[f_x\left(t, x(t), \dot{x}\left(t\right)\right)\right], \tag{5.1}$$

and apply the chain rule to obtain Hilbert's (second order) differential equation

$$\begin{aligned} \ddot{x}\left(t\right) \cdot \left[f_{uu}\left(t, x(t), \dot{x}\left(t\right)\right)\right] = {} & \left[f_x\left(t, x(t), \dot{x}\left(t\right)\right)\right] - \left[f_{ut}\left(t, x(t), \dot{x}\left(t\right)\right)\right] \\ & - \left[f_{ux}\left(t, x(t), \dot{x}\left(t\right)\right)\right] \cdot \dot{x}\left(t\right). \end{aligned} \tag{5.2}$$

5.1 A Field of Extremals

Assume that $x_0(\cdot)$ is a fixed smooth extremal satisfying Hilbert's equation (5.2). Also assume that there exists a one parameter family of solutions $\varphi(\cdot, \alpha)$ to (5.2) with the following properties:

(FE-i) For each parameter α in an open interval $(\alpha_0 - \gamma, \alpha_0 + \gamma)$, the function $x_\alpha(\cdot)$ defined by $x_\alpha(t) \triangleq \varphi(t, \alpha)$ is a solution to Hilbert's equation on the interval $t_0 \le t \le t_1$ and $x_0(t) = \varphi(t, \alpha_0)$.

(FE-ii) The function $\varphi(t, \alpha)$ and all the partial derivatives $\frac{\partial}{\partial t}\varphi(t, \alpha)$, $\frac{\partial}{\partial \alpha}\varphi(t, \alpha)$, $\frac{\partial^2}{\partial t \partial \alpha}\varphi(t, \alpha)$, and $\frac{\partial^2}{\partial t^2}\varphi(t, \alpha)$ exist and are continuous on the set

$$[t_0, t_1] \times (\alpha_0 - \gamma, \alpha_0 + \gamma).$$

(FE-iii) The equation

$$x - \varphi(t, \alpha) = 0 \tag{5.3}$$

implicitly defines a function $\alpha : S \longrightarrow \mathbb{R}$ on a region S in the (t, x) plane defined by

$$S = [t_0, t_1] \times \{x \colon x_0(t) - \delta < x < x_0(t) + \delta\}$$

for some $\delta > 0$. In particular, $\alpha = \alpha(t, x)$ satisfies

$$x - \varphi(t, \alpha(t, x)) = 0, \tag{5.4}$$

for all $[t \;\; x]^T \in S$.

(FE-iv) The partial derivatives $\frac{\partial}{\partial t}\alpha(t, x)$ and $\frac{\partial}{\partial x}\alpha(t, x)$ exist and are continuous on S.

It is helpful to think of the family of solutions $\varphi(\cdot, \alpha)$ as providing a "strip" of graphs of smooth extremals about the graph of $x_0(\cdot)$.

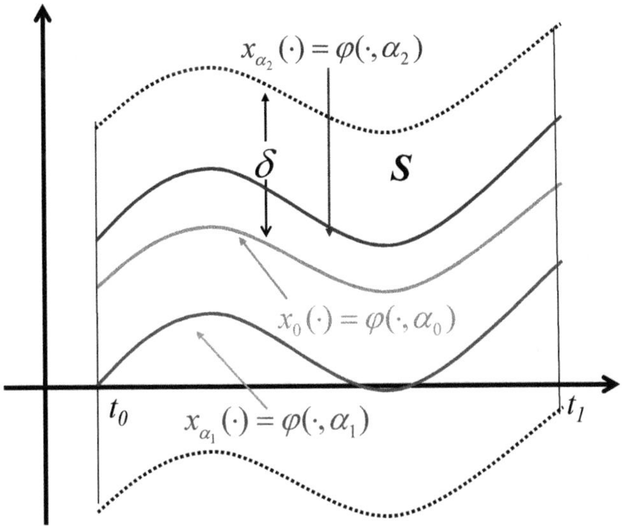

Figure 5.1: A Field of Extremals

Note that (FE-i) through (FE-iv) imply that through each fixed point $[\hat{t} \;\; \hat{x}]^T \in S$, there exists exactly one extremal

$$\hat{x}(t) = x_{\hat{\alpha}}(t) = \varphi(t, \hat{\alpha}) = \varphi(t, \alpha(\hat{t}, \hat{x})).$$

Let $p(\hat{t}, \hat{x})$ denote the slope of the **unique** extremal that goes through the point $[\hat{t} \;\; \hat{x}]^T$. Note that at a specific point $[\hat{t} \;\; \hat{x}]^T \in S$ there is a value $\hat{\alpha} = \alpha(\hat{t}, \hat{x})$ such that the value of the slope at $[\hat{t} \;\; \hat{x}]^T$ is the slope of the extremal

$$\hat{x}(t) = x_{\hat{\alpha}}(t) = \varphi(t, \hat{\alpha})$$

at $t = \hat{t}$. Thus, it follows that

$$p(\hat{t}, \hat{x}) = \dot{x}_{\hat{\alpha}}(\hat{t}) = \left.\frac{\partial}{\partial t}\varphi(t, \hat{\alpha})\right|_{t=\hat{t}} = \frac{\partial}{\partial t}\varphi(\hat{t}, \hat{\alpha}) = \frac{\partial}{\partial t}\varphi(\hat{t}, \alpha(\hat{t}, \hat{x})) \tag{5.5}$$

and since (5.5) holds at each $[\hat{t} \;\; \hat{x}]^T \in S$ one has that

$$\dot{x}_{\alpha}(t) = p(t, x) = \frac{\partial}{\partial t}\varphi(t, \alpha(t, x)) = \frac{\partial}{\partial t}\varphi(t, \alpha) \tag{5.6}$$

holds for all $[t \;\; x]^T \in S$. The function $p(t, x)$ is called the *slope function* on S.

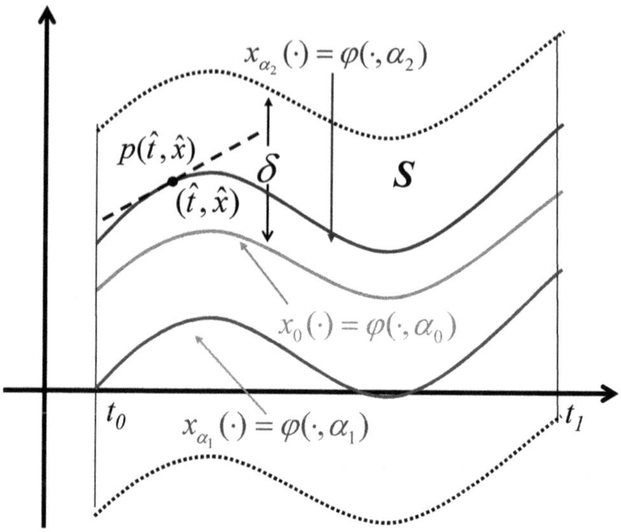

Figure 5.2: An Extremal Through a Specified Point and the Slope Function

Definition 5.1 *The pair $\mathcal{F} = (S, p(\cdot, \cdot))$ is called a **field of extremals about** $x_0(\cdot)$. Here S is the domain of the function $\alpha(t, x)$, and we say that $x_0(\cdot)$ **is embedded** in $\mathcal{F}$.*

The existence of a field of extremals will provide the basic tool for developing sufficient conditions. However, we need a few preliminary results to set the stage for the development of these results. We begin by establishing the following theorem about the slope function. Assume $x_0(\cdot)$ is embedded in the field $\mathcal{F} = (S, p(\cdot, \cdot))$ with slope function $p(\cdot, \cdot)$.

Theorem 5.1 (Basic PDE for Slope) *The slope function $p(t, x)$ satisfies the partial differential equation*

$$\begin{aligned} \{p_t(t,x) + p(t,x)p_x(t,x)\} &\cdot f_{uu}(t,x,p(t,x)) \\ &= f_x(t,x,p(t,x)) \\ &- p(t,x) \cdot f_{ux}(t,x,p(t,x)) \\ &- f_{ut}(t,x,p(t,x)) \end{aligned} \tag{5.7}$$

for all $[t \quad x]^T \in S$.

Proof: For each α in the open interval $(\alpha_0 - \gamma, \alpha_0 + \gamma)$, the extremal $x_\alpha(t) = \varphi(t, \alpha)$ satisfies Hilbert's Equation

$$\begin{aligned}\ddot{x}_\alpha(t) \cdot f_{uu}(t, x_\alpha(t), \dot{x}_\alpha(t)) = f_x(t, x_\alpha(t), \dot{x}_\alpha(t)) \\ - \dot{x}_\alpha(t) \cdot f_{ux}(t, x_\alpha(t), \dot{x}_\alpha(t)) \\ - f_{ut}(t, x_\alpha(t), \dot{x}_\alpha(t)).\end{aligned} \tag{5.8}$$

Given any "fixed" pair $[t \;\; x]^T$, let $\alpha = \alpha(t, x)$ so that

$$x_\alpha(t) = \varphi(t, \alpha) = \varphi(t, \alpha(t, x))$$

and 5.4 in (FE-iii) implies that

$$x = x_\alpha(t) = \varphi(t, \alpha) = \varphi(t, \alpha(t, x)).$$

Moreover, since $\varphi(t, \alpha) = x_\alpha(t)$ and $\varphi_t(t, \alpha) = \dot{x}_\alpha(t)$, then

$$\begin{aligned}(t, x_\alpha(t), \dot{x}_\alpha(t)) &= (t, \varphi(t, \alpha), p(t, x)) = (t, \varphi(t, \alpha(t, x)), p(t, x)) \\ &= (t, x, p(t, x)).\end{aligned} \tag{5.9}$$

Differentiating both sides of (5.6)

$$p(t, x) = p(t, \varphi(t, \alpha)) = \varphi_t(t, \alpha) = \dot{x}_\alpha(t), \tag{5.10}$$

yields

$$p_t(t, \varphi(t, \alpha)) + p_x(t, \varphi(t, \alpha)) \cdot \varphi_t(t, \alpha) = \varphi_{tt}(t, \alpha) = \ddot{x}_\alpha(t).$$

In addition, since $x = \varphi(t, \alpha)$, $p(t, x) = \varphi_t(t, \alpha)$ and $\varphi_{tt}(t, \alpha) = \ddot{x}_\alpha(t)$ it follows that

$$\{p_t(t, x) + p_x(t, x)\, p(t, x)\} = \ddot{x}_\alpha(t).$$

Multiplying both sides of this equation by $f_{uu}(t, x_\alpha(t), \dot{x}_\alpha(t)) = f_{uu}(t, x, p(t, x))$ and substituting (5.10) into (5.8) yields

$$\begin{aligned}\{p_t(t, x) &+ p_x(t, x) p(t, x)\} \cdot f_{uu}(t, x_\alpha(t), p(t, x)) \\ &= \ddot{x}_\alpha(t) \cdot f_{uu}(t, x_\alpha(t), \dot{x}_\alpha(t)) \\ &= f_x(t, x_\alpha(t), \dot{x}_\alpha(t)) \\ &\quad - \dot{x}_\alpha(t) \cdot f_{ux}(t, x_\alpha(t), \dot{x}_\alpha(t)) \\ &\quad - f_{ut}(t, x_\alpha(t), \dot{x}_\alpha(t)) \\ &= f_x(t, x, p(t, x)) \\ &\quad - p(t, x) \cdot f_{ux}(t, x, p(t, x)) \\ &\quad - f_{ut}(t, x, p(t, x))\end{aligned}$$

and this completes the proof. □

5.2 The Hilbert Integral

Assume that $\mathcal{F} = (S, p(\cdot,\cdot))$ is a field of extremals about $x_0(\cdot)$ and suppose $q(t,x)$ is any function defined on S. Given any $x(\cdot) \in PWS(t_0,t_1)$, the *Hilbert Integral* is defined by

$$J_q^*(x(\cdot)) = \int_{t_0}^{t_1} \{f(s,x(s),q(s,x(s))) + [\dot{x}(s) - q(s,x(s))] \cdot f_u(s,x(s),q(s,x(s)))\}ds$$
$$= \int_{t_0}^{t_1} \{f(s,x(s),q(s,x(s))) - q(s,x(s)) \cdot f_u(s,x(s), q(s,x(s)))\}ds + \int_{t_0}^{t_1} \{\frac{d}{ds}x(s) \cdot f_u(s,x(s),q(s,x(s)))\}ds.$$

If one defines $\mathcal{P}(t,x)$ and $\mathcal{Q}(t,x)$ by

$$\mathcal{P}(t,x) = \{f(t,x,q(t,x)) - q(t,x) \cdot f_u(t,x,q(t,x))\}, \tag{5.11}$$

and

$$\mathcal{Q}(t,x) = f_u(t,x,q(t,x)), \tag{5.12}$$

respectively, then $J_q^*(x(\cdot))$ is a line integral given by

$$J_q^*(x(\cdot)) = \int_{t_0}^{t_1} \mathcal{P}(s,x)ds + \mathcal{Q}(s,x)dx.$$

We are interested in the case where $q(s,x)$ renders the line integral $J_q^*(x(\cdot))$ independent of path. Recall from vector calculus (see Section 9.5 in [48]) that $J_q^*(x(\cdot))$ is independent of path if and only if $\mathcal{P}_x(t,x) = \mathcal{Q}_t(t,x)$. A direct calculation yields

$$\begin{aligned}\mathcal{P}_x(t,x) &= f_x(t,x,q(t,x)) + f_u(t,x,q(t,x))q_x(t,x)\\ &\quad - q_x(t,x)f_u(t,x,q(t,x)) - q(t,x)[f_{ux}(t,x,q(t,x))\\ &\quad + f_{uu}(t,x,q(t,x))q_x(t,x)]\\ &= f_x(t,x,q(t,x)) - q(t,x)f_{ux}(t,x,q(t,x))\\ &\quad - f_{uu}(t,x,q(t,x))q(t,x)q_x(t,x).\end{aligned}$$

Likewise, a direct calculation yields

$$\mathcal{Q}_t(t,x) = f_{ut}(t,x,q(t,x)) + f_{uu}(t,x,q(t,x))q_t(t,x).$$

Hence,

$$\mathcal{Q}_t(t,x) = \mathcal{P}_x(t,x),$$

if and only if

$$\begin{aligned}&f_{ut}(t,x,q(t,x)) + f_{uu}(t,x,q(t,x))q_t(t,x)\\ &\quad = f_x(t,x,q(t,x)) - q(t,x)f_{ux}(t,x,q(t,x))\\ &\qquad - f_{uu}(t,x,q(t,x))q(t,x)q_x(t,x).\end{aligned}$$

Therefore, $J_q^*(x(\cdot))$ is independent of path if and only if for all $[t \;\; x]^T \in S$, $q\,(t,x)$ satisfies the partial differential equation

$$\begin{aligned}&\{q_t(t,x) + q(t,x)q_x(t,x)\} \cdot f_{uu}(t,x,q(t,x))\\ &\quad = f_x(t,x,q(t,x))\\ &\qquad - q(t,x) \cdot f_{ux}(t,x,q(t,x))\\ &\qquad - f_{ut}(t,x,q(t,x)).\end{aligned} \tag{5.13}$$

But we know from (5.7) that the slope function satisfies this partial differential equation. Hence, if $q(t,x) = p(t,x)$, then the Hilbert Integral

$$\begin{aligned}J_p^*(x(\cdot)) = \int_{t_0}^{t_1} &\{f(s,x(s),p(s,x(s)))\\ &+ [\dot{x}(s) - p(s,x(s))] \cdot f_u(s,x(s),p(s,x(s)))\}ds,\end{aligned}$$

is independent of path. Thus we have established the following result.

Theorem 5.2 (Hilbert's Integral Theorem) *Assume that* $\mathcal{F} = (S, p(\cdot,\cdot))$ *is a field of extremals about* $x_0(\cdot)$ *and let* $x_1(\cdot)$ *and* $x_2(\cdot) \in PWS(t_0, t_1)$ *be any two functions with graphs contained in* S. *If* $x_1(t_0) = x_2(t_0)$, *and* $x_1(t_1) = x_2(t_1)$, *then*

$$J_p^*(x_1(\cdot)) = J_p^*(x_2(\cdot)).$$

5.3 Fundamental Sufficient Results

If there exists a field of extremals $\mathcal{F} = (S, p(\cdot,\cdot))$ about a smooth function $x_0(\cdot)$, then Hilbert's Integral Theorem can be exploited to yield sufficient conditions. The key result relates the cost function to the *Weierstrass Excess Function.*

Theorem 5.3 (Weierstrass-Hilbert) *Suppose* $x_0(\cdot)$ *is a smooth function embedded in a field* $\mathcal{F} = (S, p(\cdot,\cdot))$ *and the graph of* $x(\cdot) \in PWS(t_0, t_1)$ *is in* S. *If*

$$x(t_0) = x_0(t_0),$$

and

$$x(t_1) = x_0(t_1),$$

then

$$J(x(\cdot)) - J(x_0(\cdot)) = \int_{t_0}^{t_1} E(s, x(s), p(s, x(s)), \dot{x}(s))ds. \tag{5.14}$$

Proof: Consider Hilbert's integral with $q(t,x) = p(t,x)$. Evaluating $J_p^*(\cdot)$ at $x_0(\cdot)$, it follows that

$$J_p^*(x_0(\cdot)) = \int_{t_0}^{t_1} f(s, x_0(s), p(s, x_0(s))ds$$
$$+ \int_{t_0}^{t_1} [\dot{x}_0(s) - p(s, x_0(s))] \cdot [f_u(s, x_0(s), p(s, x_0(s))]ds$$

and since

$$\dot{x}_0(s) = p(s, x_0(s))$$

the Hilbert integral reduces to

$$J_p^*(x_0(\cdot)) = \int_{t_0}^{t_1} f(s, x_0(s), p(s, x_0(s))ds = J(x_0(\cdot)). \tag{5.15}$$

However, Hilbert's Integral Theorem above implies that

$$J_p^*(x_0(\cdot)) = J_p^*(x(\cdot)). \tag{5.16}$$

Therefore it follows from (5.15) and (5.16) above that

$$\begin{aligned}
&J(x(\cdot)) - J(x_0(\cdot)) \\
&\quad = J(x(\cdot)) - J_p^*(x_0(\cdot)) = J(x(\cdot)) - J_p^*(x(\cdot)) \\
&\quad = \int_{t_0}^{t_1} f(s, x(s), \dot{x}(s))ds - \int_{t_0}^{t_1} f(s, x(s), p(s, x(s))ds \\
&\qquad - \int_{t_0}^{t_1} [\dot{x}(s) - p(s, x(s))] \cdot [f_u(s, x(s), p(s, x(s))]ds \\
&\quad = \int_{t_0}^{t_1} E(s, x(s), p(s, x(s)), \dot{x}(s))ds,
\end{aligned}$$

and this completes the proof. □

We now have the following fundamental sufficiency condition based on the Weierstrass excess function. Note that we need the existence of a field of extremals.

Theorem 5.4 (Fundamental Sufficiency Theorem) *Assume that $x_0(\cdot)$ is a smooth extremal embedded in a field $\mathcal{F} = (S, p(\cdot, \cdot))$ satisfying $x_0(t_0) = x_0$ and $x_0(t_1) = x_1$. If $x(\cdot) \in \Theta$ is any other piecewise smooth function with graph in S and if for all $t \in [t_0, t_1]$*

$$E(t, x(t), p(t, x(t)), \dot{x}(t)) \geq 0,$$

then

$$J(x_0(\cdot)) \leq J(x(\cdot)).$$

In order to make use of the Fundamental Sufficiency Theorem, one needs to have the answer to the following two questions:

(A) When can a smooth function $x_0(\cdot)$ be embedded in a field $\mathcal{F} = (S, p(\cdot,\cdot))$?

and

(B) When is $E(t, x(t), p(t, x(t)), \dot{x}(t)) \geq 0$?

One answer to the first question is given by the following result. The proof of this result is outlined in Ewing's book [77] and details can be found in Bliss [27] and Bolza [31].

Theorem 5.5 (Fundamental Field Theorem) *If $x_0(\cdot)$ is smooth and satisfies*

$(F1)$ Euler's Equation

$$\frac{d}{dt}\left[f_u\left(t, x_0(t), \dot{x}_0\left(t\right)\right)\right] = \left[f_x\left(t, x_0(t), \dot{x}_0\left(t\right)\right)\right], \tag{5.17}$$

$(F2)$ the Strengthen Legendre Condition

$$f_{uu}(t, x_0(t), \dot{x}_0(t)) > 0, \quad t_0 \leq t \leq t_1, \tag{5.18}$$

$(F3)$ the Strengthen Jacobi Condition that there is no value $\hat{t}_c$ conjugate to t_0 satisfying

$$\hat{t}_c \leq t_1, \tag{5.19}$$

then there exists a field of extremals $\mathcal{F} = (S, p(\cdot,\cdot))$ about $x_0(\cdot)$.

The following result connects the Weierstrass and Legendre conditions and is the key to obtaining the basic sufficient conditions.

Theorem 5.6 (Excess Function Expansion) *If $x_0(\cdot)$ is embedded in a field with slope function $p(t,x)$, then for each $v \in \mathbb{R}$ and $[t \quad x]^T \in S$ there is a function $\theta = \theta(t,x,v)$ such that $0 < \theta(t,x,v) < 1$ and*

$$\begin{aligned} E(t,x,p(t,x),v) &= 1/2[v - p(t,x)]^2 f_{uu}(t,x,p(t,x) \\ &\quad + \theta(t,x,v)[v - p(t,x)]). \end{aligned} \tag{5.20}$$

Proof: Assume $x_0(\cdot)$ is embedded in a field of extremals $\mathcal{F} = (S, p(\cdot,\cdot))$ about $x_0(\cdot)$ with slope function $p(t,x)$. For each $[\hat{t} \ \ \hat{x}]^T \in S$ define the function

$$r(v) = f(\hat{t}, \hat{x}, v)$$

so that

$$\frac{d}{dv} r(v) = r'(v) = f_u(\hat{t}, \hat{x}, v),$$

and

$$\frac{d^2}{dv^2} r(v) = r''(v) = f_{uu}(\hat{t}, \hat{x}, v).$$

Let $\hat{p} = p(\hat{t}, \hat{x})$. Taylor's theorem with remainder implies for any v that there is a $\theta = \theta(\hat{t}, \hat{x}, v)$, with $0 < \theta < 1$ such that

$$\begin{aligned} r(v) &= r(\hat{p}) + [v - \hat{p}]r'(\hat{p}) + 1/2[v - \hat{p}]^2[r''(\hat{p} + \theta[v - \hat{p}])] \\ &= f(\hat{t}, \hat{x}, \hat{p}) + [v - \hat{p}][f_u(\hat{t}, \hat{x}, \hat{p})] \\ &\quad + 1/2[v - \hat{p}]^2[f_{uu}(\hat{t}, \hat{x}, \hat{p} + \theta[v - \hat{p}])], \end{aligned}$$

which implies

$$\begin{aligned} f(\hat{t}, \hat{x}, v) = f(\hat{t}, \hat{x}, \hat{p}) &+ [v - \hat{p}][f_u(\hat{t}, \hat{x}, \hat{p})] \\ &+ 1/2[v - \hat{p}]^2[f_{uu}(\hat{t}, \hat{x}, \hat{p} + \theta[v - \hat{p}])], \end{aligned}$$

or equivalently,

$$\begin{aligned} f(\hat{t}, \hat{x}, v) - f(\hat{t}, \hat{x}, \hat{p}) &- [v - \hat{p}][f_u(\hat{t}, \hat{x}, \hat{p})] \\ &= 1/2[v - \hat{p}]^2[f_{uu}(\hat{t}, \hat{x}, \hat{p} + \theta[v - \hat{p}])]. \end{aligned}$$

The previous equality implies that

$$E(\hat{t}, \hat{x}, \hat{p}, v) = 1/2[v - \hat{p}]^2 f_{uu}(\hat{t}, \hat{x}, \hat{p} + \theta[v - \hat{p}]).$$

In particular, since $[\hat{t} \ \ \hat{x}]^T \in S$ is arbitrary, it follows that the function $\theta = \theta(t, x, v)$ exists for all $[\hat{t} \ \ \hat{x}]^T \in S$, $0 < \theta(t, x, v) < 1$ and

$$\begin{aligned} &E(t, x, p(t,x), v) \\ &\quad = 1/2[v - p(t,x)]^2 f_{uu}(t, x, p(t,x) + \theta(t,x,v)[v - p(t,x)]). \end{aligned} \tag{5.21}$$

This completes the proof and leads to the first sufficient condition for a strong local minimum. □

Theorem 5.7 (Sufficient Condition (1)) *If the problem is regular (i.e. $f_{uu}(t,x,u) > 0$ for all (t,x,u)), $x_0(\cdot) \in \Theta$ is smooth, and satisfies*
(S1–1) Euler's Differential Equation

$$\frac{d}{dt}\left[f_u\left(t, x_0(t), \dot{x}_0\left(t\right)\right)\right] = \left[f_x\left(t, x_0(t), \dot{x}_0\left(t\right)\right)\right],$$

(S1–2) the Strengthen Jacobi Condition that there is no value $\hat{t}_c$ conjugate to t_0 satisfying

$$\hat{t}_c \leq t_1,$$

then $x_0(\cdot)$ provides a strong local minimum for $J(\cdot)$ on Θ.

Proof: By **Theorem** 5.5 it follows that the smooth function $x_0(\cdot)$ can be embedded in a field $\mathcal{F} = (S, p(\cdot,\cdot))$, $x_0(t_0) = x_0$ and $x_0(t_1) = x_1$. Also, there is a $\delta > 0$ such that if $x(\cdot)$ is any piecewise smooth function satisfying $x(t_0) = x_0$, $x(t_1) = x_1$ and $d_0(x_0(\cdot), x(\cdot)) < \delta$, then $x(\cdot) \in \Theta$ has its graph in S. Thus, (5.20) implies that there exists a $\theta(t) = \theta(t, x(t), \dot{x}(t))$ with $0 < \theta(t) < 1$ such that

$$\begin{aligned} E(t, x(t), p(t, x(t)), \dot{x}(t)) = & \, 1/2[\dot{x}(t) - p(t, x(t))]^2 \cdot [f_{uu}(t, x(t), \\ & \, p(t, x(t)) + \theta(t)[\dot{x}(t) - p(t, x(t))]])] \\ \geq & \, 0 \end{aligned}$$

for all $t \in [t_0, t_1]$ and the result follows from the Fundamental Sufficiency Condition above. □

We also have the following sufficient condition for a weak local minimum. The proof is similar to the proof to **Theorem** 5.7 above and is left as an exercise.

Theorem 5.8 (Sufficient Condition (2)) *If $x_0(\cdot) \in \Theta$ is smooth, and satisfies*
$(S2-1)$ Euler's Equation

$$\frac{d}{dt}\left[f_u\left(t, x_0(t), \dot{x}_0\left(t\right)\right)\right] = \left[f_x\left(t, x_0(t), \dot{x}_0\left(t\right)\right)\right],$$

$(S2-2)$ the Strengthen Legendre Condition

$$f_{uu}(t, x_0(t), \dot{x}_0(t)) > 0, \quad t_0 \leq t \leq t_1,$$

$(S2-3)$ the Strengthen Jacobi Condition that there is no value $\hat{t}_c$ conjugate to t_0 satisfying

$$\hat{t}_c \leq t_1,$$

then $x_0(\cdot)$ provides a weak local minimum for $J(\cdot)$ on Θ.

5.4 Problem Set for Chapter 5

Consider the **Simplest Problem in the Calculus of Variations (SPCV)**: Find $x^*(\cdot)$ to minimize the cost function

$$J(x(\cdot)) = \int_{t_0}^{t_1} f(s, x(s), \dot{x}(s))ds,$$

subject to

$$x(t_0) = x_0, \quad x(t_1) = x_1.$$

Determine if the sufficient conditions above are helpful in analyzing the following problems. Be sure to distinguish between weak, strong and global minimizers.

Problem 5.1 *Minimize the functional $J(x(\cdot)) = \int\limits_0^1 \dot{x}(s)\, ds$, subject to the endpoint conditions $x(0) = 0$ and $x(1) = 1$.*

Problem 5.2 *Minimize the functional $J(x(\cdot)) = \int\limits_0^1 x(s)\,\dot{x}(s)\, ds$, subject to the endpoint conditions $x(0) = 0$ and $x(1) = 1$.*

Problem 5.3 *Minimize the functional $J(x(\cdot)) = \int\limits_0^1 s x(s)\,\dot{x}(s)\, ds$, subject to the endpoint conditions $x(0) = 0$ and $x(1) = 1$.*

Problem 5.4 *Minimize the functional* $J(x(\cdot)) = \int_0^b [\dot{x}(s)]^3 ds$, *subject to the endpoint conditions* $x(0) = 0$ *and* $x(b) = x_1$.

Problem 5.5 *Minimize the functional*

$$J(x(\cdot)) = \int_0^1 \{[\dot{x}(s)]^2 + [x(s)]^2 + 2e^s x(s)\} ds,$$

subject to the endpoint conditions $x(0) = 0$ *and* $x(1) = e/2$.

Problem 5.6 *Minimize the functional*

$$J(x(\cdot)) = \int_1^2 s^{-3}[\dot{x}(s)]^2 ds,$$

subject to the endpoint conditions $x(1) = 1$ *and* $x(2) = 16$.

Problem 5.7 *Minimize the functional*

$$J(x(\cdot)) = \int_0^4 [\dot{x}(s) - 1]^2 [\dot{x}(s) + 1]^2 ds,$$

subject to the endpoint conditions $x(0) = 0$ *and* $x(4) = 2$.

Problem 5.8 *Minimize the functional*

$$J(x(\cdot)) = \int_0^{\pi/2} \{[\dot{x}(s)]^2 - [x(s)]^2\} ds,$$

subject to the endpoint conditions $x(0) = 0$ *and* $x(\pi/2) = 0$.

Problem 5.9 *Minimize the functional*

$$J(x(\cdot)) = \int_0^{\pi} \{[\dot{x}(s)]^2 - [x(s)]^2\} ds,$$

subject to the endpoint conditions $x(0) = 0$ *and* $x(\pi) = 0$.

Problem 5.10 *Minimize the functional*

$$J(x(\cdot)) = \int_0^{3\pi/2} \{[\dot{x}(s)]^2 - [x(s)]^2\} ds,$$

subject to the endpoint conditions $x(0) = 0$ *and* $x(3\pi/2) = 0$.

Problem 5.11 *Minimize the functional*

$$J(x(\cdot)) = \int_0^{b} x(s)\sqrt{1 + [\dot{x}(s)]^2} ds,$$

subject to the endpoint conditions $x(0) = 1$ *and* $x(b) = 2$.

Problem 5.12 *Minimize the functional* $J(x(\cdot)) = \int_0^b \sqrt{\frac{1+[\dot{x}(s)]^2}{2gx(s)}} ds$, *subject to the endpoint conditions* $x(0) = 1$ *and* $x(b) = 0$.

Problem 5.13 *Minimize the functional*

$$J(x(\cdot)) = \int_1^{2} \{[\dot{x}(s)]^2 - 2sx(s)\} ds,$$

subject to the endpoint conditions $x(1) = 0$ *and* $x(2) = -1$.

Problem 5.14 *Minimize the functional*

$$J(x(\cdot)) = \int_0^{\pi} \{[x(s)]^2(1 - [\dot{x}(s)]^2)\} ds,$$

subject to the endpoint conditions $x(0) = 0$ *and* $x(\pi) = 0$.

Problem 5.15 *Minimize the functional*

$$J(x(\cdot)) = \int_1^{3} \{[3s - x(s)]x(s)\} ds,$$

subject to the endpoint conditions $x(1) = 1$ *and* $x(3) = 9/2$.

Problem 5.16 *Minimize the functional*

$$J(x(\cdot)) = 4\pi\rho v^2 \int_0^L \left\{[\dot{x}(s)]^3 x(s)\right\} ds,$$

subject to the endpoint conditions $x(0) = 1$ *and* $x(L) = R$. *Here,* ρ, v^2, $L > 0$ *and* $R > 0$ *are all constants.*

Problem 5.17 *Minimize the functional*

$$J(x(\cdot)) = \int_1^2 \{\dot{x}(s)[1 + s^2\dot{x}(s)]\} ds,$$

subject to the endpoint conditions $x(1) = 3$ *and* $x(2) = 5$.

Advanced Problems

Problem 5.18 *Prove the* ***Fundamental Field Theorem*** *5.5.*

Problem 5.19 *Consider the functional*

$$J(x(\cdot)) = \int_0^2 \{[\dot{x}(s)]^3 + [sin(s)]^2\} ds,$$

subject to the endpoint conditions $x(0) = 1$ *and* $x(2) = 1$. *Find a field of extremals about the extremal* $x_0(t) = 1$.

Problem 5.20 *Consider the functional*

$$J(x(\cdot)) = \int_0^2 \{[\dot{x}(s)]^3 + [sin(s)]^2\} ds,$$

subject to the endpoint conditions $x(0) = 0$ *and* $x(2) = 4$. *Find a field of extremals about the extremal* $x_0(t) = 2t$.

Problem 5.21 *Consider the functional*

$$J(x(\cdot)) = \int_{-1}^{1} \{\dot{x}(s)[2t - \frac{1}{2}\dot{x}(s)]\} ds,$$

subject to the endpoint conditions $x(-1) = 0$ *and* $x(1) = \frac{1}{2}$. *Find a field of extremals about the extremal* $x_0(t) = t^2 + \frac{1}{4}t - \frac{3}{4}$.

Problem 5.22 *Prove the Sufficient Condition (2) as stated in* ***Theorem*** *5.8.*

Problem 5.23 *Consider the problem of minimizing*

$$\int_0^T \{[\dot{x}(s)]^2 - [x(s)]^2\} ds,$$

subject to $x(0) = 0$, $x(T) = 0$. *[(a)] For which* T *is it the case that* $x(t) \equiv 0$ *is a strong local minimizer? [(b)] How does your analysis extend to the more general problem of minimizing*

$$\int_0^T \{p(s)[\dot{x}(s)]^2 + q(s)[x(s)]^2\} ds,$$

subject to $x(0) = 0$ *and* $x(T) = 0$. *Here,* $p(\cdot)$ *and* $q(\cdot)$ *are real valued smooth functions defined on* $[0, \infty)$ *with* $p(t) > 0$.

Chapter 6

Summary for the Simplest Problem

This is a summary of the definitions and fundamental results for the Simplest Problem in the Calculus of Variations. Recall that $f(t, x, u)$, t_0, t_1, x_0 and x_1 are given and we assume that $f(t, x, u)$ belongs to C^2.

Let $X = PWS(t_0, t_1)$ denote the space of all real-valued piecewise smooth functions defined on $[t_0, t_1]$. For each PWS function $x : [t_0, t_1] \to \mathbb{R}$, define the *functional* $J : X \to \mathbb{R}$ by

$$J(x(\cdot)) = \int_{t_0}^{t_1} f\left(s, x(s), \dot{x}\left(s\right)\right) ds. \tag{6.1}$$

Assume that the points $[t_0 \;\; x_0]^T$ and $[t_1 \;\; x_1]^T$ are given and define the set of PWS functions Θ by

$$\Theta = \left\{x(\cdot) \in PWS(t_0, t_1) \colon x\left(t_0\right) = x_0, x\left(t_1\right) = x_1\right\}. \tag{6.2}$$

Observe that $J : X \to \mathbb{R}$ is a real valued function on X.

The Simplest Problem in the Calculus of Variations: Find $x^*(\cdot) \in \Theta$ such that

$$J(x^*(\cdot)) = \int_{t_0}^{t_1} f(s, x^*(s), \dot{x}^*(s))ds \le J(x(\cdot))$$
$$= \int_{t_0}^{t_1} f(s, x(s), \dot{x}(s))ds,$$

for all $x(\cdot) \in \Theta$.

If $x(\cdot)$ and $z(\cdot) \in PWS(t_0, t_1)$, then

$$d_0(x(\cdot), z(\cdot)) \triangleq \sup_{t_0 \le s \le t_1} \{|x(s) - z(s)|\} \tag{6.3}$$

defines the d_0 *distance between* $x(\cdot)$ and $z(\cdot)$. Given $\hat{x}(\cdot) \in PWS(t_0, t_1)$ and $\delta > 0$, the $U_0(\hat{x}(\cdot), \delta)$*-neighborhood* of $\hat{x}(\cdot)$ is defined to be the open ball

$$U_0(\hat{x}(\cdot), \delta) = \{x(\cdot) \in PWS(t_0, t_1) \colon d_0(\hat{x}(\cdot), x(\cdot)) < \delta\}.$$

For $x(\cdot)$ and $z(\cdot) \in PWS(t_0, t_1)$, there is a (finite) partition of $[t_0, t_1]$, say $t_0 = \hat{t}_0 < \hat{t}_1 < \hat{t}_2 < \cdots < \hat{t}_{p-1} < \hat{t}_p = t_1$, such that $\dot{x}(t)$ and $\dot{z}(t)$ exist and are continuous (and bounded) on each open subinterval $(\hat{t}_{i-1}, \hat{t}_i)$. The d_1 *distance between* $x(\cdot)$ and $z(\cdot)$ is defined by

$$d_1(x(\cdot), z(\cdot)) \triangleq \sup_{t_0 \le t \le t_1} \{|x(t) - z(t)|\} + \sup_{t_0 \le s \le t_1,\ s \ne \hat{t}_i} \{|\dot{x}(t) - \dot{z}(t)|\}$$
$$= d_0(x(\cdot), z(\cdot)) + \sup_{t_0 \le t \le t_1,\ s \ne \hat{t}_i} \{|\dot{x}(t) - \dot{z}(t)|\}. \tag{6.4}$$

The $U_1(\hat{x}(\cdot), \delta)$*-neighborhood* of $\hat{x}(\cdot)$ is defined to be the open ball

$$U_1(\hat{x}(\cdot), \delta) = \{x(\cdot) \in PWS(t_0, t_1) \colon d_1(\hat{x}(\cdot), x(\cdot)) < \delta\}.$$

If $x^*(\cdot) \in \Theta$ satisfies $J(x^*(\cdot)) \leq J(x(\cdot))$ for all $x(\cdot) \in \Theta$, then $x^*(\cdot)$ is called a *global minimizer* for $J(\cdot)$ on Θ.

If there is a $\delta > 0$ and a $x^*(\cdot) \in \Theta$, such that $J(x^*(\cdot)) \leq J(x(\cdot))$, for all $x(\cdot) \in U_0(x^*(\cdot), \delta) \cap \Theta$, then $x^*(\cdot)$ is called a *strong local minimizer* for $J(\cdot)$ on Θ.

Similarly, if there is a $\delta > 0$ and a $x^*(\cdot) \in \Theta$, such that $J(x^*(\cdot)) \leq J(x(\cdot))$, for all $x(\cdot) \in U_1(x^*(\cdot), \delta) \cap \Theta$, then $x^*(\cdot)$ is called a *weak local minimizer* for $J(\cdot)$ on Θ.

Theorem 6.1 (Euler Necessary Condition - (I)) *If $x^*(\cdot) \in \Theta$ provides a weak local minimum for $J(\cdot)$ on Θ, then,*

(E-1) there is a constant c such that for all $t \in [t_0, t_1]$,

$$[f_u(t, x^*(t), \dot{x}^*(t))] = c + \int_{t_0}^{t} [f_x(s, x^*(s), \dot{x}^*(s))]\, ds, \qquad (6.5)$$

(E-2) $x^*(t_0) = x_0$,

(E-3) $x^*(t_1) = x_1$.

(E-4) Between corners of $x^*(\cdot)$ the function $f_u(t, x^*(t), \dot{x}^*(t))$ is differentiable and if t is not a corner of $x^*(\cdot)$, then

$$\frac{d}{dt}[f_u(t, x^*(t), \dot{x}^*(t))] = [f_x(t, x^*(t), \dot{x}^*(t))]. \qquad (6.6)$$

Any piecewise smooth function $x(\cdot)$ satisfying (6.5) is called an *extremal*. The Euler Necessary Condition 6.1 implies that any local minimizer of $J(\cdot)$ on Θ is an extremal.

Observe that extremals do not have to satisfy the boundary conditions.

If $f_{uu}(t,x,u) \neq 0$ for all (t,x,u), $t_0 \leq t \leq t_1$, then the integrand $f(t,x,u)$ is called *non-singular.* If $f_{uu}(t,x,u) > 0$ for all (t,x,u), then the integrand $f(t,x,u)$ is said to be *regular* and the SPCV is called a *regular problem.* If $x(\cdot) \in PWS(t_0,t_1)$ is an extremal, then $x(\cdot)$ is called a *non-singular extremal* if at all points $t \in [t_0,t_1]$ where $\dot{x}(t)$ is defined, $f_{uu}(t,x(t),\dot{x}(t)) \neq 0$. If $x(\cdot) \in PWS(t_0,t_1)$ is an extremal, then $x(\cdot)$ is called a *regular extremal* if at all points $t \in [t_0,t_1]$ where $\dot{x}(t)$ is defined, $f_{uu}(t,x(t),\dot{x}(t)) > 0$.

Weierstrass-Erdmann Corner Condition. If $x(\cdot) \in PWS(t_0,t_1)$ is an extremal, then

$$f_u(\hat{t},x(\hat{t}),\dot{x}(\hat{t}^+)) = f_u(\hat{t},x(\hat{t}),\dot{x}(\hat{t}^-)) \tag{6.7}$$

for all $\hat{t} \in (t_0,t_1)$.

Theorem 6.2 (Hilbert's Differentiability Theorem) *If $x(\cdot) \in PWS(t_0,t_1)$ is an extremal, $\hat{t}$ is not a corner of $x(\cdot)$, and $f_{uu}(\hat{t},x(\hat{t}),\dot{x}(\hat{t})) \neq 0$, then there exists a $\delta > 0$ such that $x(\cdot)$ has a continuous second derivative for all $t \in (\hat{t}-\delta,\hat{t}+\delta)$ and*

$$\begin{aligned}[f_{uu}(t,x(t),\dot{x}(t))] \cdot \ddot{x}(t) + [f_{ux}(t,x(t),\dot{x}(t))] \cdot \dot{x}(t) \\ + [f_{ut}(t,x(t),\dot{x}(t))] \\ = [f_x(t,x(t),\dot{x}(t))].\end{aligned} \tag{6.8}$$

If in addition, $f(t,x,u)$ is of class C^p, $p \geq 2$, then any non-singular extremal $x(\cdot)$ is also of class C^p.

The *Weierstrass Excess Function* is defined by

$$E(t,x,u,v) = [f(t,x,v) - f(t,x,u)] - [v-u]f_u(t,x,u) \tag{6.9}$$

for all $(t,x,u,v) \in [t_0,t_1] \times \mathbb{R}^3$.

Theorem 6.3 (Weierstrass Necessary Condition - (II)) *If $x^*(\cdot) \in \Theta$ provides a strong local minimum for $J(\cdot)$ on Θ, then,*

(W-1) $E(t, x^*(t), \dot{x}^*(t), v) \geq 0$ *for all* $t \in [t_0, t_1]$ *and* $v \in \mathbb{R}$,

(W-2) $x^*(t_0) = x_0$,

(W-3) $x^*(t_1) = x_1$.

Condition (*W-1*)

$$E(t, x^*(t), \dot{x}^*(t), v) \geq 0, \tag{6.10}$$

is the essential new information in Weierstrass' Necessary Condition. Moreover, (6.10) holds at all $t \in [t_0, t_1]$, including corners. In particular, for all $v \in \mathbb{R}$

$$E(t, x^*(t), \dot{x}^*(t^+), v) \geq 0, \tag{6.11}$$

and

$$E(t, x^*(t), \dot{x}^*(t^-), v) \geq 0. \tag{6.12}$$

Assume that $x^*(\cdot) \in \Theta$ provides a strong local minimum for $J(\cdot)$ on Θ and define

$$H(t, v) \triangleq -E(t, x^*(t), \dot{x}^*(t), v), \tag{6.13}$$

and note that Weierstrass' Necessary Condition may be written as

$$H(t, v) = -E(t, x^*(t), \dot{x}^*(t), v) \leq 0$$

for all $v \in \mathbb{R}$. However, if $v = u^*(t) = \dot{x}^*(t)$, then using the definition of the excess function, one has that

$$\begin{aligned}H(t,\dot{x}^*(t)) &= H(t,u^*(t)) \\ &= -E(t,x^*(t),\dot{x}^*(t),u^*(t)) \\ &= -\{[f(t,x^*(t),u^*(t)) - f(t,x^*(t),\dot{x}^*(t))] - [u^*(t) \\ &\quad - \dot{x}^*(t)]f_u(t,x^*(t),\dot{x}^*(t)\} \\ &= -\{[f(t,x^*(t),\dot{x}^*(t)) - f(t,x^*(t),\dot{x}^*(t))] \\ &\quad - [\dot{x}^*(t) - \dot{x}^*(t)]f_u(t,x^*(t),\dot{x}^*(t))\} \\ &= 0.\end{aligned}$$

Consequently,

$$H(t,v) \le 0 = H(t,\dot{x}^*(t)) = H(t,u^*(t)),$$

for all $v \in \mathbb{R}$, and we have the following equivalent version of Weierstrass' Necessary Condition.

Theorem 6.4 (Weierstrass Maximum Principle) *If $x^*(\cdot) \in \Theta$ provides a strong local minimum for $J(\cdot)$ on Θ, then,*

(WMP-1) $v = u^*(t) = \dot{x}^*(t)$ maximizes $H(t,v)$

(WMP-2) $x^*(t_0) = x_0$,

(WMP-3) $x^*(t_1) = x_1$.

In particular,

$$H(t,u^*(t)) = H(t,\dot{x}^*(t)) = \max_{v\in\mathbb{R}} H(t,v) = 0 \tag{6.14}$$

for all $t \in [t_0,t_1]$.

Theorem 6.5 (Legendre Necessary Condition - (III)) *If $x^*(\cdot) \in \Theta$ provides a weak local minimum for $J(\cdot)$ on Θ, then,*

(L-1) $f_{uu}(t,x^*(t),\dot{x}^*(t)) \ge 0$, for all $t \in [t_0,t_1]$,

(L-2) $x^*(t_0) = x_0$,

(L-3) $x^*(t_1) = x_1$.

It is important to again note that condition $(L-1)$ holds at corners. In particular,

$$f_{uu}\left(t, x^*(t), \dot{x}^*\left(t\right)\right) \geq 0 \tag{6.15}$$

implies that

$$f_{uu}\left(t, x^*(t), \dot{x}^*\left(t^+\right)\right) \geq 0$$

for all $t \in [t_0, t_1)$, and

$$f_{uu}\left(t, x^*(t), \dot{x}^*\left(t^-\right)\right) \geq 0$$

for all $t \in (t_0, t_1]$.

If $x^*(\cdot) \in \Theta$ provides a weak local minimum for $J(\cdot)$ on Θ, then define the functions $f^*_{xx}(t)$, $f^*_{xu}(t)$, and $f^*_{uu}(t)$ by

$$f^*_{xx}(t) = f_{xx}(t, x^*(t), \dot{x}^*(t)), \qquad f^*_{xu}(t) = f_{xu}(t, x^*(t), \dot{x}^*(t)),$$

and

$$f^*_{uu}(t) = f_{uu}(t, x^*(t), \dot{x}^*(t)),$$

respectively. Also, define the function $\mathcal{F}(t, \eta, \xi)$ by

$$\mathcal{F}(t, \eta, \xi) = \frac{1}{2}[f^*_{xx}(t)\eta^2 + 2f^*_{xu}(t)\eta\xi + f^*_{uu}(t)\xi^2], \tag{6.16}$$

and consider the functional $\mathcal{J} : PWS(t_0, t_1) \longrightarrow \mathbb{R}$ given by

$$\mathcal{J}(\eta(\cdot)) = \int_{t_0}^{t_1} \mathcal{F}(s, \eta(s), \dot{\eta}(s))ds. \tag{6.17}$$

Let $\Theta_S \subset PWS(t_0, t_1)$ be defined by

$$\Theta_S = \{\eta(\cdot) \in PWS(t_0, t_1) \colon \eta\left(t_0\right) = 0, \eta\left(t_1\right) = 0\} = V_0, \tag{6.18}$$

and consider the so called **Accessory (Secondary) Minimum Problem**: Find $\eta^*(\cdot) \in \Theta_S$, such that

$$\mathcal{J}(\eta^*(\cdot)) = \int_{t_0}^{t_1} \mathcal{F}(s, \eta^*(s), \dot{\eta}^*(s))ds \leq \mathcal{J}(\eta(\cdot)) = \int_{t_0}^{t_1} \mathcal{F}(s, \eta(s), \dot{\eta}(s))ds,$$

for all $\eta(\cdot) \in \Theta_S$.

If $\eta^*(\cdot) \in \Theta_S$ is any minimizer of $\mathcal{J}(\eta(\cdot))$ on Θ_S, then there is a constant c such that $\eta^*(\cdot)$ satisfies Jacobi's Integral Equation

$$[\mathcal{F}_\xi(t, \eta^*(t), \dot{\eta}^*(t))] = c + \int_{t_0}^{t} [\mathcal{F}_\eta(s, \eta^*(s), \dot{\eta}^*(s))] ds. \quad (6.19)$$

In addition, between corners the function $\mathcal{F}_\xi(t, \eta^*(t), \dot{\eta}^*(t))$ is differentiable and

$$\frac{d}{dt}[\mathcal{F}_\xi(t, \eta^*(t), \dot{\eta}^*(t))] = [\mathcal{F}_\eta(t, \eta^*(t), \dot{\eta}^*(t))]. \quad (6.20)$$

Observe that Jacobi's Differential Equation is Euler's Differential Equation for the case where $f(t, x, u)$ is replaced by $\mathcal{F}(t, \eta, \xi)$. In particular, Jacobi's Differential Equation (6.20) has the form

$$\frac{d}{dt}[f_{xu}^*(t)\eta(t) + f_{uu}^*(t)\dot{\eta}(t)] = [f_{xx}^*(t)\eta(t) + f_{xu}^*(t)\dot{\eta}(t)]. \quad (6.21)$$

A PWS function $\eta(\cdot)$ satisfying Jacobi's equation (6.19) (or (6.20)) is called a *secondary extremal.*

A value $\hat{t}_c$ is said to be a **conjugate value to** t_0, if $t_0 < \hat{t}_c$, and there is a solution $\eta_c(\cdot)$ to Jacobi's Equation (6.19) (or (6.20)) satisfying (i) $\eta_c(t_0) = \eta_c(\hat{t}_c) = 0$ and $\eta_c(t) \neq 0$, for some $t \in (t_0, \hat{t}_c)$. In particular, $\eta_c(\cdot)$ does not vanish on $(t_0, \hat{t}_c)$. The point $[\hat{t}_c \ \ x^*(\hat{t}_c)]^T \in \mathbb{R}^2$ on the graph of $x^*(\cdot)$ is said to be a **conjugate point to the initial point** $[t_0 \ \ x^*(t_0)]^T in \mathbb{R}^2$.

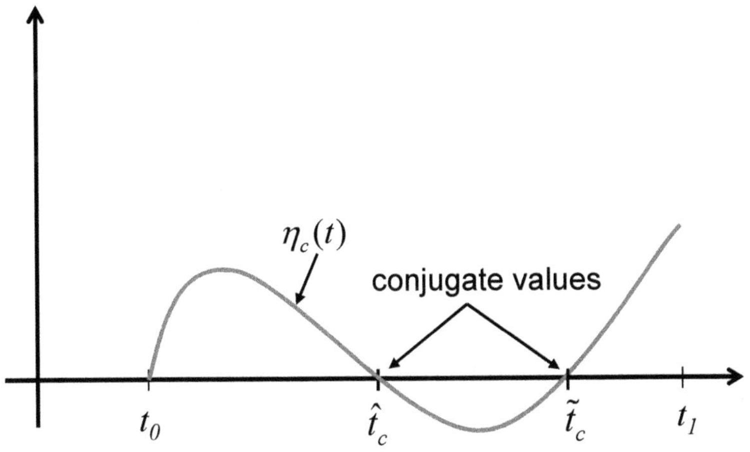

Figure 6.1: Definition of a Conjugate Value

Theorem 6.6 (Jacobi's Necessary Condition - (IV)) *Assume that $x^*(\cdot) \in \Theta$ provides a weak local minimum for $J(\cdot)$ on Θ. If $x^*(\cdot)$ is smooth and regular (i.e. $f_{uu}(t, x^*(t), \dot{x}^*(t)) > 0$), then*

(J-1) there can not be a value t_c conjugate to t_0 with

$$t_c < t_1, \tag{6.22}$$

(J-2) $x^*(t_0) = x_0$,

(J-3) $x^*(t_1) = x_1$.

Theorem 6.7 (Sufficient Condition (1)) *If the problem is regular (i.e. $f_{uu}(t, x, u) > 0$ for all (t, x, u)), $x_0(\cdot) \in \Theta$ is smooth, and satisfies*
$(S1\text{–}1)$ Euler's Equation

$$\frac{d}{dt}\left[f_u\left(t, x_0(t), \dot{x}_0(t)\right)\right] = \left[f_x\left(t, x_0(t), \dot{x}_0(t)\right)\right],$$

(S1–2) the Strengthen Jacobi Condition that there is no value $\hat{t}_c$ conjugate to t_0 satisfying

$$\hat{t}_c \leq t_1,$$

then $x_0(\cdot)$ provides a strong local minimum for $J(\cdot)$ on Θ.

Theorem 6.8 (Sufficient Condition (2)) *If $x_0(\cdot) \in \Theta$ is smooth, and satisfies*

(S2–1) Euler's Equation

$$\frac{d}{dt}\left[f_u\left(t, x_0(t), \dot{x}_0(t)\right)\right] = \left[f_x\left(t, x_0(t), \dot{x}_0(t)\right)\right],$$

(S2–2) the Strengthen Legendre Condition

$$f_{uu}(t, x_0(t), \dot{x}_0(t)) > 0, \quad t_0 \leq t \leq t_1,$$

(S2–3) the Strengthen Jacobi Condition that there is no value $\hat{t}_c$ conjugate to t_0 satisfying

$$\hat{t}_c \leq t_1,$$

then $x_0(\cdot)$ provides a weak local minimum for $J(\cdot)$ on Θ.

Chapter 7

Extensions and Generalizations

In this chapter we discuss several extensions of the Simplest Problem of the Calculus of Variations. Although we focus on the first necessary condition (Euler's Necessary Condition), there are extensions of all the Necessary conditions (I) - (IV) presented above. We shall derive the necessary conditions for global minimizers and simply note that the proofs may be modified as in Chapter 4 for weak local minimizers.

7.1 Properties of the First Variation

We recall the definition of the first variation of a functional defined on $PWS(t_0, t_1)$ at a given $x_o(\cdot) \in PWS(t_0, t_1)$. We restate Definition 2.14 in Chapter 2. If $x_o(\cdot)$ and $\eta(\cdot) \in PWS(t_0, t_1)$, then the **first variation of** $J(\cdot)$ **at** $x_o(\cdot)$ **in the direction of** $\eta(\cdot)$ is denoted by $\delta J(x_o(\cdot); \eta(\cdot))$ and is defined by

$$\delta J(x_o(\cdot); \eta(\cdot)) = \frac{d}{d\varepsilon} \left[J(x_o(\cdot) + \varepsilon \eta(\cdot)) \right] \Big|_{\varepsilon=0} . \tag{7.1}$$

As in Chapter 3 one uses Leibniz's formula (**Lemma** 3.2) to show that the first variation exists for all $\eta(\cdot) \in PWS(t_0, t_1)$. Moreover, the first variation $\delta J(x_o(\cdot); \eta(\cdot))$ of $J(\cdot)$ at $x_o(\cdot)$ in the direction of

$\eta(\cdot)$ has the form

$$\delta J(x_o(\cdot);\eta(\cdot)) = \int_{t_0}^{t_1} \{f_x(s,x_o(s),\dot{x}_o(s))\eta(s) + f_u(s,x_o(s),\dot{x}_o(s))\dot{\eta}(s)\}ds. \tag{7.2}$$

Observe that the above definition is valid for any function $x_o(\cdot) \in PWS(t_0,t_1)$. When $x_o(\cdot)$ is an extremal, the first variation has a special form. Assume now that $x_o(\cdot)$ is an extremal so that there is a constant c such that for all $t \in [t_0,t_1]$

$$[f_u(t,x_o(t),\dot{x}_o(t))] = c + \int_{t_0}^{t} [f_x(s,x_o(s),\dot{x}_o(s))]\,ds.$$

Between corners, $x_o(\cdot)$ is differentiable and

$$\frac{d}{dt}[f_u(t,x_o(t),\dot{x}_o(t))] = f_x(t,x_o(t),\dot{x}_o(t)). \tag{7.3}$$

Using the fact that $x_o(\cdot)$ satisfies the Euler equation (7.3), then the first variation (7.2) becomes

$$\begin{aligned}
\delta J(x_o(\cdot);\eta(\cdot)) &= \int_{t_0}^{t_1} \{[f_x(s,x_o(s),\dot{x}_o(s))]\cdot\eta(s) \\
&\quad + [f_u(s,x_o(s),\dot{x}_o(s))]\cdot\dot{\eta}(s)\}ds \\
&= \int_{t_0}^{t_1} \{\frac{d}{ds}[f_u(s,x_o(s),\dot{x}_o(s))]\cdot\eta(s) \\
&\quad + [f_u(s,x_o(s),\dot{x}_o(s))]\cdot\dot{\eta}(s)\}ds \\
&= \int_{t_0}^{t_1} \frac{d}{ds}[f_u(s,x_o(s),\dot{x}_o(s))\cdot\eta(s)]\,ds \\
&= [f_u(t,x_o(t),\dot{x}_o(t))\cdot\eta(t)]\big|_{t=t_0}^{t=t_1}.
\end{aligned}$$

Therefore, we have shown that if $x_o(\cdot)$ is any extremal and $\eta(\cdot) \in PWS(t_0, t_1)$, then

$$\begin{aligned}\delta J(x_o(\cdot);\eta(\cdot)) &= [f_u(t, x_o(t), \dot{x}_o(t)) \cdot \eta(t)]|_{t=t_0}^{t=t_1} \\ &= f_u(t_1, x_o(t_1), \dot{x}_o(t_1)) \cdot \eta(t_1) \\ &\quad - f_u(t_0, x_o(t_0), \dot{x}_o(t_0)) \cdot \eta(t_0).\end{aligned}$$

We summarize this as the following Lemma.

Lemma 7.1 *If $x_o(\cdot) \in PWS(t_0, t_1)$ is an extremal and $\eta(\cdot) \in PWS(t_0, t_1)$, then the first variation of $J(\cdot)$ at $x_o(\cdot)$ in the direction of $\eta(\cdot)$ exists and is given by*

$$\delta J(x_o(\cdot);\eta(\cdot)) = [f_u(t, x_o(t), \dot{x}_o(t)) \cdot \eta(t)]|_{t=t_0}^{t=t_1}. \tag{7.4}$$

7.2 The Free Endpoint Problem

Recall the river crossing problem discussed in Section 2.1.2 and Section 2.4.3. The river crossing problem is to find a smooth function $x^* : [0, 1] \longrightarrow \mathbb{R}$ that minimizes

$$J(x(\cdot)) = \int_0^1 \frac{\sqrt{c^2(1 + [\dot{x}(s)]^2) - [v(s)]^2} - v(s)\dot{x}(s)}{c^2 - [v(s)]^2} ds \tag{7.5}$$

among all smooth functions satisfying

$$x(0) = 0. \tag{7.6}$$

Observe that unlike the SPCV, there is no specified value at $t = 1$. In particular, the value of $x(1)$ is "free" and must be determined as part of finding the minimizing function $x^*(\cdot)$. This a typical example of the so-called *free-endpoint problem.*

In this section we consider the free-endpoint problem and focus on obtaining (first order) necessary conditions. As in the previous chapters, let $X = PWS(t_0, t_1)$ denote the space of all real-valued piecewise smooth functions defined on $[t_0, t_1]$. For each PWS function $x : [t_0, t_1] \to \mathbb{R}$, define the *functional* $J : X \to \mathbb{R}$ by

$$J(x(\cdot)) = \int_{t_0}^{t_1} f(s, x(s), \dot{x}(s))\, ds. \tag{7.7}$$

Assume that the interval $[t_0, t_1]$ and initial value x_0 are given (no value is assigned at t_1) and define the set of PWS functions Θ_L by

$$\Theta_L = \{x(\cdot) \in PWS(t_0, t_1) : x(t_0) = x_0\}. \tag{7.8}$$

The Free Endpoint Problem is the problem of minimizing $J(\cdot)$ on Θ_L. In particular, the goal is to find $x^*(\cdot) \in \Theta_L$ such that

$$\begin{aligned} J(x^*(\cdot)) &= \int_{t_0}^{t_1} f(s, x^*(s), \dot{x}^*(s))\, ds \le J(x(\cdot)) \\ &= \int_{t_0}^{t_1} f(s, x(s), \dot{x}(s))\, ds, \end{aligned}$$

for all $x(\cdot) \in \Theta_L$.

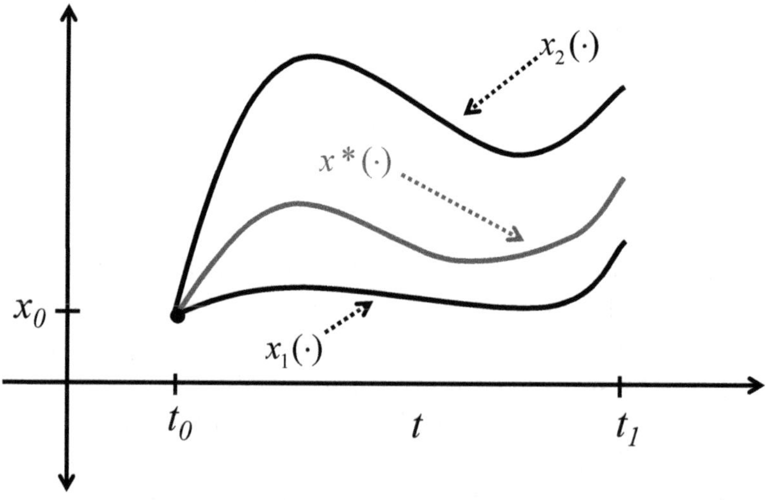

Figure 7.1: The Free Endpoint Problem

In order to derive the first order necessary condition for minimizing $J(\cdot)$ on the set Θ_L we must decide what class of variations are admissible. Assume that $x^*(\cdot) \in \Theta_L$ minimizes $J(\cdot)$ on Θ_L. We wish to make sure that for all ε (or for all ε sufficiently small) the variations $\varphi(\cdot, \varepsilon) = x^*(\cdot) + \varepsilon\eta(\cdot)$ belongs to Θ_L. In particular, we need to show that

$$\varphi(t, \varepsilon) = x^*(t) + \varepsilon\eta(t) \in \Theta_L. \tag{7.9}$$

It is clear that if

$$V_L(t_0, t_1) = \{\eta(\cdot) \in PWS(t_0, t_1) : \eta(t_0) = 0\}, \tag{7.10}$$

then (7.9) is satisfied for all ε. Observe that $V_0(t_0, t_1) \subset V_L(t_0, t_1) \subset PWS(t_0, t_1)$ so that (7.9) also holds for any $\eta(\cdot) \in V_0(t_0, t_1) \subset V_L(t_0, t_1) \subset PWS(t_0, t_1)$. The space $V_L(t_0, t_1)$ defined by (7.10) is called the space of "admissible variations" for the free endpoint problem.

We turn now to the derivation of Euler's Necessary Condition for the free endpoint problem. Although the presentation below is almost identical to the derivation for the SPCV in Chapter 3, we present the details again to re-enforce the basic idea. The main

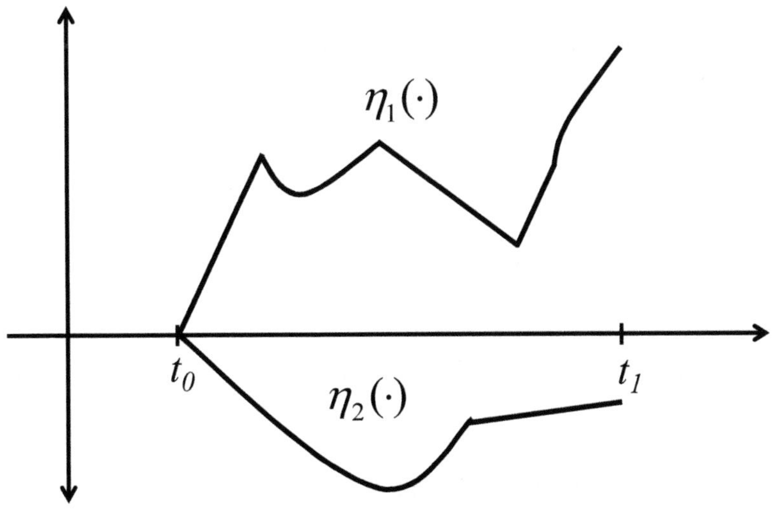

Figure 7.2: Admissible Variations for the Free Endpoint Problem

difference here is that the free endpoint condition leads to a new "*natural boundary condition*" that is an essential component of the necessary condition.

7.2.1 The Euler Necessary Condition

Assume that $x^*(\cdot) \in \Theta_L$ is a (global) minimizer for $J(\cdot)$ on Θ_L. In particular, assume that $x^*(\cdot) \in \Theta_L$ satisfies

$$J(x^*(\cdot)) = \int_{t_0}^{t_1} f(s, x^*(s), \dot{x}^*(s))\, ds \le J(x(\cdot)) = \int_{t_0}^{t_1} f(s, x(s), \dot{x}(s))\, ds, \tag{7.11}$$

for all $x(\cdot) \in \Theta_L$.

Let $\eta(\cdot) \in V_L(t_0, t_1)$ and consider the "variation"

$$\varphi(t, \varepsilon) = x^*(t) + \varepsilon\eta(t). \tag{7.12}$$

Observe that for each $\varepsilon \in \mathbb{R}$, $\varphi(t,\varepsilon) \in \Theta_L$ since $\varphi(t,\varepsilon) \in PWS(t_0, t_1)$ and satisfies $\varphi(t_0,\varepsilon) = x^*(t_0) + \varepsilon\eta(t_0) = x_0 + \varepsilon\eta(t_0) = x_0$. Hence, if $\eta(\cdot) \in V_L(t_0, t_1)$, then for all $\varepsilon \in \mathbb{R}$ the variation $\varphi(t,\varepsilon) \triangleq x^*(t) + \varepsilon\eta(t) \in \Theta_L$, i.e. it is admissible. Since $x^*(\cdot) \in \Theta_L$ minimizes $J(\cdot)$ on Θ_L, it follows that

$$J(x^*(\cdot)) \le J(x^*(\cdot) + \varepsilon\eta(\cdot)) \tag{7.13}$$

for all $\varepsilon \in (-\infty, +\infty)$. Define $F : (-\infty, +\infty) \longrightarrow \mathbb{R}$ by

$$F(\varepsilon) = J(x^*(\cdot) + \varepsilon\eta(\cdot)) = \int_{t_0}^{t_1} f(s, x^*(s) + \varepsilon\eta(s), \dot{x}^*(s) + \varepsilon\dot{\eta}(s))ds, \tag{7.14}$$

and note that (7.13) implies that

$$F(0) = J(x^*(\cdot)) \le J(x^*(\cdot) + \varepsilon\eta(\cdot)) = F(\varepsilon)$$

for all $\varepsilon \in (-\infty, +\infty)$. Therefore, $F(\cdot)$ has a minimum on $(-\infty, +\infty)$ at $\varepsilon^* = 0$ and applying **Theorem** 2.1 it follows that the first variation must be zero. That is,

$$\delta J(x^*(\cdot);\eta(\cdot)) \triangleq \frac{d}{d\varepsilon}F(\varepsilon)\Big|_{\varepsilon=0} = \frac{d}{d\varepsilon}\left[J(x^*(\cdot)+\varepsilon\eta(\cdot))\right]\Big|_{\varepsilon=0} = 0, \tag{7.15}$$

for all $\eta(\cdot) \in V_L(t_0, t_1)$. However, (7.2) implies that

$$\delta J(x^*(\cdot);\eta(\cdot)) = \int_{t_0}^{t_1} \{[f_x(s, x^*(s), \dot{x}^*(s))]\,\eta(s) + [f_u(s, x^*(s), \dot{x}^*(s))]\,\dot{\eta}(s)\}ds \tag{7.16}$$

and (7.15) yields

$$\int_{t_0}^{t_1} \{[f_x(s, x^*(s), \dot{x}^*(s))]\,\eta(s) + [f_u(s, x^*(s), \dot{x}^*(s))]\,\dot{\eta}(s)\}\,ds = 0 \tag{7.17}$$

for all $\eta(\cdot) \in V_L(t_0, t_1)$.

Since (7.17) holds for all $\eta(\cdot) \in V_L(t_0, t_1)$ and $V_0(t_0, t_1) \subset V_L(t_0, t_1)$ it follows that

$$\int_{t_0}^{t_1} \{[f_x(s, x^*(s), \dot{x}^*(s))]\,\eta(s) + [f_u(s, x^*(s), \dot{x}^*(s))]\,\dot{\eta}(s)\}\,ds = 0$$

holds for all $\eta(\cdot) \in V_0(t_0, t_1)$. Therefore, the **Fundamental Lemma of the Calculus of Variations** 3.1, Part (B) implies that there is a c such that

$$[f_u(t, x^*(t), \dot{x}^*(t))] = c + \int_{t_0}^{t} [f_x(s, x^*(s), \dot{x}^*(s))]\,ds$$

e.f. on $t \in [t_0, t_1]$ and hence $x^*(\cdot)$ is an extremal. Moreover, between corners

$$\frac{d}{dt}[f_u(t, x^*(t), \dot{x}^*(t))] = f_x(t, x^*(t), \dot{x}^*(t)). \tag{7.18}$$

Consequently, **Lemma** 7.1 implies that the first variation of $J(\cdot)$ at $x^*(\cdot)$ in the direction of $\eta(\cdot)$ exists and is given by

$$\delta J(x^*(\cdot);\eta(\cdot)) = \left[f_u\left(t, x^*(t), \dot{x}^*\left(t\right)\right) \eta\left(t\right)\right]\Big|_{t=t_0}^{t=t_1}. \tag{7.19}$$

Returning to (7.15) we have that

$$0 = \delta J(x^*(\cdot);\eta(\cdot)) = \left[f_u\left(t, x^*(t), \dot{x}^*\left(t\right)\right) \eta\left(t\right)\right]\Big|_{t=t_0}^{t=t_1}$$

must hold for all $\eta(\cdot) \in V_L(t_0, t_1)$. Hence, if $\eta(\cdot) \in V_L(t_0, t_1)$, then $\eta(t_0) = 0$ and it follows that

$$\begin{aligned} 0 &= \left[f_u\left(t, x^*(t), \dot{x}^*\left(t\right)\right) \eta\left(t\right)\right]\Big|_{t=t_0}^{t=t_1} \\ &= f_u\left(t_1, x^*\left(t_1\right), \dot{x}^*\left(t_1\right)\right) \eta\left(t_1\right) - f_u\left(t_0, x^*\left(t_0\right), \dot{x}^*\left(t_0\right)\right) \eta\left(t_0\right) \\ &= f_u\left(t_1, x^*\left(t_1\right), \dot{x}^*\left(t_1\right)\right) \eta\left(t_1\right). \end{aligned}$$

However, for $\eta(\cdot) \in V_L(t_0, t_1)$ there is no restriction on $\eta\left(t_1\right)$ so that

$$f_u\left(t_1, x^*\left(t_1\right), \dot{x}^*\left(t_1\right)\right) \eta\left(t_1\right) = 0 \tag{7.20}$$

must hold for arbitrary values of $\eta\left(t_1\right)$. Hence (7.20) holds for any value of $\eta\left(t_1\right)$ which implies that

$$f_u\left(t_1, x^*\left(t_1\right), \dot{x}^*\left(t_1\right)\right) = 0. \tag{7.21}$$

Condition (7.21) is called the *natural boundary condition* for the free endpoint problem. Thus, we have derived the (global) Euler necessary condition for the Free Endpoint Problem. It is straightforward to extend this proof to the case where we only assume that $x^*\left(\cdot\right) \in \Theta_L$ is a weak local minimizer in Θ_L. In particular, we have the following Euler Necessary Condition for the free endpoint problem.

Theorem 7.1 (Euler Necessary Condition for the Free Endpoint Problem) *If $x^*\left(\cdot\right) \in \Theta_L$ is a weak local minimizer for $J(\cdot)$ on Θ_L, then*
(EF-1) there is a constant c such that for all $t \in [t_0, t_1]$,

$$f_u\left(t, x^*(t), \dot{x}^*\left(t\right)\right) = c + \int_{t_0}^{t} f_x\left(s, x^*(s), \dot{x}^*\left(s\right)\right) ds, \tag{7.22}$$

(EF-2) $x^*(t_0) = x_0$,

(EF-3) $f_u(t_1, x^*(t_1), \dot{x}^*(t_1)) = 0$.

(EF-4) Between corners the function $f_u(t, x^*(t), \dot{x}^*(t))$ *is differentiable and if* t *is not a corner of* $x^*(\cdot)$

$$\frac{d}{dt} f_u(t, x^*(t), \dot{x}^*(t)) = f_x(t, x^*(t), \dot{x}^*(t)). \tag{7.23}$$

As in the previous chapters, equation (7.22) is called *Euler's Equation in integral form*, while equation (7.23) is called *Euler's Differential Equation*. Therefore, we have shown that a minimizer $x^*(\cdot)$ of $J(\cdot)$ on Θ_L must be an extremal. This implies that the Erdmann corner condition

$$f_u\left(\hat{t}, x^*(\hat{t}), \dot{x}^*\left(\hat{t}^+\right)\right) = f_u(\hat{t}, x^*(\hat{t}), \dot{x}^*(\hat{t}^-))$$

also holds, and if the problem is regular, then extremals can not have corners. The Hilbert Differentiability Theorem also holds and applying the chain rule to the left side of (7.23) yields

$$\begin{aligned}\frac{d}{dt}\left[f_u(t, x^*(t), \dot{x}^*(t))\right] = {} & \left[f_{ut}(t, x^*(t), \dot{x}^*(t))\right] \\ & + \left[f_{ux}(t, x^*(t), \dot{x}^*(t))\right] \cdot \dot{x}^*(t) \\ & + \left[f_{uu}(t, x^*(t), \dot{x}^*(t))\right] \cdot \ddot{x}^*(t).\end{aligned}$$

Hence, the Euler Equation (7.23) becomes the second order differential equation

$$\begin{aligned}\left[f_{uu}(t, x^*(t), \dot{x}^*(t))\right] \cdot \ddot{x}^*(t) = {} & \left[f_x(t, x^*(t), \dot{x}^*(t))\right] \\ & - \left[f_{ut}(t, x^*(t), \dot{x}^*(t))\right] \\ & - \left[f_{ux}(t, x^*(t), \dot{x}^*(t))\right] \cdot \dot{x}^*(t).\end{aligned} \tag{7.24}$$

7.2.2 Examples of Free Endpoint Problems

We shall go through a couple of simple examples to illustrate the application of the necessary condition.

Example 7.1 *Find a PWS function $x^*(\cdot)$ satisfying $x^*(0) = 0$ and such that $x^*(\cdot)$ minimizes*

$$J(x(\cdot)) = \int_0^{\pi/2} \frac{1}{2}\left([\dot{x}(s)]^2 - [x(s)]^2\right) ds.$$

We note that $t_0 = 0$, $t_1 = \pi/2$, $x_0 = 0$, and $x(\pi/2)$ is free. The integrand $f(t,x,u)$ is given by

$$f(t,x,u) = \frac{1}{2}([u]^2 - [x]^2),$$

and hence

$$\begin{aligned} f_x(t,x,u) &= -x \\ f_u(t,x,u) &= +u \\ f_{uu}(t,x,u) &= +1 > 0. \end{aligned}$$

We see that $f(t,x,u)$ is regular and hence the minimizer can not have corners. Euler's Equation

$$\frac{d}{dt}\left[f_u\left(t, x^*(t), \dot{x}^*(t)\right)\right] = \left[f_x\left(t, x^*(t), \dot{x}^*(t)\right)\right]$$

becomes

$$\frac{d}{dt}\left[\dot{x}^*(t)\right] = \left[-x^*(t)\right],$$

or equivalently,

$$\ddot{x}^*(t) + x^*(t) = 0.$$

The general solution is

$$x^*(t) = \alpha\cos(t) + \beta\sin(t),$$

and applying the boundary condition at $t = t_0 = 0$ yields

$$0 = x^*(0) = \alpha\cos(0) + \beta\sin(0) = \alpha,$$

so that

$$x^*(t) = \beta\sin(t).$$

The natural boundary at $t = t_1 = \pi/2$ becomes

$$f_u\left(\pi/2, x^*\left(\pi/2\right), \dot{x}^*\left(\pi/2\right)\right) = \dot{x}^*\left(\pi/2\right) = 0.$$

However, $\dot{x}^*\left(t\right) = \beta\cos(t)$ *so that*

$$\dot{x}^*\left(\pi/2\right) = \beta\cos(\pi/2) = 0.$$

Since $\cos(\pi/2) = 0$*, it follows that* β *can be any number and hence*

$$x^*\left(t\right) = \beta\sin(t)$$

*are **possible** minimizers. Observe that we do not know that* $x^*(\cdot)$ *minimizes* $J(\cdot)$ *on* Θ_L *for any number* β *since we only checked the necessary condition.*

Example 7.2 *Consider the functional*

$$J(x(\cdot)) = \int_0^1 [\dot{x}\left(s\right)]^2 ds$$

with $x(0) = 0$ *and the endpoint* $x(1)$ *free. The integrand is given by* $f(t,x,u) = u^2$, $f_u(t,x,u) = 2u$, $f_{uu}(t,x,u) = 2$, *and* $f_x(t,x,u) = 0$. *Since the problem is regular, all extremals are regular and Euler's Integral Equation is given by*

$$2\dot{x}^*(t) = f_u(t, x^*(t), \dot{x}^*(t)) = c + \int_0^t f_x(s, x^*(s), \dot{x}^*(s))ds$$

$$= c + \int_0^t 0 ds = c,$$

or equivalently,

$$\dot{x}^*(t) = k$$

for some constant k. *Therefore,*

$$x^*(t) = kt + b$$

and the condition $x(0) = 0$ *implies*

$$x^*(t) = kt.$$

The natural boundary condition at $t_1 = 1$ *takes the form*

$$f_u(1, x^*(1), \dot{x}^*(1)) = 2\dot{x}^*(1) = 2k = 0$$

which means that

$$x^*(t) = 0.$$

Hence, the only extremal satisfying the necessary condition **Theorem** *7.1 is* $x^*(t) = 0$. *Clearly,* $x^*(t) = 0$ *is a global minimizer.*

7.3 The Simplest Point to Curve Problem

We assume that the initial time t_0 and initial value x_0 are given and that there is a given **smooth function** $\phi(\cdot)$ defined on the interval $(t_0, +\infty)$. The problem is to minimize

$$J(x(\cdot)) = \int_{t_0}^{t_1} f(s, x(s), \dot{x}(s))\, ds,$$

subject to

$$x(t_0) = x_0$$

and

$$x(t_1) = \phi(t_1),$$

where $t_0 < t_1$ and t_1 is a point where the graph of the function $x(\cdot)$ intersects the graph of $\phi(\cdot)$. This problem is illustrated in Figure 7.3.

Note that t_1 is not fixed and different functions may intersect $\phi(\cdot)$ at different "final times". To formulate the optimization problem we first define the space of piecewise smooth functions on $[t_0, +\infty)$ by $PWS_\infty = PWS(t_0, +\infty)$, where

$$PWS_\infty = \{x(\cdot) : [t_0, +\infty) \to \mathbb{R} : x(\cdot) \in PWS(t_0, T) \text{ for all } T > t_0\}. \tag{7.25}$$

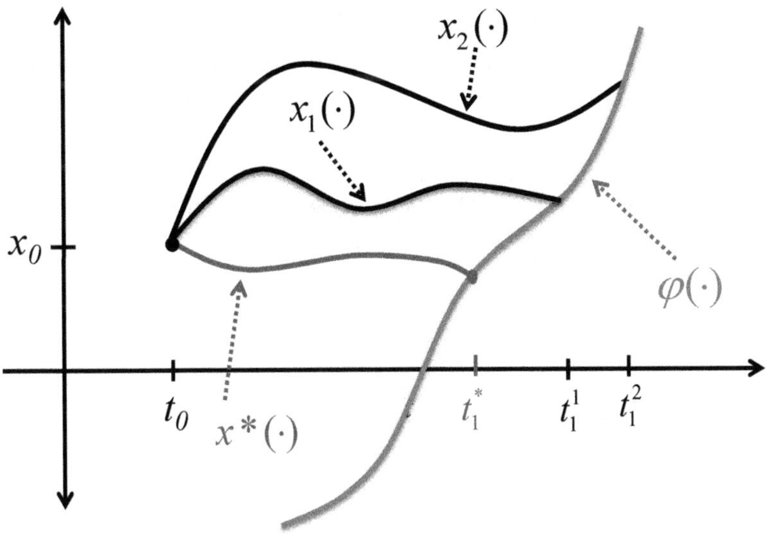

Figure 7.3: Point to Curve Problem

Note that $PWS_\infty = PWS(t_0, +\infty)$ is the set of all "locally" piecewise smooth functions defined on the half line $[t_0, +\infty)$. Assume that t_0, x_0 and $\phi(\cdot)$ are given as above and define the set of PWS functions (intersecting the graph of ϕ) Θ_ϕ by

$$\Theta_\phi = \{x(\cdot) \in PWS_\infty : x(t_0) = x_0,\ x(t_1) \\ = \phi(t_1) \text{ for some } t_1 > t_0\}. \tag{7.26}$$

The simplest **Point to Curve Problem** is the problem of minimizing $J(\cdot)$ on Θ_ϕ. In particular, the goal is to find $x^*(\cdot) \in \Theta_\phi$ **and** a $t_1^* > t_0$ such that $x^*(t_1^*) = \phi(t_1^*)$ and

$$J(x^*(\cdot)) = \int_{t_0}^{t_1^*} f(s, x^*(s), \dot{x}^*(s))\, ds \le J(x(\cdot)) \\ = \int_{t_0}^{t_1} f(s, x(s), \dot{x}(s))\, ds,$$

for all $x(\cdot) \in \Theta_\phi$.

We will derive a necessary condition for a global minimizer, but the result holds for a weak local minimizer and the following derivation is easily extended to this case. We start with a few basic lemmas.

Lemma 7.2 *If $x^*(\cdot) \in \Theta_\phi$ is piecewise smooth and minimizes $J(\cdot)$ on Θ_ϕ, with $x^*(t_1^*) = \phi(t_1^*)$, then*

(EPC-1) there is a constant c such that for all $t \in [t_0, t_1^*]$,

$$[f_u(t, x^*(t), \dot{x}^*(t))] = c + \int_{t_0}^{t} [f_x(s, x^*(s), \dot{x}^*(s))]\, ds, \qquad (7.27)$$

(EPC-2) $x^*(t_0) = x_0$,

(EPC-3) $x^*(t_1^*) = \phi(t_1^*)$.

(EPC-4) Between corners of $x^*(\cdot)$ the function $\frac{\partial}{\partial u} f(t, x^*(t), \dot{x}^*(t))$ is differentiable and

$$\frac{d}{dt}[f_u(t, x^*(t), \dot{x}^*(t))] = [f_x(t, x^*(t), \dot{x}^*(t))]. \qquad (7.28)$$

Proof. This lemma follows from the standard derivation for a fixed endpoint problem. In particular, for $\eta(\cdot) \in V_0^* = PWS_0(t_0, t_1^*)$, the variation $x^*(\cdot) + \varepsilon\eta(\cdot)$ belongs to Θ_ϕ and hence

$$\delta J(x^*(\cdot); \eta(\cdot)) = \int_{t_0}^{t_1^*} [f_x(s, x^*(s), \dot{x}^*(s))\eta(s) + f_u(s, x^*(s), \dot{x}^*(s))\dot{\eta}(s)]ds = 0. \qquad (7.29)$$

Since this holds for all $\eta(\cdot) \in V_0^* = PWS_0(t_0, t_1^*)$, the Fundamental Lemma of the Calculus of Variations implies that $x^*(\cdot)$ is an extremal and this completes the proof. □

Note that the Euler equations (7.27) or (7.28) must be solved on the interval $[t_0, t_1^*]$. However, we need an additional piece of information to determine the extra unknown parameter t_1^*. To obtain this condition, we must enlarge the space of variations.

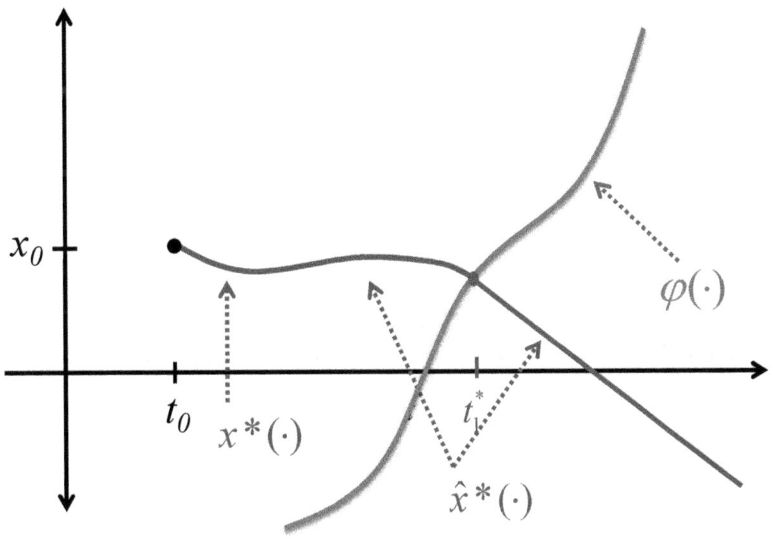

Figure 7.4: Extending the Optimal Curve

In order to derive the *transversality condition* we define another class of variations that are admissible for this problem. If $x^*(\cdot) \in \Theta_\phi$ minimizes $J(\cdot)$ on Θ_ϕ, with $x^*(t_1^*) = \phi(t_1^*)$, then we need to make sure that for all ε (or for all ε sufficiently small) the variations $x^*(t) + \varepsilon\eta(t)$ belong to Θ_ϕ. In particular, we need to find a set of admissible variations V_ϕ so that

$$x^*(t) + \varepsilon\eta(t) \in \Theta_\phi \tag{7.30}$$

for all $\eta(\cdot) \in V_\phi$ and all ε sufficiently small. Thus, we must define V_ϕ in such a way that $x^*(t) + \varepsilon\eta(t)$ intersects $\phi(\cdot)$ at some time $t_1^\varepsilon > t_0$.

Given the minimizer $x^*(\cdot)$, define $\hat{x}^*(\cdot)$ by

$$\hat{x}^*(t) = \begin{cases} x^*(t), & t_0 \le t \le t_1^*, \\ \dot{x}^*(t_1^*)(t - t_1^*) + x^*(t_1^*), & t_1^* \le t. \end{cases}$$

Also, given any $\eta(\cdot) \in V_L^* = \{\eta(\cdot) \in PWS(t_0, t_1^*) : \eta(t_0) = 0\}$ define $\hat{\eta}(\cdot)$ by

$$\hat{\eta}(t) = \begin{cases} \eta(t), & t_0 \le t \le t_1^*, \\ \dot{\eta}(t_1^*)(t - t_1^*) + \eta(t_1^*), & t_1^* \le t. \end{cases} \tag{7.31}$$

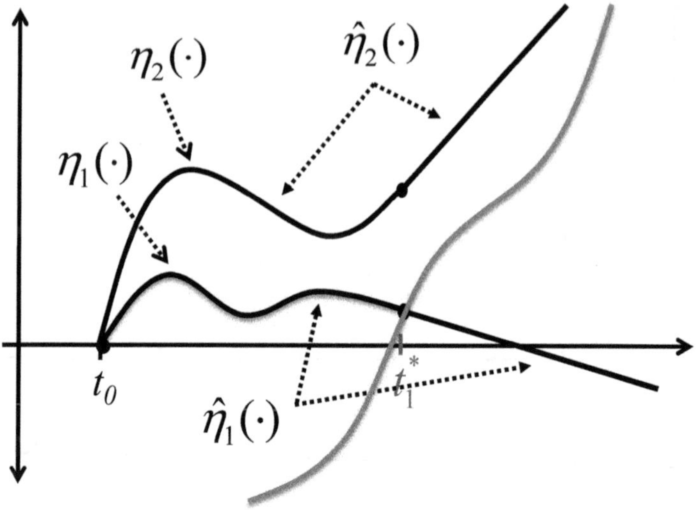

Figure 7.5: Extending $\eta(\cdot)$ to $\hat{\eta}(\cdot)$

Remark 7.1 *The functions $\hat{x}^*(\cdot)$ and $\hat{\eta}(\cdot)$ are continuously differentiable in a neighborhood of t_1^*. To see this, pick $\hat{t}_L < t_1^*$ so that neither $x^*(\cdot)$ nor $\eta(\cdot)$ has a corner $\hat{t}_c$ with $\hat{t}_L < \hat{t}_c$. By construction $\hat{x}^*(\cdot)$ and $\hat{\eta}(\cdot)$ will be continuously differentiable for all $t > \hat{t}_L$.*

Now we define V_ϕ by

$$V_\phi = \{\hat{\eta}(\cdot) \in PWS(t_0, +\infty) : \eta(\cdot) \in V_L^*\}, \tag{7.32}$$

where $\hat{\eta}(\cdot) \in V_\phi$ is defined by (7.31) above. What we want to show is that for sufficiently small ε, $x^*(\cdot) + \varepsilon\eta(\cdot)$ intersects $\phi(\cdot)$ at some time $t_1^\varepsilon > t_0$. Observe that all functions $\hat{\eta}(\cdot) \in V_\phi$ are linear on $t_1^* \leq t$. Also, $\hat{\eta}(\cdot)$ and $\hat{x}^*(\cdot)$ are smooth on an interval about t_1^*. Note that for any $\hat{\eta}(\cdot) \in V_\phi$, the function

$$\hat{x}^*(t) + \varepsilon\hat{\eta}(t),$$

is defined for all $t \geq t_0$. Since at $\varepsilon = 0$

$$\hat{x}^*(t) + \varepsilon\hat{\eta}(t) = \hat{x}^*(t),$$

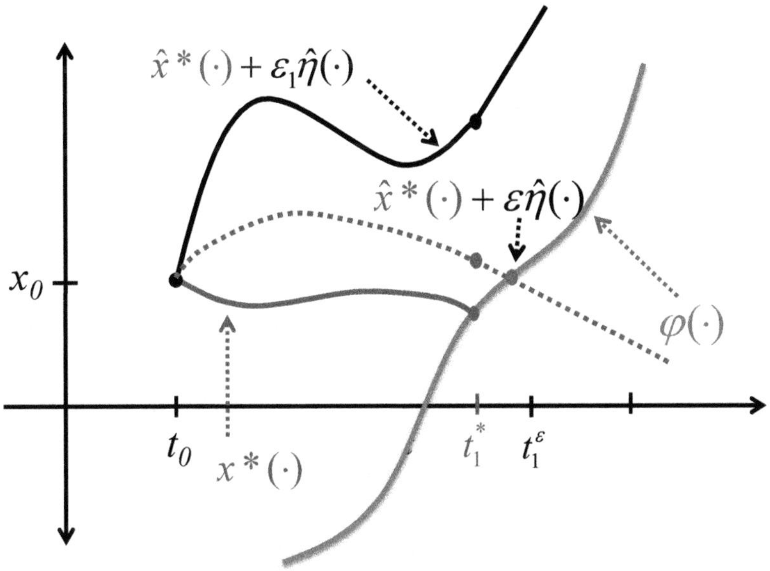

Figure 7.6: The Variation $\hat{x}^*(\cdot)+\varepsilon\hat{\eta}(\cdot)$

one might expect that for sufficiently small ε the curve $\hat{x}^*(t) + \varepsilon\hat{\eta}(t)$ would intersect $\phi(\cdot)$ at some time $t_1^\varepsilon > t_0$ (see Figure 7.6). For one important case we shall show that this is true. In order to show this we need the **Implicit Function Theorem**. The following theorem follows directly from Theorem 41.9 and Lemma 41.10 in Bartle's book [15] (see pages 382 - 391).

Theorem 7.2 (Implicit Function Theorem) *Suppose* $H : \mathbb{R}\times\mathbb{R} \to \mathbb{R}$ *satisfies (i.e.* $H(\varepsilon,\gamma)$*):*

i) $H(0,0) = 0$.

ii) *H is of class C^1 on a neighborhood of* $[0 \;\; 0]^T$.

iii) $\frac{\partial}{\partial\gamma}H(0,0) = H_\gamma(0,0) \neq 0$.

Then, there is a neighborhood of 0 (i.e. $(-\delta,+\delta)$*) such that one can solve for γ in terms of ε. In particular, there exists a function* $\gamma(\cdot) : (-\delta,+\delta) \to \mathbb{R}$ *such that for all* $-\delta < \varepsilon < \delta$

$$H(\varepsilon,\gamma(\varepsilon)) = 0,$$

and

$$\gamma(0) = 0.$$

Moreover, the function $\gamma(\varepsilon)$ *is continuously differentiable with respect to* ε *and*

$$\frac{d}{d\varepsilon}\gamma(\varepsilon) = \dot{\gamma}(\varepsilon) = -\frac{H_{\varepsilon}(\varepsilon, \gamma(\varepsilon))}{H_{\gamma}(\varepsilon, \gamma(\varepsilon))}.$$

We shall define a specific function $\hat{H}(\varepsilon, \gamma)$ and apply the Implicit Function Theorem. Again, we are trying to show that for sufficiently small ε the variation $\hat{x}^*(t) + \varepsilon\hat{\eta}(t)$ will intersect $\phi(\cdot)$ at some time $t_1^{\varepsilon} > t_0$. If $t_1^{\varepsilon} = t_1^* + \gamma(\varepsilon)$, then we are asking if for each ε on an interval $(-\delta, +\delta)$, is there a $\gamma = \gamma(\varepsilon)$ such that

$$\phi(t_1^* + \gamma) - \hat{x}^*(t_1^* + \gamma) - \varepsilon\hat{\eta}(t_1^* + \gamma) = 0?$$

Thus, we are motivated to define $\hat{H}(\varepsilon, \gamma)$ by

$$\hat{H}(\varepsilon, \gamma) = \phi(t_1^* + \gamma) - \hat{x}^*(t_1^* + \gamma) - \varepsilon\hat{\eta}(t_1^* + \gamma). \tag{7.33}$$

Observe that $\hat{H}(\varepsilon, \gamma)$ defined above has the following properties:

i) $\hat{H}(0, 0) = \phi(t_1^*) - \hat{x}^*(t_1^*) = 0.$

ii) $\hat{H}_{\varepsilon}(\varepsilon, \gamma) = -\hat{\eta}(t_1^* + \gamma)$ and $\hat{H}_{\gamma}(\varepsilon, \gamma) = \frac{d}{dt}\phi(t_1^* + \gamma) - \frac{d}{dt}\hat{x}^*(t_1^* + \gamma) - \varepsilon\frac{d}{dt}\hat{\eta}(t_1^* + \gamma)$ are continuous and hence $\hat{H}(\varepsilon, \gamma)$ is of class C^1 on $(-\infty, \infty) \times (\hat{t}_L, +\infty)$.

iii) $\hat{H}_{\gamma}(0, 0) = \frac{d}{dt}\phi(t_1^*) - \frac{d}{dt}\hat{x}^*(t_1^*) = \dot{\phi}(t_1^*) - \dot{x}^*(t_1^*).$

Therefore, if $\dot{\phi}(t_1^*) - \dot{x}^*(t_1^*) \neq 0$, then $\hat{H}(\varepsilon, \gamma)$ satisfies the assumption of the Implicit Function Theorem. In particular, if $\dot{\phi}(t_1^*) - \dot{x}^*(t_1^*) \neq 0$, then there exists a function $\gamma(\cdot) : (-\delta, +\delta) \to \mathbb{R}$ such that for all $-\delta < \varepsilon < \delta$

$$0 = \hat{H}(\varepsilon, \gamma(\varepsilon)) = \phi(t_1^* + \gamma(\varepsilon)) - \hat{x}^*(t_1^* + \gamma(\varepsilon)) - \varepsilon\hat{\eta}(t_1^* + \gamma(\varepsilon)). \tag{7.34}$$

Equation (7.34) implies that for each $\varepsilon \in (-\delta, +\delta)$, the variations $\hat{x}^*(\cdot) + \varepsilon\hat{\eta}(\cdot)$ intersects $\phi(\cdot)$ at the time $\beta(\varepsilon) = t_1^* + \gamma(\varepsilon)$. Moreover, $\dot{\beta}(\varepsilon)$ exists and equals

$$\begin{aligned}\dot{\beta}(\varepsilon) &= \dot{\gamma}(\varepsilon)\\ &= -\left[\frac{-\hat{\eta}(t_1^* + \gamma(\varepsilon))}{\frac{d}{dt}\phi(t_1^* + \gamma(\varepsilon)) - \frac{d}{dt}\hat{x}^*(t_1^* + \gamma(\varepsilon)) - \varepsilon\frac{d}{dt}\hat{\eta}(t_1^* + \gamma(\varepsilon))}\right].\end{aligned}$$

In particular,

$$\dot{\beta}(0) = \frac{\eta(t_1^*)}{\frac{d}{dt}\phi(t_1^*) - \frac{d}{dt}x^*(t_1^*)} = \frac{\eta(t_1^*)}{\dot{\phi}(t_1^*) - \dot{x}^*(t_1^*)} = \dot{\gamma}(0). \tag{7.35}$$

We now have established the following result.

Theorem 7.3 *If $x^*(\cdot)$ satisfies*

$$\begin{aligned}J(x^*(\cdot)) &= \int_{t_0}^{t_1^*} f(s, x^*(s), \dot{x}^*(s))\, ds \leq J(x(\cdot))\\ &= \int_{t_0}^{t_1} f(s, x(s), \dot{x}(s))\, ds,\end{aligned}$$

on Θ_ϕ and $\dot{\phi}(t_1^*) - \dot{x}^*(t_1^*) \neq 0$, then there exists a function $\gamma(\cdot) : (-\delta, +\delta) \to \mathbb{R}$ such that for all $-\delta < \varepsilon < \delta$ the variations $\hat{x}^*(\cdot) + \varepsilon\hat{\eta}(\cdot) \in \Theta_\phi$ and intersect $\phi(\cdot)$ at the time $\beta(\varepsilon) = t_1^* + \gamma(\varepsilon)$.

Define the function $F : (-\delta, +\delta) \to \mathbb{R}$ by

$$\begin{aligned}F(\varepsilon) &= J(\hat{x}^*(\cdot) + \varepsilon\hat{\eta}(\cdot))\\ &= \int_{t_0}^{\beta(\varepsilon)} f\left(s, \hat{x}^*(s) + \varepsilon\hat{\eta}(s), \frac{d}{ds}\hat{x}^*(s) + \varepsilon\frac{d}{ds}\hat{\eta}(s)\right) ds\\ &= \int_{t_0}^{\beta(\varepsilon)} G(s, \varepsilon)\, ds,\end{aligned}$$

where

$$G(t,\varepsilon) = f\left(t, \hat{x}^*(t) + \varepsilon\hat{\eta}(t), \frac{d}{dt}\hat{x}^*(t) + \varepsilon\frac{d}{dt}\hat{\eta}(t)\right).$$

Therefore, for all $-\delta < \varepsilon < \delta$ we have

$$\begin{aligned} F(0) &= \int_{t_0}^{\beta(0)} f(s, x^*(s), \dot{x}^*(s))\, ds \\ &\le \int_{t_0}^{\beta(\varepsilon)} f\left(s, \hat{x}^*(s) + \varepsilon\hat{\eta}(s), \frac{d}{ds}\hat{x}^*(s) + \varepsilon\frac{d}{ds}\hat{\eta}(s)\right) ds = F(\varepsilon), \end{aligned}$$

so that $\varepsilon^* = 0$ minimizes $F(\varepsilon)$ on the open interval $(-\delta, +\delta)$. Consequently, it follows that $\frac{d}{d\varepsilon}F(0) = F'(0) = 0$. In order to compute $F'(\varepsilon)$, we apply Leibniz's Formula which yields

$$F'(\varepsilon) = G(\beta(\varepsilon), \varepsilon)\dot{\beta}(\varepsilon) + \int_{t_0}^{\beta(\varepsilon)} G_\varepsilon(s,\varepsilon)\, ds.$$

In particular, since $\beta(\varepsilon) = t_1^* + \gamma(\varepsilon)$ and $\gamma(0) = 0$, it follows that

$$\begin{aligned} F'(0) &= G(\beta(0), 0)\dot{\beta}(0) + \int_{t_0}^{\beta(0)} G_\varepsilon(s, 0)\, ds \\ &= f(t_1^*, x^*(t_1^*), \dot{x}^*(t_1^*))\dot{\beta}(0) \\ &\quad + \int_{t_0}^{t_1^*} [f_x(s, x^*(s), \dot{x}^*(s))\eta(s) + f_u(s, x^*(s), \dot{x}^*(s))\dot{\eta}(s)] ds \\ &= f(t_1^*, x^*(t_1^*), \dot{x}^*(t_1^*)) \frac{\eta(t_1^*)}{[\dot{\phi}(t_1^*) - \dot{x}^*(t_1^*)]} \\ &\quad + \int_{t_0}^{t_1^*} (f_x^*(s)\eta(s) + f_u^*(s)\dot{\eta}(s))\, ds. \end{aligned}$$

Hence, for all $\eta(\cdot) \in V_\phi$ we have

$$0 = \frac{f\left(t_1^*, x^*\left(t_1^*\right), \dot{x}^*\left(t_1^*\right)\right)\eta\left(t_1^*\right)}{\left[\dot{\phi}\left(t_1^*\right) - \dot{x}^*\left(t_1^*\right)\right]} + \int_{t_0}^{t_1^*} \left(f_x^*(s)\eta(s) + f_u^*(s)\dot{\eta}(s)\right) ds.$$

In particular, (7.3) is valid for all $\eta(\cdot)$ satisfying $\eta\left(t_1^*\right) = 0$, so we obtain Euler's equation

$$\frac{d}{dt} f_u^*(t) = f_x^*(t),$$

which must hold on $[t_0, t_1^*]$.

Substituting $\frac{d}{dt} f_u^*(t) = f_x^*(t)$ into the above equation yields

$$\begin{aligned} 0 &= \frac{f\left(t_1^*, x^*\left(t_1^*\right), \dot{x}^*\left(t_1^*\right)\right)\eta\left(t_1^*\right)}{\left[\dot{\phi}\left(t_1^*\right) - \dot{x}^*\left(t_1^*\right)\right]} + \int_{t_0}^{t_1^*} \left(f_x^*(s)\eta(s) + f_u^*(s)\dot{\eta}(s)\right) ds \\ &= \frac{f^*(t_1^*)\eta\left(t_1^*\right)}{\left[\dot{\phi}\left(t_1^*\right) - \dot{x}^*\left(t_1^*\right)\right]} + \int_{t_0}^{t_1^*} \left([\frac{d}{ds} f_u^*(s)]\eta(s) + f_u^*(s)\dot{\eta}(s) \right) ds \\ &= \frac{f^*(t_1^*)\eta\left(t_1^*\right)}{\left[\dot{\phi}\left(t_1^*\right) - \dot{x}^*\left(t_1^*\right)\right]} + \int_{t_0}^{t_1^*} \left(\frac{d}{ds}[f_u^*(s)\eta(s)] \right) ds \\ &= \frac{f^*(t_1^*)\eta\left(t_1^*\right)}{\left[\dot{\phi}\left(t_1^*\right) - \dot{x}^*\left(t_1^*\right)\right]} + f_u^*(t_1^*)\eta\left(t_1^*\right) - f_u^*(t_0)\eta\left(t_0\right). \end{aligned}$$

Using the fact that $\eta\left(t_0\right) = 0$, it follows that

$$0 = \frac{f^*(t_1^*)\eta\left(t_1^*\right)}{\left[\dot{\phi}\left(t_1^*\right) - \dot{x}^*\left(t_1^*\right)\right]} + f_u^*(t_1^*)\eta\left(t_1^*\right),$$

or equivalently,

$$0 = \{f^*(t_1^*) + [\dot{\phi}\left(t_1^*\right) - \dot{x}^*\left(t_1^*\right)]f_u^*(t_1^*)\}\eta\left(t_1^*\right).$$

Since $\eta\left(t_1^*\right)$ is arbitrary, we obtain the transversality condition

$$f^*(t_1^*) + f_u^*(t_1^*)[\dot{\phi}\left(t_1^*\right) - \dot{x}^*\left(t_1^*\right)] = 0. \tag{7.36}$$

Combining **Lemma** 7.2 with the transversality condition (7.36) we have proven the following theorem.

Theorem 7.4 (Euler Necessary Condition for the Point to Curve Problem) *If* $x^*(\cdot) \in \Theta_\phi$ *is piecewise smooth and minimizes* $J(\cdot)$ *on* Θ_ϕ*, with* $x^*(t_1^*) = \phi(t_1^*)$*, then*

(EPC-1) there is a constant c *such that for all* $t \in [t_0, t_1^*]$*,*

$$[f_u(t, x^*(t), \dot{x}^*(x))] = c + \int_{t_0}^{t} [f_x(s, x^*(s), \dot{x}^*(s))]\, ds, \tag{7.37}$$

(EPC-2) $x^*(t_0) = x_0$*,*

(EPC-3) $x^*(t_1^*) = \phi(t_1^*)$*.*

(EPC-3') If $\dot{\phi}(t_1^*) \neq \dot{x}^*(t_1^*)$*, then*

$$f^*(t_1^*) + f_u^*(t_1^*)[\dot{\phi}(t_1^*) - \dot{x}^*(t_1^*)] = 0. \tag{7.38}$$

(EPC-4) Between corners of $x^*(\cdot)$ *the function* $\frac{\partial}{\partial u} f(t, x^*(t), \dot{x}^*(x))$ *is differentiable and*

$$\frac{d}{dt}[f_u(t, x^*(t), \dot{x}^*(x))] = [f_x(t, x^*(t), \dot{x}^*(x))]. \tag{7.39}$$

These results can clearly be extended to two curves, curve to free endpoint, etc. To motivate the term "transversality condition", consider the class of equations where $f(t, x, u)$ has the form

$$f(t, x, u) = g(t, x)\sqrt{1 + u^2},$$

(e.g. the brachistochrone problem, minimal surface area of revolution problem, etc.). Here,

$$\begin{aligned} f_u(t, x, u) &= g(t, x)\frac{u}{\sqrt{1+u^2}} = [g(t, x)\sqrt{1+u^2}][\frac{u}{1+u^2}] \\ &= \frac{u}{1+u^2} f(t, x, u). \end{aligned}$$

Hence,

$$
\begin{aligned}
&f^*(t) + f_u^*(t)[\dot{\phi}(t) - \dot{x}^*(t)] \\
&= f^*(t) + f^*(t)\left[\frac{\dot{x}^*(t)}{1+[\dot{x}^*(t)]^2}\right][\dot{\phi}(t) - \dot{x}^*(t)] \\
&= \frac{f^*(t)(1+[\dot{x}^*(t)]^2 + [\dot{x}^*(t)][\dot{\phi}(t) - \dot{x}^*(t)])}{[1+[\dot{x}^*(t)]^2]} \\
&= \frac{f^*(t)(1+[\dot{x}^*(t)]^2 + \dot{x}^*(t)\dot{\phi}(t) - [\dot{x}^*(t)]^2)}{[1+[\dot{x}^*(t)]^2]} \\
&= \left[\frac{f^*(t)}{1+[\dot{x}^*(t)]^2}\right][1+\dot{x}^*(t)\dot{\phi}(t)].
\end{aligned}
$$

The transversality condition at t_1^* implies that

$$f^*(t_1^*)+f_u^*(t_1^*)[\dot{\phi}\,(t_1^*)-\dot{x}^*\,(t_1^*)] = \left[\frac{f^*(t_1^*)}{1+[\dot{x}^*(t_1^*)]^2}\right][1+\dot{x}^*(t_1^*)\dot{\phi}(t_1^*)] = 0$$

and hence if $f^*(t_1^*) \neq 0$, then it follows that

$$[1+\dot{x}^*(t_1^*)\dot{\phi}(t_1^*)] = 0.$$

In this case

$$\dot{x}^*(t_1^*) = \frac{-1}{\dot{\phi}(t_1^*)}$$

which means that the slopes of the curves $\dot{x}^*(\cdot)$ and $\dot{\phi}(\cdot)$ are perpendicular at the intersecting value $t = t_1^*$. That is, the optimal trajectory $\dot{x}^*(\cdot)$ must be orthogonal (transversal) to the curve $\phi(\cdot)$ at the intersection point.

Example 7.3 *Consider the problem of minimizing*

$$J\,(x(\cdot)) = \int_{t_0}^{t_1} \sqrt{1+[\dot{x}\,(s)]^2}ds,$$

subject to

$$x\,(t_0) = \varphi\,(t_0), \quad x\,(t_1) = \psi\,(t_1),$$

where

$$\varphi\,(t) = -\,(t+1)\,, \quad \psi\,(t) = \left(t-\frac{5}{2}\right)^2.$$

Since the problem is regular, we need only consider the Euler Differential Equation

$$\frac{d}{dt}f_u(t, x^*(t), \dot{x}^*(t)) = f_x(t, x^*(t), \dot{x}^*(t))$$

which has the form

$$\frac{d}{dt}\left(\frac{\dot{x}^*(t)}{\sqrt{1+[\dot{x}^*(t)]^2}}\right) = 0,$$

or equivalently,

$$\frac{\dot{x}^*(t)}{\sqrt{1+[\dot{x}^*(t)]^2}} = c.$$

Solving this equation we get

$$[\dot{x}^*(t)]^2 = c^2\left(1+[\dot{x}^*(t)]^2\right),$$

which implies that

$$[\dot{x}^*(t)]^2 = \frac{c^2}{1-c^2},$$

or equivalently, there is a constant a such that

$$\dot{x}^*(t) = a$$

and the extremals are straight lines $x(t) = at + b$.
The transversality conditions become

$$a = \dot{x}^*\left(t_0^*\right) = \frac{-1}{\dot{\varphi}\left(t_0^*\right)} = \frac{-1}{-1} = 1,$$

and hence it follows that $a = 1$. At the other end $t_1 = t_1^*$, the transversality condition becomes

$$a = 1 = \dot{x}^*\left(t_1\right) = \frac{-1}{\dot{\psi}\left(t_1\right)} = \frac{-1}{2\left(t_1 - \frac{5}{2}\right)}$$

and solving for t_1 it follows that

$$2\left(t_1 - \frac{5}{2}\right) = -1,$$

so that

$$t_1^* = t_1 = \frac{5}{2} - \frac{1}{2} = 2.$$

Since $a = 1$,

$$x^*(t) = t + b,$$

and from above $t_1^* = 2$ *so that the relationship*

$$x^*(t_1^*) = \psi(t_1^*)$$

implies that

$$x^*(t_1^*) = 2 + b = t_1^* + b = \left(t_1^* - \frac{5}{2}\right)^2 = \left(2 - \frac{5}{2}\right)^2 = \left(-\frac{1}{2}\right)^2$$
$$= \frac{1}{4} = \psi(t_1^*).$$

Hence

$$b = -\frac{7}{4}$$

so that

$$x^*(t) = t - \frac{7}{4},$$

and $t_1^* = 2$. *To find* t_0^* *we know that*

$$x^*(t_0^*) = \varphi(t_0^*)$$

which implies

$$t_0^* - \frac{7}{4} = -(t_0^* + 1),$$

and hence

$$t_0^* = \frac{3}{8}.$$

We have found that the optimal curve is given by $x^*(t) = t - \frac{7}{4}$ *and it intersects* $\varphi(t) = -(t+1)$ *at* $t_0^* = \frac{3}{8}$ *and* $\psi(t) = \left(t - \frac{5}{2}\right)^2$ *at* $t_1^* = 2$.

7.4 Vector Formulations and Higher Order Problems

In this section we discuss extensions of the SPCV to vector systems and higher order models. Consider the problem where we look for two (or more) functions $x_1(\cdot), x_2(\cdot)$ to minimize a functional of the form

$$J(x_1(\cdot), x_2(\cdot)) = \int_{t_0}^{t_1} f(s, x_1(s), x_2(s), \dot{x}_1(s), \dot{x}_2(s))\, ds,$$

subject to

$$x_1(t_0) = x_{1,0}\ x_1(t_1) = x_{1,1},$$
$$x_2(t_0) = x_{2,0}\ x_2(t_1) = x_{2,1}.$$

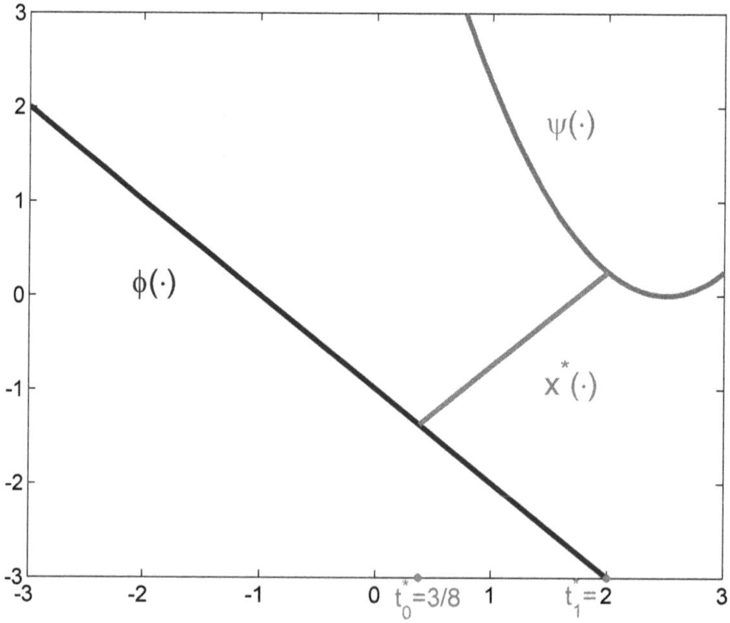

Figure 7.7: The Optimal Curve to Curve Solution

Here we assume that $f(t, x_1, x_2, u_1, u_2)$ is a C^2 function and we will apply the same ideas used in the SPCV. This will yield an Euler Necessary Condition where the Euler Differential Equation is a system of equations of the form

$$\frac{d}{dt} f_{u_1}(t, x_1(t), x_2(t), \dot{x}_1(t), \dot{x}_2(t)) = f_{x_1}(t, x_1(t), x_2(t), \dot{x}_1(t), \dot{x}_2(t)),$$

$$\frac{d}{dt} f_{u_2}(t, x_1(t), x_2(t), \dot{x}_1(t), \dot{x}_2(t)) = f_{x_2}(t, x_1(t), x_2(t), \dot{x}_1(t), \dot{x}_2(t)).$$

Higher order problems are a special case of the vector formulation. For example, consider the problem of minimizing

$$J(x(\cdot)) = \int_{t_0}^{t_1} f(s, x(s), \dot{x}(s), \ddot{x}(s))ds,$$

subject to

$$x(t_0) = x_{0,} \quad x(t_1) = x_1,$$

$$\dot{x}(t_0) = u_{0,} \quad \dot{x}(t_1) = u_1,$$

where $f(t, x, u, v)$ is a C^2 function. Here we assume that $x(\cdot)$ is smooth and $\dot{x}(\cdot)$ is piecewise smooth on $[t_0, t_1]$. In this case setting the first variation to zero and using the higher order Fundamental Lemma of the Calculus of Variations will lead to the higher order Euler Differential Equation

$$-\frac{d^2}{dt^2} f_v(t, x(t), \dot{x}(t), \ddot{x}(t)) + \frac{d}{dt} f_u(t, x(t), \dot{x}(t), \ddot{x}(t))$$
$$= f_x(t, x(t), \dot{x}(t), \ddot{x}(t)).$$

In order to formulate the vector problem we assume that t_0 and t_1 are given and fixed and $f(t, x_1, x_2, u_1, u_2)$ is a C^2 function. Let $PWS(t_0, t_1; \mathbb{R}^2)$ denote the space of all $\mathbb{R}^2$-valued piecewise smooth functions defined on $[t_0, t_1]$. In particular,

$$PWS(t_0, t_1; \mathbb{R}^2) = \Big\{ \boldsymbol{x}(\cdot) = \begin{bmatrix} x_1(\cdot) & x_2(\cdot) \end{bmatrix}^T$$
$$: x_i(\cdot) \in PWS(t_0, t_1), i = 1, 2 \Big\}. \qquad (7.40)$$

For each $\boldsymbol{x}(\cdot) = \begin{bmatrix} x_1(\cdot) & x_2(\cdot) \end{bmatrix}^T \in PWS(t_0, t_1; \mathbb{R}^2)$, define the *functional*
$J : PWS(t_0, t_1; \mathbb{R}^2) \to \mathbb{R}$ by

$$J(\boldsymbol{x}(\cdot)) = J\left(x_1(\cdot), x_2(\cdot)\right) = \int_{t_0}^{t_1} f\left(s, x_1(s), x_2(s), \dot{x}_1(s), \dot{x}_2\left(s\right)\right) ds. \tag{7.41}$$

Assume that the points $x_{1,0}$, $x_{2,0}$, $x_{1,1}$ and $x_{2,1}$ are given. Define the set of PWS functions $\boldsymbol{\Theta}_2$ by

$$\boldsymbol{\Theta}_2 = \left\{\boldsymbol{x}(\cdot) \in PWS(t_0, t_1; \mathbb{R}^2) : x_i\left(t_j\right) = x_{i,\,j}, \quad i = 1, 2, \quad j = 0, 1\right\} \tag{7.42}$$

Observe that $J : PWS(t_0, t_1; \mathbb{R}^2) \to \mathbb{R}$ is a real valued function defined on $PWS(t_0, t_1; \mathbb{R}^2)$.

The Simplest Problem in Vector Form (the fixed endpoint problem) is the problem of minimizing $J(\cdot)$ on $\boldsymbol{\Theta}_2$. In particular, the goal is to find

$$\boldsymbol{x}^*\left(\cdot\right) = [x_1^*(\cdot)\ x_2^*(\cdot)]^T \in \boldsymbol{\Theta}_2$$

such that

$$J(\boldsymbol{x}^*(\cdot)) = \int_{t_0}^{t_1} f\left(s, x_1^*(s), x_2^*(s), \dot{x}_1^*(s), \dot{x}_2^*\left(s\right)\right) ds$$

$$\leq J(\boldsymbol{x}(\cdot)) = \int_{t_0}^{t_1} f\left(s, x_1(s), x_2(s), \dot{x}_1(s), \dot{x}_2\left(s\right)\right) ds,$$

for all $\boldsymbol{x}\left(\cdot\right) \in \boldsymbol{\Theta}_2$.

In order to formulate the higher order problem we assume that t_0 and t_1 are given and fixed and $f(t, x, u, v)$ is a C^2 function. Let $PWS^2(t_0, t_1)$ denote the space of all real-valued piecewise smooth functions $x(\cdot)$ defined on $[t_0, t_1]$ such that $x(\cdot)$ is smooth and $\dot{x}(\cdot)$ is piecewise smooth on $[t_0, t_1]$. In particular,

$$PWS^2(t_0, t_1) = \left\{x : [t_0, t_1] \longrightarrow \mathbb{R} : x(\cdot) \in C^1([t_0, t_1]) \text{ and } \dot{x}(\cdot) \in PWS(t_0, t_1)\right\} \tag{7.43}$$

Note that if $x(\cdot) \in PWS^2(t_0,t_1)$, then $\ddot{x}(\cdot) \in PWC(t_0,t_1)$. For each $x(\cdot) \in PWS^2(t_0,t_1)$, define the *functional* $J : PWS^2(t_0,t_1) \to \mathbb{R}$ by

$$J(x(\cdot)) = \int_{t_0}^{t_1} f\left(s, x(s), \dot{x}(s), \ddot{x}\left(s\right)\right) ds. \tag{7.44}$$

Since $\ddot{x}(\cdot) \in PWC(t_0,t_1)$, the cost function defined by (7.44) is well defined. Assume that the points x_0, x_1, u_0 and u_1 are given and define the set of functions $\mathbf{\Theta}^2$ by

$$\mathbf{\Theta}^2 = \left\{x(\cdot) \in PWS^2(t_0,t_1) : x\left(t_i\right) = x_i,\ \dot{x}\left(t_i\right) = u_i,\ i = 0,1\right\}. \tag{7.45}$$

The Simplest Problem in Higher Order Form (the fixed endpoint problem) is the problem of minimizing $J(\cdot)$ on $\mathbf{\Theta}^2$. In particular, the goal is to find $x^*\left(\cdot\right) \in \mathbf{\Theta}^2$ such that

$$J(x^*(\cdot)) = \int_{t_0}^{t_1} f\left(s, x^*(s), \dot{x}^*(s), \ddot{x}^*\left(s\right)\right) ds$$

$$\leq J(x(\cdot)) = \int_{t_0}^{t_1} f\left(s, x(s), \dot{x}(s), \ddot{x}\left(s\right)\right) ds,$$

for all $x\left(\cdot\right) \in \mathbf{\Theta}^2$.

Remark 7.2 *Observe that the Simplest Problem in Higher Order Form is a special case of the Simplest Problem in Vector Form. To make this precise, we set*

$$x_1(\cdot) = x(\cdot) \quad \text{and} \quad x_2(\cdot) = \dot{x}(\cdot)$$

so that $\dot{x}_2(\cdot) = \ddot{x}(\cdot)$ and the cost function (7.44) becomes

$$\hat{J}(x_1(\cdot), x_2(\cdot)) = \int_{t_0}^{t_1} \hat{f}\left(s, x_1(s), x_2(s), \dot{x}_1(s), \dot{x}_2\left(s\right)\right) ds$$

$$= J(x(\cdot)) = \int_{t_0}^{t_1} f\left(s, x(s), \dot{x}(s), \ddot{x}\left(s\right)\right) ds,$$

where $\hat{f}(t, x_1, x_2, u_1, u_2)$ *does not explicitly depend* u_1 *and is defined by*

$$\hat{f}(t, x_1, x_2, u_1, u_2) = f(t, x_1, x_2, u_2).$$

Before deriving necessary conditions, we present additional lemmas that extend the Fundamental Lemma of the Calculus of Variations.

7.4.1 Extensions of Some Basic Lemmas

In order to derive first order necessary conditions for vector and higher order problems, we present a few fundamental lemmas that provide the backbone to much of variational theory. These results are natural extensions of the FLCV. In addition, we review some basic results on differentiation.

The Fundamental Lemma of the Calculus of Variations 3.1 involves only scalar functions and first order derivatives. We shall extend this lemma to higher order derivatives and to vector valued functions. To set the stage for the lemmas, we need some additional notation. Let $p \geq 1$ be a given integer and define the set

$$PWS^p(t_0, t_1) = \left\{x(\cdot) : x^{[k]}(\cdot) \in PWS(t_0, t_1),\ k = 0, 1, \ldots, p-1\right\}, \tag{7.46}$$

where

$$x^{[k]}(\cdot) \triangleq \frac{d^k x(\cdot)}{dt^k} \tag{7.47}$$

is the kth derivative of $x(\cdot)$. For notational purposes we define the zero derivative as the function $x^{[0]}(\cdot) \triangleq x(\cdot)$. Observe that if $x(\cdot) \in PWS^p(t_0, t_1)$, then the pth derivative of $x(\cdot)$ exists except at a finite number of points and $x^{[p]}(\cdot) \in PWC(t_0, t_1)$.

Likewise, we define $V_0^p = V_0^p(t_0, t_1) = PWS_0^p(t_0, t_1)$ to be the set

$$V_0^p(t_0, t_1) = \left\{\eta(\cdot) \in PWS^p(t_0, t_1) : \eta^{[k]}(t_0) = 0 = \eta^{[k]}(t_1),\right.$$
$$\left. k = 0, 1, \ldots, p-1\right\}. \tag{7.48}$$

Note that $V_0^p(t_0, t_1) \subseteq PWS^p(t_0, t_1)$ is the set of all functions $\eta(\cdot) \in PWS^p(t_0, t_1)$ with the property that $\eta(\cdot)$ and all its derivatives up to order $p - 1$ are zero at both ends of the interval $[t_0, t_1]$. Also, by definition $PWS^1(t_0, t_1) = PWS(t_0, t_1)$ and

$V_0^1(t_0, t_1) = V_0(t_0, t_1)$. The following is the basic result needed for all extensions of the FLCV.

Lemma 7.3 *Let $p \geq 1$ be any integer. If $v(\cdot)$ is piecewise continuous on $[t_0, t_1]$ and*

$$\int_{t_0}^{t_1} v(s)\,\eta(s)\,ds = 0 \tag{7.49}$$

for all $\eta(\cdot) \in V_0^p(t_0, t_1)$, then $v(\cdot) \in PWS(t_0, t_1)$ and

$$v(t) = 0, \tag{7.50}$$

except at a finite number of points. The converse is also true.

Proof: Since $v(\cdot)$ is piecewise continuous on $[t_0, t_1]$, there exist a finite set of points $t_0 < \hat{t}_1 < \hat{t}_2 < ... < \hat{t}_k < t_1$ in (t_0, t_1) such that $v(\cdot)$ is continuous on each subinterval $(\hat{t}_j, \hat{t}_{j-1})$. Assume that $v(\cdot)$ is not zero on one of these intervals. In particular, assume that there is a point $\hat{z} \in (\hat{t}_j, \hat{t}_{j-1})$ so that $v(\hat{z}) > 0$. Since $v(\cdot)$ is continuous on $(\hat{t}_j, \hat{t}_{j-1})$, there is a $\delta > 0$ and $a < b$ such that $(\hat{z}-\delta, \hat{z}+\delta) \subset [a, b] \subset (\hat{t}_j, \hat{t}_{j-1})$ and $v(t) > 0$ for all $t \in (\hat{z}-\delta, \hat{z}+\delta)$. Define the function

$$\tilde{\eta}(t) = \begin{cases} [(t-a)(b-t)]^p, & t \in [a, b] \\ 0, & t \notin [a, b] \end{cases}$$

and note that $\tilde{\eta}(t)$ has the following properties:

(i) $\tilde{\eta}(\cdot) \in V_0^p(t_0, t_1)$,

(ii) $\tilde{\eta}(t) > 0$ for all $t \in (\hat{z} - \delta, \hat{z} + \delta)$,

(iii) $v(t)\,\tilde{\eta}(t) > 0$ for all $t \in (\hat{z} - \delta, \hat{z} + \delta)$.

Consequently, it follows that

$$\int_{t_0}^{t_1} v(s)\,\tilde{\eta}(s)\,ds \geq \int_{\hat{z}-\delta}^{\hat{z}+\delta} v(s)\,\tilde{\eta}(s)\,ds > 0$$

which contradicts the assumption (7.49). Hence, $v(t) = 0$ on each of the subintervals $(\hat{t}_j, \hat{t}_{j-1})$ and this completes the proof. □

We now state a higher order form on the Fundamental Lemma of the Calculus of Variations.

Lemma 7.4 (High Order Form of the FLCV) *Let $p \geq 1$ be any integer. If $v(\cdot)$ is piecewise continuous on $[t_0, t_1]$ and*

$$\int_{t_0}^{t_1} v(s)\, \eta^{[p]}(s)\, ds = 0 \tag{7.51}$$

for all $\eta(\cdot) \in V_0^p(t_0, t_1)$, then there exist p constants a_0, a_1, a_2, ...,a_{p-1} such that

$$v(t) = a_{p-1}t^{p-1} + a_{p-2}t^{p-2} + \cdots + a_2 t^2 + a_1 t + a_0$$

except at a finite number of points. In particular, $v(\cdot) \in PWS^p(t_0, t_1)$ and at all points t where $v^{[p-1]}(\cdot)$ is continuous

$$v^{[p-1]}(t) = (p-1)!a_{p-1}.$$

The converse is also true.

Observe that this is a powerful result that implies the function $v(\cdot) \in PWS^p(t_0, t_1)$ is equal *e.f.* to a polynomial of degree $p-1$. The proof of this result is nontrivial and will not be given here (see pages 112 to 117 in Reid's book [153] and Lemma 13.1 (page 105) in Hestenes' book [101]). However, if one assumes that $v(\cdot) \in PWS^p(t_0, t_1)$, the proof follows from an easy integration by parts and applying **Lemma** 7.3 above.

Example 7.4 *Consider the case where $p = 2$.* ***Lemma*** *7.4 implies that if $v(\cdot) \in PWC(t_0, t_1)$ and*

$$\int_{t_0}^{t_1} v(s)\ddot{\eta}(s)ds = 0$$

for all $\eta(\cdot) \in V_0^2(t_0, t_1)$ where

$$V_0^2(t_0, t_1) = \{\eta(\cdot) \in PWC(t_0, t_1) : \eta(\cdot),\ \dot{\eta}(\cdot) \in PWS(t_0, t_1),$$
$$\eta(t_0) = \dot{\eta}(t_0) = \eta(t_1) = \dot{\eta}(t_1) = 0\},$$

then there are constants a_1 and a_2 such that

$$v(t) = a_1 t + a_0$$

except at a finite number of points on $[t_0, t_1]$. Thus, $v(\cdot)$ is equal to a linear function e.f. If one assumes that there are two piecewise continuous functions $\alpha(\cdot)$ and $\beta(\cdot)$ such that

$$\int_{t_0}^{t_1} \{\alpha(s)\ddot{\eta}(s) + \beta(s)\eta(s)\}\, ds = 0$$

for all $\eta(\cdot) \in V_0^2(t_0, t_1)$, then setting $\gamma(t) = \int_0^t \int_0^s \beta(\mu)d\mu$ and integrating by parts twice one has

$$0 = \int_{t_0}^{t_1} \{\alpha(s)\ddot{\eta}(s) + \beta(s)\eta(s)\}\, ds = \int_{t_0}^{t_1} \{\alpha(s)\ddot{\eta}(s) + \gamma(s)\ddot{\eta}(s)\}\, ds$$
$$= \int_{t_0}^{t_1} \{\alpha(s) + \gamma(s)\}\, \ddot{\eta}(s) ds$$

for all $\eta(\cdot) \in V_0^2(t_0, t_1)$. Consequently, it follows that

$$\alpha(t) + \gamma(t) = \alpha(t) + \int_0^t \int_0^s \beta(\mu)d\mu = a_1 t + a_0,$$

or equivalently,

$$\alpha(t) = a_1 t + a_0 - \int_0^t \int_0^s \beta(\mu)d\mu.$$

Thus, $\alpha(\cdot)$ is twice differentiable (except at a finite number of points) and

$$\ddot{\alpha}(t) = -\beta(t), \quad e.f.$$

Let $PWC(t_0, t_1; \mathbb{R}^n)$ denote the space of all $\mathbb{R}^n$-valued piecewise smooth functions defined on $[t_0, t_1]$. In particular,

$$PWC(t_0, t_1; \mathbb{R}^n) = \left\{ \begin{array}{c} \boldsymbol{x}(\cdot) = \begin{bmatrix} x_1(\cdot) & x_2(\cdot) & \cdots & x_n(\cdot) \end{bmatrix}^T : \\ \text{and } x_i(\cdot) \in PWC(t_0, t_1), \ i = 1, 2, \ldots, n \end{array} \right\}. \tag{7.52}$$

Likewise, let $PWS(t_0,t_1;\mathbb{R}^n)$ denote the space of all $\mathbb{R}^n$-valued piecewise smooth functions defined on $[t_0,t_1]$. In particular,

$$PWS(t_0,t_1;\mathbb{R}^n) = \left\{ \begin{array}{c} \boldsymbol{x}(\cdot) = \left[\ x_1(\cdot) \quad x_2(\cdot) \quad \cdots \quad x_n(\cdot)\ \right]^T : \\ \text{and } x_i(\cdot) \in PWS(t_0,t_1),\ i = 1,2,\ldots,n \end{array} \right\}. \tag{7.53}$$

Also, $PWS_0(t_0,t_1;\mathbb{R}^n)$ will denote the space of all $\mathbb{R}^n$-valued piecewise smooth functions defined on $[t_0,t_1]$ satisfying zero boundary conditions

$$\boldsymbol{x}(t_0) = \boldsymbol{0} = \boldsymbol{x}(t_1)$$

and we define $V_0(t_0,t_1;\mathbb{R}^n)$ to be

$$V_0(t_0,t_1;\mathbb{R}^n) = \left\{ \begin{array}{c} \boldsymbol{\eta}(\cdot) = \left[\ \eta_1(\cdot) \quad \eta_2(\cdot) \quad \cdots \quad \eta_n(\cdot)\ \right]^T \in PWS(t_0,t_1): \\ \text{and } \eta_i(t_j) = 0,\ i = 1,2,\ldots,n,\ j = 0,1 \end{array} \right\}. \tag{7.54}$$

Note that $V_0(t_0,t_1;\mathbb{R}^n) = PWS_0(t_0,t_1;\mathbb{R}^n)$ which leads to following extension of the FLCV Part A to the vector case.

Lemma 7.5 *Let $p \geq 1$ be any integer. If $\boldsymbol{v}(\cdot) = [v_1(\cdot)\ v_2(\cdot)\ \cdots\ v_n(\cdot)]^T \in PWC(t_0,t_1;\mathbb{R}^n)$ is piecewise continuous on $[t_0,t_1]$ and*

$$\int_{t_0}^{t_1} < \boldsymbol{v}(s), \boldsymbol{\eta}(s) > ds = 0 \tag{7.55}$$

for all $\boldsymbol{\eta}(\cdot) \in V_0(t_0,t_1;\mathbb{R}^n)$, then $\boldsymbol{v}(\cdot) \in PWS(t_0,t_1;\mathbb{R}^n)$ and

$$\boldsymbol{v}(t) = \boldsymbol{0}, \tag{7.56}$$

except at a finite number of points. The converse is also true.

Proof: Since $\boldsymbol{v}(\cdot)$ is piecewise continuous on $[t_0,t_1]$, there exists a finite set of points $t_0 < \hat{t}_1 < \hat{t}_2 < \ldots < \hat{t}_k < t_1$, in (t_0,t_1) such that $\boldsymbol{v}(\cdot)$ is continuous on each subinterval $(\hat{t}_j,\hat{t}_{j-1})$. Assume that $\boldsymbol{v}(\cdot)$ is not zero on one of these intervals. In particular, assume that there is a point $\hat{z} \in (\hat{t}_j,\hat{t}_{j-1})$ and m, with $1 \leq m \leq n$ so that $v_m(\hat{z}) > 0$. Since $v_m(\cdot)$ is continuous on $(\hat{t}_j, \hat{t}_{j-1})$, there is an $\delta > 0$ and $a < b$ such that $(\hat{z} - \delta, \hat{z} + \delta) \subset [a, b] \subset (\hat{t}_j, \hat{t}_{j-1})$ and $v_m(t) > 0$ for all $t \in (\hat{z} - \delta, \hat{z} + \delta)$. Define the function

$$\eta_m(t) = \begin{cases} [(t-a)(b-t)]^p, & t \in [a, b] \\ 0, & t \notin [a, b] \end{cases}$$

and let $\tilde{\boldsymbol{\eta}}(\cdot)$ note the function $\tilde{\boldsymbol{\eta}}(\cdot) = \begin{bmatrix} \eta_1(\cdot) & \eta_2(\cdot) & \cdots & \eta_n(\cdot) \end{bmatrix}^T \in V_0(t_0, t_1; \mathbb{R}^n)$ defined by

$$\eta_j(t) = \begin{cases} \eta_m(t), & j = m \\ 0, & j \neq m \end{cases}.$$

Observe that $\tilde{\boldsymbol{\eta}}(\cdot)$ has the following properties:

(i) $\tilde{\boldsymbol{\eta}}(\cdot) \in V_0(t_0, t_1; \mathbb{R}^n)$,

(ii) $\eta_m(t) > 0$ for all $t \in (\hat{z} - \delta, \hat{z} + \delta)$,

(iii) $v_m(t)\,\eta_m(t) > 0$ for all $t \in (\hat{z} - \delta, \hat{z} + \delta)$.

Consequently, it follows that

$$\begin{aligned} \int_{t_0}^{t_1} < \boldsymbol{v}(s), \boldsymbol{\eta}(s) > ds &= \int_{t_0}^{t_1} v_m(s)\,\eta_m(s)\,ds \\ &\geq \int_{\hat{z}-\delta}^{\hat{z}+\delta} v_m(s)\,\eta_m(s)\,ds > 0, \end{aligned}$$

which contradicts the assumption (7.55). Hence, $\boldsymbol{v}(t) = \boldsymbol{0}$ on each of the subintervals $(\hat{t}_j, \hat{t}_{j-1})$ and this completes the proof. □

Lemma 7.6 ***(FLCV: Part A in Vector Form)***

Let $p \geq 1$ be any integer. If $\boldsymbol{v}(\cdot) = \begin{bmatrix} v_1(\cdot) & v_2(\cdot) & \cdots & v_n(\cdot) \end{bmatrix}^T \in PWC(t_0, t_1; \mathbb{R}^n)$ is piecewise continuous on $[t_0, t_1]$ and

$$\int_{t_0}^{t_1} < \boldsymbol{v}(s), \dot{\boldsymbol{\eta}}(s) > ds = 0 \tag{7.57}$$

for all $\boldsymbol{\eta}(\cdot) \in V_0(t_0, t_1; \mathbb{R}^n)$, *then* $\boldsymbol{v}(\cdot) \in PWS(t_0, t_1; \mathbb{R}^n)$ *and there is a constant vector* $\mathbf{c}$ *such that*

$$\boldsymbol{v}(t) = \mathbf{c}, \tag{7.58}$$

except at a finite number of points. The converse is also true.

Proof: Assume that $\boldsymbol{v}(\cdot)$ is piecewise continuous on $[t_0, t_1]$. Let $\eta(\cdot)$ be any piecewise smooth function in $PWS_0(t_0, t_1)$ and define the function $\boldsymbol{\eta}(\cdot) \in V_0(t_0, t_1; \mathbb{R}^n)$ by

$$\eta_j(t) = \begin{cases} \eta(t), & j = m \\ 0, & j \neq m \end{cases}.$$

Observe that that $\boldsymbol{\eta}(\cdot) \in V_0(t_0, t_1; \mathbb{R}^n)$ and hence

$$\int_{t_0}^{t_1} v_m(s)\,\dot{\eta}(t)ds = \int_{t_0}^{t_1} < \boldsymbol{v}(s), \dot{\boldsymbol{\eta}}(s) > ds = 0.$$

Since $\eta(\cdot) \in PWS_0(t_0, t_1)$ is arbitrary, it follows from the Fundamental Lemma of the Calculus of Variations that there is a constant c_m such that $v_m(t) = c_m$ except at a finite number of points. This can be repeated for each $m = 1, 2, \cdots, n$ and if $\mathbf{c} = \begin{bmatrix} c_1 & c_2 & \cdots & c_n \end{bmatrix}^T$, then $\boldsymbol{v}(t) = \mathbf{c}$ except at a finite number of points and this completes the proof. □

The following lemma follows immediately from the previous result.

Lemma 7.7 ***(FLCV: Part B in Vector Form)***

Let $p \geq 1$ *be any integer. If* $\boldsymbol{\alpha}(\cdot) = \begin{bmatrix} \alpha_1(\cdot) & \alpha_2(\cdot) & \cdots & \alpha_n(\cdot) \end{bmatrix}^T$ *and*
$\boldsymbol{\beta}(\cdot) = \begin{bmatrix} \beta_1(\cdot) & \beta_2(\cdot) & \cdots & \beta_n(\cdot) \end{bmatrix}^T \in PWC(t_0, t_1; \mathbb{R}^n)$ *are piecewise continuous on* $[t_0, t_1]$ *and*

$$\int_{t_0}^{t_1} \{\langle \boldsymbol{\alpha}(s), \boldsymbol{\eta}(s) \rangle + \langle \boldsymbol{\beta}(s), \dot{\boldsymbol{\eta}}(s) \rangle\}\, ds = 0 \tag{7.59}$$

for all $\boldsymbol{\eta}(\cdot) \in V_0(t_0, t_1; \mathbb{R}^n)$, *then* $\boldsymbol{\beta}(\cdot) \in PWS(t_0, t_1; \mathbb{R}^n)$ *and there is a constant vector* $\mathbf{c}$ *such that*

$$\boldsymbol{\beta}(t) = \mathbf{c} + \int_{t_0}^{t} \boldsymbol{\alpha}(s)\, ds, \tag{7.60}$$

except at a finite number of points. The converse is also true.

We can now formulate the general vector and higher order forms of the Simplest Problem in the Calculus of Variations.

7.4.2 The Simplest Problem in Vector Form

Let t_0 and $t_1$be given and suppose the function $\boldsymbol{f} : [t_0, t_1] \times \mathbb{R}^n \times \mathbb{R}^n \longrightarrow \mathbb{R}$ is C^2. In particular, $\boldsymbol{f}(t, \boldsymbol{x}, \boldsymbol{u})$ has the form

$$\boldsymbol{f}(t, \boldsymbol{x}, \boldsymbol{u}) = \boldsymbol{f}(t, x_1, x_2, \ldots, x_n, u_1, u_2, \ldots, u_n),$$

where all the partial derivatives of order 2 exist and are continuous. Here,

$$\frac{\partial \boldsymbol{f}(t, \boldsymbol{x}, \boldsymbol{u})}{\partial t} = \boldsymbol{f}_t(t, \boldsymbol{x}, \boldsymbol{u}) = \boldsymbol{f}_t(t, x_1, x_2, \ldots, x_n, u_1, u_2, \ldots, u_n),$$

$$\frac{\partial \boldsymbol{f}(t, \boldsymbol{x}, \boldsymbol{u})}{\partial x_i} = \boldsymbol{f}_{x_i}(t, \boldsymbol{x}, \boldsymbol{u}) = \boldsymbol{f}_{x_i}(t, x_1, x_2, \ldots, x_n, u_1, u_2, \ldots, u_n)$$

and

$$\frac{\partial \boldsymbol{f}(t, \boldsymbol{x}, \boldsymbol{u})}{\partial u_i} = \boldsymbol{f}_{u_i}(t, \boldsymbol{x}, \boldsymbol{u}) = \boldsymbol{f}_{u_i}(t, x_1, x_2, \ldots, x_n, u_1, u_2, \ldots, u_n).$$

We use standard notation for the gradients

$$\nabla_{\boldsymbol{x}} \boldsymbol{f}(t, \boldsymbol{x}, \boldsymbol{u}) \triangleq \begin{bmatrix} \boldsymbol{f}_{x_1}(t, \boldsymbol{x}, \boldsymbol{u}) \\ \boldsymbol{f}_{x_2}(t, \boldsymbol{x}, \boldsymbol{u}) \\ \vdots \\ \boldsymbol{f}_{x_n}(t, \boldsymbol{x}, \boldsymbol{u}) \end{bmatrix}$$

and

$$\nabla_{\boldsymbol{u}} \boldsymbol{f}(t, \boldsymbol{x}, \boldsymbol{u}) \triangleq \begin{bmatrix} \boldsymbol{f}_{u_1}(t, \boldsymbol{x}, \boldsymbol{u}) \\ \boldsymbol{f}_{u_2}(t, \boldsymbol{x}, \boldsymbol{u}) \\ \vdots \\ \boldsymbol{f}_{u_n}(t, \boldsymbol{x}, \boldsymbol{u}) \end{bmatrix}.$$

For each $\boldsymbol{x}(\cdot) = \begin{bmatrix} x_1(\cdot) & x_2(\cdot) & \cdots & x_n(\cdot) \end{bmatrix}^T \in PWS(t_0, t_1; \mathbb{R}^n)$, define the *functional* $\boldsymbol{J} : PWS(t_0, t_1; \mathbb{R}^n) \to \mathbb{R}$ by

$$\boldsymbol{J}(\boldsymbol{x}(\cdot)) = \int_{t_0}^{t_1} \boldsymbol{f}(s, \boldsymbol{x}(s), \dot{\boldsymbol{x}}(s))ds. \tag{7.61}$$

Assume that the vectors $\boldsymbol{x}_0$ and $\boldsymbol{x}_1$ are given. Define the set of piecewise smooth vector functions $\boldsymbol{\Theta}_n$ by

$$\boldsymbol{\Theta}_n = \{\boldsymbol{x}(\cdot) \in PWS(t_0, t_1; \mathbb{R}^n) : \boldsymbol{x}(t_0) = \boldsymbol{x}_0, \quad \boldsymbol{x}(t_1) = \boldsymbol{x}_1\}. \tag{7.62}$$

The Simplest Problem in Vector Form (the fixed endpoint problem) is the problem of minimizing $\boldsymbol{J}(\cdot)$ on $\boldsymbol{\Theta}_n$. In particular, the goal is to find $\boldsymbol{x}^*(\cdot) = \begin{bmatrix} x_1^*(\cdot) & x_2^*(\cdot) & \cdots & x_n^*(\cdot) \end{bmatrix}^T \in \boldsymbol{\Theta}_n$ such that

$$\boldsymbol{J}(\boldsymbol{x}^*(\cdot)) = \int_{t_0}^{t_1} \boldsymbol{f}(s, \boldsymbol{x}^*(s), \dot{\boldsymbol{x}}^*(s))\, ds$$

$$\leq \boldsymbol{J}(\boldsymbol{x}(\cdot)) = \int_{t_0}^{t_1} \boldsymbol{f}(s, \boldsymbol{x}(s), \dot{\boldsymbol{x}}(s))\, ds$$

for all $\boldsymbol{x}(\cdot) \in \boldsymbol{\Theta}_n$.

Theorem 7.5 (Vector Form of the Euler Necessary Condition) *If $\boldsymbol{x}^*(\cdot) \in \boldsymbol{\Theta}_n$ minimizes $\boldsymbol{J}(\cdot)$ on $\boldsymbol{\Theta}_n$, then*

(1) there is a vector $\mathbf{c}$ such that

$$[\nabla_{\boldsymbol{u}} \boldsymbol{f}(t, \boldsymbol{x}^*(t), \dot{\boldsymbol{x}}^*(t))] = \mathbf{c} + \int_{t_0}^{t} [\nabla_{\boldsymbol{x}} \boldsymbol{f}(s, \boldsymbol{x}^*(s), \dot{\boldsymbol{x}}^*(s))]\, ds, \tag{7.63}$$

except at a finite number of points,

(2) $\boldsymbol{x}^*(t_0) = \boldsymbol{x}_0$,

(3) $\boldsymbol{x}^*(t_1) = \boldsymbol{x}_1$.

(4) Between corners of $\boldsymbol{x}^*(\cdot)$ *the function* $\nabla_{\boldsymbol{u}}\boldsymbol{f}\left(t, \boldsymbol{x}^*(t), \dot{\boldsymbol{x}}^*(t)\right)$ *is differentiable and*

$$\frac{d}{dt}\left[\nabla_{\boldsymbol{u}}\boldsymbol{f}\left(t, x^*(t), \dot{x}^*(t)\right)\right] = \left[\nabla_{\boldsymbol{x}}\boldsymbol{f}\left(t, \boldsymbol{x}^*(t), \dot{\boldsymbol{x}}^*(t)\right)\right]. \qquad (7.64)$$

Proof: Suppose that $\boldsymbol{x}^*(\cdot) \in \boldsymbol{\Theta}_n$ minimizes $\boldsymbol{J}(\cdot)$ on $\boldsymbol{\Theta}_n$ and $\boldsymbol{\eta}(\cdot) \in PWS(t_0, t_1; \mathbb{R}^n)$. Define the function

$$\boldsymbol{g}(t, \varepsilon) = \boldsymbol{f}\left(t, \boldsymbol{x}^*(t) + \varepsilon\boldsymbol{\eta}(t), \dot{\boldsymbol{x}}^*(t) + \varepsilon\,\dot{\boldsymbol{\eta}}(t)\right).$$

Since $\dot{x}^*(t)$ and $\dot{\eta}(t)$ are continuous on a finite partition $t_0 = \hat{t}_1 < \hat{t}_2 < \ldots < \hat{t}_k = t_1$ of $[t_0, t_1]$, it follows that $\boldsymbol{g}(t, \varepsilon)$ and $\boldsymbol{g}_\varepsilon(t, \varepsilon)$ are both continuous on each subinterval $(\hat{t}_i, \hat{t}_{i+1})$. Without loss of generality we may assume that $\boldsymbol{x}^*(\cdot)$ and $\boldsymbol{\eta}(\cdot)$ are smooth. It follows that

$$\begin{aligned}\boldsymbol{F}(\varepsilon) &= \boldsymbol{J}(\boldsymbol{x}^*(\cdot) + \varepsilon\boldsymbol{\eta}(\cdot)) \\ &= \int_{t_0}^{t_1} \boldsymbol{f}(s, \boldsymbol{x}^*(s) + \varepsilon\boldsymbol{\eta}(s), \dot{\boldsymbol{x}}^*(s) + \varepsilon\dot{\boldsymbol{\eta}}(s))ds = \int_{t_0}^{t_1} \boldsymbol{g}(s, \varepsilon)ds\end{aligned}$$

and the goal is to differentiate $\boldsymbol{F}(\varepsilon)$ at $\varepsilon = 0$. Applying the chain rule we have

$$\begin{aligned}\boldsymbol{g}_\varepsilon(t, \varepsilon) &= \frac{d}{d\varepsilon}[\boldsymbol{f}\left(t, \boldsymbol{x}^*(t) + \varepsilon\boldsymbol{\eta}(t), \dot{\boldsymbol{x}}^*(s) + \varepsilon\dot{\boldsymbol{\eta}}(t)\right)] \\ &= \sum_{i=1}^{n} \boldsymbol{f}_{x_i}\left(t, \boldsymbol{x}^*(t) + \varepsilon\boldsymbol{\eta}(t), \dot{\boldsymbol{x}}^*(t) + \varepsilon\dot{\boldsymbol{\eta}}(t)\right)\eta_i(t) \\ &\qquad + \boldsymbol{f}_{u_i}\left(t, \boldsymbol{x}^*(t) + \varepsilon\boldsymbol{\eta}(t), \dot{\boldsymbol{x}}^*(t) + \varepsilon\dot{\boldsymbol{\eta}}(t)\right)\dot{\eta}_i(t).\end{aligned}$$

Applying Leibniz's Lemma 3.2 yields

$$\left.\frac{d}{d\varepsilon}\boldsymbol{F}(\varepsilon)\right|_{\varepsilon=0} = \left.\frac{d}{d\varepsilon}\left[\boldsymbol{J}(\boldsymbol{x}^*(\cdot) + \varepsilon\boldsymbol{\eta}(\cdot))\right]\right|_{\varepsilon=0} = \int_{t_0}^{t_1} \boldsymbol{g}_\varepsilon(s, 0)ds. \quad (7.65)$$

Again, note that we really only need to compute $\boldsymbol{g}_{\varepsilon}(t,\varepsilon)$ on each subinterval $(\hat{t}_i, \hat{t}_{i+1})$ where both $\dot{\boldsymbol{x}}^*(\cdot)$ and $\dot{\boldsymbol{\eta}}(\cdot)$ are continuous and the chain rule produces an explicit formula for the first variation $\delta J(x^*(\cdot);\eta(\cdot))$ of $J(\cdot)$ at $x^*(\cdot)$ in the direction of $\eta(\cdot)$. In particular,

$$\begin{aligned}\delta \boldsymbol{J}(\boldsymbol{x}^*(\cdot);\boldsymbol{\eta}(\cdot)) = \int_{t_0}^{t_1} \sum_{i=1}^{n} &\boldsymbol{f}_{x_i}(s, \boldsymbol{x}^*(s) + \varepsilon\boldsymbol{\eta}(s), \dot{\boldsymbol{x}}^*(s) \\ &+ \varepsilon\dot{\boldsymbol{\eta}}(s))\eta_i(s)\,ds \\ &+ \int_{t_0}^{t_1} \sum_{i=1}^{n} \boldsymbol{f}_{u_i}(s, \boldsymbol{x}^*(s) + \varepsilon\boldsymbol{\eta}(s), \dot{\boldsymbol{x}}^*(s) \\ &+ \varepsilon\dot{\boldsymbol{\eta}}(s))\dot{\eta}_i(s)\,ds.\end{aligned}$$

Consequently, it follows that

$$\begin{aligned}\delta \boldsymbol{J}(\boldsymbol{x}^*(\cdot);\boldsymbol{\eta}(\cdot)) = \int_{t_0}^{t_1} &\{\langle \nabla_{\boldsymbol{x}}\boldsymbol{f}(s, \boldsymbol{x}^*(s), \dot{\boldsymbol{x}}^*(s)), \boldsymbol{\eta}(s)\rangle \\ &+ \langle \nabla_{\boldsymbol{u}}\boldsymbol{f}(s, \boldsymbol{x}^*(s), \dot{\boldsymbol{x}}^*(s)), \dot{\boldsymbol{\eta}}(s)\rangle\}ds.\end{aligned}$$

In addition, if $\boldsymbol{\eta}(\cdot) \in PWS_0(t_0,t_1;\mathbb{R}^n)$, then

$$\begin{aligned}\int_{t_0}^{t_1} &\{\langle \nabla_{\boldsymbol{x}}\boldsymbol{f}(s, \boldsymbol{x}^*(s), \dot{\boldsymbol{x}}^*(s)), \boldsymbol{\eta}(s)\rangle \\ &+ \langle \nabla_{\boldsymbol{u}}\boldsymbol{f}(s, \boldsymbol{x}^*(s), \dot{\boldsymbol{x}}^*(s)), \dot{\boldsymbol{\eta}}(s)\rangle\}ds = 0\end{aligned}$$

and the theorem follows from the Fundamental Lemma Part B in Vector Form given in **Lemma** 7.7 above. □

7.4.3 The Simplest Problem in Higher Order Form

Assume $p \geq 1$ is a given integer and the endpoints t_0 and t_1 are fixed. Let $f : [t_0, t_1] \times \mathbb{R}^{1+p} \longrightarrow \mathbb{R}$ be a C^2 real valued function of the form

$$f(t, x, u_1, u_2, \ldots, u_p),$$

where all the partial derivatives of order 2 exist and are continuous. As above, we let $PWS^p(t_0, t_1)$ denote the space of all real-valued piecewise smooth functions $x(\cdot)$ defined on $[t_0, t_1]$ such that $x^{[k]}(\cdot)$ is piecewise smooth on $[t_0, t_1]$ for all $k = 0, 1, \ldots, p-1$. In particular,

$$PWS^p(t_0, t_1) = \left\{x(\cdot) : x^{[k]}(\cdot) \in PWS(t_0, t_1),\; k = 0, 1, \ldots, p-1\right\}. \tag{7.66}$$

Observe that if $x(\cdot) \in PWS^p(t_0, t_1)$, then $x^{[p]}(\cdot) \in PWC(t_0, t_1)$. For each $x(\cdot) \in PWS^p(t_0, t_1)$, define the *functional* $J : PWS^p(t_0, t_1) \to \mathbb{R}$ by

$$J(x(\cdot)) = \int_{t_0}^{t_1} f\left(s, x(s), \dot{x}(s), \ddot{x}(s), \ldots, x^{[p]}(s)\right) ds. \tag{7.67}$$

Since $x^{[p]}(\cdot) \in PWC(t_0, t_1)$, the cost function defined by (7.67) is well defined. Assume that the points $x_{k,0}$ and $x_{k,1}$ for $k = 0, 1, 2, \ldots, p-1$ are given and define the set of functions $\mathbf{\Theta}^p$ by

$$\mathbf{\Theta}^p = \{x(\cdot) \in PWS^p(t_0, t_1) : x^{[k]}(t_i) = x_{k,i},\; k = 0, 1, 2, \ldots, p-1, \quad i = 0, 1\}. \tag{7.68}$$

The Simplest Problem in Higher Order Form (the fixed endpoint problem) is the problem of minimizing $J(\cdot)$ on $\mathbf{\Theta}^p$. In particular, the goal is to find $x^*(\cdot) \in \mathbf{\Theta}^p$ such that

$$J(x^*(\cdot)) = \int_{t_0}^{t_1} f\left(s, x^*(s), \dot{x}^*(s), \ddot{x}^*(s), \ldots, (x^*)^{[p]}(s)\right) ds$$

$$\leq J(x(\cdot)) = \int_{t_0}^{t_1} f\left(s, x(s), \dot{x}(s), \ddot{x}(s), \ldots, x^{[p]}(s)\right) ds,$$

for all $x(\cdot) \in \mathbf{\Theta}^p$.

Remark 7.3 *Note that the Simplest Problem in Higher Order Form is a special case of the Simplest Problem in Vector Form. To make this precise, we set*

$$x_1(\cdot) = x(\cdot),\; x_2(\cdot) = \dot{x}(\cdot),\; x_3(\cdot) = \ddot{x}(\cdot), \ldots,\; x_p(\cdot) = x^{[p-1]}(\cdot)$$

so that $\dot{x}_p(\cdot) = x^{[p]}(\cdot)$. *Define* $\hat{\boldsymbol{f}} : [t_0, t_1] \times \mathbb{R}^p \times \mathbb{R}^p \longrightarrow \mathbb{R}$ *by*

$$\hat{\boldsymbol{f}}(t, \boldsymbol{x}, \boldsymbol{u})) = \hat{\boldsymbol{f}}(t, x_1, x_2, \ldots, x_p, u_1, u_2, \ldots, u_p) \triangleq f(t, x, u_1, u_2, \ldots, u_p)$$

and observe that for

$$\boldsymbol{x}(\cdot) = \begin{bmatrix} x(\cdot) & \dot{x}(\cdot) & \cdots & x^{[p-1]}(\cdot) \end{bmatrix}^T = \begin{bmatrix} x_1(\cdot) & x_2(\cdot) & \cdots & x_p(\cdot) \end{bmatrix}^T,$$

$\hat{\boldsymbol{f}}(t, \boldsymbol{x}(t), \dot{\boldsymbol{x}}(t))$ *has the form*

$$\begin{aligned} \hat{\boldsymbol{f}}(t, \boldsymbol{x}(t), \dot{\boldsymbol{x}}(t)) &= \hat{\boldsymbol{f}}(t, x_1(t), x_2(t), \ldots, x_p(t), \dot{x}_1(t), \dot{x}_2(t), \ldots, \dot{x}_p(t)) \\ &= f\left(t, x(t), \dot{x}(t), \ddot{x}(t), \ldots, x^{[p]}(t)\right). \end{aligned}$$

Consequently, the cost function (7.67) becomes

$$\begin{aligned} \hat{J}(\boldsymbol{x}(t)) &= \int_{t_0}^{t_1} \hat{\boldsymbol{f}}(s, \boldsymbol{x}(s), \dot{\boldsymbol{x}}(s)) ds \\ &= \int_{t_0}^{t_1} f\left(s, x(s), \dot{x}(s), \ddot{x}(t), \ldots, x^{[p]}(s)\right) ds = J(x(\cdot)), \end{aligned}$$

where

$$f\left(t, x_1, x_2, \ldots, x_p, u_1, u_2, \ldots, u_p\right)$$

does not explicitly depend u_k *for* $k = 1, 2, \ldots, p-1$.

In order to simplify expressions, recall that we use $f^*(t)$ to denote the evaluation of $f(\cdot)$ along the optimal curve so that

$$\begin{aligned} f^*(t) &= f\left(t, x^*(t), \dot{x}^*(t), \ddot{x}^*(t), \ldots, (x^*)^{[p]}(t)\right), \\ f_x^*(t) &= f_x\left(t, x^*(t), \dot{x}^*(t), \ddot{x}^*(t), \ldots, (x^*)^{[p]}(t)\right), \\ f_{u_i}^*(t) &= f_{u_i}\left(t, x^*(t), \dot{x}^*(t), \ddot{x}^*(t), \ldots, (x^*)^{[p]}(t)\right), \quad i = 1, 2, ...p. \end{aligned}$$

Also, we set

$$f_{u_0}^*(t) = f_x^*(t).$$

Theorem 7.6 ***(High Order Form of the Euler Necessary Condition)*** *If $x^*(\cdot) \in \Theta^p$ minimizes $J(\cdot)$ on Θ^p, then*

(1) there are constants c_i, $i = 1, 2, \ldots, p-1$ such that for all $t \in [t_0, t_1]$,

$$f^*_{u_p}(t) = \sum_{i=0}^{p-1} c_i \frac{(i-t)^{p-1-i}}{(p-1-i)!} + \int_{t_0}^{t} \sum_{i=0}^{p-1} \left[\frac{(s-t)^{p-1-i}}{(p-1-i)!} f^*_{u_i}(s) \right] ds \tag{7.69}$$

except at a finite number of points,

(2) $x^{[k]}(t_0) = x_{k,0}$, $k = 0, 1, 2, \ldots, p-1$,

(3) $x^{[k]}(t_1) = x_{k,1}$, $k = 0, 1, 2, \ldots, p-1$.

(4) Between corners of $(x^*)^{[p-1]}(\cdot)$, $x^*(\cdot)$ satisfies

$$(-1)^p \frac{d^p}{dt^p}\left[f^*_{u_p}(t)\right] + \sum_{i=1}^{p-1} (-1)^{p-i} \frac{d^{p-i}}{dt^{p-i}}\left[f^*_{u_{p-i}}(t)\right] + f^*_x(t) = 0, \tag{7.70}$$

where

$$\frac{d^k}{d^k}\left[f^*_{u_k}(t)\right] = \frac{d^k}{d^k}[f^*_{u_k}(t, x^*(t), \dot{x}^*(t), \ddot{x}^*(t), \ldots, (x^*)^{[p]}(t))].$$

7.5 Problems with Constraints: Isoperimetric Problem

Here we impose a "functional" constraint on the curve $x(\cdot)$. For example, consider the problem of finding $x^*(\cdot) \in PWS(t_0, t_1)$ such that $x^*(\cdot)$ minimizes

$$J(x(\cdot)) = \int_{t_0}^{t_1} f(s, x(s), \dot{x}(s))\, ds,$$

subject constraint

$$x(t_0) = x_0, \quad x(t_1) = x_1,$$

and

$$\mathcal{G}(x(\cdot)) = \int_{t_0}^{t_1} g(s, x(s), \dot{x}(s))\, ds = 0. \tag{7.71}$$

Example 7.5 *Find the curve of length 5 that passes through the points* $[-1\ 0]^T$ *and* $[1\ 0]^T$ *and minimizes*

$$J(x(\cdot)) = \int_{-1}^{1} -x(s)\, ds$$

subject to to the length constraint

$$\int_{-1}^{1} \sqrt{1 + [\dot{x}(s)]^2} ds = 5.$$

Observe that

$$\mathcal{G}(x(\cdot)) = \int_{-1}^{1} g(s, x(s), \dot{x}(s))\, ds,$$

where

$$g(t, x, u) = \sqrt{1 + [u]^2} - 5/2.$$

In order to obtain a first order necessary condition for this constrained problem, we recall the definition of an extremal. An extremal for the functional

$$J(x(\cdot)) = \int_{t_0}^{t_1} f(s, x(s), \dot{x}(s))\, ds$$

is any function $x(\cdot) \in PWS(t_0, t_1)$ that satisfies the integral form of the Euler equation

$$f_u(t, x(t), \dot{x}(t)) = c + \int_{t_0}^{t} f_x(s, x(s), \dot{x}(s))\, ds$$

for some c. Note that because of the Fundamental Lemma of the Calculus of Variations, $x(\cdot) \in PWS(t_0, t_1)$ is an extremal for $J(\cdot)$ if and only if

$$\delta J(x(\cdot); \eta(\cdot)) = 0 \tag{7.72}$$

for all $\eta(\cdot) \in PWS_0(t_0, t_1)$. Again, $\delta J(x(\cdot); \eta(\cdot))$ is the first variation of $J(\cdot)$ at $x(\cdot)$ in the direction of $\eta(\cdot)$ and is given by

$$\delta J(x(\cdot); \eta(\cdot)) = \int_{t_0}^{t_1} \{f_x(s, x(s), \dot{x}(s))\, \eta(s) + f_u(s, x(s), \dot{x}(s))\, \dot{\eta}(s)\} ds. \tag{7.73}$$

Likewise, we say that $x(\cdot) \in PWS(t_0, t_1)$ is an extremal for the functional

$$\mathcal{G}(x(\cdot)) = \int_{t_0}^{t_1} g(s, x(s), \dot{x}(s))\, ds$$

if $x(\cdot) \in PWS(t_0, t_1)$ satisfies the Euler Integral Equation

$$g_u(t, x(t), \dot{x}(t)) = c + \int_{t_0}^{t} g_x(s, x(s), \dot{x}(s))\, ds$$

for some c. Thus, the Fundamental Lemma of the Calculus of Variations implies that $x(\cdot) \in PWS(t_0, t_1)$ is an extremal for $\mathcal{G}(\cdot)$ if and only if

$$\delta\mathcal{G}(x(\cdot); \eta(\cdot)) = 0 \tag{7.74}$$

for all $\eta(\cdot) \in PWS_0(t_0, t_1)$, where $\delta\mathcal{G}(x(\cdot); \eta(\cdot))$ is the first variation of $\mathcal{G}(\cdot)$ at $x(\cdot)$ in the direction of $\eta(\cdot)$ and is given by

$$\delta\mathcal{G}(x(\cdot); \eta(\cdot)) = \int_{t_0}^{t_1} \{g_x(s, x(s), \dot{x}(s))\, \eta(s) + g_u(s, x(s), \dot{x}(s))\, \dot{\eta}(s)\} ds. \tag{7.75}$$

We now can state a necessary condition which is an infinite dimensional form of the Lagrange Multiplier Theorem. We outline a proof in Section 7.5.1 below.

Theorem 7.7 (Multiplier Theorem for the Isoperimetric Problem) *If $x^*(\cdot) \in PWS(t_0, t_1)$ is smooth and minimizes*

$$J(x(\cdot)) = \int_{t_0}^{t_1} f(s, x(s), \dot{x}(s))\, ds,$$

subject to

$$x(t_0) = x_0, \quad x(t_1) = x_1,$$

and

$$\mathcal{G}(x(\cdot)) = \int_{t_0}^{t_1} g(s, x(s), \dot{x}(s)) = 0,$$

then there exist constants λ_0 *and* λ_1 *such that*

(i) $|\lambda_0| + |\lambda_1| \neq 0$ *and*

(ii) $x^*(\cdot)$ *is an extremal of*

$$\lambda_0 J(x(\cdot)) + \lambda_1 \mathcal{G}(x(\cdot)). \tag{7.76}$$

(iii) If in addition $x^*(\cdot)$ *is not an extremal of* $\mathcal{G}(\cdot)$*, then the constant* λ_0 *is not zero.*

To make the notation more compact we define the *Lagrangian* by

$$\mathcal{L}(\lambda_0, \lambda_1, x(\cdot)) = \lambda_0 J(x(\cdot)) + \lambda_1 \mathcal{G}(x(\cdot)). \tag{7.77}$$

If we define $l(\lambda_0, \lambda_1, t, x, u)$ by

$$l(\lambda_0, \lambda_1, t, x, u) = \lambda_0 f(t, x, u) + \lambda_1 g(t, x, u),$$

then

$$\mathcal{L}(\lambda_0, \lambda_1, x(\cdot)) = \int_{t_0}^{t_1} l(\lambda_0, \lambda_1, s, x(s), \dot{x}(s))\, ds. \tag{7.78}$$

Remark 7.4 *The statement that* $x^*(\cdot)$ *is an extremal of* $\mathcal{L}(\lambda_0, \lambda_1, x(\cdot)) = \lambda_0 J(x(\cdot)) + \lambda_1 \mathcal{G}(x(\cdot))$ *implies that* $x^*(\cdot)$ *is a solution of the corresponding Euler equation*

$$[\lambda_0 f_u(t, x(t), \dot{x}(t)) + \lambda_1 g_u(t, x(t), \dot{x}(t))] = c + \int_{t_0}^{t} [\lambda_0 f_x(s, x(s), \dot{x}(s))] ds$$
$$+ \int_{t_0}^{t} [\lambda_1 g_x(s, x(s), \dot{x}(s))] ds$$

for some c. Also, if $x^*(\cdot)$ is smooth then $x^*(\cdot)$ satisfies the differential equation

$$\frac{d}{dt}[\lambda_0 f_u(t, x(t), \dot{x}(t)) + \lambda_1 g_u(t, x(t), \dot{x}(t))]$$
$$= [\lambda_0 f_x(t, x(t), \dot{x}(t)) + \lambda_1 g_x(t, x(t), \dot{x}(t))] .$$

In the case (iii) above, when, $x^*(\cdot)$ is not an extremal of $\mathcal{G}(\cdot)$, then the minimizer $x^*(\cdot)$ is called a ***normal minimizer****. Thus, if $x^*(\cdot)$ is a normal minimizer, then $\lambda_0 \neq 0$. This definition of normality of a minimizing arc was first given by Bliss in [28] where he noted that (global) normality implied that the multiplier λ_0 was not zero. As we shall see later, this idea is true for general Lagrange multiplier theorems.*

7.5.1 Proof of the Lagrange Multiplier Theorem

In this section we prove **Theorem** 7.7. The proof is essentially the same as the proof for the necessary condition for the $2D$ Lagrange Multiplier Theorem in Section 2.2.3.

Proof of Theorem 7.7: Assume $x^*(\cdot) \in C^1(t_0, t_1)$ minimizes

$$J(x(\cdot)) = \int_{t_0}^{t_1} f(s, x(s), \dot{x}(s))\, ds,$$

subject to

$$x(t_0) = x_0, \quad x(t_1) = x_1,$$

and

$$\mathcal{G}(x(\cdot)) = \int_{t_0}^{t_1} g(s, x(s), \dot{x}(s)) = 0.$$

If $x^*(\cdot)$ is an extremal of $\mathcal{G}(\cdot)$, then

$$\delta\mathcal{G}(x^*(\cdot); \eta(\cdot)) = 0$$

for all $\eta(\cdot) \in PWS_0(t_0, t_1)$. In this case set $\lambda_0 = 0$ and $\lambda_1 = 1$. It follows that $|\lambda_0| + |\lambda_1| = 1 \neq 0$ and

$$\begin{aligned}
&\delta\left[\lambda_0 J(x^*(\cdot);\eta(\cdot))\right] + \delta\left[\lambda_1 \mathcal{G}(x^*(\cdot);\eta(\cdot))\right] \\
&\quad = \lambda_0 \delta J(x^*(\cdot);\eta(\cdot)) + \lambda_1 \delta \mathcal{G}(x^*(\cdot);\eta(\cdot)) \\
&\quad = 0\delta J(x^*(\cdot);\eta(\cdot)) + \delta \mathcal{G}(x^*(\cdot);\eta(\cdot)) \\
&\quad = \delta \mathcal{G}(x^*(\cdot);\eta(\cdot)) = 0,
\end{aligned}$$

for all $\eta(\cdot) \in PWS_0(t_0, t_1)$. Hence $x^*(\cdot)$ is an extremal of

$$\lambda_0 J\left(x(\cdot)\right) + \lambda_1 \mathcal{G}\left(x(\cdot)\right)$$

and the theorem is clearly true.

Now consider the case where $x^*(\cdot)$ is not an extremal of $\mathcal{G}(\cdot)$. Since $x^*(\cdot)$ is not extremal of $\mathcal{G}(\cdot)$, there exists a $\bar{\boldsymbol{\eta}}(\cdot) \in PWS_0(t_0, t_1)$ such that $\delta \mathcal{G}(x^*(\cdot);\bar{\boldsymbol{\eta}}(\cdot)) \neq 0$. Clearly, $\bar{\eta}(\cdot)$ is not the zero function since $\delta \mathcal{G}(x(\cdot);0(\cdot)) = 0$. Let

$$\lambda_0 = \delta \mathcal{G}(x^*(\cdot);\bar{\eta}(\cdot)) \neq 0$$

and

$$\lambda_1 = -\delta J(x^*(\cdot);\bar{\eta}(\cdot)).$$

We now show that $x^*(\cdot)$ is an extremal of

$$\lambda_0 J(x(\cdot)) + \lambda_1 \mathcal{G}(x(\cdot)) = [\delta \mathcal{G}(x^*(\cdot);\bar{\eta}(\cdot))] J(x(\cdot)) + [\delta J(x^*(\cdot);\bar{\eta}(\cdot))] \mathcal{G}(x(\cdot)).$$

Observe that

$$\begin{aligned}
&\delta[\lambda_0 J(x^*(\cdot);\eta(\cdot)) + \lambda_1 \mathcal{G}(x^*(\cdot);\eta(\cdot))] \\
&\quad = \lambda_0[\delta J(x^*(\cdot);\eta(\cdot))] + \lambda_1[\delta \mathcal{G}(x^*(\cdot);\eta(\cdot))] \\
&\quad = [\delta \mathcal{G}(x^*(\cdot);\bar{\eta}(\cdot))][\delta J(x^*(\cdot);\eta(\cdot))] \\
&\qquad - [\delta J(x^*(\cdot);\bar{\eta}(\cdot))][\delta \mathcal{G}(x^*(\cdot);\eta(\cdot))],
\end{aligned}$$

or equivalently,

$$\begin{aligned}
&\delta[\lambda_0 J(x^*(\cdot);\eta(\cdot)) + \lambda_1 \mathcal{G}(x^*(\cdot);\eta(\cdot))] \\
&\quad = \det \begin{bmatrix} [\delta J(x^*(\cdot);\eta(\cdot))] & [\delta J(x^*(\cdot);\bar{\eta}(\cdot))] \\ [\delta \mathcal{G}(x^*(\cdot);\eta(\cdot))] & [\delta \mathcal{G}(x^*(\cdot);\bar{\eta}(\cdot))] \end{bmatrix}.
\end{aligned}$$

Therefore, to establish that $x^*(\cdot)$ is an extremal of

$$\lambda_0 J(x(\cdot)) + \lambda_1 \mathcal{G}(x(\cdot)) = [\delta\mathcal{G}(x^*(\cdot);\bar{\eta}(\cdot))]J(x(\cdot)) + [\delta J(x^*(\cdot);\bar{\eta}(\cdot))]\mathcal{G}(x(\cdot))$$

we must show that

$$\det \begin{bmatrix} [\delta J(x^*(\cdot);\eta(\cdot))] & [\delta J(x^*(\cdot);\bar{\eta}(\cdot))] \\ [\delta \mathcal{G}(x^*(\cdot);\eta(\cdot))] & [\delta \mathcal{G}(x^*(\cdot);\bar{\eta}(\cdot))] \end{bmatrix} = 0 \tag{7.79}$$

for all $\eta(\cdot) \in PWS_0(t_0, t_1)$. This is accomplished by applying the **Inverse Mapping Theorem** 2.4.

Define $T : \mathbb{R}^2 \to \mathbb{R}^2$ by

$$T(\alpha, \beta) = [p(\alpha, \beta) \;\; q(\alpha, \beta)]^T, \tag{7.80}$$

where

$$p(\alpha, \beta) = J(x^*(\cdot) + \alpha\eta(\cdot) + \beta\bar{\eta}(\cdot)) \tag{7.81}$$

and

$$q(\alpha, \beta) = \mathcal{G}(x^*(\cdot) + \alpha\eta(\cdot) + \beta\bar{\eta}(\cdot)), \tag{7.82}$$

respectively.

Note that $T(\alpha, \beta)$ maps the open set $\mathbb{R}^2$ to $\mathbb{R}^2$ and $T(0,0) = [p(0,0) \;\; q(0,0)]^T = [J(x^*(\cdot)) \; 0]^T = [\hat{p} \;\; \hat{q}]^T$. Also, the Jacobian of $T(\alpha, \beta)$ at $[\hat{\alpha} \; \hat{\beta}]^T = [0 \; 0]^T$ is given by

$$\begin{bmatrix} \frac{\partial p(0,0)}{\partial \alpha} & \frac{\partial p(0,0)}{\partial \beta} \\ \frac{\partial q(0,0)}{\partial \alpha} & \frac{\partial q(0,0)}{\partial \beta} \end{bmatrix} = \begin{bmatrix} [\delta J(x^*(\cdot);\eta(\cdot))] & [\delta J(x^*(\cdot);\bar{\eta}(\cdot))] \\ [\delta \mathcal{G}(x^*(\cdot);\eta(\cdot))] & [\delta \mathcal{G}(x^*(\cdot);\bar{\eta}(\cdot))], \end{bmatrix}.$$

Assume that (7.79) is **not true**. This assumption implies that the Jacobian of $T(\alpha, \beta)$ is non-singular at $[\hat{\alpha} \; \hat{\beta}]^T = [0 \; 0]^T$ so we may apply the Inverse Mapping Theorem 2.4. Here, $T(\alpha, \beta)$ as defined by (7.80), (7.81) and (7.82) with $[\hat{\alpha} \; \hat{\beta}]^T = [0 \; 0]^T$ and

$$T(\hat{\alpha}, \hat{\beta}) = T(0,0) = [J(x^*(\cdot)) \; \mathcal{G}(x^*(\cdot))]^T = [\hat{p} \; 0]^T.$$

In particular, there is a neighborhood $\mathcal{U} = \left\{ [\alpha \; \beta]^T : \sqrt{\alpha^2 + \beta^2} < \gamma \right\}$ of $[0 \; 0]^T$ and a neighborhood $\mathcal{V}$ of $[J(x^*(\cdot)) \; 0]^T$ such that the restriction of $T(\alpha, \beta)$ to $\mathcal{U}$, $\mathcal{T}(\alpha, \beta) : \mathcal{U} \to \mathcal{V}$, has a continuous inverse $\mathcal{T}^{-1}(p, q) : \mathcal{V} \to \mathcal{U}$ belonging to C^1.

Let $[\tilde{p}\ 0]^T \in \mathcal{V}$ be any point with $\tilde{p} < J(x^*(\cdot))$ and let $[\tilde{\alpha}\ \tilde{\beta}]^T = \mathcal{T}^{-1}(\tilde{p}, 0) \in \mathcal{U}$. Observe that for $[\tilde{\alpha}\ \tilde{\beta}]^T$,

$$J(x^*(\cdot) + \tilde{\alpha}\eta(\cdot) + \tilde{\beta}\bar{\eta}(\cdot)) = p(\tilde{\alpha}, \tilde{\beta}) = \tilde{p} < J(x^*(\cdot)) \tag{7.83}$$

and

$$\mathcal{G}(x^*(\cdot) + \tilde{\alpha}\eta(\cdot) + \tilde{\beta}\bar{\eta}(\cdot)) = 0. \tag{7.84}$$

Therefore, by construction the function

$$\tilde{x}(\cdot) = x^*(\cdot) + \tilde{\alpha}\eta(\cdot) + \tilde{\beta}\bar{\eta}(\cdot)$$

satisfies all the constraints of the Isoperimetric Problem and

$$J(\tilde{x}(\cdot)) < J(x^*(\cdot))$$

which contradicts the assumption that $x^*(\cdot) \in C^1(t_0, t_1)$ minimizes

$$J(x(\cdot)) = \int_{t_0}^{t_1} f(s, x(s), \dot{x}(s))\, ds,$$

subject to

$$x(t_0) = x_0, \quad x(t_1) = x_1,$$

and

$$\mathcal{G}(x(\cdot)) = \int_{t_0}^{t_1} g(s, x(s), \dot{x}(s)) = 0.$$

Therefore the assumption that

$$\det \begin{bmatrix} \frac{\partial p(0,0)}{\partial \alpha} & \frac{\partial p(0,0)}{\partial \beta} \\ \frac{\partial q(0,0)}{\partial \alpha} & \frac{\partial q(0,0)}{\partial \beta} \end{bmatrix} = \det \begin{bmatrix} [\delta J(x^*(\cdot); \eta(\cdot))] & [\delta J(x^*(\cdot); \bar{\eta}(\cdot))] \\ [\delta \mathcal{G}(x^*(\cdot); \eta(\cdot))] & [\delta \mathcal{G}(x^*(\cdot); \bar{\eta}(\cdot))] \end{bmatrix} \neq 0$$

must be false. Hence, $x^*(\cdot)$ is an extremal of

$$\lambda_0 J(x(\cdot)) + \lambda_1 \mathcal{G}(x(\cdot)) = [\delta \mathcal{G}(x^*(\cdot); \bar{\eta}(\cdot))] J(x(\cdot)) + [\delta J(x^*(\cdot); \bar{\eta}(\cdot))] \mathcal{G}(x(\cdot))$$

which completes the proof. □

7.6 Problems with Constraints: Finite Constraints

Here we consider a non-integral constraint. Assume that the function $g(t,x)$ is C^2 and consider the problem of minimizing

$$J(x(\cdot)) = \int_{t_0}^{t_1} f(s, x(s), \dot{x}(s))\, ds,$$

subject to

$$x(t_0) = x_0, \quad x(t_1) = x_1,$$

and the "finite constraint"

$$g(t, x(t)) = 0. \tag{7.85}$$

Observe that equation (7.85) implies that the trajectory $x(\cdot)$ must lie on the curve

$$M_g = \left\{ [t\ x]^T : g(t,x) = 0 \right\}. \tag{7.86}$$

Theorem 7.8 (Multiplier Theorem for the Finite Constraint Problem) *If*
$x^*(\cdot) \in PWS(t_0, t_1)$ *is smooth and minimizes*

$$J(x(\cdot)) = \int_{t_0}^{t_1} f(s, x(s), \dot{x}(s))\, ds,$$

subject to

$$x(t_0) = x_0, \quad x(t_1) = x_1,$$

and the finite constraint

$$g(t, x(t)) = 0,$$

then there exists a constant λ_0 *and a function* $\lambda_1(\cdot)$ *such that*

(i) $|\lambda_0| + |\lambda_1(t)| \neq 0$ *for all* $t_0 \le t \le t_1$ *and*

(ii) $x^*(\cdot)$ *is an extremal of*

$$\mathcal{L}(\lambda_0, \lambda_1(\cdot), x(\cdot)) = \int_{t_0}^{t_1} (\lambda_0 f(s, x(s), \dot{x}(s)) + \lambda_1(s) g(s, x(s)))ds. \tag{7.87}$$

(iii) If in addition, $g_x(t,x) \neq 0$ *for all* $[t \quad x]^T$ *on the surface* $M_g = \{[t\ x]^T : g(t,x) = 0\}$, *then* $\lambda_0 \neq 0$. *Thus, there exists a function* $\lambda(t) = \lambda_1(t)/\lambda_0$ *such that* $x^*(\cdot)$ *is an extremal of the functional*

$$\int_{t_0}^{t_1} \{f(s,x(s),\dot{x}(s)) + \lambda(s)\, g(s,x(s))\} ds.$$

Let

$$\boldsymbol{F}(t,x,u) = \lambda_0 f(t,x,u) + \lambda_1(t) g(t,x)$$

so that

$$\boldsymbol{F}_x(t,x,u) = \lambda_0 f_x(t,x,u) + \lambda_1(t) g_x(t,x)$$

and

$$\boldsymbol{F}_u(t,x,u) = \lambda_0 f_u(t,x,u).$$

The statement that $x^*(\cdot)$ is an extremal of

$$\mathcal{L}(\lambda_0, \lambda_1(\cdot), x(\cdot)) = \int_{t_0}^{t_1} (\lambda_0 f(s,x(s),\dot{x}(s)) + \lambda_1(s) g(s,x(s))) ds$$

implies that $x^*(\cdot)$ is a solution of the corresponding Euler Integral Equation

$$\lambda_0 f_u(t,x(t),\dot{x}(t)) = c + \int_{t_0}^{t} (\lambda_0 f_x(s,x(s),\dot{x}(s)) + \lambda_1(s) g_x(s,x(s)))\, ds$$

for some c. Also, between corners $x^*(\cdot)$ satisfies

$$\frac{d}{dt}\lambda_0 f_u(t,x(t),\dot{x}(t)) = [\lambda_0 f_x(t,x(t),\dot{x}(t)) + \lambda_1(t) g_x(t,x(t))].$$

Remark 7.5 *It is worth noting that* ***Theorem*** *7.8 does not provide information about the function* $\lambda_1(\cdot)$. *In particular, there is no claim that* $\lambda_1(\cdot)$ *is piecewise smooth or even piecewise continuous. In order to obtain additional information of this kind requires mathematical techniques beyond advanced calculus and will not be considered here.*

We close this chapter with a very brief introduction to abstract optimization problems. Although this material is not required reading, it does help the reader to see how the idea of the first variation can be extended to rather general settings. It is extensions of this type that provide the foundations for the modern development of necessary and sufficient conditions for optimization with applications to optimal control.

7.7 An Introduction to Abstract Optimization Problems

In this section we provide an basic introduction to the theory of necessary conditions for general optimization problems. Generally speaking, the subjects of the calculus of variations and optimal control theory belong to the larger discipline of infinite dimensional optimization theory. Problems in the calculus of variations came from classical physics and date back 300 years. The theory of optimal control is relatively new and some mathematicians place its formal beginning in 1952, with Bushaw's Ph.D. thesis (see Hermes and La Salle [100]). However, many of the basic ideas date back into the last century.

Here we review some basic finite and infinite dimensional unconstrained and constrained optimization theory. The reader is referred to [102], [131] and [166] for details. The book by Neustadt [144] contains very general results on necessary conditions with applications to a wide variety of problems.

7.7.1 The General Optimization Problem

We shall consider problems that are special cases of the following general optimization problem.

Let $\boldsymbol{Z}$ be a vector space and assume that $\boldsymbol{\Theta} \subseteq \boldsymbol{Z}$ is given a given (constraint) set. Also, let $\boldsymbol{J} : \boldsymbol{D}(\boldsymbol{J}) \subseteq \boldsymbol{Z} \longrightarrow \mathbb{R}^1$ be a real valued function defined on the domain $\boldsymbol{D}(\boldsymbol{J})$. The *General Optimization Problem* (GOP) is defined by: Find an element $\boldsymbol{z}^* \in$

$\boldsymbol{\Theta} \cap \boldsymbol{D}(\boldsymbol{J})$ such that

$$\boldsymbol{J}(\boldsymbol{z}^*) \leq \boldsymbol{J}(\boldsymbol{z})$$

for all $\boldsymbol{z} \in \boldsymbol{\Theta} \cap \boldsymbol{D}(\boldsymbol{J})$.

Of course, specific problems have much more structure and the challenge is to formulate the problem in such a way that there is a solution and one can "compute" $\boldsymbol{z}^*$. In addition to the Simplest Problem in the Calculus of Variations, this framework includes much of finite dimensional optimization as well as problems in optimal control. The following examples will be used to motivate the discussion in this section.

Example 7.6 *Let $\boldsymbol{J}(\cdot)$ be a real-valued function defined on some interval $I = [a, b]$. In ordinary calculus one considers the problem of minimizing $\boldsymbol{J}(\cdot)$ on $[a, b]$. In this example, $\boldsymbol{Z} = \mathbb{R}^1$ and $\boldsymbol{\Theta} = [a, b]$.*

Example 7.7 *To give another elementary example, let us consider the problem of finding the point on the plane with equation*

$$2x + 3y - z = 5$$

which is nearest to the origin in $\mathbb{R}^3$. In this problem, we set

$$\boldsymbol{\Theta} = \left\{ [x \ y \ z]^T : 2x + 3y - z - 5 = 0 \right\}$$

and let $\boldsymbol{J}(\cdot)$ be the square of the distance to the origin; namely

$$\boldsymbol{J}(\boldsymbol{z}) = \boldsymbol{J}(x, y, z) = x^2 + y^2 + z^2.$$

Here, $\boldsymbol{Z} = \mathbb{R}^3$ is the vector space.

Example 7.8 *The problem is to find a piecewise continuous (control) function $u(t), 0 \leq t \leq 3$ such that the cost functional*

$$\boldsymbol{J}(x(\cdot), u(\cdot)) = \int_0^3 [1 - u(s) \cdot x(s)]\, ds$$

is minimized where $x(\cdot)$ *satisfies the differential equation*

$$\dot{x}(t) = u(t)\,x(t)$$

with initial condition

$$x(0) = 1$$

and $u(\cdot)$ *is constrained by*

$$0 \le u(t) \le 1.$$

This is the so called Farmer's Allocation Problem. For this example we set $\boldsymbol{X} = PWS(0,3) \times PWC(0,3)$ *and define the constraint set* $\boldsymbol{\Theta}$ *by*

$$\boldsymbol{\Theta} = \left\{ \begin{array}{r} \boldsymbol{z}(\cdot) = [x(\cdot) \quad u(\cdot)]^T : u(\cdot) \in PWC(0,3),\ 0 \le u(t) \le 1, \\ \dot{x}(t) = u(t)\,x(t), \quad x(0) = 1 \end{array} \right\}.$$

The cost function is defined on $\boldsymbol{J} : \boldsymbol{Z} = PWS(0,3) \times PWC(0,3) \longrightarrow \mathbb{R}^1$ *by* $\boldsymbol{J}(\boldsymbol{z}(\cdot)) = \boldsymbol{J}(x(\cdot), u(\cdot))$. *We shall see later that there is a more efficient way to describe the constraint set* $\boldsymbol{\Theta}$ *in terms of an equality and inequality constraint.*

Example 7.9 *Returning to Example 2.6 we let* $\boldsymbol{J} : \mathbb{R}^2 \longrightarrow \mathbb{R}^1$ *be defined by* $\boldsymbol{J}(x,y) = x^2 + y^2$ *and* $\boldsymbol{G} : \mathbb{R}^2 \longrightarrow \mathbb{R}^1$ *be given by* $\boldsymbol{G}(x,y) = x^2 - (y-1)^3$. *In this problem the constraint set is the "level curve" defined by*

$$\boldsymbol{\Theta} = \boldsymbol{\Theta}_{\boldsymbol{G}} = \left\{ \boldsymbol{z} = [x \quad y]^T \in \mathbb{R}^2 : \boldsymbol{G}(x,y) = 0 \right\}.$$

Recall that the minimizer is $\boldsymbol{z}^* = [x^* \ y^*]^T = [0 \ 1]^T$, *but the Lagrange Multiplier Rule is not very helpful.*

7.7.2 General Necessary Conditions

In this section we present the basic first order necessary conditions for the general optimization problem described above. The problem of finding the minimum of $\boldsymbol{J}(\cdot)$ over a constraint set $\boldsymbol{\Theta}$ can range from the trivial problem in Example (7.6) to the difficult optimal control problem in Example (7.8). The approach most often used to "solve" such problems has been the application of necessary conditions for a minimum. The basic idea behind the use of a necessary condition is really very simple. A necessary condition is used to reduce the problem of minimizing $\boldsymbol{J}(\cdot)$ over all of $\boldsymbol{\Theta}$ to a problem of minimizing $\boldsymbol{J}(\cdot)$ over a smaller set $\boldsymbol{\Theta}_S \subseteq \boldsymbol{\Theta}$. We illustrate this idea with the following trivial example.

Example 7.10 *Consider the problem of minimizing the continuous function*

$$J(z) = \frac{2z^2 - z^4 + 3}{4}$$

on the compact interval $\boldsymbol{\Theta} = [-2, 1]$. *From elementary calculus we know that there is a* $z^* \in \boldsymbol{\Theta}$ *that minimizes* $J(\cdot)$ *on the compact interval* $[-2, 1]$. *Moreover, if* $J(z^*) \leq J(z)$ *for all* $z \in \boldsymbol{\Theta} = [-2, 1]$ *then*

$$\begin{aligned} J'(z^*) &= 0 \quad if \quad -2 < z^* < 1 \\ J'(z^*) &\geq 0 \quad if \quad z^* = -2 \\ J'(z^*) &\leq 0 \quad if \quad z^* = 1. \end{aligned}$$

We use the first derivative test and solve for all $z^* \in (-2, 1)$ *satisfying*

$$0 = \frac{dJ(z)}{dz} = J'(z) = z - z^3.$$

Clearly, $z = -1$ *and* $z = 0$ *are the only solutions in the open interval* $(-2, 1)$. *At the left endpoint, we find that*

$$J'(-2) = 6 > 0$$

and at the right endpoint

$$J'(1) = 0 \leq 0.$$

Thus, at this point we have reduced the possible set of minimizers to

$$\boldsymbol{\Theta}_S = \{-2, -1, 0, 1\}.$$

*It is now trivial to find the **global** minimum, since*

$$J(-2) = -5/4, \quad J(0) = 3/4, \quad and \quad J(-1) = J(1) = 1.$$

Thus, $z^* = -2$ is the global minimum of $J(\cdot)$ on $\mathbf{\Theta} = [-2, 1]$. Note that we could have reduced the set $\mathbf{\Theta}_S$ even more by applying the second order condition that

$$\frac{d^2 J(z)}{dz^2} = J''(z^*) \geq 0 \quad if -2 < z^* < 0.$$

For in this problem

$$J''(z) = 1 - 3z^2,$$

so $J''(-1) = -2 \leq 0$, while $J''(0) = 1 \geq 0$. Thus, $z = -1$ could have been removed from $\mathbf{\Theta}_S$ leading to the problem of minimizing $J(\cdot)$ on $\mathbf{\Theta}_{SS} = \{-2, 0, 1\}$.

Recall from Section 2.2.2 that the proof of the basic necessary condition used in the previous example is rather simple. However, the idea can be extended to very general settings. Consider the problem of minimizing $\boldsymbol{J}(\cdot)$ on a set $\mathbf{\Theta} \subseteq \boldsymbol{Z}$. Assume that $\boldsymbol{z}^* \in \mathbf{\Theta} \cap \boldsymbol{D}(\boldsymbol{J})$ provides a (global) minimizer for $\boldsymbol{J} : \boldsymbol{D}(\boldsymbol{J}) \subseteq \boldsymbol{Z} \longrightarrow \mathbb{R}^1$ on $\mathbf{\Theta} \cap \boldsymbol{D}(\boldsymbol{J})$. Thus,

$$\boldsymbol{J}(\boldsymbol{z}^*) \leq \boldsymbol{J}(\boldsymbol{z}),$$

for all $\boldsymbol{z} \in \mathbf{\Theta} \cap \boldsymbol{D}(\boldsymbol{J})$. Furthermore, assume that there is a function

$$\boldsymbol{\varphi}(\cdot) : (-\delta, +\delta) \to \mathbf{\Theta} \cap \boldsymbol{D}(\boldsymbol{J}) \tag{7.88}$$

satisfying

$$\boldsymbol{\varphi}(0) = \boldsymbol{z}^*. \tag{7.89}$$

Define $\boldsymbol{F} : (-\delta, +\delta) \to \mathbb{R}^1$ by

$$\boldsymbol{F}(\varepsilon) = \boldsymbol{J}(\boldsymbol{\varphi}(\varepsilon)),$$

and observe that, since $\boldsymbol{\varphi}(\varepsilon) \in \mathbf{\Theta} \cap \boldsymbol{D}(\boldsymbol{J})$ for all $\varepsilon \in (-\delta, +\delta)$,

$$\boldsymbol{F}(0) = \boldsymbol{J}(\boldsymbol{\varphi}(0)) = \boldsymbol{J}(\boldsymbol{z}_*) \leq \boldsymbol{J}(\boldsymbol{\varphi}(\varepsilon)) = \boldsymbol{F}(\varepsilon),$$

for all $\varepsilon \in (-\delta, +\delta)$. If the derivative

$$\boldsymbol{F}'(0) = \left. \frac{d}{d\varepsilon} \boldsymbol{F}(\varepsilon) \right|_{\varepsilon=0} = \left. \frac{d}{d\varepsilon} \boldsymbol{J}(\boldsymbol{\varphi}(\varepsilon)) \right|_{\varepsilon=0}$$

exists, then

$$F'(0) = \frac{d}{d\varepsilon} J(\varphi(\varepsilon))\Big|_{\varepsilon=0} = 0. \tag{7.90}$$

Note that this is the basic necessary condition and to be "useful" one must construct functions $\varphi(\cdot)$ and have enough freedom in the choice of these functions so that the condition (7.90) implies something about the minimizer $z^* \in \Theta \cap D(J)$. In the following sections we show that this general framework can be applied to a wide variety of problems.

Note that in all of the above examples Θ is a subset of a vector space Z. Assume that $z^* \in \Theta \cap D(J)$ provides a (global) minimizer for $J : D(J) \subseteq Z \longrightarrow \mathbb{R}^1$ on $\Theta \cap D(J)$. We focus on variations $\varphi(\cdot) : (-\delta, +\delta) \to \Theta \cap D(J)$ of the form $\varphi(\varepsilon) = z^* + \varepsilon\eta$ where $\eta \in Z$ is "limited" to a specific subset of Z. The key issue is to make sure that, for a suitable class of η, the variations $\varphi(\varepsilon) = z^* + \varepsilon\eta$ are admissible in the following sense.

Definition 7.1 *The vector $\varphi(\varepsilon) = z^* + \varepsilon\eta$ is called an* ***admissible variation*** *of z^* if $z^* + \varepsilon\eta \in \Theta \cap D(J)$ on some nontrivial interval containing $\varepsilon = 0$.*

It is clear that some "directions" η may lead to admissible variations and others may not. Also, sometimes it is sufficient to assume that $\varphi(\cdot) : [0, +\delta) \to \Theta \cap D(J)$ and consider only one sided derivatives at $\varepsilon = 0$. Thus, we are led to the following concepts.

Definition 7.2 *Suppose $\mathcal{A}$ is a subset of the vector space Z with $z \in \mathcal{A}$ and $\eta \in Z$. We say that $z \in \mathcal{A}$ is an* ***internal point of $\mathcal{A}$ in the direction of $\eta \in Z$****, if there is a $\delta = \delta(\eta) > 0$ such that $(z + \varepsilon\eta) \in \mathcal{A}$ for all ε in the open interval $-\delta(\eta) < \varepsilon < \delta(\eta)$. The point $z \in \mathcal{A}$ is called a* ***radial point of $\mathcal{A}$ in the direction of η****, if there is an $\delta = \delta(\eta) > 0$ such that $(z + \varepsilon\eta) \in \mathcal{A}$ for all ε in the interval $0 \le \varepsilon < \delta(\eta)$. The point $z \in \mathcal{A}$ is called a* ***core point of $\mathcal{A}$*** *if z is an internal point of $\mathcal{A}$ for all directions $\eta \in Z$. Observe that the concept of internal, core and radial points are valid in a general vector space and require no notion of distance. In particular, there is no requirement that Z be a topological vector space.*

7.7.3 Abstract Variations

Let $\boldsymbol{J} : \boldsymbol{D}(\boldsymbol{J}) \subseteq \boldsymbol{Z} \longrightarrow \mathbb{R}^1$. If $\boldsymbol{z}$ is an internal point of $\boldsymbol{D}(\boldsymbol{J})$ in the direction of $\boldsymbol{\eta}$ then the function

$$\boldsymbol{\varphi}(\varepsilon) = \boldsymbol{z} + \varepsilon\boldsymbol{\eta}$$

maps the interval $(-\delta(\boldsymbol{\eta}), \delta(\boldsymbol{\eta}))$ into $\boldsymbol{D}(\boldsymbol{J})$. On the other hand, if $\boldsymbol{z} \in \boldsymbol{D}(\boldsymbol{J})$ is a radial point of $\boldsymbol{D}(\boldsymbol{J})$ in the direction of η then

$$\boldsymbol{\varphi}(\varepsilon) = \boldsymbol{z} + \varepsilon\boldsymbol{\eta}$$

maps $[0, \delta(\boldsymbol{\eta}))$ into $\boldsymbol{D}(\boldsymbol{J})$. If $z \in \boldsymbol{D}(\boldsymbol{J})$, $\boldsymbol{\eta} \in \boldsymbol{Z}$ and $\boldsymbol{z}$ is either an internal point or radial point of $\boldsymbol{D}(\boldsymbol{J})$ in the direction of $\boldsymbol{\eta}$, then we define the **first variation of $\boldsymbol{J}(\cdot)$ *at* $\boldsymbol{z}$ in the direction $\boldsymbol{\eta}$** by

$$\delta\boldsymbol{J}(\boldsymbol{z};\boldsymbol{\eta}) \triangleq \left.\frac{d}{d\varepsilon}\boldsymbol{J}(\boldsymbol{z} + \varepsilon\boldsymbol{\eta})\right|_{\epsilon=0}, \tag{7.91}$$

provided the derivative exists. Here the derivative is two-sided if $\boldsymbol{z}$ is an internal point and one-sided if $\boldsymbol{z}$ is a radial point of $\boldsymbol{D}(\boldsymbol{J})$.

If the two-sided or one-sided second derivatives

$$\left.\frac{d^2}{d\varepsilon^2}\boldsymbol{J}(\boldsymbol{z} + \varepsilon\boldsymbol{\eta})\right|_{\epsilon=0}$$

exist, then we say that $\boldsymbol{J}(\cdot)$ has a **second variation at $\boldsymbol{z}$ in the direction $\boldsymbol{\eta}$** and denote this by

$$\delta^2\boldsymbol{J}(\boldsymbol{z};\boldsymbol{\eta}) \triangleq \left.\frac{d^2}{d\varepsilon^2}\boldsymbol{J}(\boldsymbol{z} + \varepsilon\boldsymbol{\eta})\right|_{\epsilon=0}. \tag{7.92}$$

If $\boldsymbol{z}$ is a core point of $\boldsymbol{D}(\boldsymbol{J})$ and the first variation of $\boldsymbol{J}(\cdot)$ at $\boldsymbol{z}$ in the direction $\boldsymbol{\eta}$ exists for all $\boldsymbol{\eta} \in \boldsymbol{Z}$, then we say that $\boldsymbol{J}(\cdot)$ has a **Gâteaux variation at $\boldsymbol{z}$**. In other words, if $\delta\boldsymbol{J}(\boldsymbol{z};\boldsymbol{\eta})$ exists for all $\boldsymbol{\eta} \in \boldsymbol{Z}$, then $\delta\boldsymbol{J}(\boldsymbol{z};\boldsymbol{\eta})$ is called the Gâteaux variation (or **weak differential**) **of $\boldsymbol{J}(\cdot)$ at $\boldsymbol{z}$ in the direction $\boldsymbol{\eta}$**. Likewise, if $\delta^2\boldsymbol{J}(\boldsymbol{z};\boldsymbol{\eta})$ exists for all $\boldsymbol{\eta} \in \boldsymbol{Z}$, then $\delta^2\boldsymbol{J}(\boldsymbol{z};\boldsymbol{\eta})$ is called the **second Gâteaux variation** (or **weak second differential**) **of $\boldsymbol{J}(\cdot)$** at **$\boldsymbol{z}$ in the direction $\boldsymbol{\eta}$** (see [143] for details).

Observe that $\boldsymbol{J}(\cdot)$ has Gâteaux variation at $\boldsymbol{z} \in \boldsymbol{D}(\boldsymbol{J})$ if and only if

1. $\boldsymbol{z}$ is a core point of $\boldsymbol{D}(\boldsymbol{J})$ and

2. $\delta \boldsymbol{J}(\boldsymbol{z};\boldsymbol{\eta})$ exists for all $\boldsymbol{\eta} \in \boldsymbol{Z}$.

Moreover, if $\boldsymbol{J}(\cdot)$ has Gâteaux variation at $\boldsymbol{z} \in \boldsymbol{D}(\boldsymbol{J})$, then the derivative (7.91) is two-sided since $\boldsymbol{\eta} \in \boldsymbol{Z}$ implies that $-\boldsymbol{\eta} \in \boldsymbol{Z}$.

It is now rather easy to establish a very general necessary condition for the abstract problem. In particular, assume that $\boldsymbol{z}^* \in \boldsymbol{\Theta} \cap \boldsymbol{D}(\boldsymbol{J})$ provides a (global) minimizer for $\boldsymbol{J} : \boldsymbol{D}(\boldsymbol{J}) \subseteq \boldsymbol{Z} \longrightarrow \mathbb{R}^1$ on $\boldsymbol{\Theta} \cap \boldsymbol{D}(\boldsymbol{J})$. In order for the necessary condition to be useful, the set of all $\boldsymbol{\eta}$ for which $\boldsymbol{z}^* \in \boldsymbol{\Theta}$ is an internal point or radical point of $\mathcal{A} \triangleq \boldsymbol{\Theta} \cap \boldsymbol{D}(\boldsymbol{J})$ in the direction $\boldsymbol{\eta}$ needs to be sufficiently "large". Thus, we define two spaces of admissible variations given by

$$V_{\mathcal{A}}^{I} = V_{\mathcal{A}}^{I}(\boldsymbol{z}^*)$$
$$\triangleq \{\boldsymbol{\eta} \in \boldsymbol{Z} : \boldsymbol{z}^* \text{ is an internal point of } \mathcal{A} \text{ in the direction of } \boldsymbol{\eta}\}$$

and

$$V_{\mathcal{A}}^{R} = V_{\mathcal{A}}^{R}(\boldsymbol{z}^*)$$
$$\triangleq \{\boldsymbol{\eta} \in \boldsymbol{Z} : \boldsymbol{z}^* \text{ is an radial point of } \mathcal{A} \text{ in the direction of } \boldsymbol{\eta}\}.$$

The following theorem provides the starting point for much of the theory of general necessary conditions in optimization.

Theorem 7.9 (Fundamental Abstract Necessary Condition) *Let $\boldsymbol{J} : \boldsymbol{D}(\boldsymbol{J}) \subseteq \boldsymbol{Z} \longrightarrow \mathbb{R}^1$ be a real valued function defined on a vector space $\boldsymbol{Z}$ and assume that $\boldsymbol{\Theta} \subseteq \boldsymbol{Z}$ is a given subset of $\boldsymbol{Z}$. Let $\mathcal{A} \triangleq \boldsymbol{\Theta} \cap \boldsymbol{D}(\boldsymbol{J})$ and assume that $\boldsymbol{z}^* \in \mathcal{A}$ satisfies $\boldsymbol{J}(\boldsymbol{z}^*) \leq \boldsymbol{J}(\boldsymbol{z})$ for all $\boldsymbol{z} \in \mathcal{A}$.*

(A) *If $\boldsymbol{\eta} \in V_{\mathcal{A}}^{I}$ and $\delta \boldsymbol{J}(\boldsymbol{z}^*;\boldsymbol{\eta})$ exists, then*

$$\delta \boldsymbol{J}(\boldsymbol{z}^*;\boldsymbol{\eta}) = 0. \tag{7.93}$$

Moreover, if $\delta^2 \boldsymbol{J}(\boldsymbol{z}^*;\boldsymbol{\eta})$ exists, then

$$\delta^2 \boldsymbol{J}(\boldsymbol{z}^*;\boldsymbol{\eta}) \geq 0. \tag{7.94}$$

(B) *If $\boldsymbol{\eta} \in V_{\mathcal{A}}^{R}$ and $\delta \boldsymbol{J}(\boldsymbol{z}^*;\boldsymbol{\eta})$ exists, then*

$$\delta \boldsymbol{J}(\boldsymbol{z}^*;\boldsymbol{\eta}) \geq 0. \tag{7.95}$$

7.7.4 Application to the SPCV

To apply the abstract necessary condition given by **Theorem** 7.9 to the Simplest Problem in the Calculus of Variations, first note that $\boldsymbol{Z} = PWS(t_0, t_1)$ is a vector space. Also, $J : \boldsymbol{Z} \to \mathbb{R}^1$ is defined on the domain $D(J) = \boldsymbol{Z}$ by

$$J(x(\cdot)) = \int_{t_0}^{t_1} f(s, x(s), \dot{x}(s))ds.$$

Since the constraint set is

$$\boldsymbol{\Theta} = \{x(\cdot) \in PWS(t_0, t_1) : x(t_0) = x_0, x(t_1) = x_1\},$$

it follows that $\hat{x}(\cdot) \in \mathcal{A} \triangleq \boldsymbol{\Theta} \cap D(J)$ is an internal point of $\mathcal{A} \triangleq \boldsymbol{\Theta} \cap D(J) = \boldsymbol{\Theta}$ in the direction of $\eta(\cdot)$ if $\eta(\cdot) \in V_0$, where V_0 is the set of "admissible variations" given by

$$V_0 = \{\eta(\cdot) \in PWS(t_0, t_1) : \boldsymbol{\eta}(t_0) = 0, \boldsymbol{\eta}(t_1) = 0\}.$$

Thus, if $\hat{x}(\cdot) \in \mathcal{A} \triangleq \boldsymbol{\Theta} \cap D(J)$, then

$$V_0 \subset V_{\mathcal{A}}^I(\hat{\boldsymbol{x}}(\cdot)) \triangleq \{\eta(\cdot) \in \boldsymbol{Z} : \hat{x}(\cdot) \text{ is an internal point of } \mathcal{A} \text{ in the direction of } \eta(\cdot)\}$$

and the first variation $\delta J(\hat{x}(\cdot); \eta(\cdot))$ exists. However, in general it is sometimes neither obvious nor easy to pick a "good" set of admissible variations.

Note that for the Simplest Problem in the Calculus of Variations, the constraint set

$$\boldsymbol{\Theta} = \{x(\cdot) \in PWS(t_0, t_1) : x(t_0) = x_0, x(t_1) = x_1\},$$

has no core points! For example, given any $\hat{x}(\cdot) \in \boldsymbol{\Theta}$, let $\bar{\eta}(\cdot)$ be the constant function $\bar{\eta}(t) \equiv 1$. Since

$$\hat{x}(t_0) + \varepsilon\bar{\eta}(t_0) = x_0 + \varepsilon \neq x_0$$

for any $\varepsilon \neq 0$, it follows that $[\hat{\boldsymbol{x}}(\cdot) + \varepsilon\bar{\eta}(\cdot)] \notin \boldsymbol{\Theta}$, and hence $\hat{x}(\cdot)$ can not be a core point of $\mathcal{A} \triangleq \boldsymbol{\Theta} \cap D(J)$. Therefore, requiring that $J(\cdot)$ have Gâteaux variation at a minimizer $x^*(\cdot) \in \boldsymbol{\Theta}$ is too strong for even the Simplest Problem of the Calculus of Variations.

7.7.5 Variational Approach to Linear Quadratic Optimal Control

Although the second part of this book is devoted to optimal control problems, some of these problems may be formulated as abstract optimization problems and the variational approach as set out in **Theorem** 7.9 may be applied. To illustrate this we consider a simple linear quadratic control problem governed by the linear system

$$(\mathcal{LS}) \qquad \dot{\boldsymbol{x}}(t) = \boldsymbol{A}\boldsymbol{x}(t), +\boldsymbol{B}\boldsymbol{u}(t), \quad 0 < t \leq 1, \tag{7.96}$$

where $\boldsymbol{A}$ is an $n \times n$ constant matrix and $\boldsymbol{B}$ is an $n \times m$ constant matrix. The initial conditions are given by

$$\boldsymbol{x}(0) = \boldsymbol{x}_0 \in \mathbb{R}^m. \tag{7.97}$$

The function $\boldsymbol{u}(\cdot) \in PWC(0,1;\mathbb{R}^m)$ is called the control and the solution $\boldsymbol{x}(\cdot) = \boldsymbol{x}(\cdot;\boldsymbol{u}(\cdot))$ to the initial value problem (7.96) - (7.97) is called the state. Given a control $\boldsymbol{u}(\cdot) \in PWC(0,1;\mathbb{R}^m)$, the (quadratic) cost functional is defined by

$$J(\boldsymbol{x}(s),\boldsymbol{u}(\cdot)) = \frac{1}{2}\int_0^1 \left\{\|\boldsymbol{x}(s)\|^2 + \|\boldsymbol{u}(s)\|^2\right\} ds, \tag{7.98}$$

where $\boldsymbol{x}(\cdot) = \boldsymbol{x}(\cdot;\boldsymbol{u}(\cdot)) \in PWS(0,1;\mathbb{R}^n)$ is the piecewise smooth solution to the initial value problem (7.96) - (7.97). In particular,

$$\boldsymbol{x}(t;\boldsymbol{u}(\cdot)) = e^{\boldsymbol{A}t}\boldsymbol{x}_0 + \int_0^t e^{\boldsymbol{A}(t-\tau)}\boldsymbol{B}\boldsymbol{u}(\tau)d\tau. \tag{7.99}$$

Therefore,

$$J(\boldsymbol{u}(\cdot)) = \frac{1}{2}\int_0^1 \left\{\|\boldsymbol{x}(s)\|^2 + \|\boldsymbol{u}(s)\|^2\right\}$$

$$= \frac{1}{2}\int_0^1 \left\{\left\| e^{\boldsymbol{A}s}\boldsymbol{x}_0 + \int_0^s e^{\boldsymbol{A}(s-\tau)}\boldsymbol{B}\boldsymbol{u}(\tau)d\tau \right\|^2 + \|\boldsymbol{u}(s)\|^2\right\} ds$$

depends only on $\boldsymbol{u}(\cdot) \in PWC(0,1;\mathbb{R}^m)$.

If one defines $\boldsymbol{Z} = PWC(0,1;\mathbb{R}^m)$ and $\boldsymbol{J} : \boldsymbol{Z} \to \mathbb{R}$ by $\boldsymbol{J}(\boldsymbol{u}(\cdot)) = J(\boldsymbol{u}(\cdot))$ and the set of all admissible controllers by

$$\boldsymbol{\Theta} = \{\boldsymbol{u}(\cdot) \in PWC(0,1;\mathbb{R}^m)\}, \tag{7.100}$$

then the optimal control problem is equivalent to the general problem of minimizing $\boldsymbol{J}(\cdot)$ on the set of all admissible controllers $\boldsymbol{\Theta}$. One can apply the abstract necessary condition given by **Theorem** 7.9 to this problem and obtain the standard necessary conditions. Observe that the domain of $J(\cdot)$, is the entire space $D(J) = PWC(0,1;\mathbb{R}^m)$.

Remark 7.6 *This variational approach is valid because there are no constraints on the control. In particular,* $\boldsymbol{\Theta} = PWC(0,1;\mathbb{R}^m)$ *is a vector space and any vector* $\mathbf{u}^*(\cdot) \in PWC(0,1;\mathbb{R}^m)$ *is a core point of* $\mathcal{A} = \boldsymbol{D}(\boldsymbol{J}) \cap PWC(0,1;\mathbb{R}^m) = PWC(0,1;\mathbb{R}^m)$. *We note that the classical variational approach for optimal control was the first method used in optimal control of both ODE systems and systems governed by partial differential equations. Also, as long as the control is not constrained, as is the case above, this variational method works for nonlinear systems as well (see Lee and Markus [119], page 18 to 22).*

7.7.6 An Abstract Sufficient Condition

The derivation of sufficient conditions is not as easy, nor is it a well developed subject. However, for certain convex problems (which almost never occur in the calculus of variations) there is a simple result. Again we assume that $\boldsymbol{J} : \boldsymbol{D}(\boldsymbol{J}) \subseteq \boldsymbol{Z} \longrightarrow \mathbb{R}^1$ is a real valued function defined on a vector space $\boldsymbol{Z}$ and that $\boldsymbol{\Theta} \subseteq \boldsymbol{Z}$ is a given subset of $\boldsymbol{Z}$.

Definition 7.3 *The set* $\boldsymbol{\Theta}$ *is called* ***convex*** *if*

$$\boldsymbol{z}_1 \quad \text{and} \quad \boldsymbol{z}_2 \in \boldsymbol{\Theta}$$

implies that for all $\lambda \in [0,1]$

$$[\lambda \boldsymbol{z}_1 + (1-\lambda)\boldsymbol{z}_2] \in \boldsymbol{\Theta}.$$

In particular, the line segment between $\boldsymbol{z}_1$ *and* $\boldsymbol{z}_2$ *lies inside* $\boldsymbol{\Theta}$. *Figure 7.8 illustrates a convex set. On the other hand, the set shown in Figure 7.9 is not convex.*

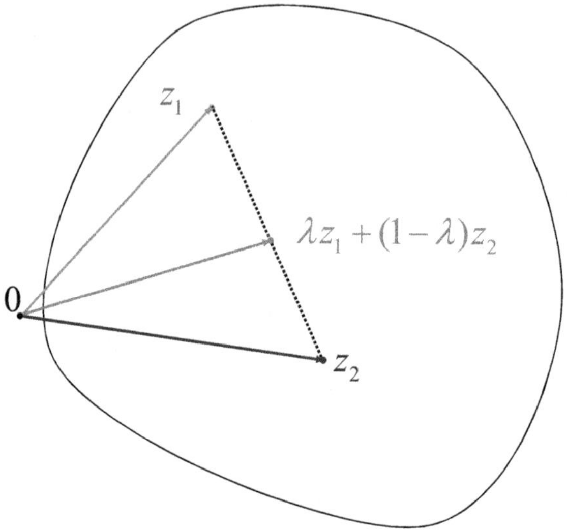

Figure 7.8: A Convex Set

Definition 7.4 *If* $\boldsymbol{\Theta}$ *is convex, then a function* $\boldsymbol{J} : \boldsymbol{D}(\boldsymbol{J}) \subseteq \boldsymbol{Z} \longrightarrow \mathbb{R}^1$ *is said to be a* ***convex function*** *on* $\boldsymbol{\Theta}$ *if* $\boldsymbol{\Theta} \cap \boldsymbol{D}(\boldsymbol{J})$ *is convex and for all* $\boldsymbol{z}_1, \boldsymbol{z}_2 \in \boldsymbol{\Theta} \cap \boldsymbol{D}(\boldsymbol{J})$ *and* $0 \leq \lambda \leq 1$,

$$\boldsymbol{J}(\lambda \boldsymbol{z}_1 + (1-\lambda)\,\boldsymbol{z}_2) \leq \lambda \boldsymbol{J}(\boldsymbol{z}_1) + (1-\lambda)\,\boldsymbol{J}(\boldsymbol{z}_2).$$

The function $\boldsymbol{J}(\cdot)$ *is said to be a* ***strictly convex function*** *on* $\boldsymbol{\Theta}$ *if for all* $\boldsymbol{z}_1, \boldsymbol{z}_2 \in \boldsymbol{\Theta} \cap \boldsymbol{D}(\boldsymbol{J})$ *and* $0 < \lambda < 1$,

$$\boldsymbol{J}(\lambda \boldsymbol{z}_1 + (1-\lambda)\,\boldsymbol{z}_2) < \lambda \boldsymbol{J}(\boldsymbol{z}_1) + (1-\lambda)\,\boldsymbol{J}(\boldsymbol{z}_2).$$

The following theorem applies to general abstract convex problems. Also, the proof is a rather simple extension of the ideas found in Section 2.2.2.

Theorem 7.10 (Abstract Sufficient Condition for Convex Problems) *Let*

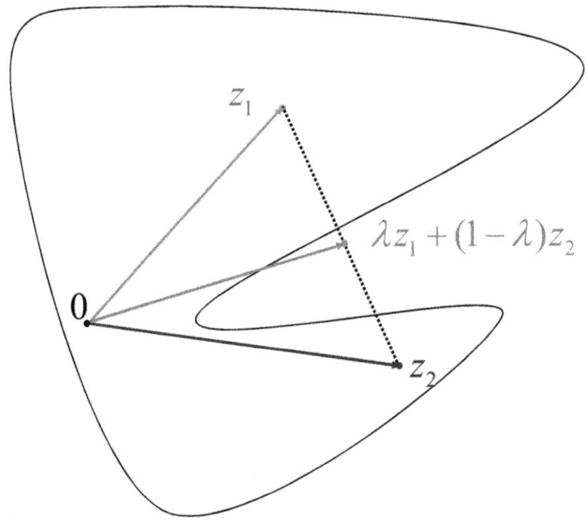

Figure 7.9: A Non-convex Set

$\boldsymbol{\Theta}$ be convex and $\boldsymbol{J}(\cdot)$ be a convex function on $\boldsymbol{\Theta}$. If $\boldsymbol{z}_0 \in \boldsymbol{\Theta} \cap \boldsymbol{D}(\boldsymbol{J})$, $\delta \boldsymbol{J}(\boldsymbol{z}_0; \boldsymbol{\eta})$ exists and $\delta \boldsymbol{J}(\boldsymbol{z}_0; \boldsymbol{\eta}) \geq 0$ for all $\boldsymbol{\eta}$ such that $(\boldsymbol{z}_0 + \boldsymbol{\eta}) \in \boldsymbol{\Theta} \cap \boldsymbol{D}(\boldsymbol{J})$, then $\boldsymbol{J}(\cdot)$ has a minimum on $\boldsymbol{\Theta}$ at $\boldsymbol{z}_0$. Moreover, if $\boldsymbol{J}(\cdot)$ is strictly convex, $\boldsymbol{z}_0$ is unique.

Proof: If $\boldsymbol{z} \in \boldsymbol{\Theta}$ and $0 < \lambda \leq 1$, then $\lambda \boldsymbol{z} + (1-\lambda)\boldsymbol{z}_0 \in \boldsymbol{\Theta}$ and

$$\boldsymbol{J}([\lambda \boldsymbol{z} + (1-\lambda)\boldsymbol{z}_0]) \leq \lambda \boldsymbol{J}(\boldsymbol{z}) + (1-\lambda)\boldsymbol{J}(\boldsymbol{z}_0),$$

or equivalently,

$$\boldsymbol{J}([\lambda \boldsymbol{z} + (1-\lambda)\boldsymbol{z}_0]) - \boldsymbol{J}(\boldsymbol{z}_0) \leq \lambda[\boldsymbol{J}(\boldsymbol{z}) - \boldsymbol{J}(\boldsymbol{z}_0)].$$

Hence,

$$\frac{\boldsymbol{J}([\lambda \boldsymbol{z} + (1-\lambda)\boldsymbol{z}_0]) - \boldsymbol{J}(\boldsymbol{z}_0)}{\lambda} \leq [\boldsymbol{J}(\boldsymbol{z}) - \boldsymbol{J}(\boldsymbol{z}_0)],$$

and it follows that

$$\begin{aligned}\delta \boldsymbol{J}(\boldsymbol{z}_0; [\boldsymbol{z} - \boldsymbol{z}_0]) &= \lim_{\lambda \to 0}[\frac{\boldsymbol{J}([\lambda \boldsymbol{z} + (1-\lambda)\boldsymbol{z}_0]) - \boldsymbol{J}(\boldsymbol{z}_0)}{\lambda} \\ &\leq [\boldsymbol{J}(\boldsymbol{z}) - \boldsymbol{J}(\boldsymbol{z}_0)].\end{aligned}$$

However, $\boldsymbol{z}_0 + [\boldsymbol{z} - z_0] = \boldsymbol{z} \in \boldsymbol{\Theta}$, so that the assumption that

$$0 \leq \delta \boldsymbol{J}(\boldsymbol{z}_0; [\boldsymbol{z} - \boldsymbol{z}^*])$$

implies that

$$0 \leq \delta \boldsymbol{J}(\boldsymbol{z}_0; [\boldsymbol{z} - \boldsymbol{z}_0]) \leq [\boldsymbol{J}(\boldsymbol{z}) - \boldsymbol{J}(\boldsymbol{z}_0)],$$

Consequently, it follows that

$$\boldsymbol{J}(\boldsymbol{z}_0) \leq \boldsymbol{J}(\boldsymbol{z}) \tag{7.101}$$

and since (7.101) holds for any $\boldsymbol{z} \in \boldsymbol{\Theta}$, $\boldsymbol{z}_0$ minimizes $\boldsymbol{J}(\cdot)$ on $\boldsymbol{\Theta}$. The second part of the theorem is trivial. □

Remark 7.7 *Although control of PDE systems lies outside of the scope of this book, we note that the abstract variational approach represented by* **Theorem** *7.9 above was the basic method used in this field. As noted in the* 1971 *review paper by A. C. Robinson [156], most of the early work on necessary conditions for PDE control systems was in fact done by variational methods (see page 374 and the cited references). The idea of using abstract variational calculus to derive the necessary conditions for PDE systems was the main theme of the 1960's and 1970's literature. J. L. Lions certainly used this approach in his fundamental book [124]. This abstract approach was popular in the early Russian school as illustrated in [144] and [151].*

7.8 Problem Set for Chapter 7

Apply the Euler Necessary Condition to the following free endpoint problems.

Problem 7.1 *Minimize the functional*

$$J(x(\cdot)) = \int_0^1 \frac{\sqrt{c^2(1 + [\dot{x}(s)]^2) - [v(s)]^2} - v(s)\dot{x}(s)}{c^2 - [v(s)]^2} ds$$

among all piecewise smooth functions satisfying

$$x(0) = 0, \text{ and } x(1) \text{ is free.}$$

Problem 7.2 *Minimize the functional*

$$J(x(\cdot)) = \int_0^1 \{[\dot{x}(s)]^2 + [x(s)]^2 + 2e^s x(s)\}ds,$$

subject to the endpoint conditions $x(0) = 0$ *and* $x(1)$ *is free.*

Problem 7.3 *Minimize the functional*

$$J(x(\cdot)) = \int_0^\pi \{[\dot{x}(s)]^2 - [x(s)]^2\}ds,$$

subject to the endpoint conditions $x(0) = 0$ *and* $x(\pi)$ *is free.*

Problem 7.4 *Minimize the functional*

$$J(x(\cdot)) = \int_0^{3\pi/2} \{[\dot{x}(s)]^2 - [x(s)]^2\}ds,$$

subject to the endpoint conditions $x(0) = 0$ *and* $x(3\pi/2)$ *is free.*

Problem 7.5 *Minimize the functional*

$$J(x(\cdot)) = \int_0^b x(s)\sqrt{1 + [\dot{x}(s)]^2}ds,$$

subject to the endpoint conditions $x(0) = 1$ *and* $x(b)$ *is free.*

Problem 7.6 *Minimize the functional*

$$J(x(\cdot)) = \int_1^2 \{[\dot{x}(s)]^2 - 2sx(s)\}ds,$$

subject to the endpoint conditions $x(1) = 0$ *and* $x(2)$ *is free.*

Problem 7.7 *Minimize the functional*

$$J(x(\cdot)) = \int_0^{\pi} \{[x(s)]^2(1 - [\dot{x}(s)]^2)\}ds,$$

subject to the endpoint conditions $x(0) = 0$ *and* $x(\pi)$ *is free.*

Problem 7.8 *Minimize the functional*

$$J(x(\cdot)) = \int_1^{3} \{[3s - x(s)]x(s)\}ds,$$

subject to the endpoint conditions $x(1) = 1$ *and* $x(3)$ *is free.*

Problem 7.9 *Minimize the functional*

$$J(x(\cdot)) = \int_1^{2} \{\dot{x}(s)[1 + s^2\dot{x}(s)]\}ds,$$

subject to the endpoint conditions $x(1)$ *is free and* $x(2) = 5$.

Problem 7.10 *Minimize the functional*

$$J(x(\cdot)) = \int_0^{1} \{\frac{1}{2}[\dot{x}(s)]^2 + x(s)\dot{x}(s) + \dot{x}(s) + x(s)\}ds,$$

subject to the endpoint conditions $x(0)$ *is free and* $x(1)$ *is free.*

Problem 7.11 *Consider the point to curve problem of minimizing the functional*

$$J(x(\cdot)) = \int_0^{t_1} \left\{ \frac{\sqrt{1 + [\dot{x}(s)]^2}}{x(s)} \right\} ds,$$

subject to the endpoint condition $x(0) = 0$ *and* $x(t_1) = \varphi(t_1)$, *where* $.\varphi(t) = t - 5$. *Use the necessary condition to find all possible minimizers.*

Problem 7.12 *Minimize the functional*

$$J(x(\cdot)) = \int_0^1 [\ddot{x}(s)]^2 ds,$$

subject to the endpoint conditions $x(0) = 1$, $\dot{x}(0) = 0$, $x(1) = 2$ *and* $\dot{x}(1) = 0$.

Problem 7.13 *Minimize the functional*

$$J(x(\cdot)) = \int_0^1 \{1 + [\ddot{x}(s)]^2\} ds,$$

subject to the endpoint conditions $x(0) = 0$, $\dot{x}(0) = 1$, $x(1) = 1$ *and* $\dot{x}(1) = 1$.

Problem 7.14 *Minimize the functional* $J(x(\cdot)) = \int_{-1}^{1} [\ddot{x}(s)]^2 ds$, *subject to the endpoint conditions* $x(-1) = 0$, $x(0) = 1$, *and* $x(1) = 0$. *Hint: Note that this requires an extension of the necessary condition for higher order problems.*

Problem 7.15 *Minimize the functional*

$$J(x_1(\cdot), x_2(\cdot)) = \int_0^{\pi/2} \left\{[x_1(s)]^2 + 2x_1(s)x_2(s) + [x_2(s)]^2\right\} ds,$$

subject to the endpoint conditions

$$x_1(0) = 0,\ x_1(\pi/2) = 1,\ x_2(0) = 0,\ x_2(\pi/2) = 1.$$

Advanced Problems

Problem 7.16 *Prove the Euler Necessary Condition for the Free Endpoint Problem 7.1 assuming that* $x^*(\cdot)$ *provides only a weak local minimum.*

Problem 7.17 *Consider the Euler Necessary Condition for the point to curve problem as stated in* **Theorem** *7.4. Modify the proof to allow for piecewise smooth functions* $x(\cdot)$ *and* $\varphi(\cdot)$.

Problem 7.18 *Prove the High Order Form of the FLCV as stated in* **Lemma** *7.4. Hint: See Problem Set III.2 in Reid's book [153].*

Problem 7.19 *Prove* **Theorem** *7.7 assuming that* $x^*(\cdot)$ *is piecewise smooth and provides only a weak local minimum.*

Problem 7.20 *Assume one has a partition* $0 = \hat{t}_0 < \hat{t}_1 < \cdots < \hat{t}_{p-1} < \hat{t}_p = 1$ *of* $[0,1]$, *and fixed points* $\hat{x}_0$, $\hat{x}_1$, $\hat{x}_2$, $\hat{x}_{p-1}$. *Let* $J(\cdot)$ *be the functional*

$$J(x(\cdot)) = \int_0^1 [\ddot{x}(s)]^2 ds$$

and consider the problem of finding $x^*(\cdot)$ *to minimize* $J(\cdot)$ *on the set*

$$\Theta_I = \left\{x(\cdot) \in PWS^2(0,1) : x(\hat{t}_i) = \hat{x}_i, \;\; \textit{for all } i = 0,1,\ldots,p\right\}.$$

Derive the Euler Necessary Condition for this problem. Hint: Look into the theory of splines and check the references [3], [159] and [160].

Chapter 8

Applications

In this chapter we apply the theory from the previous chapters to some specific problems. We begin with the brachistochrone problem.

8.1 Solution of the Brachistochrone Problem

Recall that the brachistochrone problem leads to the minimization of

$$J(x(\cdot)) = \int_0^a \sqrt{\frac{1 + [\dot{x}(s)]^2}{2gx(s)}} ds,$$

subject to

$$x(0) = 0, \ x(a) = b.$$

Here we assume that the minimizer has no corners so that $x^*(\cdot)$ is C^2. In order to solve this problem we note that the integrand does not depend explicitly on t. In particular, the functional has the form

$$J(x(\cdot)) = \int_0^a f(x(s), \dot{x}(s)) ds,$$

where

$$f(x, u) = \sqrt{\frac{1 + u^2}{2gx}} = \frac{1}{\sqrt{2gx}} \sqrt{1 + u^2},$$

and

$$f_u(x,u) = f(x,u)\frac{u}{1+u^2}.$$

The Euler Differential Equation becomes

$$\frac{d}{dt}f_u(x^*(t),\dot{x}^*(t)) = f_x(x^*(t),\dot{x}^*(t))$$

or in second order form (Hilbert's Equation)

$$f_{ux}(x^*(t),\dot{x}^*(t))\dot{x}^*(t) + f_{uu}(x^*(t),\dot{x}^*(t))\ddot{x}^*(t) = f_x(x^*(t),\dot{x}^*(t)). \tag{8.1}$$

Again, recall that we shall use $f^*(t)$, $f_u^*(t)$ and $f_x^*(t)$, etc. to denote the functions evaluated at $(x^*(t),\dot{x}^*(t))$. Multiply both sides of (8.1) by $\dot{x}^*(t)$ to obtain

$$f_{ux}^*(t)[\dot{x}^*(t)]^2 + f_{uu}^*(t)\ddot{x}^*(t)\dot{x}^*(t) - f_x^*(t)\dot{x}^*(t) = 0. \tag{8.2}$$

However,

$$\begin{aligned}\frac{d}{dt}[f_u^*(t)\dot{x}^*(t) - f^*(t)] &= f_u^*(t)\ddot{x}^*(t)\\ &\quad + [f_{ux}^*(t)\dot{x}^*(t) + f_{uu}^*(t)\ddot{x}^*(t)]\dot{x}^*(t)\\ &\quad - [f_x^*(t)\dot{x}^*(t) + f_u^*(t)\ddot{x}^*(t)]\\ &= f_u^*(t)\ddot{x}^*(t) + f_{ux}^*(t)[\dot{x}^*(t)]^2\\ &\quad + f_{uu}^*(t)\ddot{x}^*(t)\dot{x}^*(t)\\ &\quad - f_x^*(t)\dot{x}^*(t) - f_u^*(t)\ddot{x}^*(t)\\ &= f_{ux}^*(t)[\dot{x}^*(t)]^2\\ &\quad + f_{uu}^*(t)\ddot{x}^*(t)\dot{x}^*(t)\\ &\quad - f_x^*(t)\dot{x}^*(t)\\ &= 0.\end{aligned}$$

Hence, we have that

$$\frac{d}{dt}[f_u(x^*(t),\dot{x}^*(t))\dot{x}^*(t) - f(x^*(t),\dot{x}^*(t))] = 0,$$

or equivalently,

$$[f_u(x^*(t),\dot{x}^*(t))\dot{x}^*(t) - f(x^*(t),\dot{x}^*(t))] = c. \tag{8.3}$$

Thus, equation (8.3) is equivalent to the Euler Integral Equation for the brachistochrone problem and hence the Euler Integral Equation has the equivalent form

$$\frac{[\dot{x}^*(t)]^2}{\sqrt{2gx^*(t)(1+[\dot{x}^*(t)]^2)}} - \sqrt{\frac{1+[\dot{x}^*(t)]^2}{2gx^*(t)}} = c.$$

Simplifying, we have that

$$\frac{[\dot{x}^*(t)]^2}{\sqrt{x^*(t)(1+[\dot{x}^*(t)]^2)}} - \frac{1+[\dot{x}^*(t)]^2}{\sqrt{x^*(t)(1+[\dot{x}^*(t)]^2)}} = \sqrt{2gc^2},$$

or equivalently,

$$\frac{-1}{\sqrt{x^*(t)(1+[\dot{x}^*(t)]^2)}} = \sqrt{2gc^2}.$$

Solving for $[\dot{x}^*(t)]^2$ we obtain the differential equation

$$[\dot{x}^*(t)]^2 = \frac{a+x^*(t)}{x^*(t)}, \quad a = \frac{1}{2gc^2}.$$

Therefore, since the slope of $x^*(t)$ is non-positive we obtain the differential equation

$$\dot{x}^*(t) = -\sqrt{\frac{a+x^*(t)}{x^*(t)}}.$$

Solving this differential equation is not an easy task. In fact we shall make the (not so obvious) substitution

$$x^*(t) = q\,[\sin(\theta(t)/2)]^2$$

and observe that by the chain rule

$$\dot{x}^*(t) = -2q\sin(\theta(t)/2)\cos(\theta(t)/2)\frac{\dot{\theta}(t)}{2}.$$

Thus,

$$\sqrt{\frac{q+x^*(t)}{x^*(t)}} = \sqrt{\frac{q\,[1-[\sin(\theta(t)/2)]^2]}{q[\sin(\theta(t)/2)]^2}} = \sqrt{\frac{[\cos(\theta(t)/2)]^2}{[\sin(\theta(t)/2)]^2}}$$
$$= \frac{[\cos(\theta(t)/2)]}{[\sin(\theta(t)/2)]}$$

and the differential equation for $\theta(t)$ becomes

$$2q\sin(\theta(t)/2)\cos(\theta(t)/2)\frac{\dot{\theta}(t)}{2}=\frac{[\cos(\theta(t)/2)]}{[\sin(\theta(t)/2)]}.$$

Multiplying by $\sin(\theta(t)/2)$ and dividing out $\cos(\theta(t)/2)$ yields

$$q[\sin(\theta(t)/2)]^2\dot{\theta}(t)=1.$$

However, the identity

$$2[\sin(\theta(t)/2)]^2=1-\cos\theta(t)$$

implies that

$$\frac{q}{2}(1-\cos(\theta(t))\,\dot{\theta}(t)=1,$$

or equivalently,

$$\frac{q}{2}\dot{\theta}(t)-\frac{q}{2}\cos\theta(t)\,\dot{\theta}(t)=1.$$

We write this as

$$\frac{q}{2}\frac{d\theta}{dt}-\frac{q}{2}\cos\theta\frac{d\theta}{dt}=1$$

so that

$$\frac{q}{2}d\theta-\frac{q}{2}\cos\theta d\theta=dt$$

and integrating, we obtain

$$\frac{q}{2}\theta-\frac{q}{2}\sin\theta=t+t_0$$

and since $t_0=0$ we have

$$\frac{q}{2}\theta-\frac{q}{2}\sin\theta=t.$$

Thus, we have a parametric description of the solution given by

$$t=\frac{q}{2}\theta-\sin\theta$$

and

$$x=q\,[\sin(\theta/2)]^2=\frac{q}{2}(1-\cos\theta).$$

The curve

$$t = \frac{q}{2}(\theta - \sin\theta),$$
$$x = \frac{q}{2}(1 - \cos\theta),$$

is a cycloid and the graph of $-x^*(\cdot)$ lies on this curve (recall that $x^*(\cdot)$ is the distance from the curve to $y = 0$ axis) and hence the optimal curve is part of a cycloid.

8.2 Classical Mechanics and Hamilton's Principle

What we call classical mechanics today was initiated in the late 1600's by Newton and others. Newton changed our understanding of the universe by enumerating his three "Laws of Motion":

> [First Law] Every object in a state of uniform motion tends to remain in that state of motion unless an external force is applied to it.
>
> [Second Law] The rate of change of momentum ($p =$ mass·velocity) is proportional to the impressed force and is in the direction in which the force acts:
>
> $$\vec{F}(t) = \frac{d\vec{p}(t)}{dt} = \frac{d[m(t)\vec{v}(t)]}{dt}.$$
>
> [Third Law] For every action there is an equal and opposite reaction.

Although the second law is often stated as

$$\vec{F}(t) = m\vec{a}(t), \tag{8.4}$$

which is valid under the assumption that the mass is constant, it is important to note the correct version. Also, around 300 BC Aristotle had stated a "Second Law of Motion" which, for constant mass, becomes

$$\vec{F}(t) = m\vec{v}(t). \tag{8.5}$$

Aristotle's Second Law seems to be more in accord with common sense and even today we hear reports about the "force of impact" when a 3000 pound car traveling at 50 miles per hour hits a brick wall. The point is that "Laws of Motion" are not "laws", but are mathematical models of observed behavior and often evolve and change over time as our understanding improves. In short, these are assumptions that lead to models that must be validated through experiments.

In 1744 Euler showed Newton's Second Law (8.4) could be obtained by another assumption called the Principle of Least Action and in 1788 Lagrange showed that, in the case of conservative forces, a major part of Newtonian mechanics could be derived from this principle. As we shall see below, the Principle of Least Action is not true for a general time interval (this was pointed out by Jacobi in 1842), but in 1834 and 1835 William Hamilton provided an alternative principle that extended the application to more general forces and set the stage for what we now call Hamiltonian mechanics. Suppose a particle of fixed constant mass m is located at the position $\boldsymbol{x}(t) = \begin{bmatrix} x_1(t) & x_2(t) & x_3(t) \end{bmatrix}^T$ at time t and moves under a force $\boldsymbol{F}(t) = \begin{bmatrix} f_1(t) & f_2(t) & f_3(t) \end{bmatrix}^T$. By Newton's Second Law, the force $\boldsymbol{F}(t)$ will cause the particle to move along a path in $\mathbb{R}^3$ such that

$$m\ddot{\boldsymbol{x}}(t) = \boldsymbol{F}(t),$$

i.e.

$$m\ddot{x}_i(t) = f_i(t), \quad i = 1, 2, 3. \tag{8.6}$$

The particle is said to be in a *conservative field* if there is a function $U = U(t, x_1, x_2, x_3)$ such that

$$\boldsymbol{F}(t) = -\nabla_{\boldsymbol{x}} U(t, x_1(t), x_2(t), x_3(t)), \tag{8.7}$$

or equivalently,

$$f_i(t) = -\frac{\partial}{\partial x_i} U(t, x_1(t), x_2(t), x_3(t)), \quad i = 1, 2, 3.$$

The function $U : \mathbb{R} \times \mathbb{R}^3 \to \mathbb{R}$ is called the *potential* and

$$P(t) = U(t, x_1(t), x_2(t), x_3(t)) = U(t, \boldsymbol{x}(t)) \tag{8.8}$$

is called the *potential energy* of the particle. The *kinetic energy* of the particle is defined to be

$$K(t) \triangleq \frac{1}{2}m\sum_{i=1}^{3}[\dot{x}_i(t)]^2 = \frac{1}{2}m\,\|\dot{\boldsymbol{x}}(t)\|^2. \tag{8.9}$$

The *total energy* of the particle is given by

$$H(t) = K(t) + P(t). \tag{8.10}$$

The Lagrangian $L(t, \boldsymbol{x}, \boldsymbol{u}) : \mathbb{R} \times \mathbb{R}^3 \times \mathbb{R}^3 \to \mathbb{R}$ is defined by

$$L(t, \boldsymbol{x}, \boldsymbol{u}) = \frac{1}{2}m\,\|\boldsymbol{u}\|^2 - U(t, \boldsymbol{x}) \tag{8.11}$$

so that

$$L(t) = L(t, \boldsymbol{x}(t), \dot{\boldsymbol{x}}(t)) \triangleq K(t) - P(t) \tag{8.12}$$

is the difference between the kinetic and potential energy of the particle. Finally, given two times $t_0 < t_1$, the *action of the particle* is defined to be the integral

$$\mathbb{A}(\boldsymbol{x}(\cdot)) \triangleq \int_{t_0}^{t_1} [K(s) - P(s)]\,ds = \int_{t_0}^{t_1} [L(s, \boldsymbol{x}(s), \dot{\boldsymbol{x}}(s))]\,ds \tag{8.13}$$

and given the special form of the force and kinetic energy, it follows that

$$\mathbb{A}(\boldsymbol{x}(\cdot)) = \int_{t_0}^{t_1} \left[\frac{1}{2}m\sum_{i=1}^{3}[\dot{x}_i(s)]^2 - U(s, \boldsymbol{x}(s))\right] ds.$$

The *Principle of Least Action* states that the particle will move between $\boldsymbol{x}(t_0)$ and $\boldsymbol{x}(t_1)$ along a path $\boldsymbol{x}^*(\cdot)$ that minimizes the action integral. In particular,

$$\mathbb{A}(\boldsymbol{x}^*(\cdot)) \le \mathbb{A}(\boldsymbol{x}(\cdot))$$

for all curves joining $\boldsymbol{x}^*(t_0)$ and $\boldsymbol{x}^*(t_1)$. Note that this is a 3 dimensional version of the Simplest Problem of the Calculus of

Variations and hence if $\boldsymbol{x}^*(\cdot)$ minimizes the action integral (8.13) then $\boldsymbol{x}^*(\cdot)$ must satisfy

$$\delta\mathbb{A}(\boldsymbol{x}^*(\cdot);\boldsymbol{\eta}(\cdot)) = 0 \tag{8.14}$$

for all $\boldsymbol{\eta}(\cdot) \in PWS(t_0,t_1;\mathbb{R}^3)$ satisfying $\boldsymbol{\eta}(t_0) = \boldsymbol{\eta}(t_1) = \mathbf{0}$. Consequently, by Euler's Necessary Condition $\boldsymbol{x}^*(\cdot)$ satisfies Euler's Differential Equation

$$\frac{d}{dt}\nabla_{\boldsymbol{u}}L(t,\boldsymbol{x}^*(t),\dot{\boldsymbol{x}}^*(t)) = \nabla_{\boldsymbol{x}}L(t,\boldsymbol{x}^*(t),\dot{\boldsymbol{x}}^*(t)) \tag{8.15}$$

for $t_0 < t < t_1$. Using the definition of

$$\begin{aligned} L(t,\boldsymbol{x},\boldsymbol{u}) &= \frac{1}{2}m\left\|\boldsymbol{u}\right\|^2 - U(s,\boldsymbol{x}) \\ &= \frac{1}{2}m\sum_{i=1}^{3}[u_i]^2 - U(s,\boldsymbol{x}), \end{aligned}$$

it follows that

$$\nabla_{\boldsymbol{u}}L(t,\boldsymbol{x},\boldsymbol{u}) = \begin{bmatrix} mu_1 \\ mu_2 \\ mu_3 \end{bmatrix}$$

and

$$\nabla_{\boldsymbol{x}}L(t,\boldsymbol{x},\boldsymbol{u}) = -\nabla_{\boldsymbol{x}}U(t,\boldsymbol{x}).$$

Hence, along the motion of the particle we have

$$\nabla_{\boldsymbol{u}}L(t,\boldsymbol{x}^*(t),\dot{\boldsymbol{x}}^*(t)) = \begin{bmatrix} m\dot{x}_1(t) \\ m\dot{x}_2(t) \\ m\dot{x}_3(t) \end{bmatrix} = m\begin{bmatrix} \dot{x}_1(t) \\ \dot{x}_2(t) \\ \dot{x}_3(t) \end{bmatrix}$$

and

$$\nabla_{\boldsymbol{x}}L(t,\boldsymbol{x}^*(t),\dot{\boldsymbol{x}}^*(t)) = -\nabla_{\boldsymbol{x}}U(t,\boldsymbol{x}^*(t)) = \begin{bmatrix} f_1(t) \\ f_2(t) \\ f_3(t) \end{bmatrix}$$

which becomes

$$m\frac{d}{dt}\begin{bmatrix} \dot{x}_1(t) \\ \dot{x}_2(t) \\ \dot{x}_3(t) \end{bmatrix} = \begin{bmatrix} f_1(t) \\ f_2(t) \\ f_3(t) \end{bmatrix},$$

or equivalently,

$$m\ddot{\boldsymbol{x}}(t) = \boldsymbol{F}(t)$$

which is Newton's Second Law.

Example 8.1 *Consider the one dimensional version where a particle of mass $m = 1$ moves under the force defined by the potential $U(x) = \frac{\kappa}{2}x^2$ so that $f(t) = -\kappa x(t)$. The kinetic energy is given by $K(t) = \frac{1}{2}(\dot{x}(t))^2$ and the Lagrangian is defined by*

$$L(x, u) = \frac{1}{2}[u^2 - \kappa x^2].$$

Here the action integral becomes

$$\mathbb{A}(\boldsymbol{x}(\cdot)) = \frac{1}{2}\int_{t_0}^{t_1} [(\dot{x}(s))^2 - \kappa(x(s))^2]ds \tag{8.16}$$

and Euler's equation is

$$\frac{d}{dt}[\dot{x}(t)] = -\kappa x(t).$$

Therefore, the particle will move along a path $x^*(\cdot)$ defined by the differential equation

$$\ddot{x}^*(t) + \kappa x^*(t) = 0.$$

This is the same equation that comes directly from Newton's Second Law. However, consider the case where $\kappa = 1$ and the particle starts at $x_0 = 0$ at time $t_0 = 0$ and stops at $x_1 = 0$ at time $t_1 = 2\pi$. We know from the Jacobi Necessary Condition - (IV) that the action integral (8.16) does not have a minimum on this interval. Thus, the Principle of Least Action does not always hold.

Hamilton stated a new principle which is known as *Hamilton's Principle of Stationary Action.* In particular, Hamilton's principle states that the particle will move between $\boldsymbol{x}(t_0)$ and $\boldsymbol{x}(t_1)$ along a path $\boldsymbol{x}^*(\cdot)$ that makes the action integral *stationary.* In particular, $\mathbb{A}(\boldsymbol{x}^*(\cdot))$ has first variation zero for all $\boldsymbol{\eta}(\cdot) \in PWS(t_0, t_1; \mathbb{R}^3)$

satisfying $\boldsymbol{\eta}(t_0) = \boldsymbol{\eta}(t_1) = \mathbf{0}$. Thus, Hamilton's Principle of Stationary Action is equivalent to the condition that

$$\delta\mathbb{A}(\boldsymbol{x}^*(\cdot);\boldsymbol{\eta}(\cdot)) = 0 \tag{8.17}$$

all $\boldsymbol{\eta}(\cdot) \in PWS(t_0, t_1; \mathbb{R}^3)$ satisfying $\boldsymbol{\eta}(t_0) = \boldsymbol{\eta}(t_1) = \mathbf{0}$. Observe that Newton's Second Law follows from (8.17) since the Fundamental Lemma of the Calculus of Variations implies that the Euler Differential Equation holds. Clearly, the Euler Differential Equation in this case reduces to

$$m\ddot{\boldsymbol{x}}(t) = \boldsymbol{F}(t),$$

which is Newton's Second Law.

8.2.1 Conservation of Energy

Here we again assume that a particle of mass m is moving in a conservative field with potential energy given by

$$P(t) = U(t, x_1(t), x_2(t), x_3(t)) = U(t, \boldsymbol{x}(t)) \tag{8.18}$$

and kinetic energy given by

$$K(t) \triangleq \frac{1}{2}m\sum_{i=1}^{3}[\dot{x}_i(t)]^2. \tag{8.19}$$

Also, with $L(t, \boldsymbol{x}, \boldsymbol{u}) = \frac{1}{2}m\,\|\boldsymbol{u}\|^2 - U(t, \boldsymbol{x}) = \frac{1}{2}m\sum_{i=1}^{3}[u_i]^2 - U(t, \boldsymbol{x})$, let $p_i(t)$ be defined by

$$p_i(t) \triangleq m\dot{x}_i(t) = \frac{\partial}{\partial u_i}L(t, \boldsymbol{x}(t), \dot{\boldsymbol{x}}(t)) \tag{8.20}$$

and observe that the total energy $H(t) = K(t) + P(t)$ can be written as

$$\begin{aligned} H(t) &= \frac{1}{2}m\sum_{i=1}^{3}[\dot{x}_i(t)]^2 + U(t, \boldsymbol{x}(t)) \\ &= -\frac{1}{2}m\sum_{i=1}^{3}[\dot{x}_i(t)]^2 + m\sum_{i=1}^{3}[\dot{x}_i(t)]^2 + U(t, \boldsymbol{x}(t)) \end{aligned}$$

$$= -\frac{1}{2}m\sum_{i=1}^{3}[\dot{x}_i(t)]^2 + U(t,\boldsymbol{x}(t)) + \sum_{i=1}^{3}\dot{x}_i(t)[m\dot{x}_i(t)]$$

$$= -[K(t) - U(t,\boldsymbol{x}(t))] + \sum_{i=1}^{3}\dot{x}_i(t)[\frac{\partial}{\partial u_i}L(t,\boldsymbol{x}(t),\dot{\boldsymbol{x}}(t))]$$

$$= -L(t,\boldsymbol{x}(t),\dot{\boldsymbol{x}}(t)) + \sum_{i=1}^{3}\dot{x}_i(t)p_i(t).$$

If we define the *Hamiltonian* function by

$$\mathcal{H}(t,\boldsymbol{x},\boldsymbol{u},\boldsymbol{p}) \triangleq -L(t,\boldsymbol{x},\boldsymbol{u}) + \sum_{i=1}^{3}u_i p_i, \tag{8.21}$$

then

$$\mathcal{H}(t,\boldsymbol{x}(t),\dot{\boldsymbol{x}}(t),\boldsymbol{p}(t)) = -L(t,\boldsymbol{x}(t),\dot{\boldsymbol{x}}(t)) + \sum_{i=1}^{3}\dot{x}_i(t)p_i(t) \tag{8.22}$$

is the total energy of the system.

Now consider the case where the potential is independent of time so that

$$P(t) = U(x_1(t),x_2(t),x_3(t)) = U(\boldsymbol{x}(t)) \tag{8.23}$$

and since $L(\boldsymbol{x},\boldsymbol{u}) = \frac{1}{2}m\|\boldsymbol{u}\|^2 - U(\boldsymbol{x}) = \frac{1}{2}m\sum_{i=1}^{3}[u_i]^2 - U(\boldsymbol{x})$, the Euler Differential Equation

$$\frac{d}{dt}\nabla_{\boldsymbol{u}}L(\boldsymbol{x}^*(t),\dot{\boldsymbol{x}}^*(t)) = \nabla_{\boldsymbol{x}}L(\boldsymbol{x}^*(t),\dot{\boldsymbol{x}}^*(t))$$

has the form

$$\frac{d}{dt}[m\dot{x}_i(t)] = L_{x_i}(\boldsymbol{x}^*(t),\dot{\boldsymbol{x}}^*(t)), \quad i=1,2,3.$$

Note that for $i=1,2,3,$ we have

$$\frac{\partial}{\partial x_i}\mathcal{H}(\boldsymbol{x},\boldsymbol{u},\boldsymbol{p}) = -L_{x_i}(\boldsymbol{x},\boldsymbol{u})$$

and

$$\frac{\partial}{\partial p_i}\mathcal{H}(\boldsymbol{x}, \boldsymbol{u}, \boldsymbol{p}) = u_i.$$

Therefore,

$$\dot{x}_i(t) = \frac{\partial}{\partial p_i}\mathcal{H}(\boldsymbol{x}(t), \dot{\boldsymbol{x}}(t), \boldsymbol{p}(t)) \tag{8.24}$$

and

$$\frac{d}{dt}[m\dot{x}_i(t)] = L_{x_i}(\boldsymbol{x}^*(t), \dot{\boldsymbol{x}}^*(t)) = -\frac{\partial}{\partial x_i}\mathcal{H}(\boldsymbol{x}(t), \dot{\boldsymbol{x}}(t), \boldsymbol{p}(t)),$$

so that

$$\dot{p}_i(t) = -\frac{\partial}{\partial x_i}\mathcal{H}(\boldsymbol{x}(t), \dot{\boldsymbol{x}}(t), \boldsymbol{p}(t)). \tag{8.25}$$

Combining (8.24) and (8.25) it follows that Euler's Equation can be written as the system

$$\dot{x}_i(t) = \frac{\partial}{\partial p_i}\mathcal{H}(\boldsymbol{x}(t), \dot{\boldsymbol{x}}(t), \boldsymbol{p}(t)) \tag{8.26}$$

$$\dot{p}_i(t) = -\frac{\partial}{\partial x_i}\mathcal{H}(\boldsymbol{x}(t), \dot{\boldsymbol{x}}(t), \boldsymbol{p}(t)). \tag{8.27}$$

Now we differentiate $\mathcal{H}(\boldsymbol{x}(t), \dot{\boldsymbol{x}}(t), \boldsymbol{p}(t))$ with respect to time. In particular, since

$$\mathcal{H}(\boldsymbol{x}(t), \dot{\boldsymbol{x}}(t), \boldsymbol{p}(t)) = -L(\boldsymbol{x}(t), \dot{\boldsymbol{x}}(t)) + \sum_{i=1}^{3}\dot{x}_i(t)p_i(t),$$

it follows that

$$\begin{aligned}\frac{d}{dt}\mathcal{H}(\boldsymbol{x}(t), \dot{\boldsymbol{x}}(t), \boldsymbol{p}(t)) &= \sum_{i=1}^{3}\left(\frac{\partial}{\partial x_i}\mathcal{H}(\boldsymbol{x}(t), \dot{\boldsymbol{x}}(t), \boldsymbol{p}(t))\dot{x}_i(t)\right)\\ &\quad+\sum_{i=1}^{3}\left(\frac{\partial}{\partial u_i}\mathcal{H}(\boldsymbol{x}(t), \dot{\boldsymbol{x}}(t), \boldsymbol{p}(t))\ddot{x}_i(t)\right)\\ &\quad+\sum_{i=1}^{3}\left(\frac{\partial}{\partial p_i}\mathcal{H}(\boldsymbol{x}(t), \dot{\boldsymbol{x}}(t), \boldsymbol{p}(t))\dot{p}_i(t)\right).\end{aligned}$$

However,

$$\mathcal{H}(\boldsymbol{x},\boldsymbol{u},\boldsymbol{p}) \triangleq -L(\boldsymbol{x},\boldsymbol{u}) + \sum_{i=1}^{3} u_i p_i$$

so that

$$\frac{\partial}{\partial u_i}\mathcal{H}(\boldsymbol{x},\boldsymbol{u},\boldsymbol{p}) = -L_{u_i}(\boldsymbol{x},\boldsymbol{u}) + p_i = -mu_i + p_i$$

and hence

$$\frac{\partial}{\partial u_i}\mathcal{H}(\boldsymbol{x}(t),\dot{\boldsymbol{x}}(t),\boldsymbol{p}(t)) = -m\dot{x}_i(t) + p_i(t) \equiv 0.$$

We now have proven the following result concerning the conservation of energy.

Theorem 8.1 (Conservation of Energy) *Assume the potential is independent of time so that the potential energy is given by* $P(t) = U(x_1(t), x_2(t), x_3(t)) = U(\boldsymbol{x}(t))$. *If* $\boldsymbol{x}(t)$ *satisfies the Principle of Stationary Action (i.e.* $\delta\mathbb{A}(\boldsymbol{x}^*(\cdot);\boldsymbol{\eta}(\cdot)) = 0$ *for all* $\boldsymbol{\eta}(\cdot) \in PWS(t_0, t_1; \mathbb{R}^3)$ *satisfying* $\boldsymbol{\eta}(t_0) = \boldsymbol{\eta}(t_1) = \mathbf{0}$*), then the Hamiltonian is constant along the trajectory* $\boldsymbol{x}(t)$. *In particular, there is a constant* H_0 *such that the total energy* $H(t)$ *satisfies*

$$H(t) = \mathcal{H}(\boldsymbol{x}(t),\dot{\boldsymbol{x}}(t),\boldsymbol{p}(t)) \equiv H_0$$

8.3 A Finite Element Method for the Heat Equation

In this section we discuss how the Fundamental Lemma of the Calculus of Variations can be applied to the theory and numerical solutions of systems governed by partial differential equations (PDEs). The FLCV plays two important roles here. First, as noted in Section 3.4, the FLCV provides the motivation for defining "weak derivatives" which will be used to define "weak solutions" of the PDE. Moreover, this weak formulation of the PDE is the first step in developing numerical methods based on finite element schemes. We illustrate this approach for a simple PDE describing the heat flow in a uniform rod given by

$$\frac{\partial}{\partial t}\theta(t,x) = k\frac{\partial^2}{\partial x^2}\theta(t,x) + g(x)u(t), \quad t>0, \quad 0<x<1, \quad (8.28)$$

where $u(t)$ is a heat source (control) and $g(\cdot) : (0,1) \to \mathbb{R}^1$ is a given (piecewise continuous) function. We also specify boundary conditions

$$\theta(t,0) = \theta(t,1) = 0, \quad (8.29)$$

and initial data

$$\theta(0,x) = \varphi(x), \;\; 0<x<1. \quad (8.30)$$

A *strong* (or classical) solution is a function $\theta(\cdot,\cdot)$ of t and x such that $\frac{\partial}{\partial t}\theta(t,x)$, $\frac{\partial}{\partial x}\theta(t,x)$ and $\frac{\partial^2}{\partial x^2}\theta(t,x)$ are continuous and $\theta(\cdot,\cdot)$ satisfies (8.28) at every value $0<x<1$ and all $t>0$ and the boundary condition (8.29) at $x=0$, $x=1$ and all $t>0$. We are interested in developing a numerical algorithm for approximating solutions. The finite element method can be used to approximate the PDE problem (8.28) - (8.30) in much as the same way it was employed in Section 2.4.5 above to approximate the two-point boundary value problem. Indeed, the two problems are linked and the finite element method in Section 2.4.5 can be extended to the heat equation above. Recall that the basic steps begin with multiplying both sides of (8.28) by an arbitrary function $\eta(\cdot)$ so that if $\theta(t,x)$ is a solution to (8.28), then for all $t>0$ we have

$$\frac{\partial}{\partial t}\theta(t,x)\,\eta(x) = k\frac{\partial^2}{\partial x^2}\theta(t,x)\,\eta(x) + b(x)u(t)\eta(x).$$

If $\eta(\cdot) \in PWC(0,1)$, then $\eta(\cdot)$ is integrable so one can integrate both sides to obtain

$$\begin{aligned}
&\int_0^1 \left[\frac{\partial}{\partial t}\theta(t,x)\right]\eta(x)dx \\
&\quad = \int_0^1 \left[k\frac{\partial^2}{\partial x^2}\theta(t,x)\right]\eta(x)dx + \int_0^1 \left[g(x)u(t)\right]\eta(x)dx \\
&\quad = \int_0^1 \left[k\frac{\partial^2}{\partial x^2}\theta(t,x)\right]\eta(x)dx + \left[\int_0^1 g(x)\eta(x)dx\right]u(t).
\end{aligned}$$

If in addition, $\eta(\cdot) \in PWS(0,1)$, then $\frac{d}{dx}\eta(\cdot)$ is integrable and we can use integration by parts on the first term $\int_0^1 \left[k\frac{\partial^2}{\partial x^2}\theta\left(t,x\right)\right]\eta(x)dx$ which yields

$$\int_0^1 \left[k\frac{\partial^2}{\partial x^2}\theta\left(t,x\right)\right]\eta(x)dx = \left[k\frac{\partial}{\partial x}\theta\left(t,x\right)\eta(x)\right]_{x=0}^{x=1} - \int_0^1 \left[k\frac{\partial}{\partial x}\theta\left(t,x\right)\right]\left[\frac{d}{dx}\eta(x)\right]dx.$$

Finally, if $\eta(\cdot) \in PWS_0(0,1)$ so that $\eta(0) = \eta(1) = 0$, then

$$\left[k\frac{\partial}{\partial x}\theta\left(t,x\right)\eta(x)\right]_{x=0}^{x=1} = 0$$

and it follows that

$$\int_0^1 \left[\frac{\partial}{\partial t}\theta\left(t,x\right)\right]\eta(x)dx = -\int_0^1 \left[k\frac{\partial}{\partial x}\theta\left(t,x\right)\right]\left[\frac{d}{dx}\eta(x)\right]dx + \left[\int_0^1 g(x)\eta(x)dx\right]u(t), \tag{8.31}$$

for all $\eta(\cdot) \in PWS_0(0,1)$. The equation (8.31) is called the *weak* (or variational) form of the heat equation defined by the heat equation (8.28) with boundary condition (8.29).

Definition 8.1 (Weak Solution of the Heat Equation) *We say that the function $\theta(t,\cdot)$ is a weak solution of the heat equation (8.28)-(8.29), if for each $t > 0$, $\theta(t,x)$ and $\frac{\partial}{\partial t}\theta(t,x)$ are continuous functions of t and x and*
(1) $\theta(t,\cdot) \in PWS_0(0,1)$, $\frac{\partial}{\partial t}\theta(t,\cdot) \in PWS_0(0,1)$,
(2) $\theta(t,\cdot)$ satisfies (8.31) for all $\eta(\cdot) \in PWS_0(0,1)$.

As in Section 2.4.5 above, the finite element element method actually produces an approximation of the weak form of the heat equation. Observe that we have shown that a strong solution to the heat equation (8.28) - (8.29) is always a weak solution. Following the approach in Section 2.4.5, we focus on the simplest piecewise linear approximations and hence we divide the interval $[0,1]$ into $N+1$ subintervals (called *elements*) of length $\Delta_x = 1/(N+1)$ with *nodes* $0 = \hat{x}_0 < \hat{x}_1 < \hat{x}_2 < \ldots < \hat{x}_{N-1} < \hat{x}_N < \hat{x}_{N+1} = 1$, where for $i = 0, 1, 2, \ldots, N, N+1$, $\hat{x}_i = i\Delta_x$. The approximation $\theta_N(t,x)$ will be continuous in x on all of $[0,1]$ and linear between the nodes. Since continuous piecewise linear approximating functions $\theta_N(t,x)$ are not typically differentiable in x at the nodes, it is not possible to insert this approximation directly into the equation (8.28). In particular, the piecewise smooth function $\theta_N(t,x)$ has only a piecewise continuous derivative $\frac{\partial}{\partial x}\theta_N(t,x)$ and hence $\frac{\partial^2}{\partial x^2}\theta_N(t,x)$ does not exist. In order to deal with this lack of smoothness, we use the weak form (8.31).

Define the spatial *hat functions* $h_i(\cdot)$ on $[0,1]$ as in 2.67 by

$$
\begin{aligned}
h_0(x) &= \begin{cases} (\hat{x}_1 - x)/\Delta_x, & 0 \le x \le \hat{x}_1 \\ 0, & \hat{x}_1 \le x \le 1 \end{cases}, \\
h_{N+1}(x) &= \begin{cases} (x - \hat{x}_N)/\Delta_x, & \hat{x}_N \le x \le 1 \\ 0, & 0 \le x \le \hat{x}_N \end{cases}, \\
h_i(x) &= \begin{cases} (x - \hat{x}_{i-1})/\Delta_x, & \hat{x}_{i-1} \le x \le \hat{x}_i \\ (\hat{x}_{i+1} - x)/\Delta_x, & \hat{x}_i \le x \le \hat{x}_{i+1} \\ 0, & x \notin (\hat{x}_{i-1}, \hat{x}_{i+1}) \end{cases},
\end{aligned}
\tag{8.32}
$$

for $i = 1, 2, \ldots, N$.

Plots of these hat functions are identical to the plots in Figures 2.24, 2.25 and 2.26 in Section 2.4.5 with the variable t replaced by the variable x.

These hat functions provide a basis for all continuous piecewise linear functions with (possible) corners at the internal nodes $0 < \hat{x}_1 < \hat{x}_2 < \ldots < \hat{x}_{N-1} < \hat{x}_N < 1$. Therefore, any continuous piecewise linear function $\theta_N(t,x)$ with corners only at these nodes can be written as

$$
\theta_N(t,x) = \sum_{i=0}^{N+1} \theta_i(t) h_i(x), \tag{8.33}
$$

where the functions $\theta_i(t)$ determine the value of $\theta_N(t,x)$ at $x = \hat{x}_i$. In particular, $\theta_N(t,\hat{x}_i) = \theta_i(t)$ and in order to form the function $\theta_N(t,\cdot)$ one must provide the coefficients $\theta_i(t)$ for $i = 0,1,2,\ldots,N,N+1$. Moreover, since $\theta_N(t,\cdot)$ is assumed to satisfy the Dirichlet boundary conditions (8.29), then $\theta_N(t,\hat{x}_0) = \theta_N(t,0) = \theta_0(t) = 0$ and $\theta_N(t,\hat{x}_{N+1}) = \theta_N(t,1) = \theta_{N+1}(t) = 0$ and $\theta_N(t,\cdot)$ can be written as

$$\theta_N(t,x) = \sum_{i=1}^{N} \theta_i(t) h_i(x). \tag{8.34}$$

We seek an approximate continuous piecewise linear solution $\theta_N(t,\cdot)$ of the form (8.34) to the weak form of the two-point value problem (8.31). In order to compute the functions $\theta_i(t)$ for $i = 1,2,\ldots,N$, we substitute $\theta_N(t,x) = \sum_{i=1}^{N} \theta_i(t)h_i(x)$ into the weak form of the equation given by (8.31). In particular, $\theta_N(t,x)$ is assumed to satisfy

$$\int_0^1 \left[\frac{\partial}{\partial t}\theta_N(t,x)\right]\eta(x)dx = -\int_0^1 \left[k\frac{\partial}{\partial x}\theta_N(t,x)\right]\left[\frac{d}{dx}\eta(x)\right]dx \tag{8.35}$$

$$+\left[\int_0^1 g(x)\eta(x)dx\right]u(t),$$

for all $\eta(\cdot) \in PWS_0(0,1)$.

Observe that $\frac{\partial}{\partial t}\theta_N(t,x) = \sum_{i=1}^{N} \dot{\theta}_i(t)h_i(x)$ and $\frac{\partial}{\partial x}\theta_N(t,x) = \sum_{i=1}^{N} \theta_i(t)\frac{d}{dx}h_i(x)$ is piecewise continuous so that substituting $\theta_N(t,x) = \sum_{i=1}^{N} \theta_i(t)h_i(x)$ into the weak equation (8.35) yields

$$\int_0^1 \left[\sum_{i=1}^{N} \dot{\theta}_i(t)h_i(x)\right]\eta(x)dx$$

$$= -\int_0^1 \left[k\sum_{i=1}^{N} \theta_i(t)\frac{d}{dx}h_i(x)\right]\left[\frac{d}{dx}\eta(x)\right]dx$$

$$+\left[\int_0^1 g(x)\eta(x)dx\right]u(t)$$

for all $\eta(\cdot) \in PWS_0(0,1)$. This equation can be written as

$$\sum_{i=1}^{N} \dot{\theta}_i(t) \left(\int_0^1 h_i(x)\eta(x)dx \right)$$
$$= -k \sum_{i=1}^{N} \theta_i(t) \left(\int_0^1 \left[\frac{d}{dx} h_i(x) \right] \left[\frac{d}{dx} \eta(x) \right] dx \right) \tag{8.36}$$
$$+ \left[\int_0^1 g(x)\eta(x)dx \right] u(t),$$

for all $\eta(\cdot) \in PWS_0(0,1)$. In order to use the variational equation to compute the coefficients $\theta_i(t)$ for (8.34), we note that for $i = 1, 2, \ldots, N$ the basis function $h_i(\cdot)$ belongs to $PWS_0(0,1)$. Therefore, setting $\eta(\cdot) = h_j(\cdot) \in PWS_0(0,1)$ for each index $j = 1, 2, \ldots, N$, yields N equations

$$\sum_{i=1}^{N} \dot{\theta}_i(t) \left(\int_0^1 [h_i(x)] [h_j(x)] dx \right)$$
$$= -k \sum_{i=1}^{N} \theta_i(t) \left(\int_0^1 \left[\frac{d}{dx} h_i(x) \right] \left[\frac{d}{dx} h_j(x) \right] dx \right) \tag{8.37}$$
$$+ \left[\int_0^1 g(x)h_j(x)dx \right] u(t).$$

Again, as in Section 2.4.5 we define the $N \times N$ *mass matrix* $\boldsymbol{M} = \boldsymbol{M}_N$ by

$$\boldsymbol{M} = \boldsymbol{M}_N = [m_{i,j}]_{i,j=i,2,\ldots,N},$$

where the entries $m_{i,j}$ of $\boldsymbol{M}_N$ are given by the integrals

$$m_{i,j} = \left(\int_0^1 [h_i(x)] [h_j(x)] dx \right).$$

Likewise, define the $N \times N$ *stiffness matrix* $\boldsymbol{K} = \boldsymbol{K}_N$ by

$$\boldsymbol{K} = \boldsymbol{K}_N = [k_{i,j}]_{i,j=i,2,\ldots,N},$$

where the entries $k_{i,j}$ of $\boldsymbol{K}_N$ are given by the integrals

$$k_{i,j} = \left(\int_0^1 \left[\frac{d}{dx} h_i(x) \right] \left[\frac{d}{dx} h_j(x) \right] dx \right).$$

Finally, let $\boldsymbol{g}_N$ be the $N \times 1$ (column) vector defined by

$$\boldsymbol{g}_N = \begin{bmatrix} g_1 & g_2 & \cdots & g_N \end{bmatrix}^T, \tag{8.38}$$

where entries g_j of $\boldsymbol{g}_N$ are given by the integrals

$$g_j = \left(\int_0^1 h_j(x) g(x) dx \right). \tag{8.39}$$

If $\boldsymbol{\theta}_N(t)$ is the solution vector

$$\boldsymbol{\theta}_N(t) = \begin{bmatrix} \theta_1(t) \\ \theta_2(t) \\ \vdots \\ \theta_N(t) \end{bmatrix},$$

of (8.37), then $\boldsymbol{\theta}_N(t)$ satisfies the matrix differential equation

$$\boldsymbol{M}_N \dot{\boldsymbol{\theta}}_N(t) = -\, k \boldsymbol{K}_N \boldsymbol{\theta}_N + \boldsymbol{g}_N u(t). \tag{8.40}$$

Defining

$$\boldsymbol{A}_N = -k \left[\boldsymbol{M}_N\right]^{-1} \boldsymbol{K}_N$$

and

$$\boldsymbol{B}_N = \left[\boldsymbol{M}_N\right]^{-1} \boldsymbol{g}_N$$

yields the linear control system

$$\dot{\boldsymbol{\theta}}_N(t) = \boldsymbol{A}_N \, \boldsymbol{\theta}_N + \boldsymbol{B}_N u(t). \tag{8.41}$$

which must be solved to find the coefficients $\theta_i(t)$ for $i = 1, 2, \ldots, N$.

In order to find the correct initial data for the finite element model (8.41), one approximates $\theta(0, x) = \varphi(x), \ 0 < x < 1$ by

$$\varphi_N(x) = \sum_{i=1}^{N} \varphi_i h_i(x)$$

and selects the coefficients φ_i for $i = 1, 2, \ldots, N$ so that $\varphi_N(x)$ is the "best" approximation of $\varphi(x)$. To make this precise, define the subspace $V_0^h(0,1) \subseteq PWS_0(0,1)$ by

$$V_0^h(0,1) = span\left\{h_i(x) : i = 1, 2, \ldots, N\right\} = \left\{ \psi(x) = \sum_{i=1}^{N} \alpha_i h_i(x) \right\}.$$

Thus, we seek the function $\varphi_N(x) = \sum_{i=1}^{N} \varphi_i h_i(x)$ such that $\varphi_N(\cdot)$ minimizes

$$\int_0^1 |\varphi_N(x) - \varphi(x)|^2 \, dx.$$

Elementary geometry implies that $\varphi_N(\cdot) - \varphi(\cdot)$ must be "orthogonal to $V_0^h(0,1)$" so that

$$\int_0^1 [\varphi_N(x) - \varphi(x)]\eta(x))dx = 0$$

for all $\eta(\cdot) \in V_0^h(0,1)$. In particular,

$$\int_0^1 [\varphi_N(x) - \varphi(x)]h_j(x)dx = 0$$

for all $i = 1, 2, \ldots, N$ so that

$$\int_0^1 \left[\sum_{i=1}^{N} \varphi_i h_i(x) - \varphi(x)\right] h_j(x)dx = 0$$

which implies

$$\sum_{i=1}^{N} \varphi_i \left(\int_0^1 [h_i(x)]\,[h_j(x)]\,dx\right) - \sum_{i=1}^{N} \left(\int_0^1 [\varphi(x)]\,[h_j(x)]\,dx\right) = 0.$$

In matrix form we have

$$[\boldsymbol{M}_N]\boldsymbol{\varphi}_N = \tilde{\boldsymbol{\varphi}}_N,$$

where

$$\tilde{\varphi}_j = \left(\int_0^1 [\varphi(x)]\,[h_j(x)]\,dx\right)$$

and

$$\tilde{\boldsymbol{\varphi}}_N = \begin{bmatrix} \tilde{\varphi}_1 & \tilde{\varphi}_2 & \cdots & \tilde{\varphi}_N \end{bmatrix}^T.$$

Therefore, the initial condition for the finite element model (8.41) is

$$\boldsymbol{\theta}_N(0) = \boldsymbol{\varphi}_N$$

where

$$\boldsymbol{\varphi}_N = [\boldsymbol{M}_N]^{-1}\,\tilde{\boldsymbol{\varphi}}_N \tag{8.42}$$

and the finite dimensional "finite element" system is defined by

$$\dot{\boldsymbol{\theta}}_N(t) = \boldsymbol{A}_N\,\boldsymbol{\theta}_N + \boldsymbol{B}_N u(t) \tag{8.43}$$

with initial condition

$$\boldsymbol{\theta}_N(0) = \boldsymbol{\varphi}_N. \tag{8.44}$$

The finite element ODE system (8.41) can be solved using standard numerical methods. Again, the key to developing the ODE model (8.41) is to approximate the **weak form** of the PDE model (8.28) - (8.29). This is a fundamental idea that is the basis of the finite element method.

8.4 Problem Set for Chapter 8

Problem 8.1 *Consider the problem of minimizing the functional*

$$J(x(\cdot)) = \int_0^b \left\{ [\dot{x}(s)]^2 - [x(s)]^2 + f(s)x(s) \right\} ds$$

among all piecewise smooth functions satisfying

$$x(0) = 0, \quad x(b) = 0,$$

where $f(\cdot)$ *is a given continuous function. Assume* $x^*(\cdot)$ *minimizes* $J(\cdot)$.
(A) Compute the first variation $\delta J(x^*(\cdot), \eta(\cdot))$ *for this problem.*
(B) Write out the equation $\delta J(x^*(\cdot), \eta(\cdot)) = 0$.
(C) Show that if $x^*(\cdot)$ *minimizes* $J(\cdot)$ *subject to* $x(0) = 0,\ x(1) = 0$, *then* $x^*(\cdot)$ *is a weak solution to the two point boundary value problem*

$$\ddot{x}(t) + x(t) = f(t), \quad x(0) = 0, \quad x(b) = 0. \tag{8.45}$$

(D) Show that there are solutions of (8.45) that do not minimize $J(x(\cdot))$.

Problem 8.2 *Consider the problem of minimizing the functional*

$$J(x(\cdot)) = \int_0^1 \left\{ [\dot{x}(s)]^2 + 2x(s)\dot{x}(s) + [x(s)]^2 + 2e^s x(s) \right\} ds,$$

subject to the endpoint conditions $x(0) = 0$ and $x(1) = 1$. Assume $x^*(\cdot)$ is piecewise smooth and minimizes $J(\cdot)$.
(A) Compute the first variation $\delta J(x^*(\cdot), \eta(\cdot))$ for this problem.
(B) Write out the equation $\delta J(x^*(\cdot), \eta(\cdot)) = 0$.
(C) What two point boundary value problem will $x^*(\cdot)$ satisfy?

Problem 8.3 *Consider the problem of minimizing the functional*

$$J(x(\cdot)) = \int_1^3 \{[3s - x(s)]x(s)\} \, ds,$$

subject to the endpoint conditions $x(1) = 1$ and $x(3)$ is free. Assume $x^*(\cdot)$ is piecewise smooth and minimizes $J(\cdot)$.
(A) Compute the first variation $\delta J(x^*(\cdot), \eta(\cdot))$ for this problem.
(B) Write out the equation $\delta J(x^*(\cdot), \eta(\cdot)) = 0$.
(C) What two point boundary value problem will $x^*(\cdot)$ satisfy?

Problem 8.4 *Consider the problem of minimizing the functional*

$$J(x(\cdot)) = \int_1^2 \left\{ \dot{x}(s)[1 + s^2 \dot{x}(s)] \right\} ds,$$

subject to the endpoint conditions $x(1)$ is free and $x(2) = 5$. Assume $x^*(\cdot)$ is piecewise smooth and minimizes $J(\cdot)$.
(A) Compute the first variation $\delta J(x^*(\cdot), \eta(\cdot))$ for this problem.
(B) Write out the equation $\delta J(x^*(\cdot), \eta(\cdot)) = 0$.
(C) What two point boundary value problem will $x^*(\cdot)$ satisfy?

Problem 8.5 *Apply Hamilton's Principle to the idealized double pendulum described in Section 2.1.3 to derive the equations of motion for the system. Hint: Use polar coordinates and Figure 2.3 to determine the kinetic and potential energy.*

Advanced Problems

Problem 8.6 *Let $k = .25$, $g(x) = x$ and assume the initial function is given by $\varphi(x) = 1$. Construct the finite element model (8.43) - (8.44) for $N = 4, 8, 16, 32$ and 64. Let $u(t) = e^{-t}$ and use a numerical method to solve the finite element equations (8.43) - (8.44) on the interval $0 < t \leq 2$. Hint: Since $g(x) = x = \sum_{i=1}^{N+1} (\frac{i}{N+1})h_i(x)$, it follows that*

$$g_j = \left(\int_0^1 h_j(x)g(x)dx \right)$$

can be computed exactly. In particular, note

$$\left(\int_0^1 h_i(x)h_j(x)dx \right) = \frac{\Delta_x}{6} \begin{cases} 4 & 0 < i = j < N+1 \\ 2 & i = j = 0 \text{ or } i = j = N+1 \\ 1 & |i-j| = 1 \\ 0 & \text{elsewhere} \end{cases}$$

and

$$\left(\int_0^1 [\frac{d}{dx}h_i(x)][\frac{d}{dx}h_j(x)]dx \right) = \frac{1}{\Delta_x} \begin{cases} 2 & 0 < i = j < N+1 \\ 1 & i = j = 0 \text{ or } \\ & i = j = N+1 \\ -1 & |i-j| = 1 \\ 0 & \text{elsewhere} \end{cases} .$$

Thus, the $N \times N$ mass and stiffness matrices for this problem are given by

$$\boldsymbol{M}_N = \frac{\Delta_x}{6} \begin{bmatrix} 4 & 1 & 0 & 0 & \cdots & 0 & 0 \\ 1 & 4 & 1 & 0 & \cdots & 0 & 0 \\ 0 & 1 & 4 & 1 & \cdots & 0 & 0 \\ 0 & 0 & 1 & 4 & \cdots & 0 & 0 \\ \vdots & \vdots & \vdots & \vdots & \ddots & \vdots & \vdots \\ 0 & 0 & \cdots & \cdots & 1 & 4 & 1 \\ 0 & 0 & \cdots & \cdots & 0 & 1 & 4 \end{bmatrix}$$

and

$$\boldsymbol{K}_N = \frac{1}{\Delta_x}\begin{bmatrix} 2 & -1 & 0 & 0 & \cdots & 0 & 0 \\ -1 & 2 & -1 & 0 & \cdots & 0 & 0 \\ 0 & -1 & 2 & -1 & \cdots & 0 & 0 \\ 0 & 0 & -1 & 2 & \cdots & 0 & 0 \\ \vdots & \vdots & \vdots & \vdots & \ddots & \vdots & \vdots \\ 0 & 0 & \cdots & \cdots & -1 & 2 & -1 \\ 0 & 0 & \cdots & \cdots & 0 & -1 & 2 \end{bmatrix},$$

respectively.

Problem 8.7 *Let* $k = 1$, $g(x) = 1$ *and assume the initial function is given by* $\varphi(x) = x$. *Construct the finite element model (8.43) - (8.44) for* $N = 4, 8, 16, 32$ *and* 64. *Let* $u(t) = \sin(t)$ *and use a numerical method to solve the finite element equations (8.43) - (8.44) on the interval* $0 < t \leq 2\pi$.

Part II

Optimal Control

Chapter 9

Optimal Control Problems

In this chapter we describe the mathematical formulation of a typical optimal control problem as an optimization problem over a space of admissible functions. We provide some typical examples and use these examples to motivate the Maximum Principle. We start with some standard problems in the calculus of variations and make a few modifications to illustrate where the classical necessary conditions begin to break down.

The history of the development of optimal control theory is very interesting. The reader should look at the perspectives found in the references [30], [84], [85], [91], [134], [149], [154] and [172].

9.1 An Introduction to Optimal Control Problems

Consider the optimization problem: Minimize

$$J(x(\cdot), u(\cdot)) = \int_{t_0}^{t_1} \left\{ |x(s)|^2 + |u(s)|^2 \right\} ds, \tag{9.1}$$

where $u(\cdot) \in PWC(t_0, t_1)$ and $x(\cdot) \in PWS(t_0, t_1)$ are related by

$$x(t) - x_0 - \int_{t_0}^{t} \{ax(s) + bu(s)\}\ ds = 0, \tag{9.2}$$

where $b \neq 0$. Observe that $x(\cdot) \in PWS(t_0, t_1)$, $\dot{x}(t) = ax(t)+bu(t)$ *e.f.* and $x(t_0) = x_0$. Therefore, the constraint (9.2) is equivalent to the initial value problem

$$\dot{x}(t) = ax(t) + bu(t), \quad x(t_0) = x_0 \tag{9.3}$$

Thus, we can solve for $u(t)$ in terms of $x(t)$ and $\dot{x}(t)$ so that $[\dot{x}(t) - ax(t)]/b = u(t)$ and the problem of minimizing (9.1) subject to the constraint (9.2) is equivalent to the free endpoint problem of minimizing

$$J(x(\cdot), u(\cdot)) = \int_{t_0}^{t_1} \left\{ [x(s)]^2 + \frac{1}{b^2}[\dot{x}(s) - ax(s)]^2 \right\} ds,$$

subject to

$$x(t_0) = x_0 \quad \text{with} \quad x(t_0) - \text{free}.$$

In this case we can eliminate the "control" $u(\cdot)$ and reformulate the problem as a classical problem in the calculus of variations.

Also, we could have reformulated the problem as a constrained optimization problem as follows. Let $\boldsymbol{Z} = PWS(t_0, t_1) \times PWC(t_0, t_1)$ and $\boldsymbol{Y} = PWS(t_0, t_1)$, so that an element $\boldsymbol{z} \in \boldsymbol{Z} = PWS(t_0, t_1) \times PWC(t_0, t_1)$ has the form $\boldsymbol{z} = [x(\cdot) \ \ u(\cdot)]^T$. Define the functional $\boldsymbol{J} : \boldsymbol{Z} \longrightarrow \mathbb{R}^1$ by

$$\boldsymbol{J}(\boldsymbol{z}) = J(x(\cdot), u(\cdot)) = \int_{t_0}^{t_1} \{[x(s)]^2 + [u(s)]^2\} \, ds \tag{9.4}$$

and the function $\boldsymbol{G} = \boldsymbol{Z} \longrightarrow \boldsymbol{Y}$ by

$$[\boldsymbol{G}(\boldsymbol{z})](t) = [\boldsymbol{G}(x(\cdot), u(\cdot))](t) \triangleq x(t) - x_0 - \int_{t_0}^{t} \{ax(s) + bu(s)\} \ ds, \tag{9.5}$$

respectively. In this setting, the problem is equivalent to finding $\boldsymbol{z}^* = [x^*(\cdot) \ \ u^*(\cdot)]^T \in \boldsymbol{Z}$ to minimize $\boldsymbol{J}(\cdot)$ subject to the (equality) constraint $\boldsymbol{G}(\boldsymbol{z}) = \boldsymbol{0}$.

Consider now a vector version of the problem. Let A be a $n \times n$ constant real matrix and B be a $n \times m$ real constant matrix with

$m < n$ (i.e. fewer controls that states). Consider the linear vector control system

$$\dot{\boldsymbol{x}}(t) = A\boldsymbol{x}(t) + B\boldsymbol{u}(t) \tag{9.6}$$

with initial condition

$$\boldsymbol{x}(0) = \boldsymbol{x}_0 \in \mathbb{R}^n. \tag{9.7}$$

We assume that $t_0 = 0$, $\boldsymbol{x}_0 \in \mathbb{R}^n$ and $0 < t_1$ are given and that the control $\boldsymbol{u}(\cdot)$ belongs to the space $PWC(0, t_1; \mathbb{R}^m)$. The quadratic cost functional is defined by

$$J(\boldsymbol{x}(\cdot), \boldsymbol{u}(\cdot)) = \int_0^{t_1} \left\{ \|\boldsymbol{x}(s)\|^2 + \|\boldsymbol{u}(s)\|^2 \right\} ds, \tag{9.8}$$

where $\boldsymbol{x}(t) = \boldsymbol{x}(t; \boldsymbol{u}(\cdot))$ is the solution to the system (9.6) - (9.7). The vector form of the optimization problem defined by (9.1) with the constraint (9.2) is defined by (9.6) - (9.8).

Remark 9.1 *Observe that we can no longer eliminate the control $\boldsymbol{u}(\cdot)$ since the matrix B is not invertible. Thus, it is not possible to directly reformulate this vector problem as a classical vector problem in the calculus of variations.*

Let $\boldsymbol{Z} = PWS(0, t_1; \mathbb{R}^n) \times PWC(0, t_1; \mathbb{R}^m)$ and $\boldsymbol{Y} = PWS(0, t_1; \mathbb{R}^n)$, so that an element $\boldsymbol{z} \in \boldsymbol{Z} = PWS(0, t_1; \mathbb{R}^n) \times PWC(0, t_1; \mathbb{R}^m)$ has the form $\boldsymbol{z} = [\boldsymbol{x}(\cdot) \;\; \boldsymbol{u}(\cdot)]^T$. Also, define the functional $\boldsymbol{J} : \boldsymbol{Z} \longrightarrow \mathbb{R}^1$ by

$$\boldsymbol{J}(\boldsymbol{z}) = J(\boldsymbol{x}(\cdot), \boldsymbol{u}(\cdot)) = \int_0^{t_1} \left\{ \|\boldsymbol{x}(s)\|^2 + \|\boldsymbol{u}(s)\|^2 \right\} ds, \tag{9.9}$$

and the function $\boldsymbol{G} = \boldsymbol{Z} \longrightarrow \boldsymbol{Y}$ by

$$[\boldsymbol{G}(\boldsymbol{z})](t) = [\boldsymbol{G}(\boldsymbol{x}(\cdot), \boldsymbol{u}(\cdot))](t) \triangleq \boldsymbol{x}(t) - \boldsymbol{x}_0 - \int_0^t \{A\boldsymbol{x}(s) + B\boldsymbol{u}(s)\} \, ds, \tag{9.10}$$

respectively. Again, the vector problem is equivalent to finding $\boldsymbol{z}^* = [\boldsymbol{x}^*(\cdot) \ \ \boldsymbol{u}^*(\cdot)]^T \in \boldsymbol{Z}$ to minimize $\boldsymbol{J}(\cdot)$ subject to the (equality) constraint $\boldsymbol{G}(\boldsymbol{z}) = \boldsymbol{0}$.

As we shall see later, formulating the control problem as a general optimization problem with equality (or inequality) constraints can be used to obtain necessary conditions. Moreover, some classical problems such as the isoperimetric problem in the calculus of variations naturally fall into this formulation. The general formulation of the equality constrained problem assumes there are two vector spaces $\boldsymbol{Z}$ and $\boldsymbol{Y}$ and two functions

$$\boldsymbol{J} : \boldsymbol{D}(\boldsymbol{J}) \subseteq \boldsymbol{Z} \longrightarrow \mathbb{R}^1 \tag{9.11}$$

and

$$\boldsymbol{G} : \boldsymbol{D}(\boldsymbol{G}) \subseteq \boldsymbol{Z} \longrightarrow \boldsymbol{Y}. \tag{9.12}$$

The function $\boldsymbol{J}(\cdot)$ is called the *cost function* and $\boldsymbol{G}(\cdot)$ is called the *constraint function.* Define the constraint set

$$\boldsymbol{\Theta}_G \subseteq \boldsymbol{Z}$$

by

$$\boldsymbol{\Theta}_G = \{\boldsymbol{z} \in \boldsymbol{D}(\boldsymbol{G}) : \boldsymbol{G}(\boldsymbol{z}) = \boldsymbol{0} \in \boldsymbol{Y}\} \subset \boldsymbol{D}(\boldsymbol{G}). \tag{9.13}$$

The *Equality Constrained Optimization Problem* is defined to be:

Find an element $\boldsymbol{z}^* \in \boldsymbol{\Theta}_G \cap \boldsymbol{D}(\boldsymbol{J})$ such that

$$\boldsymbol{J}(\boldsymbol{z}^*) \leq \boldsymbol{J}(\boldsymbol{z})$$

for all $\boldsymbol{z} \in \boldsymbol{\Theta}_G \cap \boldsymbol{D}(\boldsymbol{J})$.

Observe that since $\boldsymbol{\Theta}_G \subset \boldsymbol{D}(\boldsymbol{G})$, it follows that $\boldsymbol{\Theta}_G \cap \boldsymbol{D}(\boldsymbol{J}) \subset \boldsymbol{D}(\boldsymbol{G}) \cap \boldsymbol{D}(\boldsymbol{J})$. Therefore, the equality constrained optimization problem is equivalent to finding $\boldsymbol{z}^* \in \boldsymbol{D}(\boldsymbol{G}) \cap \boldsymbol{D}(\boldsymbol{J})$ such that $\boldsymbol{z}^*$ minimizes $\boldsymbol{J}(\boldsymbol{z})$ subject to $\boldsymbol{G}(\boldsymbol{z}) = \boldsymbol{0} \in \boldsymbol{Y}$. We first discuss special cases and then move to the more abstract versions.

As noted before, this formulation allows us to develop and apply a Lagrange Multiplier Theorem to obtain necessary conditions.

Luenberger employs this abstract approach in his book *Optimization by Vector Space Methods* (see [131]). However, equality constraints such as (9.10) above are not always applicable to control problems with (hard) inequality constraints on the control or states. A simple example where one needs to use inequality constraints is the so called Rocket Sled Time Optimal Control problem described below.

In the following sections we describe some typical optimal control problems and provide a preliminary discussion to motivate the Maximum Principle. We start with the Rocket Sled Control problem and then revisit some of the simplest problems in the calculus of variations. In addition, we make a few modifications of the SPCV to illustrate where the classical necessary conditions begin to break down.

9.2 The Rocket Sled Problem

The Rocket Sled problem was formulated as a mathematical optimization problem in Section 2.4.4 above. Here we present another formulation that will be typical of optimal control problems considered below. Recall that the system is defined by

$$\frac{d}{dt}\begin{bmatrix} x_1(t) \\ x_2(t) \end{bmatrix} = \begin{bmatrix} 0 & 1 \\ 0 & 0 \end{bmatrix}\begin{bmatrix} x_1(t) \\ x_2(t) \end{bmatrix} + \begin{bmatrix} 0 \\ 1/m \end{bmatrix} u(t), \tag{9.14}$$

or

$$\dot{\boldsymbol{x}}(t) = A\boldsymbol{x}(t) + Bu(t), \quad \boldsymbol{x}(0) = \boldsymbol{x}_0 \tag{9.15}$$

where the matrices A and B are defined by

$$A = \begin{bmatrix} 0 & 1 \\ 0 & 0 \end{bmatrix} \text{ and } B = \begin{bmatrix} 0 \\ 1/m \end{bmatrix},$$

respectively. Here, $\boldsymbol{x}_0 = \begin{bmatrix} x_{1,0} & x_{2,0} \end{bmatrix}^T$ and $\boldsymbol{x}(t) = \begin{bmatrix} x_1(t) & x_2(t) \end{bmatrix}^T$ provides a curve (trajectory) in the plane $\mathbb{R}^2$. To be more precise, given a control $u(\cdot)$ we let $\boldsymbol{x}(t; u(\cdot))$ denote the solution of the initial value problem (9.15) with control input $u(\cdot)$. Given that the control $u(\cdot)$ is constrained by $|u(t)| \leq 1$, the time optimal problem is to find the control $u^*(\cdot)$ such that

$$u^*(\cdot) \in [-1, +1], \tag{9.16}$$

and $u^*(t)$ steers $\boldsymbol{x}_0 = \begin{bmatrix} x_{1,0} & x_{2,0} \end{bmatrix}^T$ to $\boldsymbol{x}_1 = \begin{bmatrix} 0 & 0 \end{bmatrix}^T$ in minimum time. Here the final time t_1 is not fixed and the cost function can be represented by the integral

$$J(x(\cdot), u(\cdot)) = t_1 = \int_0^{t_1} 1 ds = \int_0^{t_1} f_0(x(s), u(s)), \tag{9.17}$$

where $f_0(x, u) \equiv 1$. Let

$$\mathrm{X}_0 = \{\boldsymbol{x}_0\} \subseteq \mathbb{R}^2 \text{ and } \mathrm{X}_1 = \{\boldsymbol{x}_1\} \subseteq \mathbb{R}^2$$

be the initial set and "target" set respectively. Also, define the control constraint set

$$\Omega = [-1, +1] \subseteq \mathbb{R}.$$

The set of admissible controllers is the subset of $PWC(t_0, +\infty)$ defined by

$$\begin{aligned} \Theta &= \left\{ \begin{array}{r} u(\cdot) \in PWC(t_0, +\infty) : u(t) \in \Omega \;\; e.f. \\ \text{and } u(\cdot) \text{ steers } \mathrm{X}_0 \text{ to } \mathrm{X}_1. \end{array} \right\} \\ &= \Theta(t_0, \mathrm{X}_0, \mathrm{X}_1, \Omega). \end{aligned} \tag{9.18}$$

The time optimal control problem is the problem of minimizing $J(\cdot)$ on the set of all admissible controllers Θ. In particular, an *optimal control* is a function $u^*(\cdot) \in \Theta$ such that $u^*(\cdot)$ steers X_0 to X_1 in time $t_1^* > t_0$ and

$$J(u^*(\cdot)) = \int_{t_0}^{t_1^*} f_0(x^*(s), u^*(s)) ds \leq \int_{t_0}^{t_1} f_0(x(s), u(s)) ds = J(u(\cdot)) \tag{9.19}$$

for all $u(\cdot) \in \Theta$. This time optimal control problem will be solved below by using some simple geometric arguments that lay the foundation for the development of the Maximum Principle.

9.3 Problems in the Calculus of Variations

Problems in the calculus of variations can be formulated as special optimal control problems. We discuss two examples to illustrate how classical problems in the calculus of variations can easily be transformed into optimal control problems. We begin with the Simplest Problem in the Calculus of Variations (**SPCV**).

9.3.1 The Simplest Problem in the Calculus of Variations

We assume that $f_0(t,x,u)$ is a C^2 smooth function of three variables. In particular, we assume that f_0 is continuous and all the partial derivatives

$$\frac{\partial f_0(t,x,u)}{\partial t}, \quad \frac{\partial f_0(t,x,u)}{\partial x}, \quad \frac{\partial f_0(t,x,u)}{\partial u},$$

$$\frac{\partial^2 f_0(t,x,u)}{\partial t^2}, \quad \frac{\partial^2 f_0(t,x,u)}{\partial x^2}, \quad \frac{\partial^2 f_0(t,x,u)}{\partial u^2}$$

exist and are continuous. Note that this implies that all the mixed derivatives are equal, i.e.

$$\frac{\partial^2 f_0(t,x,u)}{\partial t \partial x} = \frac{\partial^2 f_0(t,x,u)}{\partial x \partial t}, \quad \frac{\partial^2 f_0(t,x,u)}{\partial t \partial u} = \frac{\partial^2 f_0(t,x,u)}{\partial u \partial t}$$

and

$$\frac{\partial^2 f_0(t,x,u)}{\partial x \partial u} = \frac{\partial^2 f_0(t,x,u)}{\partial u \partial x}.$$

Let $X = PWS(t_0, t_1)$ denote the space of all real-valued piecewise smooth functions defined on $[t_0, t_1]$. For each PWS function $x : [t_0, t_1] \to \mathbb{R}^1$, define the *functional* $J : X \to \mathbb{R}^1$ by

$$J(x(\cdot)) = \int_{t_0}^{t_1} f_0\left(s, x(s), \dot{x}\left(s\right)\right) ds. \tag{9.20}$$

Assume that the points (t_0, x_0) and (t_1, x_1) are given and define the set of PWS functions Ψ by

$$\Psi = \{x(\cdot) \in PWS(t_0, t_1) : x(t_0) = x_0, x(t_1) = x_1\}. \qquad (9.21)$$

Here $\dot{x}(\cdot)$ is the derivative of $x(\cdot)$, i.e. $\dot{x}(\cdot) = \frac{d}{dt}x(\cdot)$. Observe that $J : X \to \mathbb{R}^1$ is a real valued function on X.

The Simplest Problem in the Calculus of Variations (SPCV) is the problem of minimizing $J(\cdot)$ on Ψ. In particular, the goal is to find $x^*(\cdot) \in \Psi$ such that

$$J(x^*(\cdot)) = \int_{t_0}^{t_1} f_0(s, x^*(s), \dot{x}^*(s))\, ds \le J(x(\cdot)) = \int_{t_0}^{t_1} f_0(s, x(s), \dot{x}(s))\, ds,$$

for all $x(\cdot) \in \Psi$.

We introduce some notation and reformulate the **SPCV** as an optimal control problem. In particular, let

$$\dot{x}(t) = u(t) \qquad (9.22)$$

and observe that if

$$x(t_0) = x_0, \qquad (9.23)$$

then

$$x(t) = x_0 + \int_{t_0}^{t} u(s)ds \qquad (9.24)$$

so that $x(\cdot) \in PWS(t_0, t_1)$ if and only if $u(\cdot) \in PWC(t_0, t_1)$. We say that a control $u(\cdot) \in PWC(t_0, t_1)$ *steers* x_0 *to* x_1 *at time* t_1 if the solution $x(t; u(\cdot)) = x(t)$ to (9.22) - (9.23) satisfies

$$x(t_1) = x_1.$$

Define the sets

$$\mathrm{X}_0 = \{x_0\} \subseteq \mathbb{R}^1 \quad \text{and} \quad \mathrm{X}_1 = \{x_1\} \subseteq \mathbb{R}^1,$$

and let $\Omega = \mathbb{R}^1$. The set of *admissible controllers* is the subset of $PWC(t_0, t_1)$ defined by

$$\begin{aligned}\Theta &= \left\{ \begin{array}{c} u(\cdot) \in PWC(t_0,t_1) : u(t) \in \Omega \;\; e.f. \text{ and} \\ u(\cdot) \text{ steers } \mathrm{X}_0 \text{ to } \mathrm{X}_1 \;\text{ at time } \; t_1. \end{array} \right\} \\ &= \Theta(t_0, t_1, x_0, x_1, \Omega). \end{aligned} \tag{9.25}$$

Given a control $u(\cdot) \in PWC(t_0, t_1)$, the cost functional is defined by

$$J(u(\cdot)) = \int_{t_0}^{t_1} f_0(s, x(s), u(s))ds = \int_{t_0}^{t_1} f_0(s, x(s; u(\cdot)), u(s))ds, \tag{9.26}$$

where $x(\cdot) = x(\cdot; u(\cdot))$ is the solution to the initial value problem (10.1) - (10.2).

The Simplest Problem in the Calculus of Variations (**SPCV**) can now be formulated as an equivalent problem in optimal control. In particular, the **SPCV** is equivalent to the problem of minimizing $J(\cdot)$ on the set of all admissible controllers Θ. The goal is to find an *optimal control* $u^*(\cdot) \in \Theta$ such that $u^*(\cdot)$ steers x_0 to x_1 at time $t_1 > t_0$ and

$$\begin{aligned} J(u^*(\cdot)) &= \int_{t_0}^{t_1} f_0(s, x(s; u^*(\cdot)), u^*(s))ds \\ &\leq \int_{t_0}^{t_1} f_0(s, x(s; u(\cdot)), u(s))ds = J(u(\cdot)) \end{aligned} \tag{9.27}$$

for all

$$u(\cdot) \in \Theta. \tag{9.28}$$

If $u^*(\cdot) \in \Theta$ minimizes $J(\cdot)$ on Θ, then $x^*(\cdot)$ defined by (9.24) solves the **SPCV**. Conversely, if $x^*(\cdot)$ is a solution to the **SPCV**, then $\dot{x}^*(\cdot) = u^*(\cdot)$ is a solution to the optimal control problem defined by (9.22) - (9.26).

9.3.2 Free End-Point Problem

We start as in the **SPCV**, where t_0, t_1 and x_0 are given (fixed) but x_1 is not specified (free). In particular, we are given t_0, t_1 and x_0. The cost functional is

$$J(x(\cdot)) = \int_{t_0}^{t_1} f_0\left(s, x\left(s\right), \dot{x}\left(s\right)\right) ds$$

and we define the set of PWS functions $\Psi_1 = \Psi_1(t_0, t_1, x_0)$ by

$$\Psi_1 = \left\{x(\cdot) \in PWS(t_0, t_1) : x\left(t_0\right) = x_0\right\}.$$

Observe that the final condition $x(t_1)$ is not specified. The free endpoint problem is to find $x^*\left(\cdot\right) \in \Psi_1$ such that

$$\begin{aligned} J(x^*(\cdot)) &= \int_{t_0}^{t_1} f_0\left(s, x^*(s), \dot{x}^*\left(s\right)\right) ds \leq J(x(\cdot)) \\ &= \int_{t_0}^{t_1} f_0\left(s, x(s), \dot{x}\left(s\right)\right) ds \end{aligned}$$

for all $x(\cdot) \in \Psi_1$.

Again we define the state equation by

$$\dot{x}(t) = u(t) \tag{9.29}$$

with initial condition

$$x(t_0) = x_0. \tag{9.30}$$

Define the sets

$$\mathrm{X}_0 = \{x_0\} \subseteq \mathbb{R}^1 \quad \text{and} \quad \mathrm{X}_1 = \mathbb{R}^1,$$

and let $\Omega = \mathbb{R}^1$. The set of *admissible controllers* is the subset of $PWC(t_0, t_1)$ defined by

$$\begin{aligned} \Theta_1 &= \left\{ \begin{array}{l} u(\cdot) \in PWC(t_0, t_1) : u(t) \in \Omega \;\; e.f. \text{ and} \\ \qquad\qquad u(\cdot) \text{ steers } \mathrm{X}_0 \text{ to } \mathrm{X}_1 \text{ at time } t_1. \end{array} \right\} \\ &= \Theta_1(t_0, t_1, x_0, \Omega). \end{aligned} \tag{9.31}$$

Given a control $u(\cdot) \in PWC(t_0, t_1)$, the cost functional is defined by

$$J(u(\cdot)) = \int_{t_0}^{t_1} f_0(s, x(s), u(s))ds = \int_{t_0}^{t_1} f_0(s, x(s; u(\cdot)), u(s))ds, \tag{9.32}$$

where $x(\cdot) = x(\cdot; u(\cdot))$ is the solution to the initial value problem (9.29) - (9.30).

This variational problem is equivalent to the optimal control problem of minimizing $J(\cdot)$ on the set of admissible controllers Θ_1. In particular, the goal is to find an *optimal control* $u^*(\cdot) \in \Theta_1$ such that $u^*(\cdot)$ steers x_0 to X_1 at time t_1 and

$$\begin{aligned} J(u^*(\cdot)) &= \int_{t_0}^{t_1} f_0\left(s, x^*(s), u^*\left(s\right)\right) ds \leq J(u(\cdot)) \\ &= \int_{t_0}^{t_1} f_0\left(s, x(s), u\left(s\right)\right) ds \end{aligned}$$

for all $u(\cdot) \in \Theta_1$.

9.4 Time Optimal Control

The problem of time optimal control provides a natural departure from classical calculus of variational problems in that one places a "hard" constraint on the control variable. Also, the final time t_1 is not fixed and becomes the value of the cost functional. We begin with the simplest time optimal control problem defined by the rocket sled control system.

9.4.1 Time Optimal Control for the Rocket Sled Problem

Recall that the system is defined by

$$\frac{d}{dt}\begin{bmatrix} x_1(t) \\ x_2(t) \end{bmatrix} = \begin{bmatrix} 0 & 1 \\ 0 & 0 \end{bmatrix}\begin{bmatrix} x_1(t) \\ x_2(t) \end{bmatrix} + \begin{bmatrix} 0 \\ 1/m \end{bmatrix} u(t), \tag{9.33}$$

or

$$\dot{\boldsymbol{x}}(t) = A\boldsymbol{x}(t) + Bu(t), \quad \boldsymbol{x}(0) = \boldsymbol{x}_0 \tag{9.34}$$

where the matrices A and B are defined by

$$A = \begin{bmatrix} 0 & 1 \\ 0 & 0 \end{bmatrix}, \text{ and } B = \begin{bmatrix} 0 \\ 1/m \end{bmatrix},$$

respectively.

In matrix form, using the Variation of Parameters formula we have

$$\boldsymbol{x}(t) = e^{At}\boldsymbol{x}_0 + e^{At}\int_0^t e^{-As}Bu(s)ds. \tag{9.35}$$

Let

$$\mathcal{A}(t_1, \boldsymbol{x}_0) = \left\{ \begin{bmatrix} x_1(t_1; u(\cdot)) \\ x_2(t_1; u(\cdot)) \end{bmatrix} = \boldsymbol{x}(t_1; u(\cdot)) : \boldsymbol{x}(0) \right.$$
$$\left. = \boldsymbol{x}_0 \text{ is given, } |u(t)| \leq 1 \right\}$$

denote the *Attainable Set* which is the set of all terminal points that can be "attained" from $\boldsymbol{x}_0$ at time t_1 by using a control satisfying $|u(t)| \leq 1$. If $\boldsymbol{x}_0 = \boldsymbol{0}$, then we denote $\mathcal{A}(t, \boldsymbol{0})$ by $\mathcal{A}(t)$.

One can prove that $\mathcal{A}(t_1, \boldsymbol{x}_0)$ is bounded and convex (see the problems below). Moreover, the sets are "continuous in time" as described below. In order to make this precise we need to define distances between sets. Assume $\boldsymbol{B}$ is a non-empty set in $\mathbb{R}^2$ and the vector $\boldsymbol{x} \in \mathbb{R}^2$ is fixed. The distance between the point $\boldsymbol{x}$ and the set $\boldsymbol{B}$ is defined by

$$d(\boldsymbol{x}, \boldsymbol{B}) = \inf\left\{\|\boldsymbol{x} - \boldsymbol{b}\| : \boldsymbol{b} \in \boldsymbol{B}\right\}. \tag{9.36}$$

If $\boldsymbol{A}$ and $\boldsymbol{B}$ are two non-empty sets in $\mathbb{R}^2$, define

$$d(\boldsymbol{A}, \boldsymbol{B}) = \sup\left\{d(\boldsymbol{a}, \boldsymbol{B}) : \boldsymbol{a} \in \boldsymbol{A}\right\}$$
$$= \sup\inf\left\{\|\boldsymbol{a} - \boldsymbol{b}\| : \boldsymbol{a} \in \boldsymbol{A}, \quad \boldsymbol{b} \in \boldsymbol{B}\right\}.$$

It is easy to show that if $\boldsymbol{A}$ and $\boldsymbol{B}$ are bounded, then $d(\boldsymbol{A}, \boldsymbol{B}) < +\infty$. Moreover, if $d(\boldsymbol{A}, \boldsymbol{B}) = 0$, then $\boldsymbol{A} \subseteq \boldsymbol{B}$, but it is not necessarily true that $\boldsymbol{A} = \boldsymbol{B}$. The *Hausdorff distance* between two bounded sets $\boldsymbol{A}$ and $\boldsymbol{B}$ is defined by

$$d_H(\boldsymbol{A}, \boldsymbol{B}) = max\,\{d(\boldsymbol{A}, \boldsymbol{B}), d(\boldsymbol{B}, \boldsymbol{A})\} \tag{9.37}$$

Lemma 9.1 (Basic Properties) *Assume that* $\boldsymbol{A}$ *and* $\boldsymbol{B}$ *are bounded sets.*

1. *If* $d_H(\boldsymbol{A}, \boldsymbol{B}) = 0$*, then* $\boldsymbol{A}$ *and* $\boldsymbol{B}$ *have the same closure.*
2. *If* $\boldsymbol{A}$ *and* $\boldsymbol{B}$ *are compact and* $d_H(\boldsymbol{A}, \boldsymbol{B}) = 0$*, then* $\boldsymbol{A} = \boldsymbol{B}$*.*
3. *If* $\boldsymbol{A}$*,* $\boldsymbol{B}$ *and* $\boldsymbol{C}$ *are compact sets then the triangle inequality*

$$d_H(\boldsymbol{A}, \boldsymbol{C}) \le d_H(\boldsymbol{A}, \boldsymbol{B}) + d_H(\boldsymbol{B}, \boldsymbol{C})$$

holds.

For each $0 < t$, let $\mathcal{A}(t) = \mathcal{A}(t, \boldsymbol{0})$ be the attainable set defined above. Then the following results hold.

Lemma 9.2 (Properties of the Attainable Set) *The attainable set* $\mathcal{A}(t)$ *is symmetric, bounded, convex and closed.*

Lemma 9.3 (Continuity of the Attainable Set) *Let* $0 < \hat{t}$*. If* $t \longrightarrow \hat{t}$*, then* $d_H(\mathcal{A}(t), \mathcal{A}(\hat{t})) \to 0$*.*

Remark 9.2 *Except for showing that the attainable set is closed, proofs of the previous lemmas are straightforward and are given as exercises. In general,* $\mathcal{A}(t)$ *is not closed (see [97] and [164]), but in this particular rocket car problem the special structure of the control constraint set* $\Omega = [-1, +1]$ *implies that* $\mathcal{A}(t)$ *is closed. This is a rather subtle but important issue and the core difficultly arises because we assume the controls belong to the space of piecewise continuous functions. There are two approaches to deal with this issue. One approach is to "enlarge" the class of controllers to the bigger set of (Lebesgue) integrable functions and follow the method in [100]. Another approach is to "restrict" the control constraint* *sets to be of a special type (in this case, the interval $[-1, +1]$). Although Lebesgue integration and measure theory are the foundation of modern analysis and this approach provides the most advanced theory, the mathematical background required to apply this method lies outside the scope of this book. Thus, we will simply assume the control constraint set Ω has a structure sufficient to imply that $\mathcal{A}(t)$ is closed (see the problems at the end of this chapter).*

Let t_1^* be the first time that $\mathcal{A}(t, \boldsymbol{x}_0)$ contains the target $\boldsymbol{x}_1 = \boldsymbol{0} = [0\ 0]^T$. In particular, $\boldsymbol{x}_1 \in \mathcal{A}(t_1^*, \boldsymbol{x}_0)$ and $\boldsymbol{x} \notin \mathcal{A}(t, \boldsymbol{x}_0)$ for any $0 < t < t_1^*$. When the attainable set $\mathcal{A}(t_1^*, \boldsymbol{x}_0)$ first touches $\boldsymbol{x}_1 = \boldsymbol{0}$, the state $\boldsymbol{0}$ lies on the boundary of the convex set $\mathcal{A}(t_1^*, \boldsymbol{x}_0)$. Hence, there is a support plane at $\boldsymbol{x}(t_1^*; u^*(\cdot)) = \boldsymbol{x}_1 = \boldsymbol{0}$ with an outer normal given by

$$\hat{\boldsymbol{\eta}} = \begin{bmatrix} \hat{\eta}_1 \\ \hat{\eta}_2 \end{bmatrix}$$

as shown in Figure 9.1. Let $u(\cdot)$ be any other control satisfying $|u(t)| \le 1$ with trajectory ending at $\boldsymbol{x}(t_1^*; u(\cdot)) \in \mathcal{A}(t_1^*, \boldsymbol{x}_0)$. If θ is the angle between the outer normal and $\boldsymbol{x}(t_1^*; u(\cdot))$, then $\frac{\pi}{2} \le \theta \le \frac{3\pi}{2}$ and hence one has

$$\langle \hat{\boldsymbol{\eta}}, \boldsymbol{x}(t_1^*; u(\cdot)) \rangle \le 0 = \langle \hat{\boldsymbol{\eta}}, \boldsymbol{0} \rangle = \langle \hat{\boldsymbol{\eta}}, \boldsymbol{x}(t_1^*; u^*(\cdot)) \rangle,$$

or equivalently,

$$\left\langle \begin{bmatrix} \hat{\eta}_1 \\ \hat{\eta}_2 \end{bmatrix}, \begin{bmatrix} x_1(t_1^*; u(\cdot)) \\ x_2(t_1^*; u(\cdot)) \end{bmatrix} \right\rangle \le \left\langle \begin{bmatrix} \hat{\eta}_1 \\ \hat{\eta}_2 \end{bmatrix}, \begin{bmatrix} x_1(t_1^*; u^*(\cdot)) \\ x_2(t_1^*; u^*(\cdot)) \end{bmatrix} \right\rangle = 0.$$

Therefore, we have shown that the optimal control $u^*(\cdot)$ maximizes the inner product

$$\left\langle \begin{bmatrix} \hat{\eta}_1 \\ \hat{\eta}_2 \end{bmatrix}, \begin{bmatrix} x_1(t_1^*; u(\cdot)) \\ x_2(t_1^*; u(\cdot)) \end{bmatrix} \right\rangle$$

and this maximum is 0.

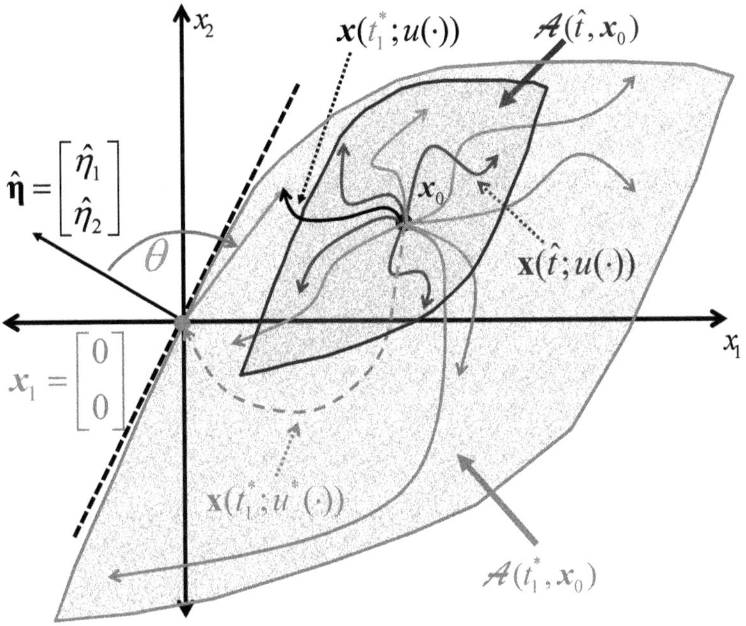

Figure 9.1: Attainable Sets at Time $\hat{t}$ and t_1^*

Consequently, it follows that

$$\langle \hat{\boldsymbol{\eta}}, \boldsymbol{x}(t_1^*; u(\cdot))\rangle = \hat{\eta}_1 x_1(t_1^*; u(\cdot)) + \hat{\eta}_2 x_2(t_1^*; u(\cdot)) \le 0,$$

for each control satisfying $|u(t)| \le 1$. Also, since

$$\langle \hat{\boldsymbol{\eta}}, \boldsymbol{x}(t_1^*; u^*(\cdot))\rangle = 0,$$

it follows that

$$\langle \hat{\boldsymbol{\eta}}, \boldsymbol{x}(t_1^*; u^*(\cdot))\rangle = \max_{|u(t)|\le 1} \langle \hat{\boldsymbol{\eta}}, \boldsymbol{x}(t_1^*; u(\cdot))\rangle = 0.$$

For this simple problem we can solve the system of equations and use this explicit solution to derive a useful necessary condition. We consider the case for $m = 1$ and note that for a given $u(\cdot)$,

$$x_1(t; u(\cdot)) = x_0 + v_0 t + \int_0^t \left[\int_0^s u(\sigma)d\sigma\right] ds \tag{9.38}$$

and

$$x_2(t;u(\cdot)) = \dot{x}(t) = v(t) = v_0 + \int_0^t u(\sigma)d\sigma. \tag{9.39}$$

For $0 \le t \le t_1^*$ define the two linear functions $\eta_1^*(\cdot)$ and $\eta_2^*(\cdot)$ by

$$\eta_1^*(t) \equiv \hat{\eta}_1, \tag{9.40}$$

$$\eta_2^*(t) = \hat{\eta}_1(t_1^* - t) + \hat{\eta}_2, \tag{9.41}$$

respectively. Note that $\eta_1^*(\cdot)$ is a constant function and $\eta_2^*(\cdot)$ is a linear function. The vector function

$$\eta^*(\cdot) = \begin{bmatrix} \eta_1^*(\cdot) \\ \eta_2^*(\cdot) \end{bmatrix}$$

has the property that

$$\eta^*(t_1^*) = \begin{bmatrix} \eta_1^*(t_1^*) \\ \eta_2^*(t_1^*) \end{bmatrix} = \begin{bmatrix} \hat{\eta}_1 \\ \hat{\eta}_2 \end{bmatrix}.$$

Also,

$$\begin{aligned}
\langle \boldsymbol{\eta}^*(t_1^*), x(t_1^*;u(\cdot))\rangle &= \langle \hat{\boldsymbol{\eta}}, x(t_1^*;u(\cdot))\rangle \\
&= \hat{\eta}_1 \left[x_0 + v_0 t_1^* + \int_0^{t_1^*}\int_0^s u(\sigma)d\sigma ds \right] \\
&\quad + \hat{\eta}_2 \left[v_0 + \int_0^{t_1^*} u(\sigma)d\sigma \right] \\
&= \hat{\eta}_1 \left[x_0 + v_0 t_1^* \right] + \left[\hat{\eta}_1 \int_0^{t_1^*}\int_0^s u(\sigma)d\sigma ds \right] \\
&\quad + \hat{\eta}_2 v_0 + \hat{\eta}_2 \int_0^{t_1^*} u(\sigma)d\sigma.
\end{aligned}$$

Since the control only appears in the integrals, we have that

$$\max_{|u(t)|\le 1} \langle \eta^*(t_1^*), x(t_1^*; u(\cdot))\rangle = \hat{\eta}_1 \left[x_0 + v_0 t_1^*\right] + \max_{|u(t)|\le 1} \left[\hat{\eta}_1 \int_0^{t_1^*}\int_0^s u(\sigma)d\sigma ds\right] + \hat{\eta}_2 v_0 + \max_{|u(t)|\le 1} \hat{\eta}_2 \int_0^{t_1^*} u(\sigma)d\sigma.$$

Therefore, to find $u^*(\cdot)$ so that

$$\langle \eta^*(t_1^*), x(t_1^*; u^*(\cdot))\rangle = \max_{|u(t)|\le 1} \langle \eta^*(t_1^*), x(t_1^*; u(\cdot))\rangle,$$

we only have to find the control $u^*(\cdot)$ so that

$$\hat{\eta}_1 \int_0^{t_1^*}\int_0^s u^*(\sigma)d\sigma ds + \hat{\eta}_2 \int_0^{t_1^*} u^*(\sigma)d\sigma = \max_{|u(t)|\le 1} \left\{\hat{\eta}_1 \int_0^{t_1^*}\int_0^s u(\sigma)d\sigma ds + \hat{\eta}_2 \int_0^{t_1^*} u(\sigma)d\sigma\right\}.$$

Observe that interchanging the order of integration and integration by parts yields

$$\hat{\eta}_1 \int_0^t\int_0^s u(\sigma)d\sigma ds = \int_0^t \hat{\eta}_1(t-\sigma)u(\sigma)d\sigma,$$

so that

$$\left\{\hat{\eta}_1 \int_0^{t_1^*}\int_0^s u(\sigma)d\sigma ds + \hat{\eta}_2 \int_0^{t_1^*} u(\sigma)d\sigma\right\} = \int_0^{t_1^*} \hat{\eta}_1(t_1^*-\sigma)u(\sigma)d\sigma + \hat{\eta}_2 \int_0^{t_1^*} u(\sigma)d\sigma$$

$$= \int_0^{t_1^*} \{\hat{\eta}_1(t_1^* - \sigma) + \hat{\eta}_2\} u(\sigma) d\sigma$$

$$= \int_0^{t_1^*} \{\eta_2^*(\sigma)\} u(\sigma) d\sigma,$$

where $\eta_2^*(\cdot)$ is defined by (9.41) above. Hence, the maximization problem is equivalent to

$$\max_{|u(t)| \le 1} \langle \eta^*(t_1^*), x(t_1^*; u(\cdot)) \rangle = \max_{|u(t)| \le 1} \int_0^{t_1^*} \eta_2^*(\sigma) u(\sigma) d\sigma.$$

Since $|u(t)| \le 1$, it follows that $u^*(\cdot)$ will maximize

$$\int_0^{t_1^*} \eta_2^*(\sigma) u(\sigma) d\sigma$$

if and only if

$$u^*(t) = sgn\{\eta_2^*(t)\},$$

where the *sgn* function is defined by

$$sgn\{z\} = \begin{cases} +1, & z > 0, \\ 0, & z = 0, \\ -1, & z < 0. \end{cases}$$

Since $\eta_2^*(\cdot)$ is linear, it follows that

$$u^*(t) = \begin{cases} +1, & \eta_2^*(t) > 0, \\ 0, & \eta_2^*(t) = 0, \\ -1, & \eta_2^*(t) < 0 \end{cases}$$

and $u^*(t) = \pm 1$ can change sign at most one time (when $\eta_2^*(t)$ crosses the axis).

To finish the problem we need to find a way to compute $\eta_2^*(\cdot)$. Note that the definitions (9.40)-(9.41) imply

$$\dot{\eta}_1^*(t) = 0,$$

and

$$\dot{\eta}_2^*(t) = \frac{d}{dt}\left\{\hat{\eta}_1(t_1^* - t) + \hat{\eta}_2\right\} = -\hat{\eta}_1 = -\eta_1^*(t).$$

We can write this as a system of the form

$$\begin{bmatrix} \dot{\eta}_1^*(t) \\ \dot{\eta}_2^*(t) \end{bmatrix} = \begin{bmatrix} 0 & 0 \\ -1 & 0 \end{bmatrix} \begin{bmatrix} \eta_1^*(t) \\ \eta_2^*(t) \end{bmatrix}, \tag{9.42}$$

with terminal condition

$$\begin{bmatrix} \eta_1^*(t_1^*) \\ \eta_2^*(t_1^*) \end{bmatrix} = \begin{bmatrix} \hat{\eta}_1 \\ \hat{\eta}_2 \end{bmatrix}. \tag{9.43}$$

It is important to observe that the system (9.42) can be written as

$$\dot{\boldsymbol{\eta}}^*(t) = -A^T\boldsymbol{\eta}^*(t), \tag{9.44}$$

where

$$A = \begin{bmatrix} 0 & 1 \\ 0 & 0 \end{bmatrix}$$

is the matrix defining the state equation. The system defined by (9.44) is called the *adjoint system*, or the *co-state system*. We shall return to this point later.

Since the optimal control is determined by the sign of the linear function $\eta_2^*(t) = \hat{\eta}_1(t_1^* - t) + \hat{\eta}_2$, we know that $u^*(t) = 1$ when $\eta_2^*(t) > 0$ and $u^*(t) = -1$ when $\eta_2^*(t) < 0$. This type of control is known as a *bang-bang* control, meaning that the optimal control takes on the extreme values of the constraint set $\Omega = [-1, 1]$. Thus, we need only consider controls that take the value of $+1$ or -1 and this allows us to "synthesize" the optimal controller by the following method.

Rather than fixing the initial condition and integrating forward in time, we fix the terminal condition at $\boldsymbol{x}_0 = \begin{bmatrix} 0 & 0 \end{bmatrix}^T$ and integrate the equation backwards in time to see what possible states can be reached by controls of this form. Thus, consider the problem

$$\dot{x}_1(t) = x_2(t),$$
$$\dot{x}_2(t) = u(t)$$

with initial values

$$x_1(0) = 0, \quad x_2(0) = 0$$

and integrate backward in time $-\infty < t < 0$. When $u(t) = -1$ the solutions are given by

$$x_1(t) = \frac{-t^2}{2} < 0,$$
$$x_2(t) = -t > 0.$$

Observe that

$$x_2(t) = +\sqrt{-2x_1(t)}$$

so that the backward trajectory lies on the parabola

$$\Gamma^- = \left\{ [x \ \ y]^T : y = +\sqrt{-2x}, \ -\infty < x \le 0 \right\}.$$

Likewise, if $u(t) = +1$, then

$$x_1(t) = \frac{t^2}{2} > 0,$$
$$x_2(t) = t < 0.$$

In this case we have

$$x_2(t) = -\sqrt{2x_1(t)}$$

and the backward trajectory lies on the parabola

$$\Gamma^+ = \left\{ [x \ \ y]^T : y = -\sqrt{2x}, \ 0 \le x < +\infty \right\}.$$

Let $\Gamma = \Gamma^- \cup \Gamma^+$ denote the union of the curves Γ^- and Γ^+ denote the *switching curve* as shown in Figure 9.2.

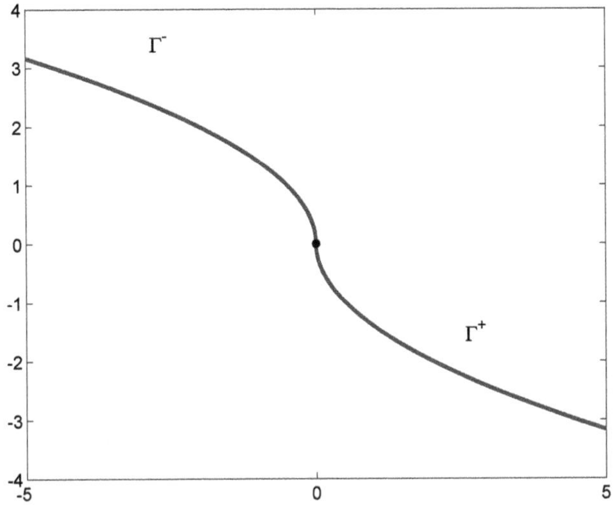

Figure 9.2: The Switching Curves

When time is positive and if $u(t) = -1$, then any initial state on the curve Γ^- will follow the parabola until it reaches $\boldsymbol{x}_1 = \mathbf{0}$ at a time t_1^*. If the control is then turned off (i.e. $u(t) = 0$ for $t \geq t_1^*$), then

$$u^*(t) = \begin{cases} -1, & 0 \leq t \leq t_1^*, \\ 0, & t_1^* < t, \end{cases}$$

is optimal. See Figure 9.3. Likewise, if the initial state is on the curve Γ^+, then

$$u^*(t) = \begin{cases} +1, & 0 \leq t \leq t_1^*, \\ 0, & t_1^* < t \end{cases}$$

will be optimal. Here, t_1^* is the time it takes for the trajectory to reach $\boldsymbol{x}_1 = \mathbf{0}$. See Figure 9.4.

When the initial data lies above Γ one applies a control of $u^*(t) = -1$ until the trajectory intersects the switching curve on Γ^+ at some time t_s^*. At this time the optimal control is switched to $u^*(t) = +1$ until the (optimal) time t_1^* when the trajectory reaches $\boldsymbol{x}_1 = \mathbf{0}$. At $t = t_1^*$ the control is set to $u^*(t) = 0$ for all time $t > t_1^*$.

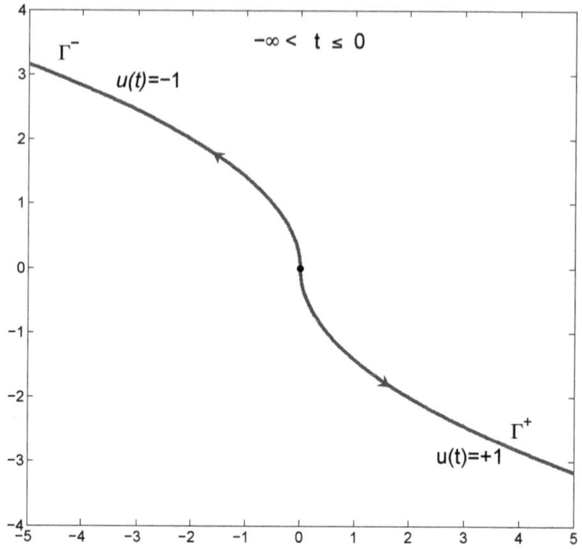

Figure 9.3: Switching Curves for Negative Times

Thus, the optimal control is

$$u^*(t) = \begin{cases} -1, & 0 \le t \le t_s^*, \\ +1, & t_s^* < t \le t_1^*, \\ 0, & t_1^* < t. \end{cases}$$

When the initial data lies below Γ one applies a control of $u^*(t) = +1$ until the trajectory intersects the switching curve Γ^- at some time t_s^*. At this time the optimal control is switched to $u^*(t) = -1$ until the (optimal) time t_1^* when the trajectory reaches $\boldsymbol{x}_1 = \mathbf{0}$. At $t = t_1^*$ the control is set to $u^*(t) = 0$ for all time $t > t_1^*$. Thus, the optimal control is

$$u^*(t) = \begin{cases} +1, & 0 \le t \le t_s^*, \\ -1, & t_s^* < t \le t_1^*, \\ 0, & t_1^* < t. \end{cases}$$

Let $W : \mathbb{R} \longrightarrow \mathbb{R}$ be defined by

$$x_2 = W(x_1) = \begin{cases} -\sqrt{2x_1}, & x_1 \ge 0, \\ +\sqrt{-2x_1}, & x_1 < 0. \end{cases} \tag{9.45}$$

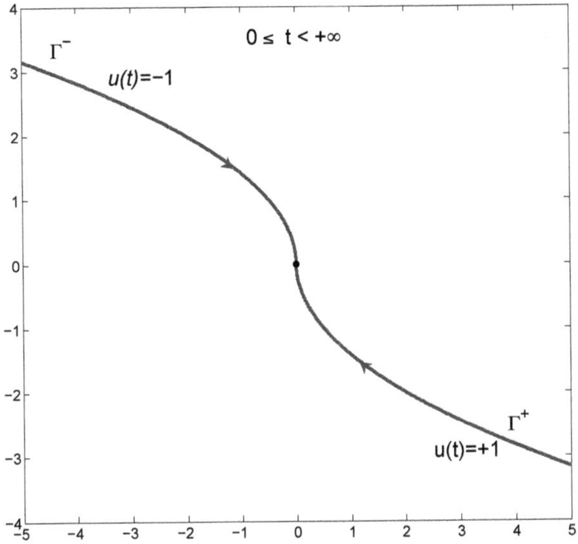

Figure 9.4: Switching Curves for Positive Times

Define the function $\Psi(x_1, x_2) : \mathbb{R}^2 \longrightarrow \mathbb{R}$ by

$$\Psi(x_1, x_2) = \begin{cases} -1, & \text{if } x_2 > W(x_1) \text{ or } (x_1, x_2) \in \Gamma^-, \\ 0, & \text{if } x_1 = x_2 = 0, \\ +1, & \text{if } x_2 < W(x_1) \text{ or } (x_1, x_2) \in \Gamma^+. \end{cases} \tag{9.46}$$

It follows that the optimal control is given by the *feedback control law*

$$u^*(t) = \Psi(x_1^*(t), x_2^*(t)) = \Psi(\boldsymbol{x}^*(t)), \tag{9.47}$$

where $\boldsymbol{x}^*(t) = \begin{bmatrix} x_1^*(t) \\ x_2^*(t) \end{bmatrix}$ is the optimal trajectory.

The previous synthesis can be summarized as follows. The *state equation* is given by

$$\dot{\boldsymbol{x}}(t) = A\boldsymbol{x}(t) + Bu(t), \quad \boldsymbol{x}(0) = \boldsymbol{x}_0 \tag{9.48}$$

where the matrices A and B are defined by

$$A = \begin{bmatrix} 0 & 1 \\ 0 & 0 \end{bmatrix} \text{ and } B = \begin{bmatrix} 0 \\ 1 \end{bmatrix},$$

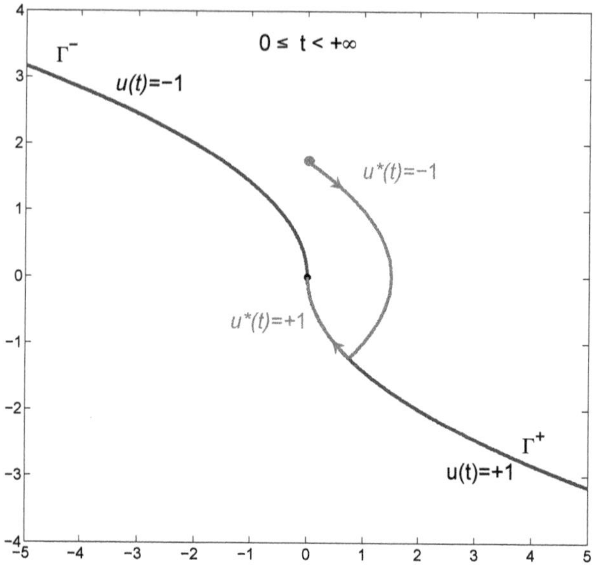

Figure 9.5: Optimal Trajectories for Initial State above Γ

respectively. Here, $\boldsymbol{x}_0 = \left[\, x_{1,0} \;\; x_{2,0} \,\right]^T$ and $\boldsymbol{x}(t) = \left[\, x_1(t) \;\; x_2(t) \,\right]^T$. The *adjoint system* (9.42) is defined by

$$\dot{\boldsymbol{\eta}}^*(t) = -A^T\boldsymbol{\eta}^*(t), \quad \boldsymbol{\eta}^*(t_1^*) = \hat{\boldsymbol{\eta}}, \tag{9.49}$$

where $\boldsymbol{\eta}^*(t) = \left[\, \eta_1^*(t) \;\; \eta_2^*(t) \,\right]^T$ satisfies the terminal boundary condition $\boldsymbol{\eta}^*(t_1^*) = \hat{\boldsymbol{\eta}} = \left[\, \hat{\eta}_1 \;\; \hat{\eta}_2 \,\right]^T$. The optimal control is given by

$$u^*(t) = sgn(\eta_2^*(t)) = sgn\left(B^T\boldsymbol{\eta}^*(t)\right). \tag{9.50}$$

If we substitute (9.50) into (9.48), then we obtain the two point boundary value problem defined by the coupled system

$$\begin{aligned} \dot{\boldsymbol{x}}^*(t) &= A\boldsymbol{x}^*(t) + B[sgn\left(B^T\boldsymbol{\eta}^*(t)\right)] \\ \dot{\boldsymbol{\eta}}^*(t) &= -A^T\boldsymbol{\eta}^*(t) \end{aligned} \tag{9.51}$$

and boundary conditions

$$\boldsymbol{x}^*(0) = \boldsymbol{x}_0 \quad \boldsymbol{\eta}^*(t_1^*) = \hat{\boldsymbol{\eta}}, \tag{9.52}$$

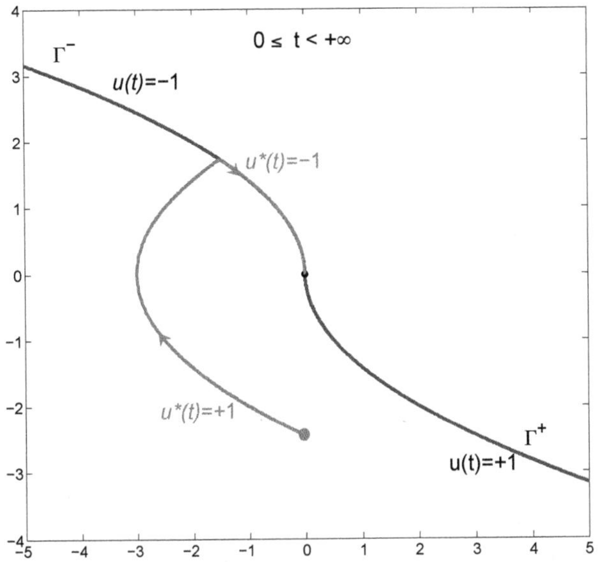

Figure 9.6: Optimal Trajectories for Initial State below Γ

where $\boldsymbol{x}^*(t)$ is the optimal trajectory. Consequently, if one can compute the outer normal $\hat{\boldsymbol{\eta}}$, then the optimal control can be computed by solving the two point boundary value problem (9.51) - (9.52) for $\boldsymbol{x}^*(t)$ and $\boldsymbol{\eta}^*(t)$ and using (9.50) to give $u^*(t) = sgn(\eta_2^*(t)) = sgn\left(B^T\boldsymbol{\eta}^*(t)\right)$.

9.4.2 The Bushaw Problem

Here we consider the so-called Bushaw Problem [52]. The general Bushaw type problem is concerned with second order control systems of the form

$$\ddot{x}(t) + 2\gamma\dot{x}(t) + \kappa x(t) = u(t), \tag{9.53}$$

where $\gamma \geq 0$ and $\kappa > 0$. This model can be thought of as a damped mechanical oscillator. We write this equation as a second order system

$$\frac{d}{dt}\begin{bmatrix} x_1(t) \\ x_2(t) \end{bmatrix} = \begin{bmatrix} 0 & 1 \\ -\kappa & -2\gamma \end{bmatrix}\begin{bmatrix} x_1(t) \\ x_2(t) \end{bmatrix} + \begin{bmatrix} 0 \\ 1 \end{bmatrix} u(t). \tag{9.54}$$

In particular, we write the system as

$$\dot{\boldsymbol{x}}(t) = A\boldsymbol{x}(t) + Bu(t) \tag{9.55}$$

with initial condition

$$\boldsymbol{x}(0) = \boldsymbol{x}_0, \tag{9.56}$$

where the matrices A and B are defined by

$$A = \begin{bmatrix} 0 & 1 \\ -\kappa & -2\gamma \end{bmatrix}, \text{ and } B = \begin{bmatrix} 0 \\ 1 \end{bmatrix},$$

respectively. Here, $\boldsymbol{x}_0 = \begin{bmatrix} x_{1,0} & x_{2,0} \end{bmatrix}^T$ where $x_{1,0}$ is the initial displacement and $x_{2,0}$ is the initial velocity of the oscillator defined by (9.53). Given a control $u(t)$, the solution to (9.55) - (9.56) is denoted by $\boldsymbol{x}(t;u(\cdot)) = \begin{bmatrix} x_1(t;u(\cdot)) & x_1(t;u(\cdot)) \end{bmatrix}^T$ and defines the state trajectory at time t with the initial state $\boldsymbol{x}(0;u(\cdot)) = \boldsymbol{x}_0 \in \mathbb{R}^2$. Again, we are interested in time optimal transfer of the initial state $\boldsymbol{x}_0$ to a fixed terminal state $\boldsymbol{x}_1 = \begin{bmatrix} 0 & 0 \end{bmatrix}^T \in \mathbb{R}^2$. We shall introduce some notation to make the statement of this problem more precise and formal.

Recall that a control $u(\cdot)$ is said to steer the state $\boldsymbol{x}_0$ to the state $\boldsymbol{x}_1$ in time t_1 if there is a solution to (9.55) - (9.56) and a **finite time** $t_1 > 0$ such that $\boldsymbol{x}(t_1;u(\cdot)) = \boldsymbol{x}_1 \in \mathbb{R}^2$. Let $\Omega = [-1,1] \subset \mathbb{R}^1$ be the *control constraint set* so that the condition $|u(t)| \leq 1$ can be written as $u(t) \in \Omega = [-1,1]$. Recall that the optimal controller for the rocket sled problem above was "bang-bang" and hence discontinuous. It turns out that many time optimal controllers are bang-bang (see [97] and [100]) and hence to ensure the existence of an optimal control we need to allow for such controls to be "admissible". Consequently, the set of *admissible controls* is a subset Θ of all *piecewise continuous functions* defined on $[0,+\infty)$ defined by

$$\Theta = \Theta(\boldsymbol{x}_0, \boldsymbol{x}_1) \triangleq \{u(\cdot) \in PWC[0,+\infty) : u(\cdot) \text{ steers } \boldsymbol{x}_0 \text{ to } \boldsymbol{x}_1, u(t) \in \Omega \; e.f.\} \tag{9.57}$$

where

$$PWC[0,+\infty) = \{u(\cdot) \in PWC[0,T] : \text{for all finite } T > 0\}. \tag{9.58}$$

Let $f_0 : \mathbb{R}^2 \times \mathbb{R}^1 \longrightarrow \mathbb{R}^1$ be defined by

$$f_0(\boldsymbol{x}, u) = f_0(x_1, x_2, u) \equiv 1$$

and define the cost functional $J : \Theta \longrightarrow \mathbb{R}^1$ by

$$J(u(\cdot)) \triangleq \int_0^{t_1} f_0(\boldsymbol{x}(s;u(s)), u(s))ds. \tag{9.59}$$

Note that if $u(\cdot) \in \Theta$, then $u(\cdot)$ steers $\boldsymbol{x}_0$ to $\boldsymbol{x}_1$ in time t_1 and $\boldsymbol{x}(s;u(s))$ is the corresponding solution of (9.55) - (9.56). Hence, $J(\cdot)$ is well defined and only depends on the control $u(\cdot) \in \Theta$. For the time optimal control problem considered here

$$J(u(\cdot)) \triangleq \int_0^{t_1} f_0(\boldsymbol{x}(s;u(s)), u(s))ds = \int_0^{t_1} 1ds = t_1 \tag{9.60}$$

is the time. If $u^*(\cdot) \in \Theta$ steers $\boldsymbol{x}_0$ to $\boldsymbol{x}_1$ in time t_1^*, then $u^*(\cdot)$ is said to be a *time optimal control* if

$$J(u^*(\cdot)) \leq J(u(\cdot))$$

for any control $u(\cdot) \in \Theta$. **The Time Optimal Control Problem is the problem of minimizing the functional $J(\cdot)$ on the set of admissible controls Θ.** In particular, the goal is to find $u^*(\cdot) \in \Theta$ such that

$$t_1^* = J(u^*(\cdot)) = \int_0^{t_1^*} f_0(\boldsymbol{x}(s;u^*(s)), u^*(s))ds \leq J(u(\cdot))$$
$$= \int_0^{t_1} f_0(\boldsymbol{x}(s;u(s)), u(s))ds = t_1.$$

for all $u(\cdot) \in \Theta$.

To obtain the optimal control we proceed as in the rocket sled problem and arrive at the same conclusion that the optimal control $u^*(\cdot)$ maximizes the inner product

$$\langle \hat{\boldsymbol{\eta}}, \boldsymbol{x}(t_1^*;u(\cdot)) \rangle$$

so that

$$\langle \hat{\boldsymbol{\eta}}, \boldsymbol{x}(t_1^*;u(\cdot))\rangle = \max_{u(\cdot)\in\Theta} \langle \hat{\boldsymbol{\eta}},\ \boldsymbol{x}(t_1^*;u(\cdot))\rangle = 0.$$

Applying the Variation of Parameters formula

$$\boldsymbol{x}(t_1^*;u(\cdot)) = e^{At_1^*}\boldsymbol{x}_0 + \int_0^{t_1^*} e^{A(t_1^*-s)}Bu(s)ds$$

we find that

$$\begin{aligned}\langle \hat{\boldsymbol{\eta}},\ \boldsymbol{x}(t_1^*;u(\cdot))\rangle &= \left\langle \hat{\boldsymbol{\eta}}, e^{At_1^*}\boldsymbol{x}_0 + \int_0^{t_1^*} e^{A(t_1^*-s)}Bu(s)ds\right\rangle \\ &= \left\langle \hat{\boldsymbol{\eta}}, e^{At_1^*}\boldsymbol{x}_0\right\rangle + \left\langle \hat{\boldsymbol{\eta}}, \int_0^{t_1^*} e^{A(t_1^*-s)}Bu(s)ds\right\rangle \\ &= \left\langle \hat{\boldsymbol{\eta}}, e^{At_1^*}\boldsymbol{x}_0\right\rangle + \int_0^{t_1^*} \left\langle \hat{\boldsymbol{\eta}}, [e^{A(t_1^*-s)}]Bu(s)\right\rangle ds.\end{aligned}$$

However,

$$\begin{aligned}\int_0^{t_1^*} \left\langle \hat{\boldsymbol{\eta}},\ [e^{A(t_1^*-s)}]Bu(s)\right\rangle ds &= \int_0^{t_1^*} \left\langle [e^{A(t_1^*-s)}]^T\hat{\boldsymbol{\eta}},\ Bu(s)\right\rangle ds \\ &= \int_0^{t_1^*} \left\langle [e^{A^T(t_1^*-s)}\hat{\boldsymbol{\eta}}],\ Bu(s)\right\rangle ds\end{aligned}$$

and if we define $\boldsymbol{\eta}^*(s)$ by

$$\boldsymbol{\eta}^*(s) = [e^{A^T(t_1^*-s)}\hat{\boldsymbol{\eta}}],$$

then

$$\begin{aligned}\langle \hat{\boldsymbol{\eta}},\ \boldsymbol{x}(t_1^*;u(\cdot))\rangle &= \left\langle \hat{\boldsymbol{\eta}},\ e^{At_1^*}\boldsymbol{x}_0\right\rangle + \int_0^{t_1^*} \langle \boldsymbol{\eta}^*(s),\ Bu(s)\rangle\, ds \\ &= \left\langle \hat{\boldsymbol{\eta}},\ e^{At_1^*}\boldsymbol{x}_0\right\rangle + \int_0^{t_1^*} \left\langle B^T\boldsymbol{\eta}^*(s),\ u(s)\right\rangle ds.\end{aligned}$$

In order to maximize

$$\langle \hat{\boldsymbol{\eta}},\ \boldsymbol{x}(t_1^*;u(\cdot))\rangle = \left\langle \hat{\boldsymbol{\eta}},\ e^{At_1^*}\boldsymbol{x}_0\right\rangle + \int_0^{t_1^*} \left\langle B^T\boldsymbol{\eta}^*(s),\ u(s)\right\rangle ds$$

over all $u(\cdot) \in \Theta$ it is clear that one needs only maximize the integral containing $u(\cdot)$. Hence, the optimal control satisfies

$$\int_0^{t_1^*} \langle B^T\boldsymbol{\eta}^*(s), u^*(s)\rangle\, ds = \max_{u(\cdot)\in\Theta} \int_0^{t_1^*} \langle B^T\boldsymbol{\eta}^*(s),\ u(s)\rangle\, ds$$

and, like in the rocket sled problem, $u^*(\cdot)$ must be

$$u^*(s) = sgn[B^T\boldsymbol{\eta}^*(s)]. \tag{9.61}$$

The only issue is how do we compute $B^T\boldsymbol{\eta}^*(s)$? Differentiating

$$\boldsymbol{\eta}^*(s) = [e^{A^T(t_1^*-s)}\hat{\boldsymbol{\eta}}]$$

it follows that

$$\dot{\boldsymbol{\eta}}^*(s) = -A^T[e^{A^T(t_1^*-s)}\hat{\boldsymbol{\eta}}] = -A^T\boldsymbol{\eta}^*(s). \tag{9.62}$$

Note that

$$\boldsymbol{\eta}^*(s) = [e^{A^T(t_1^*-s)}\ \hat{\boldsymbol{\eta}}] = [e^{-A^Ts}][e^{A^Tt_1^*}\ \hat{\boldsymbol{\eta}}] = [e^{-A^Ts}]\tilde{\boldsymbol{\eta}},$$

so that $\boldsymbol{\eta}^*(s)$ is the solution to the adjoint equation (9.62) with initial data

$$\boldsymbol{\eta}^*(0) = \tilde{\boldsymbol{\eta}} = [e^{A^Tt_1^*}\hat{\boldsymbol{\eta}}].$$

Therefore, to compute $\boldsymbol{\eta}^*(\cdot)$ from the outward normal $\hat{\boldsymbol{\eta}}$ we set $\tilde{\boldsymbol{\eta}} = [e^{A^Tt_1^*}\hat{\boldsymbol{\eta}}]$ and solve the initial value problem

$$\dot{\boldsymbol{\eta}}^*(s) = -A^T\boldsymbol{\eta}^*(s), \quad \boldsymbol{\eta}^*(0) = \tilde{\boldsymbol{\eta}}.$$

The optimal control is then given by

$$u^*(s) = sgn[B^T\boldsymbol{\eta}^*(s)] = sgn[B^Te^{-A^Ts}\tilde{\boldsymbol{\eta}}] \tag{9.63}$$

as indicated by (9.61). We will return to this problem and provide a complete solution for the special case where $\gamma = 0$ and $\kappa = 1$.

9.5 Problem Set for Chapter 9

Problem 9.1 *Consider the system*

$$\begin{aligned}\dot{x}_1(t) &= x_2(t),\\ \dot{x}_2(t) &= u(t),\end{aligned}$$

and assume $|u(t)| \leq 1$. *Compute and sketch the attainable set* $\mathcal{A}(t)$ *for* $t = 1$ *and* $t = 2$.

Problem 9.2 *Consider the system*

$$\begin{aligned}\dot{x}_1(t) &= u_1(t),\\ \dot{x}_2(t) &= u_2(t),\\ \dot{x}_3(t) &= -1.\end{aligned}$$

Show that there is a control $[u_1(t) \;\; u_2(t)]^T$ *that steers* $[0 \;\; 0 \;\; 1]^T$ *to* $[0 \;\; 0 \;\; 0]^T$ *and satisfies* $[u_1(t)]^2 + [u_2(t)]^2 \leq 1$.

Problem 9.3 *Consider the system*

$$\begin{aligned}\dot{x}_1(t) &= -x_2(t) + u_1(t),\\ \dot{x}_2(t) &= u_2(t),\end{aligned}$$

and assume $|u_1(t)| \leq 1$ *and* $|u_2(t)| \leq 1$. *Compute and sketch the attainable set* $\mathcal{A}(t)$ *for* $t = 2$.

Problem 9.4 *Consider the system*

$$\dot{x}(t) = x(t) + u(t),$$

and assume $u(t) \in \Omega$. *Compute and sketch the attainable set* $\mathcal{A}(t)$ *for* $t = 1$ *when* $\Omega = [-1, 0]$, $\Omega = [-1, 1]$ *and* $\Omega = [0, 1]$.

Problem 9.5 *Consider the system*

$$\begin{aligned}\dot{x}_1(t) &= u_1(t),\\ \dot{x}_2(t) &= u_2(t),\end{aligned}$$

and assume $|u_1(t)| \leq 1$ *and* $|u_2(t)| \leq 1$. *Show that for* $t > 0$ *the attainable set* $\mathcal{A}(t)$ *is a square.*

Problem 9.6 *Consider the system*

$$\dot{x}_1(t) = u_1(t),$$
$$\dot{x}_2(t) = u_2(t),$$

and assume $|u_1(t)|^2 + |u_2(t)|^2 \le 1$. *Compute and sketch the attainable set* $\mathcal{A}(t)$ *for* $t = 1$.

Problem 9.7 *Consider the system*

$$\dot{x}_1(t) = x_2(t),$$
$$\dot{x}_2(t) = -x_1(t) + u(t),$$

and assume $|u(t)| \le 1$. *Show that the attainable set* $\mathcal{A}(t)$ *has "corners" for* $0 < t < \pi$ *and is a circle of radius* 2 *for* $t = \pi$.

Problem 9.8 *Compute the distances* $d(A, B)$ *and* $d_H(A, B)$ *between the following sets in* $\mathbb{R}^1$:
(1) $A = \{1, 2, 3, 4\}$ *and* $B = \{2, 3\}$;
(2) $A = \{2, 3\}$ *and* $B = \{1, 2, 3, 4\}$;
(3) $A = [0, 1]$ *and* $B = [0, 2]$;
(4) $A = [0, 2]$ *and* $B = [0, 1]$.

Problem 9.9 *Let* A *be the constant matrix*

$$A = \begin{bmatrix} -1 & R \\ 0 & -2 \end{bmatrix},$$

where $R \ge 0$. *Compute the matrix exponential* e^{At}.

Problem 9.10 *Let* A *be the constant matrix*

$$A = \begin{bmatrix} 0 & 1 \\ -1 & 0 \end{bmatrix}.$$

Compute the matrix exponential e^{At}.

Problem 9.11 *Consider the system*

$$\dot{\boldsymbol{x}}(t) = A\boldsymbol{x}(t) + B\boldsymbol{u}(t),$$

where A is a $n \times n$ constant matrix and B is a $n \times m$ constant matrix. Derive the variation of parameters formula

$$\boldsymbol{x}(t) = e^{At}\boldsymbol{x}_0 + \int_0^t e^{A(t-s)} B\boldsymbol{u}(s)ds.$$

Advanced Problems

Problem 9.12 *Prove **Lemma** 9.1.*

Problem 9.13 *Prove **Lemma** 9.2.*

Problem 9.14 *Prove **Lemma** 9.3.*

Problem 9.15 *Consider the control system*

$$\dot{\boldsymbol{x}}(t) = A\boldsymbol{x}(t) + B\boldsymbol{u}(t),$$

where A is a $n \times n$ constant matrix and B is a $n \times m$ constant matrix. Assume $\Omega \subseteq \mathbb{R}^m$ is a compact and convex set. Show that the attainable set $\mathcal{A}(t)$ is convex and bounded. Also, prove that if Ω is a convex polyhedron, then $\mathcal{A}(t)$ is closed. (Hint: See [97] and [121].)

Chapter 10

Simplest Problem in Optimal Control

As in the previous chapters, let $PWC(a,b;\mathbb{R}^m)$ denote the space of all $\mathbb{R}^m$-valued piecewise continuous functions defined on $[a,b]$. Also, $PWC(a,+\infty;\mathbb{R}^m)$ denotes the set of all functions $\boldsymbol{u}(\cdot) : [a,+\infty) \to \mathbb{R}^m$ such that $\boldsymbol{u}(\cdot) \in PWC(a,T;\mathbb{R}^m)$ for all $a < T < +\infty$. A function $\boldsymbol{u}(\cdot) \in PWC(t_0,+\infty;\mathbb{R}^m)$ is called a control or control function. We start with a statement of the Simplest Problem in Optimal Control (**SPOC**) and state the basic Maximum Principle for this simplest problem.

10.1 SPOC: Problem Formulation

Assume we are given functions of class C^2

$$f : \mathbb{R}^n \times \mathbb{R}^m \to \mathbb{R}^n, \;\; f_0 : \mathbb{R}^n \times \mathbb{R}^m \to \mathbb{R},$$

initial set and time

$$\mathrm{X}_0 \subseteq \mathbb{R}^n, \;\; t_0 \in \mathbb{R},$$

terminal set (but no final time t_1)

$$\mathrm{X}_1 \subseteq \mathbb{R}^n,$$

and control constraint set

$$\Omega \subseteq \mathbb{R}^m.$$

The state equation (or control system or control process) is defined by the system of differential equations

$$(\mathcal{S}) \qquad \dot{\boldsymbol{x}}(t) = f(\boldsymbol{x}(t), \boldsymbol{u}(t)), \quad t_0 < t. \tag{10.1}$$

The initial conditions are given by

$$\boldsymbol{x}(t_0) = \boldsymbol{x}_0 \in \mathrm{X}_0. \tag{10.2}$$

Definition 10.1 *We say that a control $\boldsymbol{u}(\cdot) \in PWC(t_0, +\infty; \mathbb{R}^m)$* ***steers the state $\boldsymbol{x}_0$ to the state $\boldsymbol{x}_1$*** *if there is a solution $\boldsymbol{x}(\cdot) = \boldsymbol{x}(\cdot; \boldsymbol{x}_0; \boldsymbol{u}(\cdot))$ to the initial value problem (10.1) - (10.2) satisfying*

$$\boldsymbol{x}(t_1) = \boldsymbol{x}_1$$

for some finite time $t_0 < t_1$. A control $\boldsymbol{u}(\cdot) \in PWC(t_0, +\infty; \mathbb{R}^m)$ ***steers the set X_0 to the set X_1*** *if for some $\boldsymbol{x}_0 \in \mathrm{X}_0$, there is a solution $\boldsymbol{x}(\cdot) = \boldsymbol{x}(\cdot; \boldsymbol{x}_0; \boldsymbol{u}(\cdot))$ to the initial value problem (10.1) - (10.2) satisfying*

$$\boldsymbol{x}(t_1) = \boldsymbol{x}_1 \in \mathrm{X}_1$$

for some finite time $t_0 < t_1$.

We make the following standing assumption about the initial value problem.

> **Standing Assumption for Optimal Control:** *For each $\boldsymbol{x}_0 \in \mathrm{X}_0$ and $\boldsymbol{u}(\cdot) \in PWC(t_0, t_1; \mathbb{R}^m)$, the initial value problem (10.1) - (10.2) has a unique solution defined on $[t_0, t_1]$. We denote this solution by $\boldsymbol{x}(\cdot; \boldsymbol{x}_0; \boldsymbol{u}(\cdot))$.*

Definition 10.2 *The set of* ***admissible controllers*** *is the subset of $PWC(t_0, +\infty; \mathbb{R}^m)$ defined by*

$$\Theta = \left\{ \begin{array}{c} \boldsymbol{u}(\cdot) \in PWC(t_0, +\infty; \mathbb{R}^m) : \boldsymbol{u}(t) \in \Omega \ e.f. \\ \text{and } \boldsymbol{u}(\cdot) \text{ steers } \mathrm{X}_0 \text{ to } \mathrm{X}_1. \end{array} \right\}$$
$$= \Theta(t_0, \mathrm{X}_0, \mathrm{X}_1, \Omega). \tag{10.3}$$

Again, the abbreviation *e.f.* stands for "except at a finite number of points". Given a control $\boldsymbol{u}(\cdot) \in PWC(t_0, +\infty; \mathbb{R}^m)$ and a time $t_1 > t_0$, the **cost functional** at time $t_1 > t_0$ is defined by

$$J(\boldsymbol{u}(\cdot)) = \int_{t_0}^{t_1} f_0(\boldsymbol{x}(s), \boldsymbol{u}(s))ds = \int_{t_0}^{t_1} f_0(\boldsymbol{x}(s; \boldsymbol{u}(\cdot)), \boldsymbol{u}(s))ds, \quad (10.4)$$

where $\boldsymbol{x}(\cdot) = \boldsymbol{x}(\cdot; \boldsymbol{u}(\cdot))$ is the solution to the initial value problem (10.1) - (10.2).

The **Simplest Problem in Optimal Control** is the problem of minimizing $J(\cdot)$ on the set of all admissible controllers Θ. In particular, an **optimal control** is a function $\boldsymbol{u}^*(\cdot) \in \Theta$ such that $\boldsymbol{u}^*(\cdot)$ steers X_0 to X_1 in time $t_1^* > t_0$ and

$$\begin{aligned} J(\boldsymbol{u}^*(\cdot)) &= \int_{t_0}^{t_1^*} f_0(\boldsymbol{x}(s; \boldsymbol{u}^*(\cdot)), \boldsymbol{u}(s))ds \le \int_{t_0}^{t_1} f_0(\boldsymbol{x}(s; \boldsymbol{u}(\cdot)), \boldsymbol{u}(s))ds \\ &= J(\boldsymbol{u}(\cdot)) \end{aligned} \quad (10.5)$$

for all $\boldsymbol{u}(\cdot) \in \Theta$.

10.2 The Fundamental Maximum Principle

In this section we state the Basic Maximum Principle that applies to the SPOC stated above. We will not prove the basic result here. The proof may be found in several standard references. The books *The Calculus of Variations and Optimal Control* by Leitmann [120] and *Foundations of Optimal Control Theory* [119] by Lee and Markus both contain a complete proof of the basic result. These references provide the necessary background and alternative proofs of the theorem.

In order to state the Maximum Principle we first have to introduce some notation. Given a vector $\boldsymbol{z} = \begin{bmatrix} z_1 \ z_2 \ \dots \ z_n \end{bmatrix}^T \in \mathbb{R}^n$ and a real number z_0, define an *augmented vector* $\hat{\boldsymbol{z}} = \begin{bmatrix} z_0 \ z_1 \ z_2 \ \dots \ z_n \end{bmatrix}^T \in$

$\mathbb{R}^{n+1}$ by

$$\hat{\boldsymbol{z}} = [z_0\ z_1\ z_2\ ...\ z_n]^T = \begin{bmatrix} z_0 \\ \boldsymbol{z} \end{bmatrix} \in \mathbb{R}^{n+1}.$$

Also, given the smooth functions

$$f : \mathbb{R}^n \times \mathbb{R}^m \to \mathbb{R}^n, \;\; f_0 : \mathbb{R}^n \times \mathbb{R}^m \to \mathbb{R},$$

where $f : \mathbb{R}^n \times \mathbb{R}^m \to \mathbb{R}^n$ is given by

$$f(\boldsymbol{x}, \boldsymbol{u}) = \begin{bmatrix} f_1(\boldsymbol{x}, \boldsymbol{u}) \\ f_2(\boldsymbol{x}, \boldsymbol{u}) \\ f_3(\boldsymbol{x}, \boldsymbol{u}) \\ \vdots \\ f_n(\boldsymbol{x}, \boldsymbol{u}) \end{bmatrix} = \begin{bmatrix} f_1(x_1, x_2, ..., x_n, u_1, u_2, ..., u_m) \\ f_2(x_1, x_2, ..., x_n, u_1, u_2, ..., u_m) \\ f_3(x_1, x_2, ..., x_n, u_1, u_2, ..., u_m) \\ \vdots \\ f_n(x_1, x_2, ..., x_n, u_1, u_2, ..., u_m) \end{bmatrix},$$

it follows that all the partial derivatives

$$\frac{\partial f_i(x_1, x_2, ..., x_n, u_1, u_2, ..., u_m)}{\partial x_j}$$

and

$$\frac{\partial f_0(x_1, x_2, ..., x_n, u_1, u_2, ..., u_m)}{\partial x_j}$$

exist and are continuous. Define the augmented vector field

$$\hat{f} : \mathbb{R}^{n+1} \times \mathbb{R}^m \to \mathbb{R}^{n+1}$$

by

$$\hat{f}(\hat{\boldsymbol{x}}, \boldsymbol{u}) = \begin{bmatrix} f_0(\hat{\boldsymbol{x}}, \boldsymbol{u}) \\ f_1(\hat{\boldsymbol{x}}, \boldsymbol{u}) \\ f_2(\hat{\boldsymbol{x}}, \boldsymbol{u}) \\ \vdots \\ f_n(\hat{\boldsymbol{x}}, \boldsymbol{u}) \end{bmatrix} = \begin{bmatrix} f_0(\boldsymbol{x}, \boldsymbol{u}) \\ f_1(\boldsymbol{x}, \boldsymbol{u}) \\ f_2(\boldsymbol{x}, \boldsymbol{u}) \\ \vdots \\ f_n(\boldsymbol{x}, \boldsymbol{u}) \end{bmatrix}$$

$$= \begin{bmatrix} f_0(x_1, x_2, ..., x_n, u_1, u_2, ..., u_m) \\ f_1(x_1, x_2, ..., x_n, u_1, u_2, ..., u_m) \\ f_2(x_1, x_2, ..., x_n, u_1, u_2, ..., u_m) \\ \vdots \\ f_n(x_1, x_2, ..., x_n, u_1, u_2, ..., u_m) \end{bmatrix}. \qquad (10.6)$$

The **augmented control system** is defined by

$$\frac{d}{dt}\hat{\boldsymbol{x}}(t) = \hat{f}(\hat{\boldsymbol{x}}(t), \boldsymbol{u}(t)), \tag{10.7}$$

or equivalently, by the system

$$\frac{d}{dt}x_0(t) = f_0(\boldsymbol{x}(t), \boldsymbol{u}(t)) \tag{10.8}$$

$$\frac{d}{dt}x_i(t) = f_i(\boldsymbol{x}(t), \boldsymbol{u}(t)), \text{ for } i = 1, 2, ..., n. \tag{10.9}$$

Observe that if

$$x_0(t) = \int_{t_0}^{t} f_0(\boldsymbol{x}(s; \boldsymbol{u}(\cdot)), \boldsymbol{u}(s))ds,$$

then $x_0(\cdot)$ satisfies (10.8) with initial data $x_0(t_0) = 0$. In particular, $x_0(t)$ represents the cost of transferring the state from an initial $\boldsymbol{x}(t_0) = \boldsymbol{x}_0 \in \mathrm{X}_0$ to $\boldsymbol{x}(t; \boldsymbol{u}(\cdot))$ by the control $\boldsymbol{u}(\cdot)$.

Assume that $(\boldsymbol{x}^*(\cdot), \boldsymbol{u}^*(\cdot))$ is an optimal pair for the **SPOC**. In particular, $\boldsymbol{u}^*(\cdot) \in \Theta$ steers $\boldsymbol{x}_0^* \in \mathrm{X}_0$ to $\boldsymbol{x}_1^* \in \mathrm{X}_1$ in time t_1^* and $\boldsymbol{x}^*(\cdot) = \boldsymbol{x}(\cdot; \boldsymbol{u}^*(\cdot))$ is the corresponding optimal trajectory that satisfies the initial value problem (10.1) - (10.2) with $\boldsymbol{x}_0 = \boldsymbol{x}_0^* \in \mathrm{X}_0$. Define the $(n+1) \times (n+1)$ matrix $\hat{A}(t)$ by

$$\hat{A}(t) = \left[\frac{\partial \hat{f}_i(x_0, x_1, x_2, ..., x_n, u_1, u_2, ..., u_m)}{\partial x_j}\right]_{|(\hat{\boldsymbol{x}}^*(t), \boldsymbol{u}^*(t))}$$

$$= \left[\frac{\partial \hat{f}_i(\hat{\boldsymbol{x}}^*(t), \boldsymbol{u}^*(t))}{\partial x_j}\right], \tag{10.10}$$

so that

$$\hat{A}(t) = \begin{bmatrix} \frac{\partial f_0(\hat{\boldsymbol{x}}^*(t),\boldsymbol{u}^*(t))}{\partial x_0} & \frac{\partial f_0(\hat{\boldsymbol{x}}^*(t),\boldsymbol{u}^*(t))}{\partial x_1} & \frac{\partial f_0(\hat{\boldsymbol{x}}^*(t),\boldsymbol{u}^*(t))}{\partial x_2} & \cdots & \frac{\partial f_0(\hat{\boldsymbol{x}}^*(t),\boldsymbol{u}^*(t))}{\partial x_n} \\ \frac{\partial f_1(\hat{\boldsymbol{x}}^*(t),\boldsymbol{u}^*(t))}{\partial x_0} & \frac{\partial f_1(\hat{\boldsymbol{x}}^*(t),\boldsymbol{u}^*(t))}{\partial x_1} & \frac{\partial f_1(\hat{\boldsymbol{x}}^*(t),\boldsymbol{u}^*(t))}{\partial x_2} & \cdots & \frac{\partial f_1(\hat{\boldsymbol{x}}^*(t),\boldsymbol{u}^*(t))}{\partial x_n} \\ \vdots & \vdots & \vdots & \ddots & \vdots \\ \frac{\partial f_n(\hat{\boldsymbol{x}}^*(t),\boldsymbol{u}^*(t))}{\partial x_0} & \frac{\partial f_n(\hat{\boldsymbol{x}}^*(t),\boldsymbol{u}^*(t))}{\partial x_1} & \frac{\partial f_n(\hat{\boldsymbol{x}}^*(t),\boldsymbol{u}^*(t))}{\partial x_2} & \cdots & \frac{\partial f_n(\hat{\boldsymbol{x}}^*(t),\boldsymbol{u}^*(t))}{\partial x_n} \end{bmatrix}.$$

Moreover, since $\hat{f}_i(\hat{\boldsymbol{x}}, \boldsymbol{u}) = \hat{f}_i(\boldsymbol{x}, \boldsymbol{u}) = f_i(\boldsymbol{x}, \boldsymbol{u}) = f_i(x_1, x_2, ..., x_n, u_1, u_2, ..., u_m)$ does not depend on x_0, then

$$\frac{\partial f_i(\hat{\boldsymbol{x}}^*, \boldsymbol{u}^*)}{\partial x_0} = 0$$

for all $i = 0, 1, 2, ..., n$ and it follows that

$$\hat{A}(t) = \left[\frac{\partial \hat{f}_i(\hat{\boldsymbol{x}}^*(t), \boldsymbol{u}^*(t))}{\partial x_j}\right]$$
$$= \begin{bmatrix} 0 & \frac{\partial f_0(\hat{\boldsymbol{x}}^*(t),\boldsymbol{u}^*(t))}{\partial x_1} & \frac{\partial f_0(\hat{\boldsymbol{x}}^*(t),\boldsymbol{u}^*(t))}{\partial x_2} & \cdots & \frac{\partial f_0(\hat{\boldsymbol{x}}^*(t),\boldsymbol{u}^*(t))}{\partial x_n} \\ 0 & \frac{\partial f_1(\hat{\boldsymbol{x}}^*(t),\boldsymbol{u}^*(t))}{\partial x_1} & \frac{\partial f_1(\hat{\boldsymbol{x}}^*(t),\boldsymbol{u}^*(t))}{\partial x_2} & \cdots & \frac{\partial f_1(\hat{\boldsymbol{x}}^*(t),\boldsymbol{u}^*(t))}{\partial x_n} \\ \vdots & \vdots & \vdots & \ddots & \vdots \\ 0 & \frac{\partial f_n(\hat{\boldsymbol{x}}^*(t),\boldsymbol{u}^*(t))}{\partial x_1} & \frac{\partial f_n(\hat{\boldsymbol{x}}^*(t),\boldsymbol{u}^*(t))}{\partial x_2} & \cdots & \frac{\partial f_n(\hat{\boldsymbol{x}}^*(t),\boldsymbol{u}^*(t))}{\partial x_n} \end{bmatrix}.$$

It is important to note that the first column of $\hat{A}(t)$ is all zeros. Also, note that for $\boldsymbol{x} = [x_1\ x_2\ ...\ x_n]^T \in \mathbb{R}^n$, the gradient of $f_0(\boldsymbol{x}, \boldsymbol{u})$ with respect to $\boldsymbol{x}$ at $(\boldsymbol{x}, \boldsymbol{u})$ is defined by

$$\nabla_{\boldsymbol{x}} f_0(\boldsymbol{x}, \boldsymbol{u}) \triangleq \begin{bmatrix} \frac{\partial f_0(\boldsymbol{x},\boldsymbol{u})}{\partial x_1} \\ \frac{\partial f_0(\boldsymbol{x},\boldsymbol{u})}{\partial x_2} \\ \vdots \\ \frac{\partial f_0(\boldsymbol{x},\boldsymbol{u})}{\partial x_n} \end{bmatrix}$$

and the Jacobian matrix of $f(\boldsymbol{x}, \boldsymbol{u})$ with respect to $\boldsymbol{x} = [x_1\ x_2\ ...\ x_n]^T \in \mathbb{R}^n$ at $(\boldsymbol{x}, \boldsymbol{u})$ is given by

$$\mathbb{J}_{\boldsymbol{x}} f(\boldsymbol{x}, \boldsymbol{u}) = \begin{bmatrix} \frac{\partial f_1(\boldsymbol{x},\boldsymbol{u})}{\partial x_1} & \frac{\partial f_1(\boldsymbol{x},\boldsymbol{u})}{\partial x_2} & \cdots & \frac{\partial f_1(\boldsymbol{x},\boldsymbol{u})}{\partial x_n} \\ \frac{\partial f_2(\boldsymbol{x},\boldsymbol{u})}{\partial x_1} & \frac{\partial f_2(\boldsymbol{x},\boldsymbol{u})}{\partial x_2} & \cdots & \frac{\partial f_2(\boldsymbol{x},\boldsymbol{u})}{\partial x_n} \\ \vdots & \vdots & \ddots & \vdots \\ \frac{\partial f_n(\boldsymbol{x},\boldsymbol{u})}{\partial x_1} & \frac{\partial f_n(\boldsymbol{x},\boldsymbol{u})}{\partial x_2} & \cdots & \frac{\partial f_n(\boldsymbol{x},\boldsymbol{u})}{\partial x_n} \end{bmatrix}.$$

Consequently,

$$\hat{A}(t) = \left[\frac{\partial \hat{f}_i(\hat{\boldsymbol{x}}^*(t), \boldsymbol{u}^*(t))}{\partial x_j}\right] = \begin{bmatrix} 0 & [\nabla_{\boldsymbol{x}} f_0(\hat{\boldsymbol{x}}^*(t), \boldsymbol{u}^*(t))]^T \\ \mathbf{0} & [J_{\boldsymbol{x}} f(\hat{\boldsymbol{x}}^*(t), \boldsymbol{u}^*(t))] \end{bmatrix}$$

and the transpose matrix can be written as

$$[\hat{A}(t)]^T = \begin{bmatrix} 0 & \mathbf{0} \\ \nabla_{\boldsymbol{x}} f_0(\hat{\boldsymbol{x}}^*(t), \boldsymbol{u}^*(t)) & [J_{\boldsymbol{x}} f(\hat{\boldsymbol{x}}^*(t), \boldsymbol{u}^*(t))]^T \end{bmatrix}. \tag{10.11}$$

We now define the **adjoint equation** by

$$\frac{d}{dt}\hat{\boldsymbol{\eta}}(t) = -[\hat{A}(t)]^T \hat{\boldsymbol{\eta}}(t) \tag{10.12}$$

where $\hat{\eta}(t)$ given by

$$\hat{\boldsymbol{\eta}}(t) = [\eta_0(t) \;\; \eta_1(t) \;\; \eta_2(t) \;\; \ldots \;\; \eta_n(t)]^T = \begin{bmatrix} \eta_0(t) \\ \boldsymbol{\eta}(t) \end{bmatrix}$$

is called the **adjoint state** (or **co-state**) variable.

We need to define two more functions. The **augmented Hamiltonian** is the function $\hat{H} : \mathbb{R}^{n+1} \times \mathbb{R}^{n+1} \times \mathbb{R}^m \to \mathbb{R}^1$ defined by

$$\hat{H}(\hat{\boldsymbol{\eta}}, \hat{\boldsymbol{x}}, \boldsymbol{u}) = \hat{H}(\eta_0, \eta_1, \eta_2, \ldots, \eta_n, x_0, x_1, x_2, \ldots, x_n, u_1, u_2, \ldots, u_m) \tag{10.13}$$

$$= \eta_0 f_0(\boldsymbol{x}, \boldsymbol{u}) + \langle \boldsymbol{\eta}, \; f(\boldsymbol{x}, \boldsymbol{u}) \rangle$$

$$= \eta_0 f_0(\boldsymbol{x}, \boldsymbol{u}) + \sum_{i=1}^{n} \eta_i f_i(\boldsymbol{x}, \boldsymbol{u}).$$

Hence,

$$\hat{H}(\hat{\boldsymbol{\eta}}, \hat{\boldsymbol{x}}, \boldsymbol{u}) = \eta_0 f_0(x_0, x_1, x_2, \ldots, x_n, u_1, u_2, \ldots, u_m) + \sum_{i=1}^{n} \eta_i f_i(x_0, x_1, x_2, \ldots, x_n, u_1, u_2, \ldots, u_m).$$

is a function of $2(n+1)+m$ variables. Finally, let $\hat{M} : \mathbb{R}^{n+1} \times \mathbb{R}^{n+1} \to \mathbb{R}$ be defined by

$$\hat{M}(\hat{\boldsymbol{\eta}}, \hat{\boldsymbol{x}}) = \max_{\boldsymbol{u} \in \boldsymbol{\Omega}} \hat{H}(\hat{\boldsymbol{\eta}}, \hat{\boldsymbol{x}}, \boldsymbol{u}) \tag{10.14}$$

when $\hat{M}(\hat{\boldsymbol{\eta}}, \hat{\boldsymbol{x}})$ exists. We can now state the (Pontryagin) Maximum Principle as the following theorem.

Theorem 10.1 (Maximum Principle) *Assume that* $f : \mathbb{R}^n \times \mathbb{R}^m \to \mathbb{R}^n$, $f_0 : \mathbb{R}^n \times \mathbb{R}^m \to \mathbb{R}^1$, $\mathrm{X}_0 \subseteq \mathbb{R}^n$, $t_0 \in \mathbb{R}$, $\mathrm{X}_1 \subseteq \mathbb{R}^n$ *and* $\Omega \subseteq \mathbb{R}^m$ *are given as above and consider the control system*

$$(\mathcal{S}) \qquad \dot{\boldsymbol{x}}(t) = f(\boldsymbol{x}(t), \boldsymbol{u}(t)), \quad t_0 < t, \tag{10.15}$$

with piecewise continuous controllers $\boldsymbol{u}(\cdot) \in PWC(t_0, +\infty; \mathbb{R}^m)$ *satisfying* $\boldsymbol{u}(\cdot) \in \Omega \subseteq \mathbb{R}^m$ *e.f. If*

$$\boldsymbol{u}^*(\cdot) \in \Theta = \left\{ \begin{array}{c} \boldsymbol{u}(\cdot) \in PWC(t_0, +\infty; \mathbb{R}^m) : \boldsymbol{u}(t) \in \Omega \ e.f. \\ \text{and } \boldsymbol{u}(\cdot) \text{ steers } \mathrm{X}_0 \text{ to } \mathrm{X}_1. \end{array} \right\} \tag{10.16}$$

minimizes

$$J(\boldsymbol{u}(\cdot)) = \int_{t_0}^{t_1} f_0(\boldsymbol{x}(s), \boldsymbol{u}(s))ds \tag{10.17}$$

on the set of admissible controls Θ *with optimal response* $\boldsymbol{x}^*(\cdot)$ *satisfying* $\boldsymbol{x}^*(t_0) = \boldsymbol{x}_0^* \in \mathrm{X}_0$ *and* $\boldsymbol{x}^*(t_1^*) = \boldsymbol{x}_1^* \in \mathrm{X}_1$ *at time* $t_1^* > t_0$, *then there exists a non-trivial solution*

$$\hat{\boldsymbol{\eta}}^*(t) = [\eta_0^*(t) \ \ \eta_1^*(t) \ \ \eta_2^*(t) \ \ \ldots \ \ \eta_n^*(t)]^T = \begin{bmatrix} \eta_0^*(t) \\ \boldsymbol{\eta}^*(t) \end{bmatrix} \tag{10.18}$$

to the augmented adjoint equation

$$\frac{d}{dt}\hat{\boldsymbol{\eta}}(t) = -[\hat{A}(t)]^T \hat{\boldsymbol{\eta}}(t), \tag{10.19}$$

such that

$$\begin{aligned} \hat{H}(\hat{\boldsymbol{\eta}}^*(t), \hat{\boldsymbol{x}}^*(t), \boldsymbol{u}^*(t)) &= \hat{M}(\hat{\boldsymbol{\eta}}^*(t), \hat{\boldsymbol{x}}^*(t)) \\ &= \max_{\boldsymbol{u} \in \Omega} \hat{H}(\hat{\boldsymbol{\eta}}^*(t), \hat{\boldsymbol{x}}^*(t), \boldsymbol{u}). \end{aligned} \tag{10.20}$$

Moreover, there is a constant $\eta_0^* \leq 0$ *such that*

$$\eta_0^*(t) \equiv \eta_0^* \leq 0 \tag{10.21}$$

and for all $t \in [t_0, t_1^*]$

$$\hat{M}(\hat{\boldsymbol{\eta}}^*(t), \hat{\boldsymbol{x}}^*(t)) = \max_{\boldsymbol{u} \in \Omega} \hat{H}(\hat{\boldsymbol{\eta}}^*(t), \hat{\boldsymbol{x}}^*(t), \boldsymbol{u}) \equiv 0. \tag{10.22}$$

Also, if $X_0 \subseteq \mathbb{R}^n$ *and* $X_1 \subseteq \mathbb{R}^n$ *are manifolds with tangent spaces* $\mathbb{T}_0$ *and* $\mathbb{T}_1$ *at* $\boldsymbol{x}^*(t_0) = \boldsymbol{x}_0^* \in X_0$ *and* $\boldsymbol{x}^*(t_1^*) = \boldsymbol{x}_1^* \in X_1$*, respectively, then*

$$\hat{\boldsymbol{\eta}}^*(t) = [\eta_0^*(t) \;\; \eta_1^*(t) \;\; \eta_2^*(t) \;\; \ldots \;\; \eta_n^*(t)]^T = \left[\; \eta_0^*(t) \;\; \boldsymbol{\eta}^*(t) \;\right]^T$$

can be selected to satisfy the transversality conditions

$$\boldsymbol{\eta}^*(t_0) \perp \mathbb{T}_0 \tag{10.23}$$

and

$$\boldsymbol{\eta}^*(t_1^*) \perp \mathbb{T}_1. \tag{10.24}$$

Remark 10.1 *We have not discussed the precise definitions of "manifolds" and "tangent spaces". A rigorous presentation of the definitions of manifolds and tangent spaces would require a background in differential geometry and topology which is outside the*

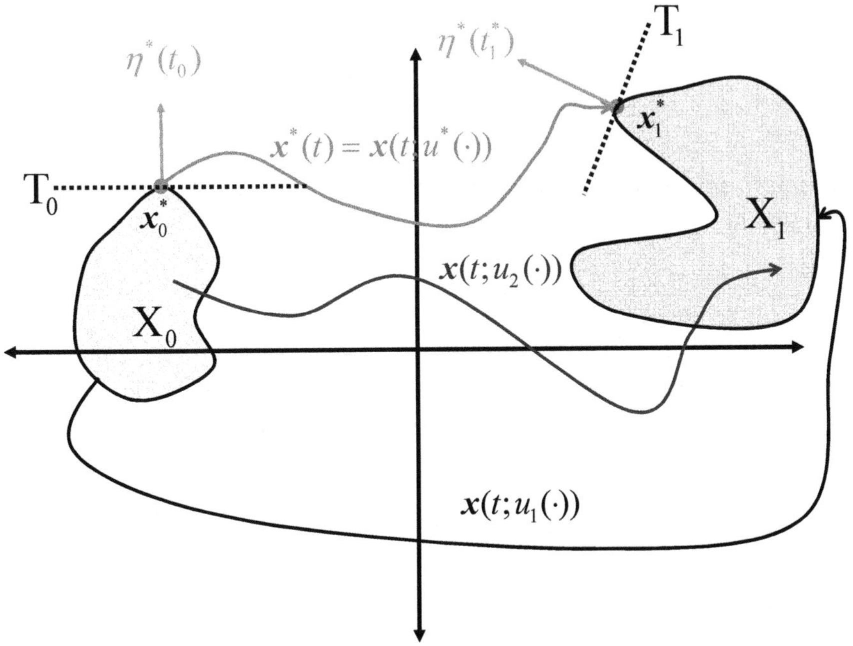

Figure 10.1: The Maximum Principle

scope of this book. However, roughly speaking a manifold of dimension k is a topological space that near each point "resembles" k-dimensional Euclidean space. Two dimensional manifolds are called surfaces. Examples of manifolds are the entire space $\mathbb{R}^n$, k-dimensional subspaces of $\mathbb{R}^n$, hyperplanes and 0-dimensional manifolds which are points and may be viewed as a translation of the zero subspace. Also, given $p \leq n$ and p smooth functions $g_1(\cdot)$, $g_2(\cdot)$, ... ,$g_p(\cdot)$ satisfying $g_i : \mathbb{R}^n \longrightarrow \mathbb{R}$, define the set

$$X_g = \{\boldsymbol{x} \in \mathbb{R}^n : g_i(\boldsymbol{x}) = 0, \; for \; i = 1, 2, ..., p\}. \qquad (10.25)$$

If the vectors

$$\nabla g_i(\boldsymbol{x}), \; i = 1, 2, ..., p$$

are linearly independent, then X_g is a (smooth) manifold of dimension $k = n - p$. Note that this condition is equivalent to the statement that the matrix

$$\begin{bmatrix} \frac{\partial g_1(\boldsymbol{x})}{\partial x_1} & \frac{\partial g_1(\boldsymbol{x})}{\partial x_2} & \cdots & \cdots & \frac{\partial g_1(\boldsymbol{x})}{\partial x_n} \\ \frac{\partial g_2(\boldsymbol{x})}{\partial x_1} & \frac{\partial g_2(\boldsymbol{x})}{\partial x_2} & \cdots & \cdots & \frac{\partial g_2(\boldsymbol{x})}{\partial x_n} \\ \vdots & \vdots & \ddots & & \vdots \\ \vdots & \vdots & & \ddots & \vdots \\ \frac{\partial g_p(\boldsymbol{x})}{\partial x_1} & \frac{\partial g_p(\boldsymbol{x})}{\partial x_2} & \cdots & \cdots & \frac{\partial g_p(\boldsymbol{x})}{\partial x_n} \end{bmatrix}$$

has maximal rank p. In this case the tangent plane $\mathbb{T}_g(\boldsymbol{x})$ at $\boldsymbol{x} \in X_g$ is given by

$$\mathbb{T}_g(\boldsymbol{x}) = \mathbb{T}_{g_1}(\boldsymbol{x}) \cap \mathbb{T}_{g_2}(\boldsymbol{x}) \cap \ldots \cap \mathbb{T}_{g_p}(\boldsymbol{x})$$

where $\mathbb{T}_{g_i}(\boldsymbol{x})$ is the tangent plane of the smooth surface $X_{g_i} = \{\boldsymbol{x} \in \mathbb{R}^n : g_i(\boldsymbol{x}) = 0\}$ with normal vector given by $\nabla g_i(\boldsymbol{x})$. Therefore, a vector $\hat{\boldsymbol{\eta}} \in \mathbb{R}^n$ is orthogonal to $\mathbb{T}_g(\boldsymbol{x})$ at $\boldsymbol{x} \in X_g$ if and only if

$$\hat{\boldsymbol{\eta}}^T \boldsymbol{z} = \sum_{j=1}^{n} \hat{\eta}_j z_z = 0$$

for $\boldsymbol{z} \in \mathbb{R}^n$ satisfying

$$\sum_{i=1}^{n} \frac{\partial g_i(\boldsymbol{x})}{\partial x_j} z_j = 0, \; for \; all \; i = 1, 2, ..., p.$$

We will use this result when the initial and target sets are defined by (10.25). The special cases where $\mathrm{X} = \mathbb{R}^n$ *and* $\mathrm{X} = \hat{\boldsymbol{x}}$ *are simple since the tangent space to any subspace is that space and the tangent space to a single point is the zero subspace. Thus, every vector* $\hat{\boldsymbol{\eta}}$ *is transversal to a single point and only the zero vector is transversal to the whole space* $\mathbb{R}^n$*.*

The references [107], [116], [147] and [184] provide introductions to differential geometry with rigorous definitions.

10.3 Application of the Maximum Principle to Some Simple Problems

Here we apply the Maximum Principle to some of the problems described above. We focus on the Simplest Problem (i.e. t_1 not fixed) and then move to other variations of this problem in later sections.

10.3.1 The Bushaw Problem

The Bushaw problem is governed by the control system

$$\frac{d}{dt}\begin{bmatrix} x_1(t) \\ x_2(t) \end{bmatrix} = \begin{bmatrix} 0 & 1 \\ -\kappa & -2\gamma \end{bmatrix}\begin{bmatrix} x_1(t) \\ x_2(t) \end{bmatrix} + \begin{bmatrix} 0 \\ 1 \end{bmatrix} u(t), \tag{10.26}$$

which has the form

$$\dot{\boldsymbol{x}}(t) = A\boldsymbol{x}(t) + Bu(t) \tag{10.27}$$

with initial condition

$$\boldsymbol{x}(0) = \boldsymbol{x}_0 \in \mathbb{R}^2, \tag{10.28}$$

where the matrices A and B are defined by

$$A = \begin{bmatrix} 0 & 1 \\ -\kappa & -2\gamma \end{bmatrix} \text{ and } B = \begin{bmatrix} 0 \\ 1 \end{bmatrix},$$

respectively. Here, $\boldsymbol{x}_0 = \begin{bmatrix} x_{1,0} & x_{2,0} \end{bmatrix}^T$ is the initial condition. Given a control $u(\cdot)$, the solution to (10.27) - (10.28) is denoted by $\boldsymbol{x}(\cdot; u(\cdot))$. We are interested in time optimal transfer of the initial state $\boldsymbol{x}_0$ to a fixed terminal state $\boldsymbol{x}_1 = \begin{bmatrix} 0 & 0 \end{bmatrix}^T \in \mathbb{R}^2$ by a control satisfying the condition $|u(t)| \leq 1$. We formulate this time optimal control problem as a **SPOC** described above.

Formulation as a Simplest Problem in Optimal Control: In order to set up the optimal control problem as a **SPOC** and apply the Maximum Principle, we need to:

1. Identify the initial time t_0;
2. Identify the initial set $X_0 \subseteq \mathbb{R}^2$, the terminal set $X_1 \subseteq \mathbb{R}^2$;
3. Identify the control constraint set $\Omega \subseteq \mathbb{R}^1$;
4. Identify the functions $f : \mathbb{R}^2 \times \mathbb{R}^1 \longrightarrow \mathbb{R}^1$, $f_0 : \mathbb{R}^2 \times \mathbb{R}^1 \longrightarrow \mathbb{R}^1$;
5. Define the augmented Hamiltonian $\hat{H}(\hat{\boldsymbol{\eta}}, \hat{\boldsymbol{x}}, u)$;
6. Form the augmented adjoint system matrix $\hat{A}(t)$.

Remark 10.2 *We will work through these steps for every example we discuss and each time the above list will be repeated. Although this may seem like overkill, it helps the reader focus on the essential steps required to formulate and apply the Maximum Principle and its extensions. Moreover, many problems require additional steps before one can formulate the problem as an optimal control problem and explicitly listing these six steps can be helpful in determining the proper formulation (see the point-to-curve problem in Section 11.2.3).*

Recall that the time optimal cost functional can be written as

$$J(u(\cdot)) = \int_0^{t_1} 1 ds = t_1$$

and the initial time is $t_0 = 0$. Since there are two state equations and one control, the state space is $\mathbb{R}^2$ and the control space is $\mathbb{R}^1$

so that $n = 2$ and $m = 1$. Let $\Omega = [-1, 1] \subset \mathbb{R}^1$ be the control constraint set; so that the condition $|u(t)| \leq 1$ can be written as $u(t) \in \Omega = [-1, 1]$. The initial and terminal sets are single points so that

$$\mathrm{X}_0 = \{\boldsymbol{x}_0\} \subseteq \mathbb{R}^2 \text{ and } \mathrm{X}_1 = \{\mathbf{0}\} \subseteq \mathbb{R}^2$$

define these sets. The functions $f : \mathbb{R}^2 \times \mathbb{R}^1 \longrightarrow \mathbb{R}^2$ and $f_0 : \mathbb{R}^2 \times \mathbb{R}^1 \longrightarrow \mathbb{R}^1$ are defined by

$$f(\boldsymbol{x}, u) = f(x_1, x_2, u) = A\boldsymbol{x} + Bu = \begin{bmatrix} x_2 \\ -\kappa x_1 - 2\gamma x_2 + u \end{bmatrix}$$

and

$$f_0(\boldsymbol{x}, u) = f_0(x_1, x_2, u) \equiv 1,$$

respectively. Hence,

$$f(\boldsymbol{x}, u) = f(x_1, x_2, u) = \begin{bmatrix} f_1(x_1, x_2, u) \\ f_2(x_1, x_2, u) \end{bmatrix} = \begin{bmatrix} x_2 \\ -\kappa x_1 - 2\gamma x_2 + u \end{bmatrix}$$

defines the functions $f_1(x_1, x_2, u)$ and $f_2(x_1, x_2, u)$.

Now to set up the Maximum Principle we start by letting $\hat{\boldsymbol{x}} = \begin{bmatrix} x_0 & x_1 & x_2 \end{bmatrix}^T$ denote the augmented state so that the augmented function $\hat{f} : \mathbb{R}^3 \times \mathbb{R}^1 \longrightarrow \mathbb{R}^3$ is given by

$$\hat{f}(\hat{\boldsymbol{x}}, u) = \hat{f}(x_0, x_1, x_2, u) = \hat{f}(x_1, x_2, u) = \begin{bmatrix} f_0(x_1, x_2, u) \\ f_1(x_1, x_2, u) \\ f_2(x_1, x_2, u) \end{bmatrix}$$

$$= \begin{bmatrix} 1 \\ x_2 \\ -\kappa x_1 - 2\gamma x_2 + u \end{bmatrix}.$$

We also need to compute the augmented Hamiltonian $\hat{H} : \mathbb{R}^3 \times \mathbb{R}^3 \times \mathbb{R}^1 \longrightarrow \mathbb{R}^1$ defined by

$$\hat{H}(\hat{\boldsymbol{\eta}}, \hat{\boldsymbol{x}}, u) = \eta_0[f_0(x_1, x_2, u)] + \eta_1[f_1(x_1, x_2, u)] + \eta_2[f_2(x_1, x_2, u)],$$

where $\hat{\boldsymbol{\eta}} = \begin{bmatrix} \eta_0 & \boldsymbol{\eta} \end{bmatrix}^T = \begin{bmatrix} \eta_0 & \eta_1 & \eta_2 \end{bmatrix}^T$. By direct substitution we have that

$$\begin{aligned} \hat{H}(\hat{\boldsymbol{\eta}}, \hat{\boldsymbol{x}}, u) &= \eta_0[1] + \eta_1[x_2] + \eta_2[-\kappa x_1 - 2\gamma x_2 + u] \\ &= \eta_0 + \eta_1 x_2 + \eta_2[-\kappa x_1 - 2\gamma x_2] + \eta_2 u. \end{aligned}$$

To set up the adjoint equation we need to compute the matrix

$$\mathbb{J}_{\hat{\boldsymbol{x}}}(\hat{\boldsymbol{x}}, u) = \mathbb{J}_{\hat{\boldsymbol{x}}}(\boldsymbol{x}, u) = \begin{bmatrix} \frac{\partial f_0(x_1,x_2,u)}{\partial x_0} & \frac{\partial f_0(x_1,x_2,u)}{\partial x_1} & \frac{\partial f_0(x_1,x_2,u)}{\partial x_2} \\ \frac{\partial f_1(x_1,x_2,u)}{\partial x_0} & \frac{\partial f_1(x_1,x_2,u)}{\partial x_1} & \frac{\partial f_1(x_1,x_2,u)}{\partial x_2} \\ \frac{\partial f_2(x_1,x_2,u)}{\partial x_0} & \frac{\partial f_2(x_1,x_2,u)}{\partial x_1} & \frac{\partial f_2(x_1,x_2,u)}{\partial x_2} \end{bmatrix},$$

which is easily seen to be

$$\mathbb{J}_{\hat{\boldsymbol{x}}}(\hat{\boldsymbol{x}}, u) = \begin{bmatrix} 0 & 0 & 0 \\ 0 & 0 & 1 \\ 0 & -\kappa & -2\gamma \end{bmatrix} = \begin{bmatrix} 0 & \mathbf{0} \\ \mathbf{0} & A \end{bmatrix}.$$

Now assume that $(\boldsymbol{x}^*(\cdot), u^*(\cdot))$ is an optimal pair so that the matrix $\hat{A}(t)$ can be computed by

$$\hat{A}(t) = \mathbb{J}_{\hat{\boldsymbol{x}}}(\hat{\boldsymbol{x}}, u)_{|(\boldsymbol{x}^*(t),u^*())} = \mathbb{J}_{\hat{\boldsymbol{x}}}(\boldsymbol{x}^*(t), u^*(t)) = \begin{bmatrix} 0 & 0 & 0 \\ 0 & 0 & 1 \\ 0 & -\kappa & -2\gamma \end{bmatrix}$$

$$= \begin{bmatrix} 0 & \mathbf{0} \\ \mathbf{0} & A \end{bmatrix}.$$

Observe that in this special case

$$\hat{A}(t) = \hat{A} = \begin{bmatrix} 0 & 0 & 0 \\ 0 & 0 & 1 \\ 0 & -\kappa & -2\gamma \end{bmatrix} = \begin{bmatrix} 0 & \mathbf{0} \\ \mathbf{0} & A \end{bmatrix}$$

is a constant matrix and does not depend on $(\boldsymbol{x}^*(\cdot), u^*(\cdot))$.

The augmented adjoint equation

$$\frac{d}{dt}\hat{\boldsymbol{\eta}}(t) = -[\hat{A}(t)]^T \hat{\boldsymbol{\eta}}(t) \tag{10.29}$$

has the form

$$\frac{d}{dt}\begin{bmatrix} \eta_0(t) \\ \eta_1(t) \\ \eta_2(t) \end{bmatrix} = -\begin{bmatrix} 0 & 0 & 0 \\ 0 & 0 & -\kappa \\ 0 & 1 & -2\gamma \end{bmatrix}\begin{bmatrix} \eta_0(t) \\ \eta_1(t) \\ \eta_2(t) \end{bmatrix} = \begin{bmatrix} 0 & 0 & 0 \\ 0 & 0 & \kappa \\ 0 & -1 & 2\gamma \end{bmatrix}\begin{bmatrix} \eta_0(t) \\ \eta_1(t) \\ \eta_2(t) \end{bmatrix}.$$

Observe that this system is given by

$$\frac{d}{dt}\begin{bmatrix} \eta_0(t) \\ \eta_1(t) \\ \eta_2(t) \end{bmatrix} = \begin{bmatrix} 0 \\ \kappa\eta_2(t) \\ -\eta_1(t) + 2\gamma\eta_2(t) \end{bmatrix}$$

and the first equation implies

$$\frac{d}{dt}\eta_0(t) = 0$$

which means that all solutions to the augmented adjoint equations have

$$\eta_0(t) \equiv \eta_0.$$

Also for this special case, the non-augmented adjoint state

$$\boldsymbol{\eta}(t) = \begin{bmatrix} \eta_1(t) \\ \eta_2(t) \end{bmatrix}$$

satisfies the system

$$\frac{d}{dt}\boldsymbol{\eta}(t) = -A^T\boldsymbol{\eta}(t) \tag{10.30}$$

where the matrix A is the state matrix in (10.27).

Application of The Maximum Principle: We are now ready to apply the Maximum Principle. Assume $(\boldsymbol{x}^*(\cdot), u^*(\cdot))$ is an optimal pair. Then there exists a non-trivial solution

$$\hat{\boldsymbol{\eta}}^*(t) = \begin{bmatrix} \eta_0^*(t) \\ \boldsymbol{\eta}^*(t) \end{bmatrix} = \begin{bmatrix} \eta_0^*(t) \\ \eta_1^*(t) \\ \eta_2^*(t) \end{bmatrix}$$

to the augmented adjoint equation (10.29) such that $\eta_0^*(t) \equiv \eta_0^* \leq 0$ and

$$\hat{H}(\hat{\boldsymbol{\eta}}^*(t), \hat{\boldsymbol{x}}^*(t), u^*(t)) = \max_{u\in\Omega} \hat{H}(\hat{\boldsymbol{\eta}}^*(t), \hat{\boldsymbol{x}}^*(t), u) \equiv 0.$$

Since

$$\begin{aligned}\hat{H}(\hat{\boldsymbol{\eta}}^*(t), \hat{\boldsymbol{x}}^*(t), u) =& \eta_0^* + \eta_1^*(t)x_2^*(t) + \eta_2^*(t)[-\kappa x_1^*(t) - 2\gamma x_2^*(t)] \\ &+ \eta_2^*(t)u,\end{aligned}$$

it follows that in order to maximize $\hat{H}(\hat{\boldsymbol{\eta}}^*(t), \hat{\boldsymbol{x}}^*(t), u)$ on $[-1, 1]$ one needs only to maximize the last term in this expression. In particular, the optimal control must maximize the expression

$$\eta_2^*(t)u$$

for all t where $-1 \leq u \leq +1$. Clearly, this occurs when

$$u^*(t) = sgn[\eta_2^*(t)]$$

and the problem now reduces to the case of computing $\eta_2^*(t)$.

Let's see what the transversality conditions tell us. Since $\mathrm{X}_0 = \{\boldsymbol{x}_0\} \subseteq \mathbb{R}^2$ and $\mathrm{X}_1 = \{\mathbf{0}\} \subseteq \mathbb{R}^2$ and the tangent plane to a point is the "zero plane", i.e. $\mathbb{T}_0 = \{\mathbf{0}\} \subseteq \mathbb{R}^2$ and $\mathbb{T}_1 = \{\mathbf{0}\} \subseteq \mathbb{R}^2$, the conditions

$$\boldsymbol{\eta}^*(0) \perp \mathbb{T}_0 \text{ and } \boldsymbol{\eta}^*(t_1^*) \perp \mathbb{T}_1$$

are satisfied by any vectors $\boldsymbol{\eta}^*(0)$ and $\boldsymbol{\eta}^*(t_1^*)$. Thus, the transversality conditions give us no additional information. However, we do know that

$$\boldsymbol{x}^*(0) = \boldsymbol{x}_0 \text{ and } \boldsymbol{x}^*(t_1^*) = \mathbf{0}$$

and

$$\frac{d}{dt}\boldsymbol{\eta}^*(t) = -A^T\boldsymbol{\eta}^*(t) \tag{10.31}$$

so that

$$\boldsymbol{\eta}^*(t) = e^{-A^T(t-t_1^*)}\boldsymbol{\eta}^*(t_1^*) = e^{-A^Tt}e^{A^Tt_1^*}\boldsymbol{\eta}^*(t_1^*) = e^{-A^Tt}\tilde{\boldsymbol{\eta}}^*$$

where

$$\tilde{\boldsymbol{\eta}}^* \triangleq e^{A^Tt_1^*}\boldsymbol{\eta}^*(t_1^*).$$

Hence,

$$\eta_2^*(t) = \begin{bmatrix} 0 & 1 \end{bmatrix} \begin{bmatrix} \eta_1^*(t) \\ \eta_2^*(t) \end{bmatrix} = B^T e^{-A^Tt}\tilde{\boldsymbol{\eta}}^*$$

and

$$u^*(t) = sgn[\eta_2^*(t)] = sgn[B^T e^{-A^Tt}\tilde{\boldsymbol{\eta}}^*]$$

where $\tilde{\boldsymbol{\eta}}^* \triangleq e^{A^Tt_1^*}\boldsymbol{\eta}^*(t_1^*)$ is not known since (at this point) neither t_1^* nor $\boldsymbol{\eta}^*(t_1^*)$ are known.

Observe that the Maximum Principle says that $\boldsymbol{\eta}^*(t_1^*)$ **can be selected** so that $\boldsymbol{\eta}^*(t_1^*) \perp \mathbb{T}_1$. However, any vector satisfies this condition. We might try the simple case where $\boldsymbol{\eta}^*(t_1^*) = \mathbf{0}$ and see what happens. In this case, $\boldsymbol{\eta}^*(t)$ satisfies the linear differential equation $\frac{d}{dt}\boldsymbol{\eta}^*(t) = -A^T\boldsymbol{\eta}^*(t)$ with zero value at $\boldsymbol{\eta}^*(t_1^*)$. Therefore, $\boldsymbol{\eta}^*(t)$ must be identically zero for all t i.e. $\boldsymbol{\eta}^*(t) \equiv \mathbf{0}$. However, this would imply that $u^*(t) = sgn(\eta_2^*(t)) \equiv \mathbf{0}$ and the augmented Hamiltonian has the form

$$\begin{aligned}\hat{H}(\hat{\boldsymbol{\eta}}^*(t), \hat{\boldsymbol{x}}^*(t), u) &= \eta_0^* + \eta_1^*(t)x_2^*(t) + \eta_2^*(t)[-\kappa x_1^*(t) - 2\gamma x_2^*(t)] \\ &\quad + \eta_2^*(t)u \\ &= \eta_0^* + 0x_2^*(t) + 0[-\kappa x_1^*(t) - 2\gamma x_2^*(t)] + 0u \\ &\equiv \eta_0^*.\end{aligned}$$

On the other hand, the Maximum Principle yields

$$\eta_0^* = \hat{H}(\hat{\boldsymbol{\eta}}^*(t), \hat{\boldsymbol{x}}^*(t), u^*(t)) = \max_{u\in\Omega} \hat{H}(\hat{\boldsymbol{\eta}}^*(t), \hat{\boldsymbol{x}}^*(t), u) \equiv 0$$

and the constant η_0^* would be zero also. Thus, $\hat{\boldsymbol{\eta}}^*(t) \equiv \mathbf{0}$ and this contradicts the statement (in the Maximum Principle) that $\hat{\boldsymbol{\eta}}^*(t)$ is a non-trivial solution to the augmented adjoint equation. Consequently, we know that $\boldsymbol{\eta}^*(t_1^*) \neq \mathbf{0}$, and hence $\tilde{\boldsymbol{\eta}}^* \triangleq e^{A^T t_1^*}\boldsymbol{\eta}^*(t_1^*) \neq \mathbf{0}$.

Therefore, we know that if $u^*(t)$ is a time optimal controller, then the following conditions hold. There is a non-trivial solution to the adjoint equation

$$\frac{d}{dt}\boldsymbol{\eta}^*(t) = -A^T\boldsymbol{\eta}^*(t)$$

with initial data

$$\boldsymbol{\eta}^*(t_1^*) \neq \mathbf{0}$$

and the optimal control has the form

$$u^*(t) = sgn(B^T e^{-A^T t}\tilde{\boldsymbol{\eta}}^*).$$

The optimal trajectories $\boldsymbol{x}^*(t)$ satisfy

$$\frac{d}{dt}\boldsymbol{x}^*(t) = A\boldsymbol{x}^*(t) + Bsgn(B^T e^{-A^T t}\tilde{\boldsymbol{\eta}}^*),$$

along with the boundary conditions

$$\boldsymbol{x}^*(0) = \boldsymbol{x}_0, \ \boldsymbol{x}^*(t_1^*) = \mathbf{0}.$$

Moreover,

$$
\begin{aligned}
0 &= \max_{u\in\Omega} \hat{H}(\hat{\boldsymbol{\eta}}^*(t), \hat{\boldsymbol{x}}^*(t), u) \equiv \hat{H}(\hat{\boldsymbol{\eta}}^*(t), \hat{\boldsymbol{x}}^*(t), u^*(t)) \\
&= \eta_0^* + \eta_1^*(t)x_2^*(t) + \eta_2^*(t)[-\kappa x_1^*(t) - 2\gamma x_2^*(t)] + \eta_2^*(t)(sgn[\eta_2^*(t)]) \\
&= \eta_0^* + \eta_1^*(t)x_2^*(t) + \eta_2^*(t)[-\kappa x_1^*(t) - 2\gamma x_2^*(t)] + |\eta_2^*(t)|
\end{aligned}
$$

holds for all $0 \le t \le t_1^*$. In addition, since $\eta_0^* \le 0$, it follows that

$$\eta_1^*(t)[x_2^*(t)] + \eta_2^*(t)[-\kappa x_1^*(t) - 2\gamma x_2^*(t)] + |\eta_2^*(t)| \ge 0$$

and this is all we can say at this point about this problem.

10.3.2 The Bushaw Problem: Special Case $\gamma = 0$ and $\kappa = 1$

Here we synthesize the optimal controller for a special case of Bushaw's problem. Since

$$A = \begin{bmatrix} 0 & 1 \\ -1 & 0 \end{bmatrix}$$

it follows that

$$-A^T = \begin{bmatrix} 0 & 1 \\ -1 & 0 \end{bmatrix} = A,$$

and hence

$$e^{-A^T s} = e^{As} = \begin{bmatrix} \cos(s) & \sin(s) \\ -\sin(s) & \cos(s) \end{bmatrix}.$$

Therefore

$$
\begin{aligned}
B^T e^{-A^T s}\tilde{\boldsymbol{\eta}} &= \begin{bmatrix} 0 & 1 \end{bmatrix} \begin{bmatrix} \cos(s) & \sin(s) \\ -\sin(s) & \cos(s) \end{bmatrix} \tilde{\boldsymbol{\eta}} \\
&= \begin{bmatrix} -\sin(s) & \cos(s) \end{bmatrix} \begin{bmatrix} \tilde{\eta}_1 \\ \tilde{\eta}_2 \end{bmatrix} \\
&= -\tilde{\eta}_1 \sin(s) + \tilde{\eta}_2 \cos(s).
\end{aligned}
$$

Using the standard identity

$$-\tilde{\eta}_1 \sin(s) + \tilde{\eta}_2 \cos(s) = \sqrt{[\tilde{\eta}_1]^2 + [\tilde{\eta}_2]^2}\, \sin(s + \tilde{\phi})$$

where

$$\tilde{\phi} = \begin{cases} \arcsin\left(\tilde{\eta}_2/\sqrt{[\tilde{\eta}_1]^2 + [\tilde{\eta}_2]^2}\right) & \text{if } \tilde{\eta}_1 \le 0, \\ \pi - \arcsin\left(\tilde{\eta}_2/\sqrt{[\tilde{\eta}_1]^2 + [\tilde{\eta}_2]^2}\right) & \text{if } \tilde{\eta}_1 > 0, \end{cases}$$

it follows from (9.63) that

$$\begin{aligned} u^*(s) &= sgn\left[B^T\boldsymbol{\eta}^*(s)\right] = sgn\left[-\tilde{\eta}_1\sin(s) + \tilde{\eta}_2\cos(s)\right] \\ &= sgn\left[\sqrt{[\tilde{\eta}_1]^2 + [\tilde{\eta}_2]^2}\sin(s + \tilde{\phi})\right]. \end{aligned}$$

Since $\sqrt{[\tilde{\eta}_1]^2 + [\tilde{\eta}_2]^2} > 0$, we have that $\sqrt{[\tilde{\eta}_1]^2 + [\tilde{\eta}_2]^2}\sin(s+\tilde{\phi})$ and $\sin(s + \tilde{\phi})$ have the same sign. Hence the optimal control is given by

$$u^*(t) = sgn\left[\sqrt{[\tilde{\eta}_1]^2 + [\tilde{\eta}_2]^2}\sin(t + \tilde{\phi})\right] = sgn\left[\sin(t + \tilde{\phi})\right]. \tag{10.32}$$

The expression (10.32) provides considerable information about the optimal control. In particular:

- **The optimal control $u^*(\cdot)$ is bang-bang just like the rocket sled problem.**
- **Switches (if they occur) occur exactly π seconds apart.**

To complete the synthesis we solve the equations for controls $u(\cdot)$ that are bang-bang with switching time intervals of length π. Thus, we focus on the cases where $u(t) = \pm 1$. If $u(t) = +1$, then the state equation has the form

$$\begin{aligned} \dot{x}_1(t) &= x_2(t) \\ \dot{x}_2(t) &= -x_1(t) + 1. \end{aligned}$$

Observe that

$$\frac{dx_1}{dx_2} = \frac{x_2}{-x_1 + 1}$$

so that

$$(-x_1 + 1)dx_1 = x_2 dx_2$$

and hence integration yields

$$c + \int (-x_1 + 1)dx_1 = \int x_2 dx_2.$$

Thus,

$$c + \frac{-1}{2}[-x_1 + 1]^2 = \frac{1}{2}[x_2]^2$$

which implies that

$$2c = [-x_1 + 1]^2 + [x_2]^2 > 0,$$

or equivalently,

$$[-x_1 + 1]^2 + [x_2]^2 = \alpha^2 > 0.$$

Therefore, the trajectories are circles centered at $[1 \;\; 0]^T$ with radius α. Likewise, if $u(t) = -1$, then

$$[-x_1 - 1]^2 + [x_2]^2 = \alpha^2 > 0$$

and trajectories are circles centered at $[-1 \;\; 0]^T$ with radius α.

Since an optimal control is bang-bang, optimal trajectories consist of portions of circles centered about the two points $[1 \;\; 0]^T$ and $[-1 \;\; 0]^T$ with trajectories moving clockwise. If time is reversed the backward trajectories move counter clockwise. Moreover, since the switching times are multiples of π only semi-circles can be used to construct optimal trajectories. First assume that $0 < \tilde{\phi} \leq \pi$, set $u(t) = 1$ and integrate backwards from $[0 \;\; 0]^T$. The solution is given by

$$x_1(t) = 1 - \cos(t) > 0$$
$$x_2(t) = \sin(t) < 0$$

for $-\pi \leq t \leq 0$. To be an optimal control, $u(t) = sgn[\sin(t + \tilde{\phi})] = 1$ must switch from $+1$ to -1 at $t_1^s = -\tilde{\phi}$. Thus, $u(t) = +1$ until $t_1^s = -\tilde{\phi}$ and then one starts applying $u(t) = -1$ with initial data

$$x_1(-\tilde{\phi}) = 1 - \cos(-\tilde{\phi}) > 0$$
$$x_2(-\tilde{\phi}) = \sin(-\tilde{\phi}) < 0.$$

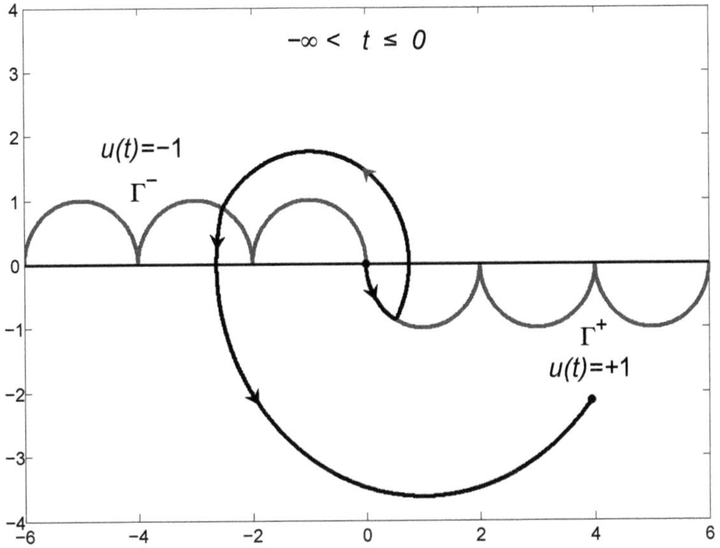

Figure 10.2: Switching Curves for Negative Time

However, now $u(t) = -1$ and the trajectory moves counter clockwise along the circle centered about $[-1 \;\; 0]^T$ with radius

$$\alpha_1 = \sqrt{[x_1(-\tilde{\phi}) + 1]^2 + [x_2(-\tilde{\phi})]^2}$$

until $t_2^s = -(\tilde{\phi} + \pi)$ when the control takes the value $u(t) = 1$ again. The trajectory will now move counter clockwise along the circle centered about $[1 \;\; 0]^T$ with radius

$$\alpha_2 = \sqrt{[x_1(-(\tilde{\phi} + \pi)) - 1]^2 + [x_2(-\tilde{\phi} + \pi)]^2}.$$

Continuing this procedure and reversing time to positive values we arrive at the switching curves as shown in the Figures 10.2 and 10.3 below.

Let $W = \Gamma^- \cup \Gamma^+ \cup H^- \cup H^+$ where H^- is the curve defined by the positive semicircles to the left of Γ^- and H^+ is the curve defined by the negative semicircles to the right of Γ^+. The curve

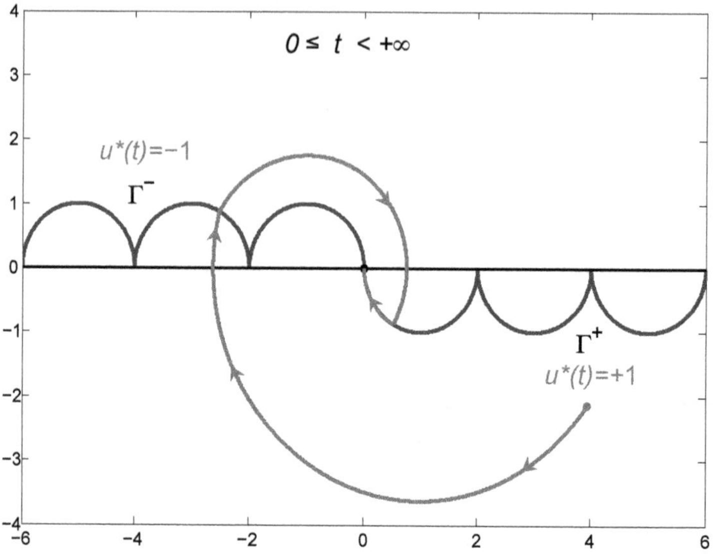

Figure 10.3: Switching Curves for Positive Time

W is the *switching curve* for this problem. Therefore, the optimal control has the feedback law given by

$$u^*(t) = \Psi(x_1^*(t), x_2^*(t)) = \Psi(\boldsymbol{x}^*(t)), \tag{10.33}$$

where

$$\Psi(x_1, x_2) = \begin{cases} -1, & \text{if } (x_1, x_2) \text{ is above } W, \\ & \quad \text{or } (x_1, x_2) \in H^- \text{ or } (x_1, x_2) \in \Gamma^-, \\ 0, & \text{if } x_1 = x_2 = 0, \\ +1, & \text{if } (x_1, x_2) \text{ is below } W, \\ & \quad \text{or } (x_1, x_2) \in H^+ \text{ or } (x_1, x_2) \in \Gamma^+. \end{cases} \tag{10.34}$$

10.3.3 A Simple Scalar Optimal Control Problem

Consider the linear control system

$$\dot{x}(t) = -x(t) + 1 - u(t) \tag{10.35}$$

with initial data

$$x(0) = 1 \in \mathbb{R}^1 \tag{10.36}$$

and target

$$x(t_1) = 0 \in \mathbb{R}^1. \tag{10.37}$$

Here $t_0 = 0$, $x_0 = 1 \in \mathbb{R}^1$ and $x_1 = 0 \in \mathbb{R}^1$ are given and the controls belong to the space $PWC(0, +\infty; \mathbb{R}^1)$. The quadratic cost functional is defined by

$$J(x(\cdot), u(\cdot)) = \int_0^{t_1} \frac{1}{2}\left\{1 + [u(s)]^2\right\} ds, \tag{10.38}$$

where $x(\cdot)$ is the solution to the system (10.35) - (10.37). The optimal control problem is to find $u^*(\cdot) \in PWC(0, +\infty; \mathbb{R}^1)$ so that

$$\begin{aligned} J(x^*(\cdot), u^*(\cdot)) &= \int_0^{t_1^*} \frac{1}{2}\left\{1 + [u^*(s)]^2\right\} ds \leq \int_0^{t_1} \frac{1}{2}\left\{1 + [u(s)]^2\right\} ds \\ &= J(x(\cdot), u(\cdot)) \end{aligned}$$

for all $u(\cdot) \in PWC(0, +\infty; \mathbb{R}^1)$ that steer $x_0 = 1$ to $x_1 = 0$, where $x^*(s) \triangleq x^*(s; u^*(\cdot))$ is the optimal trajectory. Observe that this is a Simplest Problem in Optimal Control since the final time t_1 is not fixed.

Formulation as a Simplest Problem in Optimal Control: In order to set up the optimal control problem as a **SPOC** and apply the Maximum Principle, we need to:

1. Identify the initial time t_0;
2. Identify the initial set $\mathrm{X}_0 \subseteq \mathbb{R}^1$, the terminal set $\mathrm{X}_1 \subseteq \mathbb{R}^1$;
3. Identify the control constraint set $\Omega \subseteq \mathbb{R}^1$;
4. Identify the functions $f : \mathbb{R}^1 \times \mathbb{R}^1 \longrightarrow \mathbb{R}^1$, $f_0 : \mathbb{R}^1 \times \mathbb{R}^1 \longrightarrow \mathbb{R}^1$;
5. Define the augmented Hamiltonian $\hat{H}(\hat{\boldsymbol{\eta}}, \hat{\boldsymbol{x}}, u)$;

6. Form the augmented adjoint system matrix $\hat{A}(t)$.

It is clear that $X_0 = \{1\}$ and $X_1 = \{0\}$. Also, since there is no bound on the control, it follows that $\Omega = \mathbb{R}^1$. The functions $f : \mathbb{R}^1 \times \mathbb{R}^1 \longrightarrow \mathbb{R}^1$ and $f_0 : \mathbb{R}^1 \times \mathbb{R}^1 \longrightarrow \mathbb{R}^1$ are given by

$$f(x, u) = -x + 1 - u$$

and

$$f_0(x, u) = \frac{1}{2}\{1 + [u]^2\},$$

respectively. The augmented Hamiltonian is thus defined by

$$\hat{H}(\hat{\boldsymbol{\eta}}, \hat{\boldsymbol{x}}, u) = \eta_0 f_0(x, u) + \eta f(x, u) = \eta_0 \frac{1}{2}\{1 + [u]^2\} + \eta(-x + 1 - u).$$

Moreover, since the augmented right-hand side is given by

$$\hat{f}(\hat{\boldsymbol{x}}, u) = \begin{bmatrix} f_0(x, u) \\ f(x, u) \end{bmatrix} = \begin{bmatrix} \frac{1}{2}\{1 + [u]^2\} \\ -x + 1 - u \end{bmatrix},$$

it follows that

$$\hat{A}(t) = \begin{bmatrix} \frac{\partial f_0(x,u)}{\partial x_0} & \frac{\partial f_0(x,u)}{\partial x} \\ \frac{\partial f(x,u)}{\partial x_0} & \frac{\partial f(x,u)}{\partial x} \end{bmatrix}_{|(x,u)=(x^*(t),u^*(t))}$$

$$= \begin{bmatrix} 0 & 0 \\ 0 & -1 \end{bmatrix}_{|(x,u)=(x^*(t),u^*(t))} = \begin{bmatrix} 0 & 0 \\ 0 & -1 \end{bmatrix}.$$

The augmented adjoint equation is defined by

$$\frac{d}{dt}\begin{bmatrix} \eta_0(t) \\ \eta(t) \end{bmatrix} = \frac{d}{dt}\hat{\boldsymbol{\eta}}(t) = -[\hat{A}(t)]^T \hat{\boldsymbol{\eta}}(t) = \begin{bmatrix} 0 & 0 \\ 0 & 1 \end{bmatrix}\begin{bmatrix} \eta_0(t) \\ \eta(t) \end{bmatrix}, \tag{10.39}$$

so that

$$\dot{\eta}_0(t) = 0$$

and

$$\dot{\eta}(t) = \eta(t).$$

Thus, as stated in the Maximum Principle, $\eta_0(t) \equiv \eta_0 \leq 0$ is a non-positive constant and hence

$$\eta(t) = ce^t$$

for some constant c. If $(x^*(\cdot), u^*(\cdot))$ is an optimal pair, then it follows that there is a non-trival solution $\hat{\boldsymbol{\eta}}^*(t) = \begin{bmatrix} \eta_0^*(t) \\ \eta^*(t) \end{bmatrix}$ to the augmented adjoint equation (10.39) such that $u = u^*(t)$ maximizes the augmented Hamiltonian

$$\hat{H}(\hat{\boldsymbol{\eta}}^*(t), \hat{\boldsymbol{x}}^*(t), u) = \eta_0^* \frac{1}{2}\{1 + [u]^2\} + \eta^*(t)(-x^*(t) + 1 - u).$$

Since $\Omega = \mathbb{R}^1$ it follows that

$$\frac{\partial}{\partial u}\hat{H}(\hat{\boldsymbol{\eta}}^*(t), \hat{\boldsymbol{x}}^*(t), u)|_{u=u^*(t)} = \frac{\partial}{\partial u}\hat{H}(\hat{\boldsymbol{\eta}}^*(t), \hat{\boldsymbol{x}}^*(t), u^*(t)) = 0.$$

However, a direct calculation yields

$$\frac{\partial}{\partial u}\hat{H}(\hat{\eta}^*(t), \hat{x}^*(t), u) = \eta_0^* u - \eta^*(t)$$

so that

$$0 = [\eta_0^* u - \eta^*(t)]|_{u=u^*(t)} = [\eta_0^* u^*(t) - \eta^*(t)]. \tag{10.40}$$

Note that η_0^* can not be zero since if $\eta_0^* = 0$, then (10.40) would imply that $0 = [\eta_0^* u^*(t) - \eta^*(t)] = \eta^*(t)$ which contradicts the statement that $\hat{\boldsymbol{\eta}}^*(t)$ is non-trival. Hence we can solve (10.40) for the optimal control

$$u^*(t) = \frac{\eta^*(t)}{\eta_0^*} = -\left[\frac{\eta^*(t)}{-\eta_0^*}\right].$$

If we define

$$\lambda^*(t) \triangleq \left[\frac{\eta^*(t)}{-\eta_0^*}\right],$$

then

$$\dot{\lambda}^*(t) = \lambda^*(t)$$

and

$$\lambda^*(t) = Ae^t,$$

for some constant A. Thus,

$$u^*(t) = -\lambda^*(t) = -Ae^t,$$

and since t_1 is not given it follows that

$$\max_{u\in\mathbb{R}^1} \hat{H}(\hat{\boldsymbol{\eta}}^*(t), \hat{\boldsymbol{x}}^*(t), u) = \hat{H}(\hat{\boldsymbol{\eta}}^*(t), \hat{\boldsymbol{x}}^*(t), u^*(t)) \equiv 0.$$

At $t = 0$ we have

$$\hat{H}(\hat{\boldsymbol{\eta}}^*(o), \hat{\boldsymbol{x}}^*(0), u^*(0)) = \eta_0^* \frac{1}{2}\{1 + [u^*(0)]^2\} \\ + \eta^*(0)(-x^*(0) + 1 - u^*(0)) = 0.$$

Consequently,

$$-\frac{1}{2}\{1 + [u^*(0)]^2\} + \left[\frac{\eta^*(0)}{-\eta_0^*}\right](-x^*(0) + 1 - u^*(0)) = 0,$$

and

$$-\frac{1}{2}\{1 + [u^*(0)]^2\} + \lambda^*(0)(-x^*(0) + 1 - u^*(0)) = 0$$

so that

$$-\frac{1}{2}\{1 + [-A]^2\} + A(-1 + 1 + A) = 0.$$

This equation becomes

$$-\frac{1}{2}\{1 + A^2\} + A^2 = -\frac{1}{2} + \frac{1}{2}A^2 = 0$$

or

$$A^2 = 1.$$

There are two possible roots $A = 1$ and $A = -1$. In any case, the optimal state is given by the Variation of Parameters Formula

$$\begin{aligned} x^*(t) &= e^{-t} + \int_0^t e^{-(t-s)}[1 - u^*(s)]ds \\ &= e^{-t} + \int_0^t e^{-(t-s)}[1 + Ae^s]ds \\ &= e^{-t} + \int_0^t e^{-(t-s)}[1]ds + \int_0^t e^{-(t-s)}[Ae^s]ds \\ &= e^{-t} + \left[e^{-(t-s)}\right]\Big|_{s=0}^{s=t} + A\int_0^t e^{-t+2s}ds \\ &= e^{-t} + \left[1 - e^{-t}\right] + \frac{A}{2}\left[e^{-t+2s}\right]\Big|_{s=0}^{s=t} \\ &= e^{-t} + \left[1 - e^{-t}\right] + \frac{A}{2}\left[e^{t} - e^{-t}\right]. \end{aligned}$$

Therefore,

$$x^*(t) = 1 + \frac{1}{2}Ae^t - \frac{1}{2}Ae^{-t}$$

and using the end condition

$$x^*(t_1) = 0$$

it follows that

$$1 = -A[\frac{1}{2}e^{t_1} - \frac{1}{2}e^{-t_1}] = -A\sinh(t_1).$$

Thus, at $t_1 = t_1^*$

$$\sinh(t_1^*) = \frac{1}{-A},$$

where from above $A = \pm 1$. Clearly, $A = -1$ and the optimal final time is

$$t_1^* = \sinh^{-1}(1).$$

Completing the problem, we have that the optimal control is given by

$$u^*(t) = -Ae^t = e^t, \quad 0 \le t \le t_1^* = \sinh^{-1}(1)$$

and the optimal trajectory is

$$x^*(t) = 1 - \frac{1}{2}e^t + \frac{1}{2}e^{-t}.$$

10.4 Problem Set for Chapter 10

Problem 10.1 *Consider the control system*

$$\dot{x}(t) = u(t),$$

with initial condition

$$x(0) = 1,$$

and terminal condition

$$x(t_1) = 0.$$

Let the control constraint set be $\Omega = [-1, 0]$. Investigate the optimal control problem for the cost functional

$$J(u(\cdot)) = t_1 = \int_0^{t_1} 1 ds.$$

Problem 10.2 *Consider the control system*

$$\dot{x}_1(t) = x_2(t),$$
$$\dot{x}_2(t) = u(t),$$

with initial condition

$$x_1(0) = 1, \quad x_2(0) = 1$$

and terminal condition

$$x_1(t_1) = 0, \quad x_2(t_1) = 0.$$

Let the control constraint set be $\Omega = [-1, +1]$. Investigate the optimal control problem for the cost functional

$$J(u(\cdot)) = t_1 = \int_0^{t_1} 1 ds.$$

Problem 10.3 *Consider the control system*

$$\dot{x}_1(t) = -x_1(t) + 10x_2(t) + u(t),$$
$$\dot{x}_2(t) = -x_2(t) + u(t),$$

with initial condition

$$x_1(0) = 1, \quad x_2(0) = 0,$$

and terminal condition

$$x_1(t_1) = 0, \quad x_2(t_1) = 1.$$

Let the control constraint set be $\Omega = [-1, +1]$. Investigate the optimal control problem for the cost functional

$$J(u(\cdot)) = t_1 = \int_0^{t_1} 1 ds.$$

Problem 10.4 *Consider the control system*

$$\dot{x}_1(t) = u_1(t) + u_2(t),$$
$$\dot{x}_2(t) = u_1(t) - u_2(t),$$

with initial condition

$$x_1(0) = 1, \quad x_2(0) = 1,$$

and terminal condition

$$x_1(t_1) = 0, \quad x_2(t_1) = 0.$$

Let the controls be constrained by $|u_1(t)| \leq 1$ *and* $|u_2(t)| \leq 1$. *Investigate the optimal control problem for the cost functional*

$$J(u(\cdot)) = t_1 = \int_0^{t_1} 1ds.$$

Problem 10.5 *Consider the control system*

$$\dot{x}(t) = u(t),$$

with initial condition

$$x(0) = 0,$$

and terminal condition

$$x(t_1) = \varphi(t_1),$$

where $\varphi(t) = t - 5$. *Let the control constraint set be* $\Omega = \mathbb{R}^1$. *Investigate the optimal control problem for the cost functional*

$$J(u(\cdot)) = \int_0^{t_1} \left\{ \frac{\sqrt{1 + [u(s)]^2}}{x(s)} \right\} ds.$$

Problem 10.6 *Consider the control system*

$$\dot{x}_1(t) = -x_1(t) + x_2(t),$$
$$\dot{x}_2(t) = u(t),$$

with initial condition

$$x_1(0) = 0, \quad x_2(0) = 0,$$

and terminal condition

$$[x(t_1)]^2 + [x(t_1)]^2 = \varphi(t_1),$$

where $\varphi(t) = t^2 + 1$. *Let the control constraint set be* $\Omega = \mathbb{R}^1$. *Investigate the optimal control problem for the cost functional*

$$J(u(\cdot)) = \int_0^{t_1} [u(s)]^2 ds.$$

Problem 10.7 *Consider the control system*

$$\dot{x}(t) = -x(t) + u(t),$$

with initial condition

$$x(0) = 1,$$

and terminal condition

$$x(t_1) = 0.$$

Let the control constraint set be $\Omega = \mathbb{R}^1$. *Investigate the optimal control problem for the cost functional*

$$J(u(\cdot)) = \int_0^{t_1} \left\{1 + [x(t)]^2 + [u(t)]^2\right\} ds.$$

Problem 10.8 *Consider the control system*

$$\dot{x}_1(t) = x_2(t),$$
$$\dot{x}_2(t) = u(t),$$

with initial condition

$$x_1(0) = 1, \quad x_2(0) = 1$$

and terminal condition

$$x_1(t_1) = \varphi(t_1), \quad x_2(t_1) = 0,$$

where $\varphi(t) = -t^2$. *Let the control constraint set be* $\Omega = \mathbb{R}^1$. *Investigate the optimal control problem for the cost functional*

$$J(u(\cdot)) = \int_0^{t_1} [u(s)]^2 ds.$$

Chapter 11

Extensions of the Maximum Principle

We now modify the SPOC by changing the problem data. In this chapter we treat the cases where the final time t_1 is fixed and consider the case where the system is nonautonomous. In addition, we discuss two other formulations of the optimal control problem and show that all the formulations are equivalent.

11.1 A Fixed-Time Optimal Control Problem

We start with a statement of an optimal control problem defined on a given fixed time interval $[t_0, t_1]$ with $t_0 < t_1$ and t_1 **is fixed and given**. Let $PWC(t_0, t_1; \mathbb{R}^m)$ denote the space of all $\mathbb{R}^m$-valued piecewise continuous functions defined on $[t_0, t_1]$. A function $\boldsymbol{u}(\cdot) \in PWC(t_0, t_1; \mathbb{R}^m)$ is called a *control* or *control function.*

Therefore, we are given $[t_0, t_1]$, smooth functions

$$f : \mathbb{R}^1 \times \mathbb{R}^n \times \mathbb{R}^m \to \mathbb{R}^n, \; f_0 : \mathbb{R}^1 \times \mathbb{R}^n \times \mathbb{R}^m \to \mathbb{R},$$

an initial set

$$\mathrm{X}_0 \subseteq \mathbb{R}^n,$$

a terminal set

$$\mathrm{X}_1 \subseteq \mathbb{R}^n,$$

and control constraint set

$$\Omega \subseteq \mathbb{R}^m.$$

As for the **SPOC**, the *state equation* is defined by the system of differential equations

$$(\mathcal{S}) \qquad \dot{\boldsymbol{x}}(t) = f(t, \boldsymbol{x}(t), \boldsymbol{u}(t)), \quad t_0 < t < t_1. \tag{11.1}$$

The initial conditions are given by

$$\boldsymbol{x}(t_0) = \boldsymbol{x}_0 \in \mathrm{X}_0. \tag{11.2}$$

Again, we make the following standing assumption about the initial value problem.

> **Standing Assumption for Optimal Control**: *For each $\boldsymbol{x}_0 \in \mathrm{X}_0$ and $\boldsymbol{u}(\cdot) \in PWC(t_0, t_1; \mathbb{R}^m)$, the initial value problem (11.1) - (11.2) has a unique solution defined on $[t_0, t_1]$. We denote this solution by $\boldsymbol{x}(\cdot; \boldsymbol{x}_0; \boldsymbol{u}(\cdot))$.*

The set of *admissible controllers* is the subset of $PWC(t_0, t_1; \mathbb{R}^m)$ defined by

$$\begin{aligned} \Theta &= \left\{ \begin{array}{c} \boldsymbol{u}(\cdot) \in PWC(t_0, t_1; \mathbb{R}^m) : \boldsymbol{u}(t) \in \Omega \ e.f. \\ \text{and } \boldsymbol{u}(\cdot) \text{ steers } \mathrm{X}_0 \text{ to } \mathrm{X}_1 \text{ at time } t_1. \end{array} \right\} \\ &= \Theta(t_0, t_1, \mathrm{X}_0, \mathrm{X}_1, \Omega). \end{aligned} \tag{11.3}$$

Given a control $\boldsymbol{u}(\cdot) \in PWC(t_0, t_1; \mathbb{R}^m)$, the cost functional is defined by

$$J(\boldsymbol{u}(\cdot)) = \int_{t_0}^{t_1} f_0(s, \boldsymbol{x}(s), \boldsymbol{u}(s))ds = \int_{t_0}^{t_1} f_0(s, \boldsymbol{x}(s; \boldsymbol{x}_0; \boldsymbol{u}(\cdot)), \boldsymbol{u}(s))ds, \tag{11.4}$$

where $\boldsymbol{x}(\cdot) = \boldsymbol{x}(\cdot; \boldsymbol{x}_0; \boldsymbol{u}(\cdot))$ is the solution to the initial value problem (11.1) - (11.2).

Consider the optimal control problem of minimizing $J(\cdot)$ on the set of all admissible controllers Θ. In particular, an *optimal control* is a function $\boldsymbol{u}^*(\cdot) \in \Theta$ such that $\boldsymbol{u}^*(\cdot)$ steers X_0 to X_1 at time $t_1 > t_0$ and

$$\begin{aligned} J(\boldsymbol{u}^*(\cdot)) &= \int_{t_0}^{t_1} f_0(s, \boldsymbol{x}(s; \boldsymbol{x}_0^*; \boldsymbol{u}^*(\cdot)), \boldsymbol{u}(s))ds \\ &\le \int_{t_0}^{t_1} f_0(s, \boldsymbol{x}(s; \boldsymbol{x}_0; \boldsymbol{u}(\cdot)), \boldsymbol{u}(s))ds = J(\boldsymbol{u}(\cdot)) \end{aligned} \tag{11.5}$$

for all $\boldsymbol{u}(\cdot) \in \Theta$.

Remark 11.1 *As before, if $\boldsymbol{u}^*(\cdot) \in \Theta$ is an optimal control that steers $\boldsymbol{x}_0^* \in \mathrm{X}_0$ to $\boldsymbol{x}_1^* \in \mathrm{X}_1$ at time $t_1 > t_0$, then the corresponding optimal trajectory $\boldsymbol{x}(\cdot; \boldsymbol{x}_0^*; \boldsymbol{u}^*(\cdot))$ will be denoted by $\boldsymbol{x}^*(\cdot)$. In particular, $\boldsymbol{x}^*(\cdot)$ satisfies the two point boundary value problem*

$$\begin{array}{ll} (\mathcal{S}^*) & \dot{\boldsymbol{x}}^*(t) = f(t, \boldsymbol{x}^*(t), \boldsymbol{u}^*(t)), \quad t_0 < t < t_1, \\ (\mathcal{B}\mathcal{C}s^*) & \boldsymbol{x}^*(t_0) = \boldsymbol{x}_0^* \in \mathrm{X}_0, \quad \boldsymbol{x}^*(t_1) = \boldsymbol{x}_1^* \in \mathrm{X}_1. \end{array} \tag{11.6}$$

This notation is consistent since we have assumed that the initial value problem (11.1) - (11.2) has a unique solution.

Remark 11.2 *For the Simplest Problem in Optimal Control the final time $t_1 > t_0$ is free and finding the optimal time $t_1^* > t_0$ is part of the problem. The time optimal control problem falls into this category. In the fixed-time optimal control problem the final time $t_1 > t_0$ is specified. Later we shall consider optimal control problems defined on the interval $0 \le t < +\infty$ which is a "special" fixed time problem.*

11.1.1 The Maximum Principle for Fixed t_1

When the final time t_1 is given at a fixed value $t_0 < t_1$, the Maximum Principle is the same except that the condition (10.22) is replaced by

$$\hat{M}(\hat{\boldsymbol{\eta}}^*(t), \hat{\boldsymbol{x}}^*(t)) = \max_{\boldsymbol{u} \in \Omega} \hat{H}(\hat{\boldsymbol{\eta}}^*(t), \hat{\boldsymbol{x}}^*(t), \boldsymbol{u}) \equiv c, \tag{11.7}$$

where the constant c is not necessarily 0. In particular, the following theorem is valid in this case.

Theorem 11.1 (Maximum Principle for t_1 Fixed) *Assume that $f : \mathbb{R}^n \times \mathbb{R}^m \to \mathbb{R}^n$, $f_0 : \mathbb{R}^n \times \mathbb{R}^m \to \mathbb{R}$, $\mathrm{X}_0 \subseteq \mathbb{R}^n$, $t_0 \in \mathbb{R}$, $t_1 \in \mathbb{R}$, $\mathrm{X}_1 \subseteq \mathbb{R}^n$ and $\Omega \subseteq \mathbb{R}^m$ are given as above and consider the control system*

$$(\mathcal{S}) \qquad \dot{\boldsymbol{x}}(t) = f(\boldsymbol{x}(t), \boldsymbol{u}(t)), \quad t_0 < t \leq t_1, \tag{11.8}$$

with piecewise continuous controllers $\boldsymbol{u}(\cdot) \in PWC(t_0, t_1; \mathbb{R}^m)$ satisfying $\boldsymbol{u}(t) \in \Omega \subseteq \mathbb{R}^m$ e.f. If

$$\boldsymbol{u}^*(\cdot) \in \Theta = \left\{ \begin{array}{c} \boldsymbol{u}(\cdot) \in PWC(t_0, t_1; \mathbb{R}^m) : \boldsymbol{u}(t) \in \Omega \;\; e.f. \\ \text{and } \boldsymbol{u}(\cdot) \text{ steers } \mathrm{X}_0 \text{ to } \mathrm{X}_1 \text{ at time } t_1. \end{array} \right\} \tag{11.9}$$

minimizes

$$J(\boldsymbol{u}(\cdot)) = \int_{t_0}^{t_1} f_0(\boldsymbol{x}(s), \boldsymbol{u}(s))ds \tag{11.10}$$

on the set of admissible controls Θ with optimal response $\boldsymbol{x}^*(\cdot)$ satisfying $\boldsymbol{x}^*(t_0) = \boldsymbol{x}_0^* \in \mathrm{X}_0$ and $\boldsymbol{x}^*(t_1) = \boldsymbol{x}_1^* \in \mathrm{X}_1$, then there exists a non-trivial solution

$$\hat{\boldsymbol{\eta}}^*(t) = [\eta_0^*(t) \;\; \eta_1^*(t) \;\; \eta_2^*(t) \;\; \dots \;\; \eta_n^*(t)]^T = \begin{bmatrix} \eta_0^*(t) \\ \boldsymbol{\eta}^*(t) \end{bmatrix} \tag{11.11}$$

to the augmented adjoint equation

$$\frac{d}{dt}\hat{\boldsymbol{\eta}}(t) = -[\hat{A}(t)]^T \hat{\boldsymbol{\eta}}(t), \tag{11.12}$$

such that

$$\hat{H}(\hat{\boldsymbol{\eta}}^*(t), \hat{\boldsymbol{x}}^*(t), \boldsymbol{u}^*(t)) = \hat{M}(\hat{\boldsymbol{\eta}}^*(t), \hat{\boldsymbol{x}}^*(t)) = \max_{\boldsymbol{u} \in \Omega} \hat{H}(\hat{\boldsymbol{\eta}}^*(t), \hat{\boldsymbol{x}}^*(t), \boldsymbol{u}). \tag{11.13}$$

Moreover, there are constants $\eta_0^* \leq 0$ and c such that

$$\eta_0^*(t) \equiv \eta_0^* \leq 0 \tag{11.14}$$

and for all $t \in [t_0, t_1]$,

$$\hat{M}(\hat{\boldsymbol{\eta}}^*(t), \hat{\boldsymbol{x}}^*(t)) = \max_{\boldsymbol{u}\in\Omega} \hat{H}(\hat{\boldsymbol{\eta}}^*(t), \hat{\boldsymbol{x}}^*(t), \boldsymbol{u}) \equiv c. \tag{11.15}$$

Also, if $\mathrm{X}_0 \subseteq \mathbb{R}^n$ *and* $\mathrm{X}_1 \subseteq \mathbb{R}^n$ *are manifolds with tangent spaces* $\mathbb{T}_0$ *and* $\mathbb{T}_1$ *at*
$\boldsymbol{x}^*(t_0) = \boldsymbol{x}_0^* \in \mathrm{X}_0$ *and* $\boldsymbol{x}^*(t_1) = \boldsymbol{x}_1^* \in \mathrm{X}_1$, *respectively, then*

$$\hat{\boldsymbol{\eta}}^*(t) = [\eta_0^*(t) \;\; \eta_1^*(t) \;\; \eta_2^*(t) \;\; \ldots \;\; \eta_n^*(t)]^T = [\; \eta_0^*(t) \;\; \boldsymbol{\eta}^*(t) \;]^T$$

can be selected to satisfy the transversality conditions

$$\boldsymbol{\eta}^*(t_0) \perp \mathbb{T}_0 \tag{11.16}$$

and

$$\boldsymbol{\eta}^*(t_1) \perp \mathbb{T}_1. \tag{11.17}$$

11.2 Application to Problems in the Calculus of Variations

We start with some classical problems in the calculus of variations. First we assume the cost function does not explicitly depend on time. In particular, the cost functional is given by

$$J(x(\cdot)) = \int_{t_0}^{t_1} f_0\left(x(s), \dot{x}\left(s\right)\right) ds$$

where $f_0(x, u)$ is independent of time. We will address the time dependent case later.

11.2.1 The Simplest Problem in the Calculus of Variations

Assume t_0, t_1, x_0, and x_1 are given and the cost functional is of the form

$$J(x(\cdot)) = \int_{t_0}^{t_1} f_0\left(x(s), \dot{x}\left(s\right)\right) ds.$$

Let $\Psi_1 = \Psi(t_0, t_1, x_0, x_1)$ be the set of PWS functions defined by

$$\Psi = \{x(\cdot) \in PWS(t_0, +\infty) : x\,(t_0) = x_0, \quad x\,(t_1) = x_1\}$$

and consider the **SPCV** where the objective is to find $x^*\,(\cdot) \in \Psi$ such that

$$J(x^*(\cdot)) = \int_{t_0}^{t_1} f_0\,(x^*(s), \dot{x}^*\,(s))\,ds \leq J(x(\cdot))$$
$$= \int_{t_0}^{t_1} f_0\,(x(s), \dot{x}\,(s))\,ds = J(x(\cdot))$$

for all $x(\cdot) \in \Psi$.

In order to formulate an equivalent Simplest Problem in Optimal Control (with fixed t_1), we define the state equation by

$$\dot{x}(t) = u(t) \tag{11.18}$$

with initial condition

$$x(t_0) = x_0. \tag{11.19}$$

Define the sets

$$X_0 = \{x_0\} \subseteq \mathbb{R}^1 \quad \text{and} \quad X_1 = \{x_1\} \subseteq \mathbb{R}^1,$$

and let the control constraint set be $\Omega = \mathbb{R}^1$. The set of *admissible controllers* is the subset of $PWC(t_0, t_1)$ defined by

$$\Theta = \left\{ \begin{matrix} u(\cdot) \in PWC(t_0, t_1) : u(t) \in \Omega \ e.f. \text{ and} \\ u(\cdot) \text{ steers } X_0 \text{ to } X_1 \text{ at time } t_1. \end{matrix} \right\}$$
$$= \Theta(t_0, t_1, x_0, x_1, \Omega). \tag{11.20}$$

Given a control $u(\cdot) \in PWC(t_0, t_1)$, the cost functional is defined by

$$J(u(\cdot)) = \int_{t_0}^{t_1} f_0(x(s), u(s))ds, \tag{11.21}$$

where $x(\cdot) = x(\cdot; u(\cdot))$ is the solution to the initial value problem (11.18) - (11.19). This variational problem is equivalent to the optimal control problem of minimizing $J(\cdot)$ on the set of admissible controllers Θ. In particular, the goal is to find an *optimal control* $u^*(\cdot) \in \Theta$ such that $u^*(\cdot)$ steers X_0 to X_1 at time $t_1 > t_0$ and

$$J(u^*(\cdot)) = \int_{t_0}^{t_1} f_0(x^*(s), u^*(s))\,ds \le \int_{t_0}^{t_1} f_0(x(s), u(s))\,ds$$

for all $u(\cdot) \in \Theta$.

Formulation as an Optimal Control Problem: In order to set up the Simplest Problem in the Calculus of Variations as a fixed time optimal control problem and apply the Maximum Principle, we need to:

1. Identify the initial time t_0 and final time t_1;
2. Identify the initial set $X_0 \subseteq \mathbb{R}^1$, the terminal set $X_1 \subseteq \mathbb{R}^1$;
3. Identify the control constraint set $\Omega \subseteq \mathbb{R}^1$;
4. Identify the functions $f : \mathbb{R}^1 \times \mathbb{R}^1 \longrightarrow \mathbb{R}^1$, $f_0 : \mathbb{R}^1 \times \mathbb{R}^1 \longrightarrow \mathbb{R}^1$;
5. Define the augmented Hamiltonian $\hat{H}(\hat{\boldsymbol{\eta}}, \hat{\boldsymbol{x}}, u)$;
6. Form the augmented adjoint system matrix $\hat{A}(t)$.

Clearly, $\mathrm{X}_0 = \{x_0\} \subseteq \mathbb{R}^1$, $\mathrm{X}_1 = \{x_1\} \subseteq \mathbb{R}^1$ and $\Omega = \mathbb{R}^1$. Moreover,

$$f(x, u) = u$$

and

$$f_0(x, u) = f_0(x, u).$$

Therefore, the augmented function $\hat{f} : \mathbb{R}^2 \times \mathbb{R}^1 \longrightarrow \mathbb{R}^2$ is defined by

$$\hat{f}(\hat{\boldsymbol{x}}, u) = \hat{f}(x_0, x, u) = \begin{bmatrix} f_0(x, u) \\ u \end{bmatrix},$$

where $\hat{\boldsymbol{x}} = [x_0 \;\; x]^T$ and the augmented Hamiltonian $\hat{H}(\hat{\boldsymbol{\eta}}, \hat{\boldsymbol{x}}, u) : \mathbb{R}^2 \times \mathbb{R}^2 \times \mathbb{R}^1 \longrightarrow \mathbb{R}^1$ is given by

$$\hat{H}(\hat{\boldsymbol{\eta}}, \hat{\boldsymbol{x}}, u) = \eta_0 f_0(x, u) + \eta f(x, u) = \eta_0 f_0(x, u) + \eta u$$

where $\hat{\boldsymbol{\eta}} = [\eta_0 \;\; \eta]^T$. The augmented Jacobian is given by

$$\mathbb{J}_{\hat{\boldsymbol{x}}}\hat{f}(\hat{\boldsymbol{x}}, u) = \mathbb{J}_{\hat{\boldsymbol{x}}}\hat{f}(x, u) = \begin{bmatrix} 0 & \frac{\partial f_0(x,u)}{\partial x} \\ 0 & 0 \end{bmatrix}$$

so that if $(x^*(\cdot), u^*(\cdot)) = (x^*(\cdot), \dot{x}^*(\cdot))$ is an optimal pair, then the augmented matrix $\hat{A}(t)$ is given by

$$\hat{A}(t) = \mathbb{J}_{\hat{\boldsymbol{x}}}\hat{f}(\hat{\boldsymbol{x}}, u)|_{(\boldsymbol{x}^*(t), u^*(t))} = \mathbb{J}_{\hat{\boldsymbol{x}}}\hat{f}(x^*(t), u^*(t)) = \begin{bmatrix} 0 & \frac{\partial f_0(x^*(t),u^*(t))}{\partial x} \\ 0 & 0 \end{bmatrix}.$$

Therefore,

$$-[\hat{A}(t)]^T = -\begin{bmatrix} 0 & \frac{\partial f_0(x^*(t),u^*(t))}{\partial x} \\ 0 & 0 \end{bmatrix}^T = \begin{bmatrix} 0 & 0 \\ -\frac{\partial f_0(x^*(t),u^*(t))}{\partial x} & 0 \end{bmatrix}$$

and the augmented adjoint equation becomes

$$\frac{d}{dt}\begin{bmatrix} \eta_0(t) \\ \eta(t) \end{bmatrix} = -[\hat{A}(t)]^T \begin{bmatrix} \eta_0(t) \\ \eta(t) \end{bmatrix} = \begin{bmatrix} 0 & 0 \\ -\frac{\partial f_0(x^*(t),u^*(t))}{\partial x} & 0 \end{bmatrix} \begin{bmatrix} \eta_0(t) \\ \eta(t) \end{bmatrix}. \tag{11.22}$$

Consequently,

$$\frac{d}{dt}\eta_0(t) = 0$$
$$\frac{d}{dt}\eta(t) = -\left[\frac{\partial f_0(x^*(t), u^*(t))}{\partial x}\right]\eta_0$$

which implies (as always) that the zero adjoint state is a constant $\eta_0(t) \equiv \eta_0$.

If $(x^*(\cdot), u^*(\cdot)) = (x^*(\cdot), \dot{x}^*(\cdot))$ is an optimal pair for the **SPCV**, then the Maximum Principle implies that there is a **non-trivial** solution

$$\hat{\boldsymbol{\eta}}^*(t) = \begin{bmatrix} \eta_0^*(t) \\ \eta^*(t) \end{bmatrix} \in \mathbb{R}^2$$

to the augmented adjoint equation (11.22) such that $\eta_0^*(t) \equiv \eta_0^* \leq 0$. Observe that

$$\frac{d}{dt}\eta^*(t) = -\left[\frac{\partial f_0(x^*(t), u^*(t))}{\partial x}\right]\eta_0^* \tag{11.23}$$

so that

$$\eta^*(t) = \eta^*(t_0) - \eta_0^* \int_{t_0}^{t} \left[\frac{\partial f_0(x^*(s), u^*(s))}{\partial x}\right] ds. \tag{11.24}$$

The Maximum Principle implies that

$$\begin{aligned} \hat{H}(\hat{\boldsymbol{\eta}}^*(t), \hat{\boldsymbol{x}}^*(t), u^*(t)) &= \max_{u\in\mathbb{R}^1} \hat{H}(\hat{\boldsymbol{\eta}}^*(t), \hat{\boldsymbol{x}}^*(t), u) \\ &= \max_{u\in\mathbb{R}^1} [\eta_0^* f_0(x^*(t), u) + \eta^*(t)u] \equiv c \quad (11.25) \end{aligned}$$

and since $\Omega = \mathbb{R}^1$, the derivative of $\hat{H}(\hat{\boldsymbol{\eta}}^*(t), \hat{\boldsymbol{x}}^*(t), u)$ with respect to u at $u = u^*(t)$ must be zero. In particular,

$$D_u\hat{H}(\hat{\boldsymbol{\eta}}^*(t), \hat{\boldsymbol{x}}^*(t), u)_{|u=u^*(t)} = D_u\hat{H}(\hat{\boldsymbol{\eta}}^*(t), \hat{\boldsymbol{x}}^*(t), u^*(t)) = 0,$$

where

$$\begin{aligned} D_u\hat{H}(\hat{\boldsymbol{\eta}}^*(t), \hat{\boldsymbol{x}}^*(t), u) &= \frac{\partial}{\partial u}[\eta_0^* f_0(x^*(t), u) + \eta^*(t)u] \\ &= \eta_0^*\left[\frac{\partial f_0(x^*(s), u)}{\partial u}\right] + \eta^*(t). \end{aligned}$$

Thus, when $u = u^*(t)$

$$D_u\hat{H}(\hat{\boldsymbol{\eta}}^*(t), \hat{\boldsymbol{x}}^*(t), u)_{|u=u^*(t)} = \eta_0^*\left[\frac{\partial f_0(x^*(t), u^*(t))}{\partial u}\right] + \eta^*(t) = 0 \tag{11.26}$$

and we have

$$\eta^*(t) = -\eta_0^*\left[\frac{\partial f_0(x^*(t), u^*(t))}{\partial u}\right]. \tag{11.27}$$

Observe that (11.27) implies that $\eta_0^* \neq 0$ because if $\eta_0^* = 0$, then we would also have $\eta^*(t) = 0$ which contradicts the statement that $\hat{\boldsymbol{\eta}}^*(t)$ is nontrivial. Combining (11.24) and (11.27) we find that

$$-\eta_0^*\left[\frac{\partial f_0(x^*(s), u^*(t))}{\partial u}\right] = \eta^*(t_0) - \eta_0^* \int_{t_0}^{t} \left[\frac{\partial f_0(x^*(s), u^*(s))}{\partial x}\right] ds$$

and since $\eta_0^* < 0$, we divide both sides of this equation by $-\eta_0^*$ to obtain

$$\left[\frac{\partial f_0(x^*(s), u^*(t))}{\partial u}\right] = c + \int_{t_0}^{t} \left[\frac{\partial f_0(x^*(s), u^*(s))}{\partial x}\right] ds \qquad (11.28)$$

where the constant c is given by

$$c = \frac{\eta^*(t_0)}{-\eta_0^*}.$$

Note that (11.28) is the Euler Integral Equation and hence we have shown that the optimal pair $(x^*(\cdot), u^*(\cdot))$ with $\dot{x}^*(t) = u^*(t)$ satisfies the Euler Integral Equation

$$\left[\frac{\partial f_0(x^*(s), \dot{x}^*(t))}{\partial u}\right] = c + \int_{t_0}^{t} \left[\frac{\partial f_0(x^*(s), \dot{x}^*(s))}{\partial x}\right] ds, \quad t_0 < t < t_1. \qquad (11.29)$$

Also, between corners of $x^*(\cdot)$,

$$\frac{d}{dt}\left[\frac{\partial f_0(x^*(s), \dot{x}^*(t))}{\partial u}\right] = \left[\frac{\partial f_0(x^*(s), \dot{x}^*(s))}{\partial x}\right], \quad t_0 < t < t_1 \qquad (11.30)$$

and

$$x^*(t_0) = x_0 \quad \text{and} \quad x^*(t_1) = x_1$$

which is the Euler Differential Equation for the **SPCV**. Consequently, we have used the Maximum Principle to derive the classical Euler Necessary Condition.

The Weierstrass Necessary Condition

One can also use the Maximum Principle to derive the Weierstrass Necessary Condition. Observe that for the **SPCV**

$$\hat{H}(\hat{\boldsymbol{\eta}}, \hat{\boldsymbol{x}}, u) = [\eta_0 f_0(x, u) + \eta u] \qquad (11.31)$$

and the Maximum Principle (11.25) implies that

$$\hat{H}(\hat{\boldsymbol{\eta}}^*(t), \hat{\boldsymbol{x}}^*(t), u^*(t)) = \max_{u \in \mathbb{R}^1} \hat{H}(\hat{\boldsymbol{\eta}}^*(t), \hat{\boldsymbol{x}}^*(t), u) \geq \hat{H}(\hat{\boldsymbol{\eta}}^*(t), \hat{\boldsymbol{x}}^*(t), u) \qquad (11.32)$$

for all $u \in \mathbb{R}^1$. Using the form of the augmented Hamiltonian $\hat{H}(\hat{\boldsymbol{\eta}}, \hat{\boldsymbol{x}}, u) : \mathbb{R}^2 \times \mathbb{R}^2 \times \mathbb{R}^1 \longrightarrow \mathbb{R}^1$ given by

$$\hat{H}(\hat{\boldsymbol{\eta}}, \hat{\boldsymbol{x}}, u) = \eta_0 f_0(x, u) + \eta f(x, u) = \eta_0 f_0(x, u) + \eta u,$$

it follows that

$$[\eta_0^* f_0(x^*(t), u^*(t)) + \eta^*(t) u^*(t)] \geq [\eta_0^* f_0(x^*(t), u) + \eta^*(t) u]. \quad (11.33)$$

Consequently, dividing by $\eta_0^* < 0$ yields

$$f_0(x^*(t), u^*(t)) + \frac{\eta^*(t)}{\eta_0^*} u^*(t) \leq f_0(x^*(t), u) + \frac{\eta^*(t)}{\eta_0^*} u.$$

However, from equation (11.27) above, it follows that

$$\frac{\eta^*(t)}{\eta_0^*} = -\frac{\partial f_0(x^*(t), u^*(t))}{\partial u}.$$

Therefore, substituting this into the previous inequality yields

$$\begin{aligned} f_0(x^*(t), u^*(t)) - \frac{\partial f_0(x^*(t), u^*(t))}{\partial u} u^*(t) \leq & f_0(x^*(t), u) \\ & - \frac{\partial f_0(x^*(t), u^*(t))}{\partial u} u \end{aligned}$$

which implies

$$0 \leq f_0(x^*(t), u) - f_0(x^*(t), u^*(t)) - [u - u^*(t)] \frac{\partial f_0(x^*(t), u^*(t))}{\partial u}$$

for all $u \in \mathbb{R}^1$.

Recalling the definition of the Weierstrass Excess Function we have

$$\begin{aligned} E(t, x^*(t), u^*(t), u) = & f_0(x^*(t), u) - f_0(x^*(t), u^*(t)) \\ & - [u - u^*(t)] \frac{\partial f_0(x^*(t), u^*(t))}{\partial u} \end{aligned}$$

and hence it follows that

$$0 \leq E(t, x^*(t), u^*(t), u)$$

for all $u \in \mathbb{R}^1$. This is the Weierstrass Necessary Condition - (II) for the **SPCV**.

11.2.2 Free End-Point Problems

Formulation as an Optimal Control Problem: In order to set up the End-Point Problem as an optimal control problem and apply the Maximum Principle, we need to:

1. Identify the initial time t_0 and final time t_1;
2. Identify the initial set $\mathrm{X}_0 \subseteq \mathbb{R}^1$, the terminal set $\mathrm{X}_1 \subseteq \mathbb{R}^1$;
3. Identify the control constraint set $\Omega \subseteq \mathbb{R}^1$;
4. Identify the functions $f : \mathbb{R}^1 \times \mathbb{R}^1 \longrightarrow \mathbb{R}^1$, $f_0 : \mathbb{R}^1 \times \mathbb{R}^1 \longrightarrow \mathbb{R}^1$;
5. Define the augmented Hamiltonian $\hat{H}(\hat{\boldsymbol{\eta}}, \hat{\boldsymbol{x}}, u)$;
6. Form the augmented adjoint system matrix $\hat{A}(t)$.

This problem is the same as the previous problem except that x_1 is not given. Therefore, $\mathrm{X}_0 = \{x_0\} \subseteq \mathbb{R}^1$ and $\mathrm{X}_1 = \mathbb{R}^1$ and the tangent spaces are given by

$$\mathbb{T}_0 = \{0\} \quad \text{and} \quad \mathbb{T}_1 = \mathbb{R}^1,$$

respectively. Therefore, the transversality condition at t_1 implies that

$$\eta^*(t_1) = 0. \tag{11.34}$$

All the analysis above for the **SPCV** holds. In particular,

$$\frac{d}{dt}\eta^*(t) = -\left[\frac{\partial f_0(x^*(t), u^*(t))}{\partial x}\right]\eta_0^*$$

and

$$D_u\hat{H}(\hat{\boldsymbol{\eta}}^*(t), \hat{\boldsymbol{x}}^*(t), u)_{|u=u^*(t)} = \eta_0^*\left[\frac{\partial f_0(x^*(t), u^*(t))}{\partial u}\right] + \eta^*(t) = 0,$$

which implies

$$\eta^*(t) = -\eta_0^*\left[\frac{\partial f_0(x^*(t), u^*(t))}{\partial u}\right]. \tag{11.35}$$

Also, (11.35) implies that $\eta_0^* \neq 0$.

Therefore, we have shown the optimal pair $(x^*(\cdot), u^*(\cdot))$ with $\dot{x}^*(t) = u^*(t)$ satisfies Euler's equation in integral form

$$\left[\frac{\partial f_0(x^*(s), \dot{x}^*(t))}{\partial u}\right] = c + \int_{t_0}^{t} \left[\frac{\partial f_0(x^*(s), \dot{x}^*(s))}{\partial x}\right] ds, \quad t_0 < t < t_1,$$

with initial condition

$$x^*(t_0) = x_0.$$

In order to obtain the "natural boundary condition", we apply the transversality condition (11.34) to (11.35) and obtain

$$0 = \eta^*(t_1) = -\eta_0^* \left[\frac{\partial f_0(x^*(t_1), u^*(t_1))}{\partial u}\right].$$

Again, since $\eta_0^* \neq 0$, it follows that the natural boundary condition is given by

$$\frac{\partial f_0(x^*(t_1), \dot{x}^*(t_1))}{\partial u} = \frac{\partial f_0(x^*(t_1), u^*(t_1))}{\partial u} = 0$$

which matches the classical results.

11.2.3 Point-to-Curve Problems

Recall that in this problem we are given t_0, x_0, and a "curve" defined by the graph of a smooth function $\varphi : \mathbb{R}^1 \longrightarrow \mathbb{R}^1$. The point-to-curve problem is to find $x^*(\cdot) \in PWS(t_0, +\infty)$ that minimizes

$$J(x(\cdot)) = \int_{t_0}^{t_1} f_0(s, x(s), \dot{x}(s))ds$$

on the set

$$\Psi_\varphi = \{x(\cdot) \in PWS(t_0, +\infty) : x(t_0) = x_0, \\ x(t_1) = \varphi(t_1) \text{ for a finite time } t_1 > t_0\}.$$

In particular, one must find t_1^* and $x^*(\cdot) \in PWS(t_0, t_1^*)$ so that

$$J(x^*(\cdot)) = \int_{t_0}^{t_1^*} f_0\left(s, x^*(s), \dot{x}^*\left(s\right)\right) ds$$
$$\leq \int_{t_0}^{t_1} f_0\left(s, x(s), \dot{x}\left(s\right)\right) ds = J(x(\cdot))$$

for all $x^*(\cdot) \in \Psi_\varphi$.

Remark 11.3 *In order to properly formulate this problem we note that the cost functional integrand $f_0\left(s, x, u\right)$ is time dependent and does not allow a direct application of the previous Maximum Principles. In order to overcome this issue, again use the idea of an "augmented state" to reformulate the problem as a time independent optimal control problem. This basic idea is also used to extend the Maximum Principle for the* ***SPOC*** *to the case where the integrands and differential equations defining the state equations are time dependent (see Section 11.7).*

The basic idea is to treat time t as a "state" and "augment" the system with this new state. In particular, we define

$$x_1(t) = t$$

and

$$x_2(t) = x(t).$$

Observe that

$$\dot{x}_1(t) = 1 \quad \text{and} \quad x_1(t_0) = t_0 \in \mathbb{R}^1,$$

and

$$\dot{x}_2(t) = \dot{x}(t) = u(t) \quad \text{and} \quad x_2(t_0) = x_0 \in \mathbb{R}^1, \quad x_2(t_1) = x_1 \in \mathbb{R}^1.$$

Therefore, the control problem is governed by the two dimensional control system

$$\frac{d}{dt}\begin{bmatrix} x_1(t) \\ x_2(t) \end{bmatrix} = \begin{bmatrix} 1 \\ u(t) \end{bmatrix},$$

with initial condition

$$\begin{bmatrix} x_1(t_0) \\ x_2(t_0) \end{bmatrix} = \begin{bmatrix} t_0 \\ x_0 \end{bmatrix}$$

and terminal condition

$$\begin{bmatrix} x_1(t_1) \\ x_2(t_1) \end{bmatrix} = \begin{bmatrix} t_1 \\ \varphi(t_1) \end{bmatrix},$$

where t_1 is not specified. Note that because the integrand $f_0(\cdot)$ is a time dependent function $f_0(t, x, u)$, the t variable becomes an additional "state" so the control problem is a 2-dimensional problem.

Formulation as a Simplest Problem in Optimal Control: In order to set up the Point-to-Curve Problem as an **SPOC** and apply the Maximum Principle, we need to:

1. Identify the initial time t_0;
2. Identify the initial set $\mathrm{X}_0 \subseteq \mathbb{R}^2$, the terminal set $\mathrm{X}_1 \subseteq \mathbb{R}^2$;
3. Identify the control constraint set $\Omega \subseteq \mathbb{R}^1$;
4. Identify the functions $f : \mathbb{R}^2 \times \mathbb{R}^1 \longrightarrow \mathbb{R}^1$, $f_0 : \mathbb{R}^2 \times \mathbb{R}^1 \longrightarrow \mathbb{R}^1$;
5. Define the augmented Hamiltonian $\hat{H}(\hat{\boldsymbol{\eta}}, \hat{\boldsymbol{x}}, u)$;
6. Form the augmented adjoint system matrix $\hat{A}(t)$.

Define the sets $\mathrm{X}_0 \subseteq \mathbb{R}^2$ and $\mathrm{X}_0 \subseteq \mathbb{R}^2$ by

$$\mathrm{X}_0 = \left\{ \begin{bmatrix} t_0 \\ x_0 \end{bmatrix} \right\} \subseteq \mathbb{R}^2$$

and

$$\mathrm{X}_1 = \left\{ \begin{bmatrix} x_1 \\ x_2 \end{bmatrix} = \begin{bmatrix} t_1 \\ x_2 \end{bmatrix} \in \mathbb{R}^2 : \; x_2 = \varphi(x_1) = \varphi(t_1) \right\} \subseteq \mathbb{R}^2,$$

respectively.

Observe that t_1, and hence $x_1(t_1) = t_1$ is not specified (other than $t_0 \le t_1$). Moreover, if we let $G : \mathbb{R}^2 \to \mathbb{R}^1$ be defined by $G(x_1, x_2) = \varphi(x_1) - x_2$, then

$$\mathrm{X}_1 = \left\{ \begin{bmatrix} x_1 \\ x_2 \end{bmatrix} \in \mathbb{R}^2 : G(x_1, x_2) = \varphi(x_1) - x_2 = 0 \right\}.$$

If $\begin{bmatrix} x_1^* \\ x_2^* \end{bmatrix} \in \mathrm{X}_1$, then the tangent plane $\mathbb{T}_1$ at $\begin{bmatrix} x_1^* \\ x_2^* \end{bmatrix}$ is defined by the gradient of $G(x_1, x_2)$ at $\begin{bmatrix} x_1^* \\ x_2^* \end{bmatrix}$. In particular,

$$\mathbb{T}_1 = \left\{ \boldsymbol{x} = \begin{bmatrix} x_1 \\ x_2 \end{bmatrix} : \boldsymbol{x} \perp \nabla G(x_1^*, x_2^*) \right\}$$

and hence a non-zero vector $\begin{bmatrix} \eta_1 \\ \eta_2 \end{bmatrix}$ is orthogonal to $\mathbb{T}_1$ if and only if

$$\begin{bmatrix} \eta_1 \\ \eta_2 \end{bmatrix} = k \nabla G(x_1^*, x_2^*) = k \begin{bmatrix} \dot{\varphi}(x_1^*) \\ -1 \end{bmatrix} \tag{11.36}$$

for some constant $k \neq 0$. We use this transversality condition several places below. Finally, note that $\Omega = \mathbb{R}^1$ since there are no bounds on the control $u(\cdot)$. Observe that $f : \mathbb{R}^2 \times \mathbb{R}^1 \longrightarrow \mathbb{R}^2$ is defined by

$$f(x_1, x_2, u) = \begin{bmatrix} f_1(x_1, x_1, u) \\ f_2(x_1, x_1, u) \end{bmatrix} = \begin{bmatrix} 1 \\ u \end{bmatrix},$$

and $f_0 : \mathbb{R}^2 \times \mathbb{R}^1 \longrightarrow \mathbb{R}^1$ is given as the integrand $f_0(x_1, x_2, u)$.

Hence, the augmented function $\hat{f} : \mathbb{R}^3 \times \mathbb{R}^1 \longrightarrow \mathbb{R}^3$ is defined by

$$\hat{f}(\hat{\boldsymbol{x}}, u) = \hat{f}(x_0, \boldsymbol{x}, u) = \begin{bmatrix} f_0(x_1, x_2, u) \\ f_1(x_1, x_2, u) \\ f_2(x_1, x_2, u) \end{bmatrix} = \begin{bmatrix} f_0(x_1, x_2, u) \\ 1 \\ u \end{bmatrix}$$

and the augmented Hamiltonian $\hat{H}(\hat{\boldsymbol{\eta}}, \hat{\boldsymbol{x}}, u) : \mathbb{R}^3 \times \mathbb{R}^3 \times \mathbb{R}^1 \longrightarrow \mathbb{R}^1$ is given by

$$\hat{H}(\hat{\boldsymbol{\eta}}, \hat{\boldsymbol{x}}, u) = \eta_0 f_0(x_1, x_2, u) + \eta_1 f_1(x_1, x_2, u) + \eta_2 f_2(x_1, x_2, u).$$

In particular, the augmented Hamiltonian has the form

$$\hat{H}(\hat{\boldsymbol{\eta}}, \hat{\boldsymbol{x}}, u) = \eta_0 f_0(x_1, x_2, u) + \eta_1 + \eta_2 u,$$

where $\hat{\boldsymbol{\eta}} = [\eta_0 \ \ \boldsymbol{\eta}]^T$. The augmented Jacobian is given by

$$\mathbb{J}_{\hat{\boldsymbol{x}}}\hat{f}(\hat{\boldsymbol{x}}, u) = \mathbb{J}_{\hat{\boldsymbol{x}}}\hat{f}(\boldsymbol{x}, u) = \begin{bmatrix} \frac{\partial f_0(x_1,x_2,u)}{\partial x_0} & \frac{\partial f_0(x_1,x_2,u)}{\partial x_1} & \frac{\partial f_0(x_1,x_2,u)}{\partial x_2} \\ 0 & 0 & 0 \\ 0 & 0 & 0 \end{bmatrix}$$

so that if $(x^*(\cdot), u^*(\cdot)) = (x^*(\cdot), \dot{x}^*(\cdot))$ is an optimal pair, then the augmented matrix $\hat{A}(t)$ is given by

$$\hat{A}(t) = \mathbb{J}_{\hat{\boldsymbol{x}}}\hat{f}(\hat{\boldsymbol{x}}, u)_{|(\boldsymbol{x}^*(t),u^*(t))} = \begin{bmatrix} 0 & \frac{\partial f_0(x_1^*(t),x_2^*(t),u^*(t))}{\partial x_1} & \frac{\partial f_0(x_1^*(t),x_2^*(t),u^*(t))}{\partial x_2} \\ 0 & 0 & 0 \\ 0 & 0 & 0 \end{bmatrix}.$$

Therefore,

$$-[\hat{A}(t)]^T = \begin{bmatrix} 0 & 0 & 0 \\ -\frac{\partial f_0(x_1^*(t),x_2^*(t),u^*(t))}{\partial x_1} & 0 & 0 \\ -\frac{\partial f_0(x_1^*(t),x_2^*(t),u^*(t))}{\partial x_2} & 0 & 0 \end{bmatrix}$$

and the augmented adjoint equation becomes

$$\frac{d}{dt}\begin{bmatrix} \eta_0(t) \\ \eta_1(t) \\ \eta_2(t) \end{bmatrix} = -[\hat{A}(t)]^T \begin{bmatrix} \eta_0(t) \\ \eta_1(t) \\ \eta_2(t) \end{bmatrix}. \tag{11.37}$$

Consequently,

$$\begin{aligned} \frac{d}{dt}\eta_0(t) &= 0 \\ \frac{d}{dt}\eta_1(t) &= -\left[\frac{\partial f_0(x_1^*(t), x_2^*(t), u^*(t))}{\partial x_1}\right]\eta_0 \\ \frac{d}{dt}\eta_2(t) &= -\left[\frac{\partial f_0(x_1^*(t), x_2^*(t), u^*(t))}{\partial x_2}\right]\eta_0 \end{aligned}$$

which implies (as always) that the zero adjoint state is a constant $\eta_0(t) \equiv \eta_0$.

If $(x^*(\cdot), u^*(\cdot)) = (x^*(\cdot), \dot{x}^*(\cdot))$ is an optimal pair for the **SPCV**, then the Maximum Principle implies that there is a **non-trivial** solution

$$\hat{\boldsymbol{\eta}}^*(t) = \begin{bmatrix} \eta_0^*(t) \\ \eta_1^*(t) \\ \eta_2^*(t) \end{bmatrix} \in \mathbb{R}^3$$

to the augmented adjoint equation (11.37) such that $\eta_0^*(t) \equiv \eta_0^* \leq 0$. Observe that

$$\frac{d}{dt}\eta_2^*(t) = -\left[\frac{\partial f_0(x_1^*(t), x_2^*(t), u^*(t))}{\partial x_2}\right]\eta_0 \tag{11.38}$$

so that

$$\eta_2^*(t) = \eta_2^*(t_0) - \eta_0^* \int_{t_0}^{t} \left[\frac{\partial f_0(x_1^*(s), x_2^*(s), u^*(s))}{\partial x_2}\right] ds. \tag{11.39}$$

Since t_1 is free, the Maximum Principle implies that

$$\hat{H}(\hat{\boldsymbol{\eta}}^*(t), \hat{\boldsymbol{x}}^*(t), u^*(t)) = \max_{u\in\mathbb{R}^1} \hat{H}(\hat{\boldsymbol{\eta}}^*(t), \hat{\boldsymbol{x}}^*(t), u) = 0$$

and for this problem we have

$$\begin{aligned} \max_{u\in\mathbb{R}^1} \hat{H}(\hat{\boldsymbol{\eta}}^*(t), \hat{\boldsymbol{x}}^*(t), u) &= \max_{u\in\mathbb{R}^1}[\eta_0^* f_0(x_1^*(t), x_2^*(t), u) + \eta_1^*(t) + \eta_2^*(t)u] \\ &\equiv 0. \end{aligned} \tag{11.40}$$

Since $\Omega = \mathbb{R}^1$, the derivative of $\hat{H}(\hat{\boldsymbol{\eta}}^*(t), \hat{\boldsymbol{x}}^*(t), u)$ with respect to u at $u = u^*(t)$ must be zero. In particular, since t_1 is free we have

$$D_u\hat{H}(\hat{\boldsymbol{\eta}}^*(t), \hat{\boldsymbol{x}}^*(t), u)_{|u=u^*(t)} = D_u\hat{H}(\hat{\boldsymbol{\eta}}^*(t), \hat{\boldsymbol{x}}^*(t), u^*(t)) = 0,$$

where

$$\begin{aligned} D_u\hat{H}(\hat{\boldsymbol{\eta}}^*(t), \hat{\boldsymbol{x}}^*(t), u) &= \frac{\partial}{\partial u}[\eta_0^* f_0(x_1^*(t), x_2^*(t), u) + \eta_1^*(t) + \eta_2^*(t)u] \\ &= \eta_0^*\left[\frac{\partial f_0(x_1^*(t), x_2^*(t), u)}{\partial u}\right] + \eta_2^*(t). \end{aligned}$$

Thus, when $u = u^*(t)$

$$D_u\hat{H}(\hat{\boldsymbol{\eta}}^*(t), \hat{\boldsymbol{x}}^*(t), u)_{|_{u=u^*(t)}} = \eta_0^* \left[\frac{\partial f_0(x_1^*(t), x_2^*(t), u^*(t))}{\partial u}\right] + \eta_2^*(t) = 0 \tag{11.41}$$

and we have

$$\eta_2^*(t) = -\eta_0^* \left[\frac{\partial f_0(x_1^*(t), x_2^*(t), u^*(t))}{\partial u}\right]. \tag{11.42}$$

Observe that (11.42) and (11.40) together imply that $\eta_0^* \neq 0$. To see this assume that $\eta_0^* = 0$. Equation (11.42) then implies that $\eta_2^*(t) \equiv 0$. The Maximum Principle (11.40) yields

$$\hat{H}(\hat{\boldsymbol{\eta}}^*(t), \hat{\boldsymbol{x}}^*(t), u^*(t)) = \eta_0^* f_0(x_1^*(t), x_2^*(t), u^*(t)) + \eta_1^*(t) + \eta_2^*(t)u^*(t) \equiv 0$$

so that if $\eta_0^* = 0$, then $\eta_2^*(t) \equiv 0$ and

$$\eta_0^* f_0(x_1^*(t), x_2^*(t), u^*(t)) + \eta_1^*(t) + \eta_2^*(t)u^*(t) = \eta_1^*(t) \equiv 0.$$

Hence, it would follow that $\hat{\boldsymbol{\eta}}^*(t) \equiv 0$ which contradicts the statement that $\hat{\boldsymbol{\eta}}^*(t)$ is nontrivial. Consequently, it follows that $\eta_0^* < 0$. Combining (11.39) and (11.42) we find that

$$-\eta_0^* \left[\frac{\partial f_0(x_1^*(t), x_2^*(t), u^*(t))}{\partial u}\right] = \eta_2^*(t_0) - \eta_0^* \int_{t_0}^{t} \left[\frac{\partial f_0(x_1^*(s), x_2^*(s), u^*(s))}{\partial x_2}\right] ds$$

and since $\eta_0^* < 0$, we divide both sides of this equation by $-\eta_0^*$ to obtain

$$\left[\frac{\partial f_0(x_1^*(t), x_2^*(t), u^*(t))}{\partial u}\right] = c + \int_{t_0}^{t} \left[\frac{\partial f_0(x_1^*(s), x_2^*(s), u^*(s))}{\partial x_2}\right] ds \tag{11.43}$$

where the constant c is given by

$$c = \frac{\eta_2^*(t_0)}{-\eta_0^*}.$$

Observe that (11.43) is Euler's equation in integral form. Therefore, we have shown that the optimal solution $(x_1^*(\cdot), x_2^*(\cdot), u^*(\cdot))$ with $\dot{x}_2^*(t) = \dot{x}^*(t) = u^*(t)$ and $x_1^*(\cdot) = t$ satisfies Euler's equation in integral form

$$\left[\frac{\partial f_0(t, x^*(t), \dot{x}^*(t))}{\partial u}\right] = c + \int_{t_0}^{t} \left[\frac{\partial f_0(t, x^*(t), \dot{x}^*(t))}{\partial x}\right] ds, \quad t_0 < t < t_1^*. \tag{11.44}$$

Also, we have that between corners of $x^*(\cdot)$,

$$\frac{d}{dt}\left[\frac{\partial f_0(t, x^*(t), \dot{x}^*(t))}{\partial u}\right] = \left[\frac{\partial f_0(t, x^*(t), \dot{x}^*(t))}{\partial x}\right], \quad t_0 < t < t_1^* \tag{11.45}$$

and

$$x^*(t_0) = x_0 \quad \text{and} \quad x^*(t_1^*) = \varphi(t_1^*).$$

To complete the problem we need to use the transversality condition at t_1^*. The transversality condition (11.36) implies that

$$\begin{bmatrix} \eta_1^*(t_1^*) \\ \eta_2^*(t_1^*) \end{bmatrix} = k\nabla G(x_1^*(t_1^*), x_2^*(t_1^*)) = k \begin{bmatrix} \dot{\varphi}(x_1^*(t_1^*)) \\ -1 \end{bmatrix} = k \begin{bmatrix} \dot{\varphi}(t_1^*) \\ -1 \end{bmatrix}$$

and hence it follows that

$$\eta_1^*(t_1^*) = k\dot{\varphi}(t_1^*) \tag{11.46}$$

and

$$\eta_2^*(t_1^*) = -\eta_0^* \left[\frac{\partial f_0(x_1^*(t_1^*), x_2^*(t_1^*), u^*(t_1^*))}{\partial u}\right] = -k. \tag{11.47}$$

Again, the Maximum Principle (11.40) implies that at $t = t_1^*$

$$\begin{aligned} 0 &= \hat{H}(\hat{\boldsymbol{\eta}}^*(t_1^*), \hat{\boldsymbol{x}}^*(t_1^*), u^*(t_1^*)) \\ &= \eta_0^* f_0(x_1^*(t_1^*), x_2^*(t_1^*), u^*(t_1^*)) + \eta_1^*(t_1^*) + \eta_2^*(t_1^*)u^*(t_1^*) \\ &= \eta_0^* f_0(t_1^*, x^*(t_1^*), \dot{x}^*(t_1^*)) + \eta_1^*(t_1^*) + \eta_2^*(t_1^*)\dot{x}^*(t_1^*), \end{aligned}$$

and applying (11.46) - (11.47) we have

$$
\begin{aligned}
0 &= \eta_0^* f_0(t_1^*, x^*(t_1^*), \dot{x}^*(t_1^*)) + k\dot{\varphi}(t_1^*) \\
&\qquad - \eta_0^* \left[\frac{\partial f_0(x_1^*(t_1^*), x_2^*(t_1^*), u^*(t_1^*))}{\partial u}\right] \dot{x}^*(t_1^*) \\
&= \eta_0^* f_0(t_1^*, x^*(t_1^*), \dot{x}^*(t_1^*)) + \eta_0^* \left[\frac{\partial f_0(t_1^*, x^*(t_1^*), \dot{x}^*(t_1^*))}{\partial u}\right] \dot{\varphi}(t_1^*) \\
&\qquad - \eta_0^* \left[\frac{\partial f_0(t_1^*, x^*(t_1^*), \dot{x}^*(t_1^*))}{\partial u}\right] \dot{x}^*(t_1^*).
\end{aligned}
$$

Dividing by $\eta_0^* < 0$ yields

$$
\begin{aligned}
0 &= f_0(t_1^*, x^*(t_1^*), \dot{x}^*(t_1^*)) + \left[\frac{\partial f_0(t_1^*, x^*(t_1^*), \dot{x}^*(t_1^*))}{\partial u}\right] \dot{\varphi}(t_1^*) \\
&\qquad - \left[\frac{\partial f_0(t_1^*, x^*(t_1^*), \dot{x}^*(t_1^*))}{\partial u}\right] \dot{x}^*(t_1^*) \\
&= f_0(t_1^*, x^*(t_1^*), \dot{x}^*(t_1^*)) + \left[\frac{\partial f_0(t_1^*, x^*(t_1^*), \dot{x}^*(t_1^*))}{\partial u}\right] \left[\dot{\varphi}(t_1^*) - \dot{x}^*(t_1^*)\right],
\end{aligned}
$$

which implies the classical natural transversality condition

$$
f_0(t_1^*, x^*(t_1^*), \dot{x}^*(t_1^*)) + \left[\frac{\partial f_0(t_1^*, x^*(t_1^*), \dot{x}^*(t_1^*))}{\partial u}\right] \left[\dot{\varphi}(t_1^*) - \dot{x}^*(t_1^*)\right] = 0. \tag{11.48}
$$

11.3 Application to the Farmer's Allocation Problem

Here we apply the Maximum Principle to the Farmer's Allocation Problem first presented as **Example** 7.8 in Section 7.7.1 above. We refer to Luenburger's book ([131]) for details. The basic problem is that a farmer is assumed to produce a single crop (wheat, rice, etc.) and when he sells his crop he stores the crop or else sells the crop and reinvests the money into his business to increase his production rate. The farmer's goal is to maximize the total amount of crop stored up to two years. Luenberger formulates this as the following optimal control problem.

Given the initial time $t_0 = 0$, the final time $t_1 = 2$ and the initial state $1 > 0$, find a control $u^*(\cdot)$ to minimize

$$J(u(\cdot)) = \int_0^2 (u(s) - 1)x(s)ds,$$

subject to

$$\dot{x}(t) = u(t)x(t) \tag{11.49}$$

with initial condition

$$x(0) = 1. \tag{11.50}$$

The control constraint is given by

$$u(t) \in [0, 1].$$

Here, $u(t)$ is the fraction of the production rate that is reinvested at time t. It is important to note that since $u(t) \geq 0$ and $\dot{x}(t) = u(t)x(t)$, all solutions to the system (11.49) - (11.50) satisfy

$$x(t) \geq 1. \tag{11.51}$$

Formulation as an Optimal Control Problem: In order to set up the Farmer's Allocation Problem as a fixed time optimal control problem and apply the Maximum Principle, we need to:

1. Identify the initial time t_0 and final time t_1;
2. Identify the initial set $X_0 \subseteq \mathbb{R}^1$, the terminal set $X_1 \subseteq \mathbb{R}^1$;
3. Identify the control constraint set $\Omega \subseteq \mathbb{R}^1$;
4. Identify the functions $f : \mathbb{R}^1 \times \mathbb{R}^1 \longrightarrow \mathbb{R}^1$, $f_0 : \mathbb{R}^1 \times \mathbb{R}^1 \longrightarrow \mathbb{R}^1$;
5. Define the augmented Hamiltonian $\hat{H}(\hat{\boldsymbol{\eta}}, \hat{\boldsymbol{x}}, u)$;
6. Form the augmented adjoint system matrix $\hat{A}(t)$.

It is obvious that $t_0 = 0$, $t_1 = 2$, $\mathrm{X}_0 = \{x_0\} = \{1\} \subseteq \mathbb{R}^1$ and $\mathrm{X}_1 = \mathbb{R}^1$. Note that the corresponding tangent spaces are given by

$$\mathbb{T}_0 = \{0\} \text{ and } \mathbb{T}_1 = \mathbb{R}^1$$

and hence the transversality condition on the adjoint equation at $t = 2$ is

$$\eta^*(2) = 0. \tag{11.52}$$

The control constraint set is

$$\Omega = [0, 1] \subseteq \mathbb{R}^1,$$

the functions $f : \mathbb{R}^1 \times \mathbb{R}^1 \longrightarrow \mathbb{R}^1$, $f_0 : \mathbb{R}^1 \times \mathbb{R}^1 \longrightarrow \mathbb{R}^1$ are given by

$$f(x, u) = ux$$

and

$$f_0(x, u) = (u - 1)x,$$

respectively. The set of *admissible controllers* is the subset of $PWC(0, 2)$ defined by

$$\Theta = \left\{ \begin{array}{l} u(\cdot) \in PWC(0,2) : u(t) \in \Omega \ e.f. \text{ and} \\ \qquad\qquad\qquad\qquad u(\cdot) \text{ steers } \mathrm{X}_0 \text{ to } \mathrm{X}_1 \text{ at time } 2. \end{array} \right\}$$
$$= \Theta(0, 2, x_0 = 1, \Omega). \tag{11.53}$$

Given a control $u(\cdot) \in PWC(0, 2)$, the cost functional is defined by

$$J(u(\cdot)) = \int_0^2 (u(s) - 1)x(s)ds = \int_0^2 f_0(x(s), u(s))ds, \tag{11.54}$$

where $x(\cdot) = x(\cdot; u(\cdot))$ is the solution to the initial value problem (11.18) - (11.19). This optimal control problem is equivalent to the problem of minimizing $J(\cdot)$ on the set of admissible controllers Θ. In particular, the goal is to find an *optimal control* $u^*(\cdot) \in \Theta$ such that $u^*(\cdot)$ steers X_0 to X_1 at time 2 and

$$J(x^*(\cdot)) = \int_0^2 f_0\left(x^*(s), u^*\left(s\right)\right) ds \leq \int_0^2 f_0\left(x(s), u\left(s\right)\right) ds = J(x(\cdot))$$

for all $u(\cdot) \in \Theta$.

The augmented function $\hat{f} : \mathbb{R}^2 \times \mathbb{R}^1 \longrightarrow \mathbb{R}^2$ is defined by

$$\hat{f}(\hat{\boldsymbol{x}}, u) = \hat{f}(x_0, x, u) = \left[\begin{array}{c} f_0(x,u) \\ f(x,u) \end{array} \right] = \left[\begin{array}{c} (u-1)x \\ ux \end{array} \right],$$

where $\hat{\boldsymbol{x}} = [x_0 \;\; x]^T$. The augmented Hamiltonian $\hat{H}(\hat{\boldsymbol{\eta}}, \hat{\boldsymbol{x}}, u)$: $\mathbb{R}^2 \times \mathbb{R}^2 \times \mathbb{R}^1 \longrightarrow \mathbb{R}^1$ is given by

$$\hat{H}(\hat{\boldsymbol{\eta}}, \hat{\boldsymbol{x}}, u) = \eta_0 f_0(x,u) + \eta f(x,u) = \eta_0 (u-1)x + \eta u x,$$

where $\hat{\boldsymbol{\eta}} = [\eta_0 \;\; \eta]^T$. The augmented Jacobian is given by

$$\mathbb{J}_{\hat{\boldsymbol{x}}} \hat{f}(\hat{\boldsymbol{x}}, u) = \mathbb{J}_{\hat{\boldsymbol{x}}} \hat{f}(x,u) = \left[\begin{array}{cc} 0 & \frac{\partial f_0(x,u)}{\partial x} \\ 0 & \frac{\partial f(x,u)}{\partial x} \end{array} \right] = \left[\begin{array}{cc} 0 & (u-1) \\ 0 & u \end{array} \right]$$

so that if $(x^*(\cdot), u^*(\cdot))$ is an optimal pair, then the augmented matrix $\hat{A}(t)$ is given by

$$\hat{A}(t) = \mathbb{J}_{\hat{\boldsymbol{x}}} \hat{f}(\hat{\boldsymbol{x}}, u)_{|(\boldsymbol{x}^*(t), u^*(t))} = \mathbb{J}_{\hat{\boldsymbol{x}}} \hat{f}(x^*(t), u^*(t)) = \begin{bmatrix} 0 & (u^*(t)-1) \\ 0 & u^*(t) \end{bmatrix}.$$

Therefore,

$$-[\hat{A}(t)]^T = -\left[\begin{array}{cc} 0 & (u^*(t)-1) \\ 0 & u^*(t) \end{array} \right]^T = \left[\begin{array}{cc} 0 & 0 \\ -(u^*(t)-1) & -u^*(t) \end{array} \right]$$

and the augmented adjoint equation becomes

$$\frac{d}{dt} \begin{bmatrix} \eta_0(t) \\ \eta(t) \end{bmatrix} = -[\hat{A}(t)]^T \begin{bmatrix} \eta_0(t) \\ \eta(t) \end{bmatrix} = \begin{bmatrix} 0 & 0 \\ -(u^*(t)-1) & -u^*(t) \end{bmatrix} \left[\begin{array}{c} \eta_0(t) \\ \eta(t) \end{array} \right]. \tag{11.55}$$

Consequently,

$$\frac{d}{dt}\eta_0(t) = 0$$
$$\frac{d}{dt}\eta(t) = -\eta_0 \left[u^*(t) - 1\right] - u^*(t)\eta(t)$$

which implies (as always) that the zero adjoint state is a constant $\eta_0(t) \equiv \eta_0$.

If $(x^*(\cdot), u^*(\cdot))$ is an optimal pair for the Farmer's Allocation Problem, then the Maximum Principle implies that is a **nontrivial** solution

$$\hat{\boldsymbol{\eta}}^*(t) = \begin{bmatrix} \eta_0^*(t) \\ \eta^*(t) \end{bmatrix} \in \mathbb{R}^2$$

to the augmented adjoint equation (11.55) such that $\eta_0^*(t) \equiv \eta_0^* \leq 0$. Observe that the transversality condition (11.52) implies that $\eta^*(t)$ satisfies

$$\frac{d}{dt}\eta^*(t) = -u^*(t)\eta^*(t) + \eta_0^*\left[1 - u^*(t)\right], \quad \eta^*(2) = 0. \tag{11.56}$$

It again follows that $\eta_0^* \neq 0$ because if $\eta_0^* = 0$, then $\eta^*(t)$ would be a solution to the homogenous linear initial value problem with zero initial data

$$\frac{d}{dt}\eta^*(t) = -u^*(t)\eta^*(t), \quad \eta^*(2) = 0, \tag{11.57}$$

which would imply that $\eta^*(t) = 0$. Dividing (11.56) by $-\eta_0^* > 0$ yields

$$\frac{d}{dt}\left[\frac{\eta^*(t)}{-\eta_0^*}\right] = -u^*(t)\left[\frac{\eta^*(t)}{-\eta_0^*}\right] - \left[1 - u^*(t)\right], \quad \left[\frac{\eta^*(2)}{-\eta_0^*}\right] = 0. \tag{11.58}$$

Defining the normalized adjoint variable $\lambda^*(t)$ by

$$\lambda^*(t) = \left[\frac{\eta^*(t)}{-\eta_0^*}\right]$$

yields

$$\frac{d}{dt}\lambda^*(t) = -u^*(t)\lambda^*(t) - \left[1 - u^*(t)\right], \quad \lambda^*(2) = 0. \tag{11.59}$$

Also, since $-\eta_0^* > 0$ it follows that maximizing the augmented Hamiltonian

$$\hat{H}(\hat{\boldsymbol{\eta}}^*(t), \hat{\boldsymbol{x}}^*(t), u) = \left[\eta_0^*(u-1)x^*(t) + \eta^*(t)x^*(t)u\right]$$

is equivalent to maximizing

$$\frac{1}{-\eta_0^*}[\eta_0^*(u-1)x^*(t)+\eta^*(t)x^*(t)u]$$
$$=[-(u-1)x^*(t)+\left[\frac{\eta^*(t)}{-\eta_0^*}\right]x^*(t)u]$$
$$=[-(u-1)x^*(t)+\lambda^*(t)x^*(t)u]$$

where $\lambda^*(t)$ satisfies (11.59).

Note that

$$\max_{0\le u\le 1}[-(u-1)x^*(t)+\lambda^*(t)ux^*(t)]=\max_{0\le u\le 1}[(1-u)x^*(t)+\lambda^*(t)ux^*(t)]$$

and since $x^*(t)\ge 1$ it follows that

$$\max_{0\le u\le 1}[(1-u)x^*(t)+\lambda^*(t)ux^*(t)]=\left\{\max_{0\le u\le 1}[(1-u)+\lambda^*(t)u]\right\}x^*(t)$$
$$=\left\{\max_{0\le u\le 1}[(1+(\lambda^*(t)-1)u]\right\}x^*(t).$$

Thus, the optimal control must maximize the term

$$(\lambda^*(t)-1)u$$

on the interval $[0,1]$ which implies that

$$u^*(t)=\begin{cases}1, & \text{if } \lambda^*(t)>1,\\ 0, & \text{if } \lambda^*(t)<1.\end{cases} \tag{11.60}$$

We need only compute $\lambda^*(t)$ using the adjoint equation

$$\frac{d}{dt}\lambda^*(t)=-u^*(t)\lambda^*(t)-[1-u^*(t)],\quad \lambda^*(2)=0. \tag{11.61}$$

However, at $t=2$ we know that $\lambda^*(2)=0$ and since $\lambda^*(t)$ is continuous, there is an interval $[T,2]$ with $T<2$ such that

$$\lambda^*(t)<1 \text{ for all } t\in(T,2].$$

On this interval the optimal control must be $u^*(t) = 0$ and the corresponding adjoint equation (11.61) has the form

$$\frac{d}{dt}\lambda^*(t) = -1, \quad \lambda^*(2) = 0.$$

Thus,

$$\lambda^*(t) = -t + 2$$

until the time where

$$1 = \lambda^*(T) = -T + 2$$

and then the optimal control switches to $u^*(t) = 1$. Clearly, $T = 1$ and hence the optimal control on the interval $[0, T] = [0, 1]$ is given by $u^*(t) = 1$. Again, returning to the adjoint equation (11.61) on the interval $[0, 1]$ yields

$$\frac{d}{dt}\lambda^*(t) = -u^*(t)\lambda^*(t) - [1 - u^*(t)] = -\lambda^*(t), \quad \lambda^*(1) = 1 \tag{11.62}$$

which has the solution

$$\lambda^*(t) = e^{1-t}.$$

Therefore, we know that

$$\lambda^*(t) = \begin{cases} e^{1-t}, & \text{if } 0 \le t \le 1, \\ -t + 2, & \text{if } 1 \le t \le 2 \end{cases}$$

and

$$u^*(t) = \begin{cases} 1, & \text{if } 0 \le t \le 1, \\ 0, & \text{if } 1 \le t \le 2 \end{cases} \tag{11.63}$$

is the corresponding optimal control.

Finally, the corresponding optimal trajectory is given by

$$x^*(t) = \begin{cases} e^t, & \text{if } 0 \le t \le 1, \\ e, & \text{if } 1 \le t \le 2 \end{cases} \tag{11.64}$$

and **if the optimal control problem has a solution**, then the optimal controller is given by (11.63) with corresponding optimal trajectory (11.64).

11.4 Application to a Forced Oscillator Control Problem

Suppose we have a forced oscillator with its equilibrium position at the origin and driven by the force $u(t)$. The equation of motion is given by the second order equation

$$\ddot{x}(t) + x(t) = u(t)$$

and we consider the cost function given by

$$J(u(\cdot)) = \int_0^{t_1} \frac{1}{2} u^2(s)\, ds, \tag{11.65}$$

where $t_1 < \frac{1}{2}\pi$ is fixed. Note that if the terminal time t_1 was free or if $t_1 \geq \frac{1}{2}\pi$, then one can reach the target without applying any control in a quarter-period of the oscillation and the cost is then zero. Thus, the problem is only interesting if $t_1 < \frac{1}{2}\pi$ is fixed.

If $x_1(t) = x(t)$ denotes the position, then $x_2(t) = \dot{x}(t)$ is the velocity and the state equations have the form

$$\frac{d}{dt}\begin{bmatrix} x_1(t) \\ x_2(t) \end{bmatrix} = \begin{bmatrix} 0 & 1 \\ -1 & 0 \end{bmatrix}\begin{bmatrix} x_1(t) \\ x_2(t) \end{bmatrix} + \begin{bmatrix} 0 \\ 1 \end{bmatrix} u(t). \tag{11.66}$$

Suppose that initially at time $t_0 = 0$, the system is at rest at a distance 2 (miles) from the equilibrium position $\mathbf{0} = [0 \;\; 0]^T$ and that we wish to move to the zero position at a prescribed fixed time t_1, without requiring that we arrive there with zero velocity.

Formulation as an Optimal Control Problem: In order to set up the forced oscillator control problem as a fixed time optimal control problem and apply the Maximum Principle, we need to:

1. Identify the initial time t_0 and the final time t_1 ;
2. Identify the initial set $\mathrm{X}_0 \subseteq \mathbb{R}^2$, the terminal set $\mathrm{X}_1 \subseteq \mathbb{R}^2$;
3. Identify the control constraint set $\Omega \subseteq \mathbb{R}^1$;
4. Identify the functions $f : \mathbb{R}^2 \times \mathbb{R}^1 \longrightarrow \mathbb{R}^1$, $f_0 : \mathbb{R}^2 \times \mathbb{R}^1 \longrightarrow \mathbb{R}^1$;

5. Define the augmented Hamiltonian $\hat{H}(\hat{\boldsymbol{\eta}}, \hat{\boldsymbol{x}}, u)$;
6. Form the augmented adjoint system matrix $\hat{A}(t)$.

The initial and final states are given by

$$\begin{bmatrix} x_1(0) \\ x_2(0) \end{bmatrix} = \begin{bmatrix} 2 \\ 0 \end{bmatrix}$$

and

$$\begin{bmatrix} x_1(t_1) \\ x_2(t_1) \end{bmatrix} \in \left\{ \begin{bmatrix} 0 \\ x_2 \end{bmatrix} : x_2 \in \mathbb{R}^1 \right\} = \{0\} \times \mathbb{R}^1,$$

respectively. Therefore, the initial and final sets are given by

$$\mathrm{X}_0 = \left\{ \begin{bmatrix} 2 \\ 0 \end{bmatrix} \right\} \text{ and } \mathrm{X}_1 = \{0\} \times \mathbb{R}^1, \tag{11.67}$$

respectively. Since since there is no hard constraint on the control, the control constraint set is $\Omega = \mathbb{R}^1$. The functions $f_0 : \mathbb{R}^2 \times \mathbb{R}^1 \longrightarrow \mathbb{R}^1$ and $f : \mathbb{R}^2 \times \mathbb{R}^1 \longrightarrow \mathbb{R}^1$ are defined by

$$f(x_1, x_2, u) = \frac{1}{2}u^2$$

and

$$f(x_1, x_2, u) = \begin{bmatrix} x_2 \\ (-x_1 + u) \end{bmatrix} = \begin{bmatrix} f_1(x_1, x_2, u) \\ f_2(x_1, x_2, u) \end{bmatrix},$$

so that the augmented function $\hat{f} : \mathbb{R}^3 \times \mathbb{R}^1 \to \mathbb{R}^3$ is given by

$$\hat{f}(x_0, x_1, x_2, u) = \begin{bmatrix} \frac{1}{2}u^2 \\ x_2 \\ (-x_1 + u) \end{bmatrix}.$$

The augmented Hamiltonian is given by

$$\hat{H}(\hat{\boldsymbol{\eta}}, \hat{\boldsymbol{x}}, u) = \eta_0 \frac{1}{2}u^2 + \eta_1 x_2 + \eta_2(-x_1 + u).$$

If $(\boldsymbol{x}^*(\cdot), u^*(\cdot))$ is an optimal pair, then

$$\hat{A}(t) = \begin{bmatrix} 0 & 0 & 0 \\ 0 & 0 & 1 \\ 0 & -1 & 0 \end{bmatrix}_{|\boldsymbol{x}=\boldsymbol{x}^*(t), u=u^*(t)} = \begin{bmatrix} 0 & 0 & 0 \\ 0 & 0 & 1 \\ 0 & -1 & 0 \end{bmatrix}$$

and the adjoint equations become

$$\frac{d}{dt}\hat{\boldsymbol{\eta}}(t) = \begin{bmatrix} 0 & 0 & 0 \\ 0 & 0 & 1 \\ 0 & -1 & 0 \end{bmatrix} \hat{\boldsymbol{\eta}}(t). \tag{11.68}$$

The Maximum Principle implies that there exists a nonzero function

$$\hat{\boldsymbol{\eta}}^*(\cdot) = [\eta_0^*(\cdot)\ \eta_1^*(\cdot)\ \eta_2^*(\cdot)]^T$$

which is a solution to the adjoint equation (11.68) such that $\eta_0^*(t) \equiv \eta_0^* \leq 0$ and

$$\hat{H}(\hat{\boldsymbol{\eta}}^*(t), \hat{\boldsymbol{x}}^*(t), u^*(t)) = \max_{u \in \mathbb{R}^1} \hat{H}(\hat{\boldsymbol{\eta}}^*(t), \hat{\boldsymbol{x}}^*(t), u) \equiv c.$$

Since $\Omega = \mathbb{R}^1$ it follows that

$$\frac{\partial}{\partial u}\hat{H}(\hat{\boldsymbol{\eta}}^*(t), \hat{\boldsymbol{x}}^*(t), u)_{|u=u^*(t)} \equiv 0$$

and hence we have

$$\eta_0^* u^*(t) + \eta_2^*(t) \equiv 0. \tag{11.69}$$

Moreover, the transversality condition states that $\boldsymbol{\eta}^*(t_1)$ can be taken to be normal to the tangent plane for the target set X_1 and since $X_1 = \{0\} \times \mathbb{R}^1$ is a subspace it follows that $\mathbb{T}_1 = \{0\} \times \mathbb{R}^1$. Consequently, if

$$\boldsymbol{\eta}^*(t_1) = \begin{bmatrix} \eta_1^*(t_1) \\ \eta_2^*(t_1) \end{bmatrix} \perp \{0\} \times \mathbb{R}^1,$$

then $\eta_2^*(t_1) = 0$, so that there is a constant A so that

$$\eta_1^*(t) = A\cos(t_1 - t)$$

and

$$\eta_2^*(t) = A\sin(t_1 - t).$$

Observe that if $\eta_0^* = 0$, then (11.69) implies that $\eta_2^*(t) \equiv 0$ so that $A = 0$. Hence, it would follow that $\eta_1^*(t) = A\cos(t_1 - t) \equiv 0$

which can not happen since $\hat{\boldsymbol{\eta}}^*(\cdot)$ is nonzero. Therefore, $\eta_0^* < 0$ and we can solve (11.69) for $u^*(t)$ to yield

$$u^*(t) = -\frac{\eta_2^*(t)}{\eta_0^*} = \tilde{A}\sin(t_1 - t) \tag{11.70}$$

where $\tilde{A} = -A/\eta_0^*$. When $u^*(t) = \tilde{A}\sin(t_1 - t)$, the solution to the state equation (11.66) is easily found to be given by

$$x_1^*(t) = \frac{1}{2}\tilde{A}t\cos(t_1 - t) + 2\cos t + \tilde{B}\sin t, \tag{11.71}$$

$$x_2^*(t) = \frac{1}{2}\tilde{A}\cos(t_1 - t) + \frac{1}{2}\tilde{A}t\sin(t_1 - t) - 2\sin t + \tilde{B}\cos t, \tag{11.72}$$

for some constant $\tilde{B}$. However, $x_2^*(0) = 0$ and $x_1^*(t_1) = 0$ so that

$$0 = x_2^*(0) = \frac{1}{2}\tilde{A}\cos(t_1) + \tilde{B}$$

and

$$0 = x_1^*(t_1) = \frac{1}{2}\tilde{A}t_1 + 2\cos t_1 + \tilde{B}\sin t_1.$$

Solving these two equations for $\tilde{A}$ and $\tilde{B}$ we find

$$\tilde{A} = \frac{-4\cos t_1}{t_1 - \sin t_1 \cos t_1}$$

and

$$\tilde{B} = -\frac{1}{2}\tilde{A}\cos t_1.$$

Finally, the optimal cost can be found by substituting

$$u^*(t) = \tilde{A}\sin(t_1 - t)$$

into the cost function (11.65) and integrating we find

$$J(u^*(\cdot)) = \frac{4\cos^2 t_1}{t_1 - \sin t_1 \cos t_1}. \tag{11.73}$$

Note that if $t_1 = \frac{1}{2}\pi$, then $\tilde{A} = 0$ and the optimal cost is zero. For each $t_1 < \frac{1}{2}\pi$, the optimal cost is positive and as $t_1 \to 0$, $J(u^*(\cdot)) \to +\infty$ so that the cost grows to infinity as the time interval shrinks.

11.5 Application to the Linear Quadratic Control Problem

We apply the Maximum Principle to a Linear Quadratic (LQ) optimal control problem. Let A be a $n \times n$ constant real matrix and B be a $n \times m$ real constant matrix. Consider the linear control system

$$\dot{\boldsymbol{x}}(t) = A\boldsymbol{x}(t) + B\boldsymbol{u}(t) \tag{11.74}$$

with initial data

$$\boldsymbol{x}(0) = \boldsymbol{x}_0 \in \mathbb{R}^n. \tag{11.75}$$

We assume that $t_0 = 0$, $\boldsymbol{x}_0 \in \mathbb{R}^n$ and $0 < t_1$ are given. Also, let Q be a $n \times n$ constant symmetric real matrix and R be a $m \times m$ constant symmetric real matrix such that

$$Q^T = Q \geq 0 \text{ and } R^T = R > 0.$$

Here, the inequality $Q \geq 0$ is equivalent to the condition that Q is non-negative

$$\langle Q\boldsymbol{x}, \boldsymbol{x}\rangle \geq 0 \text{ for all } \boldsymbol{x} \in \mathbb{R}^n,$$

and $R > 0$ is equivalent to the condition that R is positive definite

$$\langle R\boldsymbol{u}, \boldsymbol{u}\rangle > 0 \text{ for all } \boldsymbol{u} \in \mathbb{R}^m \text{ with } \boldsymbol{u} \neq \boldsymbol{0}.$$

We assume that the controls belong to the space $PWC(0, t_1; \mathbb{R}^m)$. The quadratic cost functional is defined by

$$J(\boldsymbol{u}(\cdot)) = \frac{1}{2}\int_0^{t_1} \{\langle Q\boldsymbol{x}(s), \boldsymbol{x}(s)\rangle + \langle R\boldsymbol{u}(s), \boldsymbol{u}(s)\rangle\}\, ds, \tag{11.76}$$

where $\boldsymbol{x}(t) = \boldsymbol{x}(t; \boldsymbol{u}(\cdot))$ is the solution to the system (11.74) - (11.75).The **Linear Quadratic (LQ) Optimal Control problem** is to find $\boldsymbol{u}^*(\cdot) \in PWC(0, t_1; \mathbb{R}^m)$ so that

$$\begin{aligned} J(\boldsymbol{u}^*(\cdot)) &= \frac{1}{2}\int_{t_0}^{t_1} \{\langle Q\boldsymbol{x}(s; \boldsymbol{u}^*(\cdot)), \boldsymbol{x}(s; \boldsymbol{u}^*(\cdot))\rangle + \langle R\boldsymbol{u}^*(s), \boldsymbol{u}^*(s)\rangle\}\, ds \\ &\leq J(\boldsymbol{u}(\cdot)) = \frac{1}{2}\int_{t_0}^{t_1} \{\langle Q\boldsymbol{x}(s; \boldsymbol{u}(\cdot)), \boldsymbol{x}(s; \boldsymbol{u}(\cdot))\rangle \\ &\qquad + \langle R\boldsymbol{u}(s), \boldsymbol{u}(s)\rangle\} ds, \end{aligned}$$

for all $\boldsymbol{u}(\cdot) \in PWC(0, t_1; \mathbb{R}^m)$.

Formulation as an Optimal Control Problem: In order to set up the Quadratic Optimal Control Problem as a fixed time optimal control problem and apply the Maximum Principle, we need to:

1. Identify the initial time t_0 and final time t_1;
2. Identify the initial set $X_0 \subseteq \mathbb{R}^n$, the terminal set $X_1 \subseteq \mathbb{R}^n$;
3. Identify the control constraint set $\Omega \subseteq \mathbb{R}^m$;
4. Identify the functions $f : \mathbb{R}^n \times \mathbb{R}^m \longrightarrow \mathbb{R}^n$, $f_0 : \mathbb{R}^n \times \mathbb{R}^m \longrightarrow \mathbb{R}^1$;
5. Define the augmented Hamiltonian $\hat{H}(\hat{\boldsymbol{\eta}}, \hat{\boldsymbol{x}}, \boldsymbol{u})$;
6. Form the augmented adjoint system matrix $\hat{A}(t)$.

The initial time is $t_0 = 0$ and the final time t_1 is given and fixed. The initial set is the single initial vector $\mathrm{X}_0 = \{\boldsymbol{x}_0\}$ and since there is no terminal constraint, the terminal set is the whole state space $\mathrm{X}_1 = \mathbb{R}^n$. Also, there is no constraint on the control $\boldsymbol{u}(\cdot) \in PWC(0, t_1; \mathbb{R}^m)$, hence the control constraint set is all of $\mathbb{R}^m$ and $\Omega = \mathbb{R}^m$. The functions $f : \mathbb{R}^n \times \mathbb{R}^m \longrightarrow \mathbb{R}^n$ and $f_0 : \mathbb{R}^n \times \mathbb{R}^m \longrightarrow \mathbb{R}^1$ are given by

$$f(\boldsymbol{x}, \boldsymbol{u}) = A\boldsymbol{x} + B\boldsymbol{u},$$

and

$$f_0(\boldsymbol{x}, \boldsymbol{u}) = \frac{1}{2}\langle Q\boldsymbol{x}, \boldsymbol{x}\rangle + \frac{1}{2}\langle R\boldsymbol{u}, \boldsymbol{u}\rangle \geq 0,$$

respectively.

The augmented Hamiltonian $\hat{H} : \mathbb{R}^{n+1} \times \mathbb{R}^{n+1} \times \mathbb{R}^m \longrightarrow \mathbb{R}^1$ is defined by

$$\begin{aligned}
\hat{H}(\hat{\boldsymbol{\eta}}, \hat{\boldsymbol{x}}, \boldsymbol{u}) &= \frac{\eta_0}{2}\{\langle Q\boldsymbol{x}, \boldsymbol{x}\rangle + \langle R\boldsymbol{u}, \boldsymbol{u}\rangle\} + \langle \boldsymbol{\eta},\ A\boldsymbol{x} + B\boldsymbol{u}\rangle \qquad (11.77)\\
&= \frac{\eta_0}{2}\boldsymbol{x}^T Q\, \boldsymbol{x} + \frac{\eta_0}{2}\boldsymbol{u}^T R\boldsymbol{u} + [A\boldsymbol{x}]^T\boldsymbol{\eta} + [B\boldsymbol{u}]^T\boldsymbol{\eta}\\
&= \frac{\eta_0}{2}\boldsymbol{x}^T Q\, \boldsymbol{x} + \frac{\eta_0}{2}\boldsymbol{u}^T R\boldsymbol{u} + \boldsymbol{x}^T A^T\boldsymbol{\eta} + \boldsymbol{u}^T B^T\boldsymbol{\eta},
\end{aligned}$$

where $\hat{\boldsymbol{x}} = \begin{bmatrix} x_0 & \boldsymbol{x} \end{bmatrix}^T \in \mathbb{R}^{n+1}$, $\hat{\boldsymbol{\eta}} = \begin{bmatrix} \eta_0 & \boldsymbol{\eta} \end{bmatrix}^T \in \mathbb{R}^{n+1}$ and $\boldsymbol{u} \in \mathbb{R}^m$. The augmented function $f : \mathbb{R}^{n+1} \times \mathbb{R}^m \longrightarrow \mathbb{R}^{n+1}$ is defined by

$$\hat{f}(\hat{\boldsymbol{x}}, \boldsymbol{u}) = \begin{bmatrix} \frac{1}{2}\langle Q\boldsymbol{x}, \boldsymbol{x}\rangle + \frac{1}{2}\langle R\boldsymbol{u}, \boldsymbol{u}\rangle \\ A\boldsymbol{x} + B\boldsymbol{u}, \end{bmatrix} = \begin{bmatrix} \frac{1}{2}\boldsymbol{x}^T Q\boldsymbol{x} + \frac{1}{2}\boldsymbol{u}^T R\boldsymbol{u} \\ A\boldsymbol{x} + B\boldsymbol{u}, \end{bmatrix}, \tag{11.78}$$

so that the Jacobian is given by

$$\mathbb{J}_{\hat{\boldsymbol{x}}}\hat{f}(\hat{\boldsymbol{x}}, \boldsymbol{u}) = \mathbb{J}_{\hat{\boldsymbol{x}}}\hat{f}(\boldsymbol{x}, u) = \begin{bmatrix} 0 & [Q\boldsymbol{x}]^T \\ \boldsymbol{0} & A \end{bmatrix}. \tag{11.79}$$

Now assume that $(\boldsymbol{x}^*(\cdot), \boldsymbol{u}^*(\cdot))$ is an optimal pair so that the matrix $\hat{A}(t)$ is given by

$$\begin{aligned} \hat{A}(t) &= \mathbb{J}_{\hat{\boldsymbol{x}}}\hat{f}(\hat{\boldsymbol{x}}, \boldsymbol{u})_{|(\boldsymbol{x}^*(t), \boldsymbol{u}^*(t))} \\ &= \mathbb{J}_{\hat{\boldsymbol{x}}}\hat{f}(\boldsymbol{x}^*(t), \boldsymbol{u}^*(t)) = \begin{bmatrix} 0 & [Q\boldsymbol{x}^*(t)]^T \\ \boldsymbol{0} & A \end{bmatrix} \end{aligned} \tag{11.80}$$

and

$$-[\hat{A}(t)]^T = -\begin{bmatrix} 0 & \boldsymbol{0} \\ [Q\boldsymbol{x}^*(t)] & A^T \end{bmatrix} = \begin{bmatrix} 0 & \boldsymbol{0} \\ -[Q\boldsymbol{x}^*(t)] & -A^T \end{bmatrix}.$$

The augmented adjoint equation has the form

$$\begin{aligned} \frac{d}{dt}\begin{bmatrix} \eta_0(t) \\ \boldsymbol{\eta}(t) \end{bmatrix} &= -[\hat{A}(t)]^T \begin{bmatrix} \eta_0(t) \\ \boldsymbol{\eta}(t) \end{bmatrix} \\ &= \begin{bmatrix} 0 & \boldsymbol{0} \\ -[Q\boldsymbol{x}^*(t)] & -A^T \end{bmatrix} \begin{bmatrix} \eta_0(t) \\ \boldsymbol{\eta}(t) \end{bmatrix} \end{aligned} \tag{11.81}$$

which is equivalent to the system

$$\begin{aligned} \frac{d}{dt}\eta_0(t) &= 0 \\ \frac{d}{dt}\boldsymbol{\eta}(t) &= -\eta_0(t)Q\boldsymbol{x}^*(t) - A^T\eta(t), \end{aligned}$$

where again $\boldsymbol{x}^*(\cdot)$ is the optimal trajectory. Again, the first equation above implies that the zero adjoint state is a constant $\eta_0(t) \equiv$

η_0. The second equation is coupled to the state equation by the term $-\eta_0(t)Q\boldsymbol{x}^*(t)$.

If $(\boldsymbol{x}^*(\cdot), \boldsymbol{u}^*(\cdot))$ is an optimal pair for the LQ optimal control problem, then the Maximum Principle implies that is a nontrivial solution

$$\hat{\boldsymbol{\eta}}^*(t) = \begin{bmatrix} \eta_0^*(t) \\ \boldsymbol{\eta}^*(t) \end{bmatrix} \in \mathbb{R}^{n+1}$$

to the augmented adjoint equation (11.81) such that $\eta_0^*(t) \equiv \eta_0^* \leq 0$,

$$\frac{d}{dt}\boldsymbol{\eta}^*(t) = -\eta_0^* Q\boldsymbol{x}^*(t) - A^T\boldsymbol{\eta}^*(t) \tag{11.82}$$

and

$$\hat{H}(\hat{\boldsymbol{\eta}}^*(t), \hat{\boldsymbol{x}}^*(t), \boldsymbol{u}^*(t)) = \max_{\boldsymbol{u}\in\mathbb{R}^m} \hat{H}(\hat{\boldsymbol{\eta}}^*(t), \hat{\boldsymbol{x}}^*(t), \boldsymbol{u}) \equiv c.$$

Since $\Omega = \mathbb{R}^m$ is open, then

$$D_{\boldsymbol{u}}\hat{H}(\hat{\boldsymbol{\eta}}^*(t), \hat{\boldsymbol{x}}^*(t), \boldsymbol{u})_{|\boldsymbol{u}=\boldsymbol{u}^*(t)} = D_{\boldsymbol{u}}\hat{H}(\hat{\boldsymbol{\eta}}^*(t), \hat{\boldsymbol{x}}^*(t), \boldsymbol{u}^*(t)) = \mathbf{0},$$

where

$$\begin{aligned} D_{\boldsymbol{u}}\hat{H}(\hat{\boldsymbol{\eta}}^*(t), \hat{\boldsymbol{x}}^*(t), \boldsymbol{u}) &= \frac{\partial}{\partial \boldsymbol{u}}\left\{\frac{\eta_0^*}{2}[\boldsymbol{x}^*(t)]^T Q\, \boldsymbol{x} + \frac{\eta_0^*}{2}\boldsymbol{u}^T R\boldsymbol{u} \right. \\ &\qquad \left. + [\boldsymbol{x}^*(t)]^T A^T\boldsymbol{\eta}^*(t) + \boldsymbol{u}^T B^T\boldsymbol{\eta}^*(t)\right\} \\ &= \frac{\partial}{\partial \boldsymbol{u}}\left\{\frac{\eta_0^*}{2}\boldsymbol{u}^T R\boldsymbol{u} + \boldsymbol{u}^T B^T\boldsymbol{\eta}^*(t)\right\} \\ &= \eta_0^* R\boldsymbol{u} + B^T\boldsymbol{\eta}^*(t), \end{aligned}$$

so that when $\boldsymbol{u} = \boldsymbol{u}^*(t)$

$$\begin{aligned} D_{\boldsymbol{u}}\hat{H}(\hat{\boldsymbol{\eta}}^*(t), \hat{\boldsymbol{x}}^*(t), \boldsymbol{u})_{|\boldsymbol{u}=\boldsymbol{u}^*(t)} &= \eta_0^* R\boldsymbol{u} + B^T\boldsymbol{\eta}^*(t)\,|_{u=u^*(t)} \\ &= \eta_0^* R\boldsymbol{u}^*(t) + B^T\boldsymbol{\eta}^*(t) = 0. \end{aligned} \tag{11.83}$$

Applying the transversality condition at $\boldsymbol{x}^*(0) = \boldsymbol{x}_0 \in \mathrm{X}_0 = \{\boldsymbol{x}_0\}$ yields that $\eta^*(0)$ can be any vector since $\mathbb{T}_0 = \{\boldsymbol{x}_0\}$. However, at t_1 we have that $\boldsymbol{x}^*(t_1) \in \mathrm{X}_1 = \mathbb{R}^n$ and since $\mathbb{T}_1 = \mathrm{X}_1 = \mathbb{R}^n$, the transversality condition

$$\boldsymbol{\eta}^*(t_1^*) \perp \mathbb{T}_1 = \mathbb{R}^n$$

implies that

$$\boldsymbol{\eta}^*(t_1) = \mathbf{0}. \tag{11.84}$$

This boundary condition in turn implies that $\eta_0^* < 0$. To see this assume that $\eta_0^* = 0$ and observe that the adjoint equation (11.82) reduces to the linear system

$$\begin{aligned}\frac{d}{dt}\boldsymbol{\eta}^*(t) &= -\eta_0^* Q\boldsymbol{x}^*(t) - A^T\boldsymbol{\eta}^*(t)\\ &= -A^T\boldsymbol{\eta}^*(t).\end{aligned}$$

Therefore, $\boldsymbol{\eta}^*(\cdot)$ would be a solution of the homogenous linear initial value problem

$$\frac{d}{dt}\boldsymbol{\eta}^*(t) = -A^T\boldsymbol{\eta}^*(t), \quad \boldsymbol{\eta}^*(t_1) = \mathbf{0}$$

and hence it follows that $\boldsymbol{\eta}^*(t) \equiv \mathbf{0}$. Consequently, we have shown that if $\eta_0^* = 0$, then $\boldsymbol{\eta}^*(t) \equiv \mathbf{0}$ and hence

$$\hat{\boldsymbol{\eta}}^*(t) = \begin{bmatrix} \eta_0^*(t) \\ \boldsymbol{\eta}^*(t) \end{bmatrix} \equiv \mathbf{0}$$

which contradicts the statement that $\hat{\boldsymbol{\eta}}^*(t)$ is a nontrivial solution of the augmented adjoint equation (11.81).

Since $\eta_0^* < 0$ we can solve (11.83) for the optimal control. We have

$$\eta_0^* R\boldsymbol{u}^*(t) + B^T\boldsymbol{\eta}^*(t) = 0$$

which yields

$$\eta_0^* R\boldsymbol{u}^*(t) = -B^T\boldsymbol{\eta}^*(t)$$

and

$$R\boldsymbol{u}^*(t) = \frac{-1}{\eta_0^*}B^T\boldsymbol{\eta}^*(t).$$

By assumption, the matrix $R = R^T > 0$ is nonsingular and hence we have the following expression for the optimal control

$$\boldsymbol{u}^*(t) = R^{-1}B^T \left[\frac{\boldsymbol{\eta}^*(t)}{-\eta_0^*}\right]. \tag{11.85}$$

Summarizing, it follows that the optimal trajectory can be obtained by solving the two point boundary value problem defined by the coupled state and adjoint equations

$$\begin{array}{ll} \frac{d}{dt}\boldsymbol{x}^*(t) = A\boldsymbol{x}^*(t) - \frac{1}{\eta_0^*}BR^{-1}B^T\boldsymbol{\eta}^*(t), & \boldsymbol{x}^*(0) = \boldsymbol{x}_0, \\ \frac{d}{dt}\boldsymbol{\eta}^*(t) = -\eta_0^* Q\boldsymbol{x}^*(t) - A^T\boldsymbol{\eta}^*(t), & \boldsymbol{\eta}^*(t_1) = \boldsymbol{0}, \end{array} \tag{11.86}$$

and setting

$$\boldsymbol{u}^*(t) = R^{-1}B^T \left[\frac{\eta^*(t)}{-\eta_0^*}\right].$$

To eliminate the η_0^* term, we divide the adjoint equation above by $-\eta_0^*$ which yields

$$\frac{d}{dt}\left[\frac{\boldsymbol{\eta}^*(t)}{-\eta_0^*}\right] = Q\boldsymbol{x}^*(t) - A^T\left[\frac{\boldsymbol{\eta}^*(t)}{-\eta_0^*}\right], \quad \left[\frac{\boldsymbol{\eta}^*(t_1)}{-\eta_0^*}\right] = \boldsymbol{0}.$$

Defining the normalized adjoint state $\boldsymbol{\lambda}^*(t)$ by

$$\boldsymbol{\lambda}^*(t) \triangleq \frac{\boldsymbol{\eta}^*(t)}{-\eta_0^*}$$

produces the optimality conditions

$$\begin{array}{ll} \frac{d}{dt}\boldsymbol{x}^*(t) = A\boldsymbol{x}^*(t) + BR^{-1}B^T\boldsymbol{\lambda}^*(t), & \boldsymbol{x}^*(0) = \boldsymbol{x}_0, \\ \frac{d}{dt}\boldsymbol{\lambda}^*(t) = Q\boldsymbol{x}^*(t) - A^T\boldsymbol{\lambda}^*(t), & \boldsymbol{\lambda}^*(t_1) = \boldsymbol{0}, \end{array} \tag{11.87}$$

where the optimal control is defined by

$$\boldsymbol{u}^*(t) = R^{-1}B^T\boldsymbol{\lambda}^*(t). \tag{11.88}$$

We can write the optimality system as

$$\frac{d}{dt}\begin{bmatrix} \boldsymbol{x}^*(t) \\ \boldsymbol{\lambda}^*(t) \end{bmatrix} = \begin{bmatrix} A & BR^{-1}B^T \\ Q & -A^T \end{bmatrix}\begin{bmatrix} \boldsymbol{x}^*(t) \\ \boldsymbol{\lambda}^*(t) \end{bmatrix}, \tag{11.89}$$

with boundary conditions

$$\boldsymbol{x}^*(0) = \begin{bmatrix} I_{n\times n} & 0_{n\times n} \end{bmatrix} \begin{bmatrix} \boldsymbol{x}^*(0) \\ \boldsymbol{\lambda}^*(0) \end{bmatrix} = \boldsymbol{x}_0, \tag{11.90}$$

and

$$\boldsymbol{\lambda}^*(t_1) = \begin{bmatrix} 0_{n\times n} & I_{n\times n} \end{bmatrix} \begin{bmatrix} \boldsymbol{x}^*(t_1) \\ \boldsymbol{\lambda}^*(t_1) \end{bmatrix} = \mathbf{0}. \tag{11.91}$$

Thus, if one solves the two point boundary value problem (11.89) - (11.91), the optimal control is defined by (11.88).

11.5.1 Examples of LQ Optimal Control Problems

Here we apply the Maximum Principle to specific LQ optimal control problem. We consider a problem with fixed final $\boldsymbol{x}_1$ and a problem where $\boldsymbol{x}_1$ is not specified.

LQ Optimal Control Problem: Example 1

Consider the linear control system

$$\dot{x}(t) = x(t) + u(t) \tag{11.92}$$

with initial data

$$x(0) = 1/2 \in \mathbb{R}^1 \tag{11.93}$$

and terminal condition

$$x(1) = 0 \in \mathbb{R}^1. \tag{11.94}$$

We assume that $t_0 = 0$, $0 < t_1 = 1$, $x_0 = 1/2 \in \mathbb{R}^1$ and $x_1 = 0 \in \mathbb{R}^1$ are given. The quadratic cost functional is defined by

$$J(u(\cdot)) = \int_0^1 \frac{1}{2}[u(s)]^2 ds. \tag{11.95}$$

The Linear Quadratic (LQ) optimal control problem is to find $u^*(\cdot) \in PWC(0,1)$ so that

$$J(u^*(\cdot)) = \int_0^1 \frac{1}{2}[u^*(s)]^2 ds \le J(u(\cdot)) = \int_0^1 \frac{1}{2}[u(s)]^2 ds$$

for all $u(\cdot) \in PWC(0,1)$ that steer $x_0 = 1/2$ to $x_1 = 0$.

Formulation as an Optimal Control Problem: In order to set up the fixed time optimal control problem and apply the Maximum Principle, we need to:

1. Identify the initial time t_0 and final time t_1;
2. Identify the initial set $X_0 \subseteq \mathbb{R}^1$, the terminal set $X_1 \subseteq \mathbb{R}^1$;
3. Identify the control constraint set $\Omega \subseteq \mathbb{R}^1$;
4. Identify the functions $f : \mathbb{R}^1 \times \mathbb{R}^1 \longrightarrow \mathbb{R}^1$, $f_0 : \mathbb{R}^1 \times \mathbb{R}^1 \longrightarrow \mathbb{R}^1$;
5. Define the augmented Hamiltonian $\hat{H}(\hat{\boldsymbol{\eta}}, \hat{\boldsymbol{x}}, u)$;
6. Form the augmented adjoint system matrix $\hat{A}(t)$.

The initial time is $t_0 = 0$ and the final time $t_1 = 1$ is given and fixed. The initial set is $X_0 = \{1/2\}$ and the terminal constraint set is $X_1 = \{0\}$. Also, there is no constraint on the control $u(\cdot) \in PWC(0,1)$, hence the control constraint set is all of $\mathbb{R}^1$ and $\Omega = \mathbb{R}^1$. The functions $f : \mathbb{R}^1 \times \mathbb{R}^1 \longrightarrow \mathbb{R}^1$ and $f_0 : \mathbb{R}^1 \times \mathbb{R}^1 \longrightarrow \mathbb{R}^1$ are given by

$$f(x,u) = x + u,$$

and

$$f_0(x,u) = \frac{1}{2}u^2 \ge 0,$$

respectively.

The augmented Hamiltonian $\hat{H} : \mathbb{R}^2 \times \mathbb{R}^2 \times \mathbb{R}^1 \longrightarrow \mathbb{R}^1$ defined by

$$\hat{H}(\hat{\boldsymbol{\eta}}, \hat{\boldsymbol{x}}, u) = \eta_0 \frac{1}{2}u^2 + \eta(x+u) \tag{11.96}$$

where $\hat{\boldsymbol{x}} = \begin{bmatrix} x_0 & x \end{bmatrix}^T \in \mathbb{R}^2$, $\hat{\boldsymbol{\eta}} = \begin{bmatrix} \eta_0 & \eta \end{bmatrix}^T \in \mathbb{R}^2$ and $u \in \mathbb{R}^1$. The augmented function $\hat{f} : \mathbb{R}^2 \times \mathbb{R}^1 \longrightarrow \mathbb{R}^2$ is defined by

$$\hat{f}(\hat{x}, u) = \begin{bmatrix} \frac{1}{2}u^2 \\ x + u \end{bmatrix}, \tag{11.97}$$

so that the Jacobian is given by

$$\mathbb{J}_{\hat{\boldsymbol{x}}}\hat{f}(\hat{\boldsymbol{x}}, u) = \mathbb{J}_{\hat{\boldsymbol{x}}}\hat{f}(x, u) = \begin{bmatrix} 0 & 0 \\ 0 & 1 \end{bmatrix}. \tag{11.98}$$

Now assume that $(x^*(\cdot), u^*(\cdot))$ is an optimal pair so that the matrix $\hat{A}(t)$ is given by

$$\hat{A}(t) = \mathbb{J}_{\hat{\boldsymbol{x}}}\hat{f}(\hat{\boldsymbol{x}}, u)_{|(x^*(t),u^*(t))} = \mathbb{J}_{\hat{\boldsymbol{x}}}\hat{f}(x^*(t), u^*(t)) = \begin{bmatrix} 0 & 0 \\ 0 & 1 \end{bmatrix} \tag{11.99}$$

and

$$-[\hat{A}(t)]^T = -\begin{bmatrix} 0 & 0 \\ 0 & 1 \end{bmatrix} = \begin{bmatrix} 0 & 0 \\ 0 & -1 \end{bmatrix}.$$

The augmented adjoint equation has the form

$$\frac{d}{dt}\begin{bmatrix} \eta_0(t) \\ \eta(t) \end{bmatrix} = -[\hat{A}(t)]^T \begin{bmatrix} \eta_0(t) \\ \eta(t) \end{bmatrix} = \begin{bmatrix} 0 & 0 \\ 0 & -1 \end{bmatrix}\begin{bmatrix} \eta_0(t) \\ \eta(t) \end{bmatrix} \tag{11.100}$$

which is equivalent to the system

$$\frac{d}{dt}\eta_0(t) = 0$$
$$\frac{d}{dt}\eta(t) = -\eta(t),$$

where again $x^*(t)$ is the optimal trajectory. The first equation above implies that the zero adjoint state is a constant $\eta_0(t) \equiv \eta_0$. Observe that

$$\eta(t) = ke^{-t}$$

for some k. We can now apply the Maximum Principle.

If $(x^*(\cdot), u^*(\cdot))$ is an optimal pair for the LQ optimal control problem, then the Maximum Principle implies that there is a **non-trivial** solution

$$\hat{\boldsymbol{\eta}}^*(t) = \begin{bmatrix} \eta_0^*(t) \\ \eta^*(t) \end{bmatrix} \in \mathbb{R}^2$$

to the augmented adjoint equation (11.100) such that $\eta_0^*(t) \equiv \eta_0^* \leq 0$,

$$\frac{d}{dt}\eta^*(t) = -\eta^*(t) \tag{11.101}$$

and

$$\hat{H}(\hat{\boldsymbol{\eta}}^*(t), \hat{\boldsymbol{x}}^*(t), u^*(t)) = \max_{u \in \mathbb{R}^1} \hat{H}(\hat{\boldsymbol{\eta}}^*(t), \hat{\boldsymbol{x}}^*(t), u) \equiv c.$$

Since $\Omega = \mathbb{R}^1$ is open and $\hat{H}(\hat{\boldsymbol{\eta}}^*(t), \hat{\boldsymbol{x}}^*(t), u)$ is maximized on the open set $\Omega = \mathbb{R}^1$, then the derivative of $\hat{H}(\hat{\boldsymbol{\eta}}^*(t), \hat{\boldsymbol{x}}^*(t), u)$ with respect to u at $u = u^*(t)$ must be zero. In particular,

$$D_u\hat{H}(\hat{\boldsymbol{\eta}}^*(t), \hat{\boldsymbol{x}}^*(t), u)_{|u=u^*(t)} = D_u\hat{H}(\hat{\boldsymbol{\eta}}^*(t), \hat{\boldsymbol{x}}^*(t), u^*(t)) = 0,$$

where

$$\begin{aligned} D_u\hat{H}(\hat{\boldsymbol{\eta}}^*(t), \hat{\boldsymbol{x}}^*(t), u) &= \frac{\partial}{\partial u}[\eta_0^* \frac{1}{2}u^2 + \eta^*(t)(x^*(t) + u)] \\ &= [\eta_0^* u + \eta^*(t)], \end{aligned}$$

so that when $u = u^*(t)$

$$D_u\hat{H}(\hat{\boldsymbol{\eta}}^*(t), \hat{\boldsymbol{x}}^*(t), u)_{|u=u^*(t)} = [\eta_0^* u^*(t) + \eta^*(t)] = 0.$$

Thus,

$$\eta_0^* u^*(t) = -\eta^*(t). \tag{11.102}$$

Note that if $\eta_0^* = 0$, then $0 = \eta_0^* u^*(t) = -\eta^*(t)$ which would imply that $\eta^*(t) \equiv 0$. However, this in turn implies that

$$\hat{\boldsymbol{\eta}}^*(t) = \begin{bmatrix} \eta_0^*(t) \\ \eta^*(t) \end{bmatrix} \equiv \begin{bmatrix} 0 \\ 0 \end{bmatrix} \in \mathbb{R}^2$$

which would contradict the fact that $\hat{\boldsymbol{\eta}}^*(t)$ is a nontrivial solution. Hence, $\eta_0^* < 0$ and we can solve (11.102) for $u^*(t)$ yielding

$$u^*(t) = \frac{\eta^*(t)}{-\eta_0^*} \triangleq \lambda^*(t),$$

where we normalize the adjoint variable by defining

$$\lambda^*(t) = \frac{\eta^*(t)}{-\eta_0^*}.$$

Note that by (11.101)

$$\dot{\lambda}^*(t) = \frac{\dot{\eta}^*(t)}{-\eta_0^*} = \frac{\eta^*(t)}{\eta_0^*} = -\lambda^*(t),$$

so that

$$\lambda^*(t) = Ae^{-t}$$

for some constant A and

$$\dot{x}^*(t) = x^*(t) + u^*(t) = x^*(t) + Ae^{-t}. \tag{11.103}$$

Solving (11.103) and applying the boundary conditions at $x^*(0) = 1/2$ and $x^*(1) = 0$ yields that

$$x^*(t) = \frac{1}{2(e^{-1} - e)}[e^{t-1} - e^{1-t}]$$

and

$$A = \frac{1}{1 - e^{-2}},$$

so the optimal control is given by

$$u^*(t) = [\frac{1}{1 - e^{-2}}]e^{-t}.$$

The second example is Linear Quadratic (LQ) optimal control problem with hard constraint on the control. In this case, the Maximum Principle does not reduce to taking the derivative of the augmented Hamiltonian and setting the resulting equation to zero. One must compute a solution to a constrained optimization problem.

LQ Optimal Control Problem: Example 2

In this problem we allow the final state to be free and place a bound on the control. Thus, we consider the linear control system

$$\dot{x}(t) = x(t) + u(t) \tag{11.104}$$

with initial data

$$x(0) = 1 \in \mathbb{R}^1. \tag{11.105}$$

We assume that $t_0 = 0$, $x_0 = 1 \in \mathbb{R}^1$ and $0 < t_1 = 1$ are given. The control is required to be bounded

$$|u(t)| \leq 1 \tag{11.106}$$

and the quadratic cost functional is defined by

$$J(u(\cdot)) = \int_0^1 \frac{1}{2}[x(s; u(\cdot))]^2 ds = \int_0^1 \frac{1}{2}[x(s)]^2 ds, \tag{11.107}$$

where $x(\cdot) = x(\cdot; u(\cdot))$ is the solution to initial value problem (11.104) - (11.105). The Linear Quadratic (LQ) optimal control problem is to find $u^*(\cdot) \in PWC(0,1)$ satisfying (11.106) so that

$$J(u^*(\cdot)) = \int_0^1 \frac{1}{2}[x^*(s; u(\cdot))]^2 ds \leq J(u(\cdot)) = \int_0^1 \frac{1}{2}[x(s; u(\cdot))]^2 ds$$

for all $u(\cdot) \in PWC(0,1)$ that steer $x_0 = 1/2$ to $x_1 \in \mathbb{R}^1$.

Formulation as an Optimal Control Problem: In order to set up the fixed time optimal control problem and apply the Maximum Principle, we need to:

1. Identify the initial time t_0 and final time t_1;
2. Identify the initial set $X_0 \subseteq \mathbb{R}^1$, the terminal set $X_1 \subseteq \mathbb{R}^1$;
3. Identify the control constraint set $\Omega \subseteq \mathbb{R}^1$;
4. Identify the functions $f : \mathbb{R}^1 \times \mathbb{R}^1 \longrightarrow \mathbb{R}^1$, $f_0 : \mathbb{R}^1 \times \mathbb{R}^1 \longrightarrow \mathbb{R}^1$;

5. Define the augmented Hamiltonian $\hat{H}(\hat{\boldsymbol{\eta}}, \hat{\boldsymbol{x}}, u)$;

6. Form the augmented adjoint system matrix $\hat{A}(t)$.

The initial time is $t_0 = 0$ and the final time $t_1 = 1$ is given and fixed. The initial set is the single initial vector $\mathrm{X}_0 = \{1\}$ and since there is no terminal constraint, the terminal set is the whole state space $\mathrm{X}_1 = \mathbb{R}^1$. The constraint on the control $u(\cdot) \in PWC(0,1)$ is $|u(t)| \leq 1$ which implies that $\Omega = [-1, 1]$. The functions $f : \mathbb{R}^1 \times \mathbb{R}^1 \longrightarrow \mathbb{R}^1$ and $f_0 : \mathbb{R}^1 \times \mathbb{R}^1 \longrightarrow \mathbb{R}^1$ are given by

$$f(x, u) = x + u,$$

and

$$f_0(x, u) = \frac{1}{2}x^2 \geq 0,$$

respectively.

The augmented Hamiltonian $\hat{H} : \mathbb{R}^2 \times \mathbb{R}^2 \times \mathbb{R}^1 \longrightarrow \mathbb{R}^1$ defined by

$$\hat{H}(\hat{\boldsymbol{\eta}}, \hat{\boldsymbol{x}}, u) = \eta_0 \frac{1}{2}x^2 + \eta(x + u) \tag{11.108}$$

where $\hat{\boldsymbol{x}} = \begin{bmatrix} x_0 & x \end{bmatrix}^T \in \mathbb{R}^2$, $\hat{\boldsymbol{\eta}} = \begin{bmatrix} \eta_0 & \eta \end{bmatrix}^T \in \mathbb{R}^2$ and $u \in \mathbb{R}^1$. The augmented function $\hat{f} : \mathbb{R}^2 \times \mathbb{R}^1 \longrightarrow \mathbb{R}^2$ is defined by

$$\hat{f}(\hat{\boldsymbol{x}}, u) = \begin{bmatrix} \frac{1}{2}x^2 \\ x + u \end{bmatrix}, \tag{11.109}$$

so that the Jacobian is given by

$$\mathbb{J}_{\hat{\boldsymbol{x}}} \hat{f}(\hat{\boldsymbol{x}}, u) = \mathbb{J}_{\hat{\boldsymbol{x}}} \hat{f}(x, u) = \begin{bmatrix} 0 & x \\ 0 & 1 \end{bmatrix}. \tag{11.110}$$

Now assume that $(x^*(\cdot), u^*(\cdot))$ is an optimal pair so that the matrix $\hat{A}(t)$ is given by

$$\begin{aligned} \hat{A}(t) &= \mathbb{J}_{\hat{\boldsymbol{x}}} \hat{f}(\hat{\boldsymbol{x}}, u)_{|(x^*(t), u^*(t))} \\ &= \mathbb{J}_{\hat{\boldsymbol{x}}} \hat{f}(x^*(t), u^*(t)) = \begin{bmatrix} 0 & x^*(t) \\ 0 & 1 \end{bmatrix} \end{aligned} \tag{11.111}$$

and

$$-[\hat{A}(t)]^T = -\begin{bmatrix} 0 & 0 \\ 0 & 1 \end{bmatrix} = \begin{bmatrix} 0 & 0 \\ -x^*(t) & -1 \end{bmatrix}.$$

The augmented adjoint equation has the form

$$\frac{d}{dt}\begin{bmatrix} \eta_0(t) \\ \eta(t) \end{bmatrix} = -[\hat{A}(t)]^T \begin{bmatrix} \eta_0(t) \\ \eta(t) \end{bmatrix} = \begin{bmatrix} 0 & 0 \\ -x^*(t) & -1 \end{bmatrix}\begin{bmatrix} \eta_0(t) \\ \eta(t) \end{bmatrix} \tag{11.112}$$

which is equivalent to the system

$$\frac{d}{dt}\eta_0(t) = 0$$
$$\frac{d}{dt}\eta(t) = -\eta_0(t)x^*(t) - \eta(t),$$

where again $x^*(t)$ is the optimal trajectory. The first equation above implies (as always) that the zero adjoint state is a constant $\eta_0(t) \equiv \eta_0$. Observe that

$$\frac{d}{dt}\eta(t) = -\eta(t) - \eta_0 x^*(t) \tag{11.113}$$

is a linear nonhomogeneous equation. We can now apply the Maximum Principle.

If $(x^*(\cdot), u^*(\cdot))$ is an optimal pair for the LQ optimal control problem, then the Maximum Principle implies that there is a **non-trivial** solution

$$\hat{\boldsymbol{\eta}}^*(t) = \begin{bmatrix} \eta_0^*(t) \\ \boldsymbol{\eta}^*(t) \end{bmatrix} \in \mathbb{R}^2$$

to the augmented adjoint equation (11.112) such that $\eta_0^*(t) \equiv \eta_0^* \leq 0$,

$$\frac{d}{dt}\eta^*(t) = -\eta^*(t) - \eta_0^* x^*(t) \tag{11.114}$$

and

$$\hat{H}(\hat{\boldsymbol{\eta}}^*(t), \hat{\boldsymbol{x}}^*(t), u^*(t)) = \max_{u \in [-1,1]} \hat{H}(\hat{\boldsymbol{\eta}}^*(t), \hat{\boldsymbol{x}}^*(t), u) \equiv c.$$

In particular,

$$\hat{H}(\hat{\boldsymbol{\eta}}^*(t), \hat{\boldsymbol{x}}^*(t), u) = \eta_0^* \frac{1}{2}x^2 + \eta^*(t)(x^*(t) + u),$$

so that

$$\begin{aligned}\max_{u\in[-1,1]} \hat{H}(\hat{\boldsymbol{\eta}}^*(t), \hat{\boldsymbol{x}}^*(t), u) &= \max_{u\in[-1,1]} \{\eta_0^* \frac{1}{2}x^2 + \eta^*(t)(x^*(t)+u)\} \\ &= \max_{u\in[-1,1]} \{\eta^*(t)u)\} \\ &\quad + \eta_0^* \frac{1}{2}x^2 + \eta^*(t)x^*(t) \equiv c.\end{aligned}$$

Thus, $u = u^*(t)$ must be selected to maximize

$$\{\eta^*(t)u)\}$$

on the interval $-1 \le u \le 1$. Clearly, this implies that

$$u^*(t) = sgn[\eta^*(t)]. \tag{11.115}$$

Applying the transversality condition at $t_1 = 1$, we find that $\mathbb{T}_1 = \mathbb{R}^1$ so $\eta^*(t_1) \perp \mathbb{T}_1$ implies that

$$\eta^*(t_1) = \eta^*(1) = 0. \tag{11.116}$$

Returning to the adjoint equation (11.113) we see that $\eta^*(t)$ satisfies the terminal value problem

$$\frac{d}{dt}\eta^*(t) = -\eta^*(t) - \eta_0^* x^*(t), \quad \eta^*(1) = 0. \tag{11.117}$$

One can see that $\eta_0^* \neq 0$ since $\eta_0^* = 0$ would imply that $\eta^*(t)$ solves the linear equation problem

$$\frac{d}{dt}\eta^*(t) = -\eta^*(t), \quad \eta^*(1) = 0,$$

and hence $\eta^*(t) \equiv 0$. Thus, if $\eta_0^* = 0$, then $\eta^*(t) \equiv 0$ which contradicts the statement that $\hat{\boldsymbol{\eta}}^*(t) = \begin{bmatrix} \eta_0^* & \eta^*(t) \end{bmatrix}^T$ is nonzero.

Hence, $\eta_0^* < 0$ and we can simplify the adjoint equation (11.117). Divide both sides of (11.117) by the positive number $-\eta_0^*$ to obtain

$$\frac{d}{dt}\left[\frac{\eta^*(t)}{-\eta_0^*}\right] = -\left[\frac{\eta^*(t)}{-\eta_0^*}\right] + x^*(t), \quad \left[\frac{\eta^*(1)}{-\eta_0^*}\right] = 0.$$

Let $\lambda^*(t)$ be defined by

$$\lambda^*(t) \triangleq \left[\frac{\eta^*(t)}{-\eta_0^*}\right]$$

and note that since $-\eta_0^* > 0$, $\lambda^*(t)$ and $\eta^*(t)$ have the same sign. Also,

$$\dot{\lambda}^*(t) = -\lambda^*(t) + x^*(t), \quad \lambda^*(1) = 0 \tag{11.118}$$

and

$$u^*(t) = sgn[\eta^*(t)] = sgn[\lambda^*(t)]. \tag{11.119}$$

Combining (11.118) - (11.119) with the state equation, we find that the system

$$\begin{aligned} \dot{x}^*(t) &= x^*(t) + sgn[\lambda^*(t)], \quad & x^*(0) &= 1, \\ \dot{\lambda}^*(t) &= -\lambda^*(t) + x^*(t), & \lambda^*(1) &= 0, \end{aligned} \tag{11.120}$$

needs to be solved in order to compute $u^*(t) = sgn(\eta^*(t))$. This is a nonlinear two-point boundary value problem that must be solved numerically.

At this point it is helpful to review numerical methods for solving two point boundary value problems. The following references [11], [25], [103], [104], [112], [162], [169] and [185] contain some useful results on this topic.

11.5.2 The Time Independent Riccati Differential Equation

We return to the general LQ optimal control problem and focus on the optimality system defined by the two point boundary value problem (11.89) - (11.91). We show that one can transform the state variable to the adjoint variable by a matrix that satisfies a Riccati differential equation. This transformation is a key step in connecting the theory of Riccati equations with the optimality conditions for linear quadratic optimal control. Also, the resulting Riccati equation provides one method for developing numerical methods for solving LQ optimal control problems.

First write (11.89) as the linear system

$$\frac{d}{dt}\begin{bmatrix} \boldsymbol{x}(t) \\ \boldsymbol{\lambda}(t) \end{bmatrix} = \begin{bmatrix} A & BR^{-1}B^T \\ Q & -A^T \end{bmatrix}\begin{bmatrix} \boldsymbol{x}(t) \\ \boldsymbol{\lambda}(t) \end{bmatrix} \triangleq F\begin{bmatrix} \boldsymbol{x}(t) \\ \boldsymbol{\lambda}(t) \end{bmatrix} \quad (11.121)$$

where $\boldsymbol{\lambda}(t_1) = 0$ and

$$F = \begin{bmatrix} A & BR^{-1}B^T \\ Q & -A^T \end{bmatrix}.$$

The solution to (11.121) has the form

$$\begin{bmatrix} \boldsymbol{x}(t) \\ \boldsymbol{\lambda}(t) \end{bmatrix} = e^{F(t-t_1)}\begin{bmatrix} \boldsymbol{x}(t_1) \\ \boldsymbol{\lambda}(t_1) \end{bmatrix} = e^{F(t-t_1)}\begin{bmatrix} \boldsymbol{x}(t_1) \\ \mathbf{0} \end{bmatrix}.$$

Let

$$\Psi(t) = e^{Ft} = \begin{bmatrix} \psi_{11}(t) & \psi_{12}(t) \\ \psi_{21}(t) & \psi_{22}(t) \end{bmatrix},$$

where $\psi_{ij}(t),\ i,j = 1,2$ are $n \times n$ square matrix functions. It follows that

$$\begin{bmatrix} \boldsymbol{x}(t) \\ \boldsymbol{\lambda}(t) \end{bmatrix} = e^{F(t-t_1)}\begin{bmatrix} \boldsymbol{x}(t_1) \\ \mathbf{0} \end{bmatrix} = \begin{bmatrix} \psi_{11}(t-t_1) & \psi_{12}(t-t_1) \\ \psi_{21}(t-t_1) & \psi_{22}(t-t_1) \end{bmatrix}\begin{bmatrix} \boldsymbol{x}(t_1) \\ \mathbf{0} \end{bmatrix}$$

so that

$$\boldsymbol{x}(t) = \psi_{11}(t-t_1)\boldsymbol{x}(t_1), \quad (11.122)$$

and

$$\boldsymbol{\lambda}(t) = \psi_{21}(t-t_1)\boldsymbol{x}(t_1). \quad (11.123)$$

If $\psi_{11}(t-t_1)$ is non-singular for $0 \le t \le t_1$, then we can solve (11.122) for $\boldsymbol{x}(t_1)$. In particular,

$$\boldsymbol{x}(t_1) = [\psi_{11}(t-t_1)]^{-1}\boldsymbol{x}(t)$$

which, when substituted into (11.123), yields

$$\boldsymbol{\lambda}(t) = [\psi_{21}(t-t_1)(\psi_{11}(t-t_1))^{-1}]\boldsymbol{x}(t).$$

If $P(t)$ is the $n \times n$ matrix defined by

$$P(t) \triangleq -[\psi_{21}(t-t_1)(\psi_{11}(t-t_1))^{-1}], \quad (11.124)$$

then we have that $\boldsymbol{\lambda}(t)$ and $\boldsymbol{x}(t)$ are linearly related by the matrix $P(t)$ and the relationship is given by

$$\boldsymbol{\lambda}(t) = -P(t)\boldsymbol{x}(t).$$

The choice of the negative sign in defining $P(\cdot)$ is made to be consistent with much of the existing literature. In order to make this step rigorous, one needs to prove that $\psi_{11}(t-t_1)$ is non-singular for $0 \le t \le t_1$. On the other hand, we could simply ask the question:

> *Is there a matrix $P(t)$ so that $\boldsymbol{\lambda}(t) = -P(t)\boldsymbol{x}(t)$ and how can $P(t)$ be computed?*

We will address the issue of the existence of $P(t)$ later. However, assume for the moment that $\boldsymbol{x}(\cdot)$ and $\boldsymbol{\lambda}(t)$ satisfying (11.89) - (11.91) and

$$\boldsymbol{\lambda}(t) = -P(t)\boldsymbol{x}(t), \tag{11.125}$$

with $P(t)$ differentiable. Differentiating the equation (11.125) one obtains

$$\begin{aligned}
\frac{d}{dt}\boldsymbol{\lambda}(t) &= -\left[\frac{d}{dt}P(t)\right]\boldsymbol{x}(t) - P(t)\left[\frac{d}{dt}\boldsymbol{x}(t)\right] \\
&= -\left[\frac{d}{dt}P(t)\right]\boldsymbol{x}(t) - P(t)\left[A\boldsymbol{x}(t) + BR^{-1}B^T\boldsymbol{\lambda}(t)\right] \\
&= -\left[\frac{d}{dt}P(t)\right]\boldsymbol{x}(t) - P(t)\left[A\boldsymbol{x}(t) - BR^{-1}B^TP(t)\boldsymbol{x}(t)\right] \\
&= -\left[\frac{d}{dt}P(t)\right]\boldsymbol{x}(t) - P(t)A\boldsymbol{x}(t) + P(t)BR^{-1}B^TP(t)\boldsymbol{x}(t).
\end{aligned}$$

However, from (11.121) it follows that

$$\begin{aligned}
\frac{d}{dt}\boldsymbol{\lambda}(t) &= Q\boldsymbol{x}(t) - A^T\boldsymbol{\lambda}(t) \\
&= Q\boldsymbol{x}(t) + A^TP(t)\boldsymbol{x}(t)
\end{aligned}$$

so that

$$\begin{aligned}
Q\boldsymbol{x}(t) + A^TP(t)\boldsymbol{x}(t) = &-\left[\frac{d}{dt}P(t)\right]\boldsymbol{x}(t) - P(t)A\boldsymbol{x}(t) \\
&+ P(t)BR^{-1}B^TP(t)\boldsymbol{x}(t).
\end{aligned}$$

Rearranging the terms we have

$$-\left[\frac{d}{dt}P(t)\right]\boldsymbol{x}(t) = A^T P(t)\boldsymbol{x}(t) + P(t)A\boldsymbol{x}(t) - P(t)BR^{-1}B^T P(t)\boldsymbol{x}(t) + Q\boldsymbol{x}(t),$$

or equivalently

$$-\left[\frac{d}{dt}P(t)\right]\boldsymbol{x}(t) = \left[A^T P(t) + P(t)A - P(t)BR^{-1}B^T P(t) + Q\right]\boldsymbol{x}(t). \tag{11.126}$$

Consequently, $P(t)$ satisfies (11.126) along the trajectory $\boldsymbol{x}(t)$. Observe that (11.126) is satisfied for any solution of the system (11.121) with $\boldsymbol{\lambda}(t_1) = 0$ and all values of $\boldsymbol{x}(t_1)$. Therefore, if

$$\boldsymbol{\lambda}(t) = -P(t)\boldsymbol{x}(t),$$

then $P(t)$ satisfies the matrix Riccati differential equation

$$-\dot{P}(t) = A^T P(t) + P(t)A - P(t)BR^{-1}B^T P(t) + Q, \quad 0 \le t < t_1, \tag{11.127}$$

with terminal condition

$$P(t_1) = 0_{n\times n}, \tag{11.128}$$

since

$$-P(t_1)\boldsymbol{x}(t_1) = \boldsymbol{\lambda}(t_1) = \mathbf{0}$$

and $\boldsymbol{x}(t_1)$ can be any vector in $\mathbb{R}^n$.

We shall show below that under the assumption that there is a solution $P(t)$ to the Riccati differential equation (11.127) satisfying (11.128), then the LQ optimal control problem has a solution and the optimal control is given by

$$\boldsymbol{u}^*(t) = -R^{-1}B^T P(t)\boldsymbol{x}^*(t). \tag{11.129}$$

In order to provide a rigorous treatment of this problem, we present two lemmas. These results relate the existence of a solution to the Riccati equation (11.127) to the existence of an optimal control for the LQ optimal control problem. First we note that any solution to the Riccati differential equation must be symmetric. In particular, $P(t) = [P(t)]^T$ for all t.

Lemma 11.1 *Suppose that $P(t) = [P(t)]^T$ is any $n \times n$ matrix function with $P(t)$ differentiable on the interval $[t_0, t_1]$. If $\boldsymbol{u}(\cdot) \in PWC(t_0, t_1; \mathbb{R}^m)$ and*

$$\dot{\boldsymbol{x}}(t) = A\boldsymbol{x}(t) + B\boldsymbol{u}(t), \quad t_0 \le t \le t_1,$$

then

$$\begin{aligned} \langle P(s)\boldsymbol{x}(s), \boldsymbol{x}(s)\rangle \Big|_{t_0}^{t_1} = & \int_{t_0}^{t_1} \left\langle \left[\dot{P}(s) + P(s)A + A^T P(s)\right] \boldsymbol{x}(s), \boldsymbol{x}(s)\right\rangle ds \\ & + \int_{t_0}^{t_1} \langle P(s)B\boldsymbol{u}(s), \boldsymbol{x}(s)\rangle \, ds \qquad (11.130) \\ & + \int_{t_0}^{t_1} \left\langle B^T P(s)\boldsymbol{x}(s), \boldsymbol{u}(s)\right\rangle ds. \end{aligned}$$

Proof: Observe that

$$\begin{aligned} \langle P(s)\boldsymbol{x}(s), \boldsymbol{x}(s)\rangle \Big|_{t_0}^{t_1} &= \int_{t_0}^{t_1} \frac{d}{ds} \langle P(s)\boldsymbol{x}(s), \boldsymbol{x}(s)\rangle \, ds \\ &= \int_{t_0}^{t_1} \left\langle \dot{P}(s)\boldsymbol{x}(s), \boldsymbol{x}(s)\right\rangle ds \\ &\quad + \int_{t_0}^{t_1} \langle P(s)\dot{\boldsymbol{x}}(s), \boldsymbol{x}(s)\rangle \, ds \\ &\quad + \int_{t_0}^{t_1} \langle P(s)\boldsymbol{x}(s), \dot{\boldsymbol{x}}(s)\rangle \, ds \end{aligned}$$

and by substituting $A\boldsymbol{x}(s) + B\boldsymbol{u}(s)$ for $\dot{\boldsymbol{x}}(s)$ we obtain

$$\begin{aligned} \langle P(s)\boldsymbol{x}(s), \boldsymbol{x}(s)\rangle \Big|_{t_0}^{t_1} = & \int_{t_0}^{t_1} \left\langle \dot{P}(s)\boldsymbol{x}(s), \boldsymbol{x}(s)\right\rangle ds \\ & + \int_{t_0}^{t_1} \langle P(s)\left[A\boldsymbol{x}(s) + B\boldsymbol{u}(s)\right], \boldsymbol{x}(s)\rangle \, ds \\ & + \int_{t_0}^{t_1} \langle P(s)\boldsymbol{x}(s), [A\boldsymbol{x}(s) + B\boldsymbol{u}(s)]\rangle \, ds. \end{aligned}$$

Simplifying this expression we obtain

$$
\begin{aligned}
\langle P(s)\boldsymbol{x}(s), \boldsymbol{x}(s)\rangle \Big|_{t_0}^{t_1} = & \int_{t_0}^{t_1} \left\langle \dot{P}(s)\boldsymbol{x}(s), \boldsymbol{x}(s)\right\rangle ds \\
& + \int_{t_0}^{t_1} \langle P(s)\left[A\boldsymbol{x}(s)\right], \boldsymbol{x}(s)\rangle ds \\
& + \int_{t_0}^{t_1} \langle P(s)\left[B\boldsymbol{u}(s)\right], \boldsymbol{x}(s)\rangle ds \\
& + \int_{t_0}^{t_1} \langle P(s)\boldsymbol{x}(s), \left[A\boldsymbol{x}(s)\right]\rangle ds \\
& + \int_{t_0}^{t_1} \langle P(s)\boldsymbol{x}(s), \left[B\boldsymbol{u}(s)\right]\rangle ds
\end{aligned}
$$

and collecting terms yields

$$
\begin{aligned}
\langle P(s)\boldsymbol{x}(s), \boldsymbol{x}(s)\rangle \Big|_{t_0}^{t_1} = & \int_{t_0}^{t_1} \left\langle [\dot{P}(s) + P(s)A + A^T P(s)]\boldsymbol{x}(s), \boldsymbol{x}(s)\right\rangle ds \\
& + \int_{t_0}^{t_1} \langle P(s)B\boldsymbol{u}(s), \boldsymbol{x}(s)\rangle ds \\
& + \int_{t_0}^{t_1} \left\langle B^T P(s)\boldsymbol{x}(s), \boldsymbol{u}(s)\right\rangle ds.
\end{aligned}
$$

which establishes (11.130). □

Lemma 11.2 *Assume that the Riccati differential equation (11.127) has a solution* $P(t) = [P(t)]^T$ *for* $t_0 \le t < t_1$ *and* $P(t_1) = 0_{n\times n}$. *If* $\boldsymbol{u}(\cdot) \in PWC(t_0, t_1; \mathbb{R}^m)$ *and*

$$
\dot{\boldsymbol{x}}(t) = A\boldsymbol{x}(t) + B\boldsymbol{u}(t), \quad t_0 \le t \le t_1,
$$

then the cost function $J(\cdot)$ *has the representation*

$$
\begin{aligned}
J(\boldsymbol{u}(\cdot)) = & \int_{t_0}^{t_1} \left\| R^{1/2}\boldsymbol{u}(s) + R^{-1/2}B^T P(s)\boldsymbol{x}(s)\right\|^2 ds \\
& + \langle P(t_0)\boldsymbol{x}(t_0), \boldsymbol{x}(t_0)\rangle .
\end{aligned}
$$

Proof: Let

$$N(\boldsymbol{x}(\cdot), \boldsymbol{u}(\cdot)) = \int_{t_0}^{t_1} \left\| R^{1/2}\boldsymbol{u}(s) + R^{-1/2}B^T P(s)\boldsymbol{x}(s) \right\|^2 ds$$

and expanding $N(\boldsymbol{x}(\cdot), \boldsymbol{u}(\cdot))$ we obtain

$$\begin{aligned} N(\boldsymbol{x}(\cdot), \boldsymbol{u}(\cdot)) = &\int_{t_0}^{t_1} \langle R^{1/2}\boldsymbol{u}(s) + R^{-1/2}B^T P(s)\boldsymbol{x}(s), R^{1/2}\boldsymbol{u}(s) \\ &\quad + R^{-1/2}B^T P(s)\boldsymbol{x}(s)\rangle ds \\ = &\int_{t_0}^{t_1} \left\langle R^{1/2}\boldsymbol{u}(s), R^{1/2}\boldsymbol{u}(s) \right\rangle ds \\ &+ \int_{t_0}^{t_1} \left\langle R^{1/2}\boldsymbol{u}(s), R^{-1/2}B^T P(s)\boldsymbol{x}(s) \right\rangle ds \\ &+ \int_{t_0}^{t_1} \left\langle R^{-1/2}B^T P(s)\boldsymbol{x}(s), R^{1/2}\boldsymbol{u}(s) \right\rangle ds \\ &+ \int_{t_0}^{t_1} \left\langle R^{-1/2}B^T P(s)\boldsymbol{x}(s), R^{-1/2}B^T P(s)\boldsymbol{x}(s) \right\rangle ds. \end{aligned}$$

Simplifying each term we have

$$\begin{aligned} N(\boldsymbol{x}(\cdot), \boldsymbol{u}(\cdot)) = &\int_{t_0}^{t_1} \left\langle R^{1/2}R^{1/2}\boldsymbol{u}(s), \boldsymbol{u}(s) \right\rangle ds \\ &+ \int_{t_0}^{t_1} \left\langle \boldsymbol{u}(s), R^{1/2}R^{-1/2}B^T P(s)\boldsymbol{x}(s) \right\rangle ds \\ &+ \int_{t_0}^{t_1} \left\langle R^{1/2}R^{-1/2}B^T P(s)\boldsymbol{x}(s), \boldsymbol{u}(s) \right\rangle ds \\ &+ \int_{t_0}^{t_1} \left\langle R^{-1/2}R^{-1/2}B^T P(s)\boldsymbol{x}(s), B^T P(s)\boldsymbol{x}(s) \right\rangle ds, \end{aligned}$$

which implies

$$
\begin{aligned}
N(\boldsymbol{x}(\cdot),\boldsymbol{u}(\cdot)) = & \int_{t_0}^{t_1} \langle R\boldsymbol{u}(s),\boldsymbol{u}(s)\rangle\, ds \\
& + \int_{t_0}^{t_1} \left\langle \boldsymbol{u}(s), B^T P(s)\boldsymbol{x}(s)\right\rangle ds \\
& + \int_{t_0}^{t_1} \left\langle B^T P(s)\boldsymbol{x}(s), \boldsymbol{u}(s)\right\rangle ds \\
& + \int_{t_0}^{t_1} \left\langle R^{-1}B^T P(s)\boldsymbol{x}(s), B^T P(s)\boldsymbol{x}(s)\right\rangle ds,
\end{aligned}
$$

or equivalently

$$
\begin{aligned}
N(\boldsymbol{x}(\cdot),\boldsymbol{u}(\cdot)) = & \int_{t_0}^{t_1} \langle R\boldsymbol{u}(s),\boldsymbol{u}(s)\rangle \\
& + \int_{t_0}^{t_1} \left\langle \boldsymbol{u}(s), B^T P(s)\boldsymbol{x}(s)\right\rangle ds \qquad (11.131) \\
& + \int_{t_0}^{t_1} \left\langle B^T P(s)\boldsymbol{x}(s), \boldsymbol{u}(s)\right\rangle ds \\
& + \int_{t_0}^{t_1} \left\langle P(s)BR^{-1}B^T P(s)\boldsymbol{x}(s), \boldsymbol{x}(s)\right\rangle ds.
\end{aligned}
$$

Since the matrix $P(s)$ satisfies the Riccati equation (11.127), it follows that

$$
P(s)BR^{-1}B^T P(s)\boldsymbol{x}(s) = \left[\dot{P}(s) + A^T P(s) + P(s)A + Q\right]\boldsymbol{x}(s)
$$

and the last term above becomes

$$
\int_{t_0}^{t_1} \left\langle \left[\dot{P}(s) + A^T P(s) + P(s)A + Q\right]\boldsymbol{x}(s), \boldsymbol{x}(s)\right\rangle ds.
$$

Substituting this expression into (11.131) and rearranging yields

$$\begin{aligned} N(\boldsymbol{x}(\cdot),\boldsymbol{u}(\cdot)) = &\int_{t_0}^{t_1} \langle R\boldsymbol{u}(s),\boldsymbol{u}(s)\rangle\, ds + \int_{t_0}^{t_1} \langle Q\boldsymbol{x}(s),\boldsymbol{x}(s)\rangle\, ds \\ &+ \int_{t_0}^{t_1} \left\langle \left[\dot{P}(s) + A^T P(s) + P(s)A\right]\boldsymbol{x}(s),\boldsymbol{x}(s)\right\rangle ds \\ &+ \int_{t_0}^{t_1} \left\langle \boldsymbol{u}(s), B^T P(s)\boldsymbol{x}(s)\right\rangle ds \\ &+ \int_{t_0}^{t_1} \left\langle B^T P(s)\boldsymbol{x}(s),\boldsymbol{u}(s)\right\rangle ds, \end{aligned}$$

which implies

$$\begin{aligned} N(\boldsymbol{x}(\cdot),\boldsymbol{u}(\cdot)) = &\,J(\boldsymbol{u}(\cdot)) \\ &+ \int_{t_0}^{t_1} \left\langle \left[\dot{P}(s) + A^T P(s) + P(s)A\right]\boldsymbol{x}(s),\boldsymbol{x}(s)\right\rangle ds \\ &+ \int_{t_0}^{t_1} \left\langle \boldsymbol{u}(s), B^T P(s)\boldsymbol{x}(s)\right\rangle ds \\ &+ \int_{t_0}^{t_1} \left\langle B^T P(s)\boldsymbol{x}(s),\boldsymbol{u}(s)\right\rangle ds. \end{aligned}$$

Applying (11.130) from the previous Lemma yields

$$N(\boldsymbol{x}(\cdot),\boldsymbol{u}(\cdot)) = J(\boldsymbol{u}(\cdot)) + \langle P(s)\boldsymbol{x}(s),\boldsymbol{x}(s)\rangle \big|_{t_0}^{t_1},$$

or equivalently

$$J(\boldsymbol{u}(\cdot)) = N(\boldsymbol{x}(\cdot),\boldsymbol{u}(\cdot)) - \langle P(s)\boldsymbol{x}(s),\boldsymbol{x}(s)\rangle \big|_{t_0}^{t_1}.$$

However, since $P(t_1) = 0_{n\times n}$ we conclude that

$$\begin{aligned} J(\boldsymbol{u}(\cdot)) = &\int_{t_0}^{t_1} \left\| R^{1/2}\boldsymbol{u}(s) + R^{-1/2}B^T P(s)\boldsymbol{x}(s)\right\|^2 ds \\ &+ \langle P(t_0)\boldsymbol{x}(t_0),\boldsymbol{x}(t_0)\rangle\,, \end{aligned} \tag{11.132}$$

which completes the proof. □

We now have the fundamental result on the relationship between solutions to the Riccati equation and the existence of an optimal control for the LQ optimal problem.

Theorem 11.2 (Existence of LQ Optimal Control) *If the Riccati differential equation (11.127) has a solution* $P(t) = [P(t)]^T$ *for* $0 \le t < t_1$ *and* $P(t_1) = 0_{n\times n}$*, then there is a control* $\boldsymbol{u}^*(\cdot) \in PWC(0, t_1; \mathbb{R}^m)$ *such that* $\boldsymbol{u}^*(\cdot)$ *minimizes*

$$J(\boldsymbol{u}(\cdot)) = \int_0^{t_1} \{\langle Q\boldsymbol{x}(s), \boldsymbol{x}(s)\rangle + \langle R\boldsymbol{u}(s), \boldsymbol{u}(s)\rangle\}\, ds$$

on the set $PWC(0, t_1; \mathbb{R}^m)$*, where the state equation is given by*

$$\dot{\boldsymbol{x}}(t) = A\boldsymbol{x}(t) + B\boldsymbol{u}(t) \tag{11.133}$$

with initial data

$$\boldsymbol{x}(0) = \boldsymbol{x}_0 \in \mathbb{R}^n. \tag{11.134}$$

In addition, the optimal control is a linear feedback law

$$\boldsymbol{u}^*(t) = -R^{-1}B^T P(t)\boldsymbol{x}^*(t) \tag{11.135}$$

and the minimum value of $J(\boldsymbol{u}(\cdot))$ *is*

$$J(\boldsymbol{u}^*(\cdot)) = \langle P(0)\boldsymbol{x}_0, \boldsymbol{x}_0\rangle\,. \tag{11.136}$$

Proof: Let $t_0 = 0$ and apply the identity (11.132) above. In particular, it follows that $J(\cdot)$ is minimized when the quadratic term

$$\int_0^{t_1} \left\|R^{1/2}\boldsymbol{u}(s) + R^{-1/2}B^T P(s)\boldsymbol{x}(s)\right\|^2 ds \ge 0$$

is minimized. If $\boldsymbol{u}^*(t) = -R^{-1}B^T P(t)\boldsymbol{x}^*(t)$, then

$$R^{1/2}\boldsymbol{u}^*(t) + R^{-1/2}B^T P(t)\boldsymbol{x}^*(t) = 0$$

and

$$\begin{aligned} J(\boldsymbol{u}^*(\cdot)) = \int_0^{t_1} &\left\|R^{1/2}\boldsymbol{u}^*(s) + R^{-1/2}B^T P(s)\boldsymbol{x}^*(s)\right\|^2 ds \\ &+ \langle P(0)\boldsymbol{x}^*(0), \boldsymbol{x}^*(0)\rangle \\ = \langle P(0)\boldsymbol{x}_0&, \boldsymbol{x}_0\rangle\,. \end{aligned}$$

Consequently, for any $\boldsymbol{u}(\cdot) \in PWC(0, t_1; \mathbb{R}^m)$ it follows that

$$\begin{aligned} J(\boldsymbol{u}^*(\cdot)) &= \langle P(0)\boldsymbol{x}_0, \boldsymbol{x}_0\rangle \\ &\le \langle P(0)\boldsymbol{x}_0, \boldsymbol{x}_0\rangle + \int_0^{t_1} \left\| R^{1/2}\boldsymbol{u}(s) + R^{-1/2}B^T P(s)\boldsymbol{x}(s)\right\|^2 ds \\ &= J(\boldsymbol{u}(\cdot)), \end{aligned}$$

which completes the proof. □

Later we will return to linear quadratic control problems and consider more general time dependent problems. Also, we will generalize the cost function to include a terminal penalty such as

$$\begin{aligned} J(\boldsymbol{u}(\cdot)) = \langle G\boldsymbol{x}(t_1), \boldsymbol{x}(t_1)\rangle + \frac{1}{2}\int_{t_0}^{t_1} \{\langle Q\boldsymbol{x}(s), \boldsymbol{x}(s)\rangle \\ + \langle R\boldsymbol{u}(s), \boldsymbol{u}(s)\rangle\}\, ds, \end{aligned} \tag{11.137}$$

where $G = G^T \ge 0$ is a symmetric non-negative matrix.

11.6 The Maximum Principle for a Problem of Bolza

Consider the case where there is an additional explicit cost on the terminal state

$$J(\boldsymbol{u}(\cdot)) = G(\boldsymbol{x}(t_1)) + \int_{t_0}^{t_1} f_0(\boldsymbol{x}(s), \boldsymbol{u}(s))ds \tag{11.138}$$

where $G : \mathbb{R}^n \to \mathbb{R}$ is a C^2 function. In order to keep the discussion simple, we begin with the simple case where

$$\mathrm{X}_0 = \{\boldsymbol{x}_0\} \subseteq \mathbb{R}^n$$

is a single vector and t_1 is fixed. We show that by augmenting the problem we can construct an equivalent problem without the terminal cost defined by G. Also note that the final target set is taken to be

$$\mathrm{X}_1 = \mathbb{R}^n.$$

The state equation is augmented by adding a new variable at the end of the vector. In particular, let

$$\tilde{\boldsymbol{x}} = [x_1\ x_2\ ...x_n\ x_{n+1}]^T = [\boldsymbol{x}\ \ x_{n+1}]^T \in \mathbb{R}^{n+1}$$

and define $f_{n+1} : \mathbb{R}^{n+1} \times \mathbb{R}^m \longrightarrow \mathbb{R}^1$ by

$$f_{n+1}(\tilde{\boldsymbol{x}}, \boldsymbol{u}) = 0.$$

By adding the equation

$$\dot{x}_{n+1}(t) = f_{n+1}(\tilde{\boldsymbol{x}}(t), \boldsymbol{u}(t)\ ) = 0$$

to the system

$$\dot{\boldsymbol{x}}(t) = f(\boldsymbol{x}(t), \boldsymbol{u}(t)), \quad t_0 < t \leq t_1,$$

we obtain the new state equation

$$\frac{d}{dt}\tilde{\boldsymbol{x}}(t) = \tilde{f}(\tilde{\boldsymbol{x}}(t), \boldsymbol{u}(t)) = \left[\begin{array}{c} f(\boldsymbol{x}(t), \boldsymbol{u}(t)) \\ 0 \end{array}\right], \quad t_0 < t \leq t_1. \tag{11.139}$$

Consider the new cost functional

$$\tilde{J}(\boldsymbol{u}(\cdot)) = \int_{t_0}^{t_1} \{x_{n+1}(s) + f_0(\boldsymbol{x}(s), \boldsymbol{u}(s))\}ds = \int_{t_0}^{t_1} \tilde{f}_0(\tilde{\boldsymbol{x}}(s), \boldsymbol{u}(s))ds$$

where

$$\tilde{f}_0(\tilde{\boldsymbol{x}}, \boldsymbol{u}) \triangleq x_{n+1} + f_0(\boldsymbol{x}, \boldsymbol{u}).$$

Observe that since $x_{n+1}(s)$ is a constant (recall that $\dot{x}_{n+1}(s) = 0$), then

$$\begin{aligned}\tilde{J}(\boldsymbol{u}(\cdot)) &= \int_{t_0}^{t_1} \{x_{n+1}(s) + f_0(\boldsymbol{x}(s), \boldsymbol{u}(s))\}ds \\ &= x_{n+1}(t_1)(t_1 - t_0) + \int_{t_0}^{t_1} f_0(\boldsymbol{x}(s), \boldsymbol{u}(s))ds.\end{aligned}$$

Hence, if we require that $x_{n+1}(t_1) = \frac{G(x(t_1))}{t_1 - t_0}$, it follows that

$$\tilde{J}(\boldsymbol{u}(\cdot)) = x_{n+1}(t_1)(t_1 - t_0) + \int_{t_0}^{t_1} f_0(\boldsymbol{x}(s), \boldsymbol{u}(s))ds$$

$$= G(x(t_1)) + \int_{t_0}^{t_1} f_0(\boldsymbol{x}(s), \boldsymbol{u}(s))ds.$$

Thus, we can reformulate the problem as the equivalent optimal control problem in $\mathbb{R}^{n+1}$.

Let

$$\tilde{\mathrm{X}}_0 = \{\boldsymbol{x}_0\} \times \mathbb{R}^1 = \left\{\tilde{\boldsymbol{x}} = [\boldsymbol{x}_0 \;\; y]^T : y \in \mathbb{R}^1\right\} \subseteq \mathbb{R}^{n+1} \tag{11.140}$$

and

$$\tilde{\mathrm{X}}_1 = \left\{\tilde{\boldsymbol{x}} = [\boldsymbol{x} \;\; x_{n+1}]^T : \tilde{G}(\tilde{\boldsymbol{x}}) \triangleq x_{n+1} - G(\boldsymbol{x})/(t_1 - t_0) = 0\right\} \subseteq \mathbb{R}^{n+1}. \tag{11.141}$$

Therefore, minimizing the cost functional (11.138) among all controls that steer $\mathrm{X}_0 = \{\boldsymbol{x}_0\}$ to $\mathrm{X}_1 = \mathbb{R}^n$ is equivalent to minimizing the cost functional

$$\tilde{J}(\boldsymbol{u}(\cdot)) = \int_{t_0}^{t_1} \tilde{f}_0(\boldsymbol{x}(s), \boldsymbol{u}(s))ds = \int_{t_0}^{t_1} \{x_{n+1}(s) + f_0(\boldsymbol{x}(s), \boldsymbol{u}(s))\}ds \tag{11.142}$$

among all controls that steer $\tilde{\mathrm{X}}_0 = \{\boldsymbol{x}_0\} \times \mathbb{R}^1$ to

$$\tilde{\mathrm{X}}_1 = \left\{\tilde{\boldsymbol{x}} = [\boldsymbol{x} \;\; x_{n+1}]^T : \tilde{G}(\tilde{\boldsymbol{x}}) \triangleq x_{n+1} - G(\boldsymbol{x})/(t_1 - t_0) = 0\right\}$$

with the state equation defined by (11.139) above. We now apply the Maximum Principle to this equivalent problem.

Observe that $\tilde{\mathrm{X}}_0 = \{\boldsymbol{x}_0\} \times \mathbb{R}^1$ is a subspace of $\mathbb{R}^{n+1}$ with tangent space $\tilde{\mathbb{T}}_0 = \{\mathbf{0}\} \times \mathbb{R}^1$ so that a vector $\tilde{\boldsymbol{\eta}} = [\boldsymbol{\eta}, \eta_{n+1}]^T \in \mathbb{R}^{n+1}$ is orthogonal to $\tilde{\mathbb{T}}_0 = \{\mathbf{0}\} \times \mathbb{R}^1$ if and only if $\eta_{n+1} = 0$. Since the target set

$$\tilde{\mathrm{X}}_1 = \left\{\tilde{\boldsymbol{x}} = [\boldsymbol{x} \;\; x_{n+1}]^T : \tilde{G}(\tilde{\boldsymbol{x}}) \triangleq x_{n+1} - G(\boldsymbol{x})/(t_1 - t_0) = 0\right\} \subseteq \mathbb{R}^{n+1}$$

is defined by the level set of the function $\tilde{G} : \mathbb{R}^{n+1} \to \mathbb{R}^1$, then a vector $\tilde{\boldsymbol{\eta}} = [\boldsymbol{\eta}, \eta_{n+1}]^T \in \mathbb{R}^{n+1}$ is orthogonal to $\tilde{\mathbb{T}}_1$ at $\tilde{\boldsymbol{x}}(t_1)$ if

$$\tilde{\boldsymbol{\eta}} = \begin{bmatrix} \boldsymbol{\eta} \\ \eta_{n+1} \end{bmatrix} = \alpha \nabla \tilde{G}(\tilde{\boldsymbol{x}}(t_1)) = \alpha \begin{bmatrix} -\nabla G(\boldsymbol{x}(t_1))/(t_1 - t_0) \\ 1 \end{bmatrix} \tag{11.143}$$

for some nonzero α.

The augmented state and co-state are given by

$$\widehat{\tilde{\boldsymbol{x}}} = [x_0 \ \ \boldsymbol{x} \ \ x_{n+1}]^T = [x_0 \ \ \tilde{\boldsymbol{x}}]^T \in \mathbb{R}^{n+2}$$

and

$$\widehat{\tilde{\boldsymbol{\eta}}} = [\eta_0 \ \ \boldsymbol{\eta} \ \ \eta_{n+1}]^T = [\eta_0 \ \ \tilde{\boldsymbol{\eta}}]^T \in \mathbb{R}^{n+2},$$

respectively. The augmented state equation is given by

$$\begin{aligned} \frac{d}{dt}\widehat{\tilde{\boldsymbol{x}}}(t) = \widehat{\tilde{\boldsymbol{f}}}(\widehat{\tilde{\boldsymbol{x}}}(t), \boldsymbol{u}(t)) &= \begin{bmatrix} \tilde{f}_0(\tilde{\boldsymbol{x}}(t), \boldsymbol{u}(t)) \\ \boldsymbol{f}(\tilde{\boldsymbol{x}}(t), \boldsymbol{u}(t)) \\ f_{n+1}(\tilde{\boldsymbol{x}}(t), \boldsymbol{u}(t)) \end{bmatrix} \\ &= \begin{bmatrix} \tilde{f}_0(\tilde{\boldsymbol{x}}(t), \boldsymbol{u}(t)) \\ \boldsymbol{f}(\boldsymbol{x}(t), \boldsymbol{u}(t)) \\ 0 \end{bmatrix}, \quad t_0 < t \le t_1, \end{aligned}$$

so that the augmented Hamiltonian has the form

$$\begin{aligned} \widehat{H}(\widehat{\tilde{\boldsymbol{\eta}}}^*(t), \widehat{\tilde{\boldsymbol{x}}}^*(t), \boldsymbol{u}) &\triangleq \eta_0 \tilde{f}_0(\tilde{\boldsymbol{x}}, \boldsymbol{u}) + \left\langle \tilde{\boldsymbol{\eta}}, \tilde{\boldsymbol{f}}(\tilde{\boldsymbol{x}}, \boldsymbol{u}) \right\rangle \\ &= \eta_0[x_{n+1} + f_0(\boldsymbol{x}, \boldsymbol{u})] + \langle \eta, \boldsymbol{f}(\boldsymbol{x}, \boldsymbol{u}) \rangle + \eta_{n+1} \cdot 0 \\ &= \eta_0[x_{n+1} + f_0(\boldsymbol{x}, \boldsymbol{u})] + \langle \eta, \boldsymbol{f}(\boldsymbol{x}, \boldsymbol{u}) \rangle . \end{aligned} \tag{11.144}$$

To construct the corresponding adjoint system we construct the Jacobian

$$\tilde{\mathbb{J}}_{\tilde{\boldsymbol{x}}}(\tilde{\boldsymbol{x}}, \boldsymbol{u}) = \begin{bmatrix} \frac{\partial \tilde{f}_0(\tilde{\boldsymbol{x}},\boldsymbol{u})}{\partial x_0} & \frac{\partial \tilde{f}_0(\tilde{\boldsymbol{x}},\boldsymbol{u})}{\partial x_1} & \frac{\partial \tilde{f}_0(\tilde{\boldsymbol{x}},\boldsymbol{u})}{\partial x_2} & \cdots & \frac{\partial \tilde{f}_0(\tilde{\boldsymbol{x}},\boldsymbol{u})}{\partial x_n} & \frac{\partial \tilde{f}_0(\tilde{\boldsymbol{x}},\boldsymbol{u})}{\partial x_{n+1}} \\ \frac{\partial \tilde{f}_1(\tilde{\boldsymbol{x}},\boldsymbol{u})}{\partial x_0} & \frac{\partial \tilde{f}_1(\tilde{\boldsymbol{x}},\boldsymbol{u})}{\partial x_1} & \frac{\partial \tilde{f}_1(\tilde{\boldsymbol{x}},\boldsymbol{u})}{\partial x_2} & \cdots & \frac{\partial \tilde{f}_1(\tilde{\boldsymbol{x}},\boldsymbol{u})}{\partial x_n} & \frac{\partial \tilde{f}_1(\tilde{\boldsymbol{x}},\boldsymbol{u})}{\partial x_{n+1}} \\ \frac{\partial \tilde{f}_2(\tilde{\boldsymbol{x}},\boldsymbol{u})}{\partial x_0} & \frac{\partial \tilde{f}_2(\tilde{\boldsymbol{x}},\boldsymbol{u})}{\partial x_1} & \frac{\partial \tilde{f}_2(\tilde{\boldsymbol{x}},\boldsymbol{u})}{\partial x_2} & \cdots & \frac{\partial \tilde{f}_2(\tilde{\boldsymbol{x}},\boldsymbol{u})}{\partial x_n} & \frac{\partial \tilde{f}_2(\tilde{\boldsymbol{x}},\boldsymbol{u})}{\partial x_{n+1}} \\ \vdots & \vdots & \vdots & \ddots & \vdots & \vdots \\ \frac{\partial \tilde{f}_n(\tilde{\boldsymbol{x}},\boldsymbol{u})}{\partial x_0} & \frac{\partial \tilde{f}_n(\tilde{\boldsymbol{x}},\boldsymbol{u})}{\partial x_1} & \frac{\partial \tilde{f}_n(\tilde{\boldsymbol{x}},\boldsymbol{u})}{\partial x_2} & \cdots & \frac{\partial \tilde{f}_n(\tilde{\boldsymbol{x}},\boldsymbol{u})}{\partial x_n} & \frac{\partial \tilde{f}_n(\tilde{\boldsymbol{x}},\boldsymbol{u})}{\partial x_{n+1}} \\ \frac{\partial \tilde{f}_{n+1}(\tilde{\boldsymbol{x}},\boldsymbol{u})}{\partial x_0} & \frac{\partial \tilde{f}_{n+1}(\tilde{\boldsymbol{x}},\boldsymbol{u})}{\partial x_1} & \frac{\partial \tilde{f}_{n+1}(\tilde{\boldsymbol{x}},\boldsymbol{u})}{\partial x_2} & \cdots & \frac{\partial \tilde{f}_{n+1}(\tilde{\boldsymbol{x}},\boldsymbol{u})}{\partial x_n} & \frac{\partial \tilde{f}_{n+1}(\tilde{\boldsymbol{x}},\boldsymbol{u})}{\partial x_{n+1}} \end{bmatrix}$$

which becomes

$$\tilde{\mathbb{J}}_{\tilde{\boldsymbol{x}}}(\tilde{\boldsymbol{x}},\boldsymbol{u}) = \begin{bmatrix} 0 & \frac{\partial f_0(\tilde{\boldsymbol{x}},\boldsymbol{u})}{\partial x_1} & \frac{\partial f_0(\tilde{\boldsymbol{x}},\boldsymbol{u})}{\partial x_2} & \cdots & \frac{\partial f_0(\tilde{\boldsymbol{x}},\boldsymbol{u})}{\partial x_n} & 1 \\ 0 & \frac{\partial \tilde{f}_1(\tilde{\boldsymbol{x}},\boldsymbol{u})}{\partial x_1} & \frac{\partial \tilde{f}_1(\tilde{\boldsymbol{x}},\boldsymbol{u})}{\partial x_2} & \cdots & \frac{\partial \tilde{f}_1(\tilde{\boldsymbol{x}},\boldsymbol{u})}{\partial x_n} & \frac{\partial \tilde{f}_1(\tilde{\boldsymbol{x}},\boldsymbol{u})}{\partial x_{n+1}} \\ 0 & \frac{\partial \tilde{f}_2(\tilde{\boldsymbol{x}},\boldsymbol{u})}{\partial x_1} & \frac{\partial \tilde{f}_2(\tilde{\boldsymbol{x}},\boldsymbol{u})}{\partial x_2} & \cdots & \frac{\partial \tilde{f}_2(\tilde{\boldsymbol{x}},\boldsymbol{u})}{\partial x_n} & \frac{\partial \tilde{f}_2(\tilde{\boldsymbol{x}},\boldsymbol{u})}{\partial x_{n+1}} \\ \vdots & \vdots & \vdots & \ddots & \vdots & \vdots \\ 0 & \frac{\partial \tilde{f}_n(\tilde{\boldsymbol{x}},\boldsymbol{u})}{\partial x_1} & \frac{\partial \tilde{f}_n(\tilde{\boldsymbol{x}},\boldsymbol{u})}{\partial x_2} & \cdots & \frac{\partial \tilde{f}_n(\tilde{\boldsymbol{x}},\boldsymbol{u})}{\partial x_n} & \frac{\partial \tilde{f}_n(\tilde{\boldsymbol{x}},\boldsymbol{u})}{\partial x_{n+1}} \\ 0 & 0 & 0 & \cdots & 0 & 0 \end{bmatrix},$$

or equivalently,

$$\tilde{\mathbb{J}}_{\tilde{\boldsymbol{x}}}(\tilde{\boldsymbol{x}},\boldsymbol{u}) = \begin{bmatrix} 0 & [\nabla f_0(\boldsymbol{x},\boldsymbol{u})]^T & 1 \\ \mathbf{0} & \mathbb{J}_{\boldsymbol{x}} f(\boldsymbol{x},\boldsymbol{u}) & \mathbf{0} \\ 0 & \mathbf{0} & 0 \end{bmatrix},$$

where $\mathbb{J}_{\boldsymbol{x}} f(\boldsymbol{x},\boldsymbol{u})$ is the Jacobian of $f(\boldsymbol{x},\boldsymbol{u})$. Consequently, if $(\tilde{\boldsymbol{x}}^*(t),\boldsymbol{u}^*(t))$ is an optimal pair, then

$$\widehat{\tilde{A}}(t) = \begin{bmatrix} 0 & [\nabla f_0(\boldsymbol{x}^*(t),\boldsymbol{u}^*(t))]^T & 1 \\ \mathbf{0} & \mathbb{J}_{\boldsymbol{x}} f(\boldsymbol{x}^*(t),\boldsymbol{u}^*(t)) & \mathbf{0} \\ 0 & \mathbf{0} & 0 \end{bmatrix} = \begin{bmatrix} \hat{A}(t) & 1 \\ \mathbf{0} & 0 \end{bmatrix}$$

and the augmented co-state equation becomes

$$\frac{d}{dt}\widehat{\tilde{\boldsymbol{\eta}}}(t) = -[\widehat{\tilde{A}}(t)]^T \frac{d}{dt}\widehat{\tilde{\boldsymbol{\eta}}}(t),$$

which has the form

$$\frac{d}{dt}\begin{bmatrix} \eta_0(t) \\ \boldsymbol{\eta}(t) \\ \eta_{n+1}(t) \end{bmatrix}$$

$$= \begin{bmatrix} 0 & \mathbf{0} & 0 \\ -\boldsymbol{\nabla} f_0(\boldsymbol{x}^*(t),\boldsymbol{u}^*(t)) & -[\mathbb{J}_{\boldsymbol{x}} f(\boldsymbol{x}^*(t),\boldsymbol{u}^*(t))]^T & \mathbf{0} \\ -1 & \mathbf{0} & 0 \end{bmatrix} \begin{bmatrix} \eta_0(t) \\ \boldsymbol{\eta}(t) \\ \eta_{n+1}(t) \end{bmatrix}. \tag{11.145}$$

Observe that the solutions of (11.145) satisfy $\eta_0(t) \equiv \eta_0$ and

$$\eta_{n+1}(t) = -\eta_0 t + k \tag{11.146}$$

for some constant k. In addition, $\boldsymbol{\eta}(t)$ satisfies the equation

$$\dot{\boldsymbol{\eta}}(t) = -\eta_0 \boldsymbol{\nabla} f_0(\boldsymbol{x}^*(t), \boldsymbol{u}^*(t)) - [\mathbb{J}_{\boldsymbol{x}} f(\boldsymbol{x}^*(t), \boldsymbol{u}^*(t))]^T \boldsymbol{\eta}(t) \quad (11.147)$$

which is the same as the adjoint equation for the Simplest Problem in Optimal Control. With the observations above it is now straightforward to establish the following Maximum Principle for a Problem of Bolza.

Theorem 11.3 ***(Maximum Principle for the Problem of Bolza)*** *Assume that* $f : \mathbb{R}^n \times \mathbb{R}^m \to \mathbb{R}^n$, $f_0 : \mathbb{R}^n \times \mathbb{R}^m \to \mathbb{R}$, $G : \mathbb{R} \to \mathbb{R}$, $\mathrm{X}_0 = \{\boldsymbol{x}_0\} \subseteq \mathbb{R}^n$, $t_0 \in \mathbb{R}$, t_1 *is fixed,* $\mathrm{X}_1 = \mathbb{R}^n$ *and* $\Omega \subseteq \mathbb{R}^m$ *are given as above. Consider the control system*

$$(\mathcal{S}) \qquad \dot{\boldsymbol{x}}(t) = f(\boldsymbol{x}(t), \boldsymbol{u}(t)), \quad t_0 < t \leq t_1, \qquad (11.148)$$

with piecewise continuous controllers $\boldsymbol{u}(\cdot) \in PWC(t_0, t_1; \mathbb{R}^m)$ *satisfying* $\boldsymbol{u}(t) \in \Omega \subseteq \mathbb{R}^m$ *e.f. If*

$$\boldsymbol{u}^*(\cdot) \in \Theta = \left\{ \begin{array}{c} \boldsymbol{u}(\cdot) \in PWC(t_0, t_1; \mathbb{R}^m) : \boldsymbol{u}(t) \in \Omega \text{ e.f.} \\ \text{and } \boldsymbol{u}(\cdot) \text{ steers } \mathrm{X}_0 \text{ to } \mathrm{X}_1 \text{ at time } t_1. \end{array} \right\} \qquad (11.149)$$

minimizes

$$J(\boldsymbol{u}(\cdot)) = G(\boldsymbol{x}(t_1)) + \int_{t_0}^{t_1} f_0(\boldsymbol{x}(s), \boldsymbol{u}(s)) ds, \qquad (11.150)$$

on the set of admissible controls Θ *with optimal response* $\boldsymbol{x}^*(\cdot)$ *satisfying* $\boldsymbol{x}^*(t_0) = \boldsymbol{x}_0^*$, *then there exists a non-trivial solution*

$$\hat{\boldsymbol{\eta}}^*(t) = [\eta_0^*(t)\ \eta_1^*(t)\ \eta_2^*(t)\ \ldots\ \eta_n^*(t)]^T = \begin{bmatrix} \eta_0^*(t) \\ \boldsymbol{\eta}^*(t) \end{bmatrix} \qquad (11.151)$$

to the augmented adjoint equation

$$\frac{d}{dt}\hat{\boldsymbol{\eta}}(t) = -[\hat{A}(t)]^T \hat{\boldsymbol{\eta}}(t), \qquad (11.152)$$

such that

$$\hat{H}(\hat{\boldsymbol{\eta}}^*(t), \hat{\boldsymbol{x}}^*(t), \boldsymbol{u}^*(t)) = \hat{M}(\hat{\boldsymbol{\eta}}^*(t), \hat{\boldsymbol{x}}^*(t)) = \max_{\boldsymbol{u} \in \Omega} \hat{H}(\hat{\boldsymbol{\eta}}^*(t), \hat{\boldsymbol{x}}^*(t), u). \qquad (11.153)$$

Moreover, there are constants $\eta_0^* \leq 0$ *and* c *such that*

$$\eta_0^*(t) \equiv \eta_0^* \leq 0 \tag{11.154}$$

and for all $t \in [t_0, t_1]$,

$$\hat{\boldsymbol{M}}(\hat{\boldsymbol{\eta}}^*(t), \hat{\boldsymbol{x}}^*(t)) = \max_{\boldsymbol{u}\in\Omega} \hat{H}(\hat{\boldsymbol{\eta}}^*(t), \hat{\boldsymbol{x}}^*(t), \boldsymbol{u}) \equiv c. \tag{11.155}$$

Also, the transversality condition at $t = t_1$ *is given by*

$$\boldsymbol{\eta}^*(t_1) = \eta_0^* \nabla G(\boldsymbol{x}^*(t_1)). \tag{11.156}$$

Proof. To prove this result, we need only to establish the final transversality condition (11.156) above. Since the tangent space $\tilde{\mathbb{T}}_0 = \{\mathbf{0}\} \times \mathbb{R}^1$, it follows from the Maximum Principle that $\tilde{\boldsymbol{\eta}}^*(t) = [\boldsymbol{\eta}^*(t_0) \;\; \eta_{n+1}^*(t_0)]^T \in \mathbb{R}^{n+1}$ is orthogonal to $\tilde{\mathbb{T}}_0 = \{\mathbf{0}\} \times \mathbb{R}^1$ and this is true if and only if $\eta_{n+1}^*(t_0) = 0$. From (11.146) $\eta_{n+1}^*(t) = -\eta_0^* t + k$ and $\eta_{n+1}^*(t_0) = 0$ implies that

$$\eta_{n+1}^*(t) = -\eta_0^*(t - t_0).$$

Also, the condition (11.143) implies that

$$\begin{aligned}
\tilde{\boldsymbol{\eta}}^*(t_1) = \begin{bmatrix} \boldsymbol{\eta}^*(t_1) \\ \eta_{n+1}^*(t_1) \end{bmatrix} &= \begin{bmatrix} \boldsymbol{\eta}^*(t_1) \\ -\eta_0^*(t_1 - t_0) \end{bmatrix} \\
&= \alpha \nabla \tilde{G}(\tilde{\boldsymbol{x}}(t_1)) = \begin{bmatrix} -\alpha \nabla G(\boldsymbol{x}^*(t_1))/(t_1 - t_0) \\ \alpha \end{bmatrix}
\end{aligned}$$

and hence

$$-\eta_0^*(t_1 - t_0) = \alpha.$$

Therefore,

$$\begin{aligned}
\boldsymbol{\eta}^*(t_1) &= -\alpha \nabla G(\boldsymbol{x}^*(t_1))/(t_1 - t_0) = \eta_0^*(t_1 - t_0) \nabla G(\boldsymbol{x}^*(t_1))/(t_1 - t_0) \\
&= \eta_0^* \nabla G(\boldsymbol{x}^*(t_1))
\end{aligned}$$

and this completes the proof. □

11.7 The Maximum Principle for Nonautonomous Systems

When the system is nonautonomous the Maximum Principle must be modified. However, we show that the general nonautonomous problem can be reduced to the SPOC and the Maximum Principle for the **SPOC** can be used to obtain the corresponding Maximum Principle. Thus, we assume that $f : \mathbb{R} \times \mathbb{R}^n \times \mathbb{R}^m \to \mathbb{R}^n$, $f_0 : \mathbb{R} \times \mathbb{R}^n \times \mathbb{R}^m \to \mathbb{R}$, $X_0 \subseteq \mathbb{R}^n$, $t_0 \in \mathbb{R}$, $X_1 \subseteq \mathbb{R}^n$ and $\Omega \subseteq \mathbb{R}^m$ are given with $X_0 \subseteq \mathbb{R}^n$, $X_1 \subseteq \mathbb{R}^n$ and $\Omega \subseteq \mathbb{R}^m$ nonempty. Here we assume that $t_1 \in \mathbb{R}$ is free as in the SPOC. The control system is given by

$$(\mathcal{S}) \qquad \dot{\boldsymbol{x}}(t) = f(t, \boldsymbol{x}(t), \boldsymbol{u}(t)), \quad t_0 < t \leq t_1 \tag{11.157}$$

and the cost functional is defined by

$$J(\boldsymbol{u}(\cdot)) = \int_{t_0}^{t_1} f_0(s, \boldsymbol{x}(s), \boldsymbol{u}(s))ds. \tag{11.158}$$

The admissible controllers are given by

$$\Theta = \left\{ \begin{array}{c} \boldsymbol{u}(\cdot) \in PWC(t_0, +\infty; \mathbb{R}^m) : \boldsymbol{u}(t) \in \Omega \ e.f. \text{ and} \\ \boldsymbol{u}(\cdot) \text{ steers } X_0 \text{ to } X_1 \text{ at a finite time } t_1. \end{array} \right\} \tag{11.159}$$

and the problem is to find $\boldsymbol{u}^*(\cdot) \in \Theta$ such that

$$\begin{aligned} J(\boldsymbol{u}^*(\cdot)) &= \int_{t_0}^{t_1} f_0(s, \boldsymbol{x}^*(s), \boldsymbol{u}^*(s))ds \\ \leq J(\boldsymbol{u}(\cdot)) &= \int_{t_0}^{t_1} f_0(s, \boldsymbol{x}(s), \boldsymbol{u}(s))ds \end{aligned}$$

for all $\boldsymbol{u}(\cdot) \in \Theta$.

In order to state the Maximum Principle for this nonautonomous case we again have to introduce some notation. Given a vector $\boldsymbol{z} = [z_1, z_2, ..., z_n]^T \in \mathbb{R}^n$, define the $n+2$ dimensional *augmented vector* $\tilde{\boldsymbol{z}} \in \mathbb{R}^{n+2}$ by

$$\tilde{\boldsymbol{z}} = [z_0 \ z_1 \ \ z_2 \ \ \ldots \ \ z_n \ \ z_{n+1}]^T = \begin{bmatrix} z_0 \\ \boldsymbol{z} \\ z_{n+1} \end{bmatrix} = \begin{bmatrix} \hat{\boldsymbol{z}} \\ z_{n+1} \end{bmatrix} \in \mathbb{R}^{n+2},$$

where as before $\hat{\boldsymbol{z}} = \begin{bmatrix} z_0 & z_1 & z_2 & \dots & z_n \end{bmatrix}^T \in \mathbb{R}^{n+1}$. Also, given the functions

$$f : \mathbb{R}^1 \times \mathbb{R}^n \times \mathbb{R}^m \to \mathbb{R}^n, \ f_0 : \mathbb{R}^1 \times \mathbb{R}^n \times \mathbb{R}^m \to \mathbb{R},$$

where $f : \mathbb{R}^1 \times \mathbb{R}^n \times \mathbb{R}^m \to \mathbb{R}^n$ is given by

$$f(t, \boldsymbol{x}, \boldsymbol{u}) = \begin{bmatrix} f_1(t, \boldsymbol{x}, \boldsymbol{u}) \\ f_2(t, \boldsymbol{x}, \boldsymbol{u}) \\ f_3(t, \boldsymbol{x}, \boldsymbol{u}) \\ \vdots \\ f_n(t, \boldsymbol{x}, \boldsymbol{u}) \end{bmatrix} = \begin{bmatrix} f_1(t, x_1, x_2, \dots, x_n, u_1, u_2, \dots, u_m) \\ f_2(t, x_1, x_2, \dots, x_n, u_1, u_2, \dots, u_m) \\ f_3(t, x_1, x_2, \dots, x_n, u_1, u_2, \dots, u_m) \\ \vdots \\ f_n(t, x_1, x_2, \dots, x_n, u_1, u_2, \dots, u_m) \end{bmatrix},$$

it follows that all the partial derivatives

$$\frac{\partial f_i(t, x_1, x_2, \dots, x_n, u_1, u_2, \dots, u_m)}{\partial x_j},$$

$$\frac{\partial f_i(t, x_1, x_2, \dots, x_n, u_1, u_2, \dots, u_m)}{\partial u_j}$$

exist.

Remark 11.4 *For the moment we make a rather strong assumption that the partial derivatives*

$$\frac{\partial f_i(t, x_1, x_2, \dots, x_n, u_1, u_2, \dots, u_m)}{\partial t}$$

also exist. This is made to simplify the derivation of the Maximum Principle for the nonautonomous problem and is not essential.

Define the (time) augmented vector field

$$\tilde{f} : \mathbb{R}^{n+2} \times \mathbb{R}^m \to \mathbb{R}^{n+2}$$

by

$$\tilde{f}(\tilde{\boldsymbol{x}}, \boldsymbol{u}) = \begin{bmatrix} \tilde{f}_0(\tilde{\boldsymbol{x}}, \boldsymbol{u}) \\ \tilde{f}_1(\tilde{\boldsymbol{x}}, \boldsymbol{u}) \\ \tilde{f}_2(\tilde{\boldsymbol{x}}, \boldsymbol{u}) \\ \vdots \\ \tilde{f}_n(\tilde{\boldsymbol{x}}, \boldsymbol{u}) \\ \tilde{f}_{n+1}(\hat{\boldsymbol{x}}, \boldsymbol{u}) \end{bmatrix} \triangleq \begin{bmatrix} f_0(x_{n+1}, x_1, x_2, \dots, x_n, u_1, u_2, \dots, u_m) \\ f_1(x_{n+1}, x_1, x_2, \dots, x_n, u_1, u_2, \dots, u_m) \\ f_2(x_{n+1}, x_1, x_2, \dots, x_n, u_1, u_2, \dots, u_m) \\ \vdots \\ f_n(x_{n+1}, x_1, x_2, \dots, x_n, u_1, u_2, \dots, u_m) \\ 1 \end{bmatrix}. \quad (11.160)$$

The **time augmented control system** is defined by

$$\frac{d}{dt}\tilde{\boldsymbol{x}}(t) = \tilde{f}(\tilde{\boldsymbol{x}}(t), \boldsymbol{u}(t)), \tag{11.161}$$

or equivalently by the system

$$\frac{d}{dt}x_0(t) = f_0(x_{n+1}(t), \boldsymbol{x}(t), \boldsymbol{u}(t)), \tag{11.162}$$

$$\frac{d}{dt}\boldsymbol{x}(t) = f(x_{n+1}(t), \boldsymbol{x}(t), \boldsymbol{u}(t)), \tag{11.163}$$

$$\frac{d}{dt}x_{n+1}(t) = f_{n+1}(x_{n+1}(t), \boldsymbol{x}(t), \boldsymbol{u}(t)) = 1. \tag{11.164}$$

Observe that

$$x_0(t) = \int_{t_0}^{t} f_0(x_{n+1}(s), \boldsymbol{x}(s), \boldsymbol{u}(s))ds,$$

and $x_{n+1}(t_0) = t_0$ so that the initial state for (11.161) is

$$\tilde{\boldsymbol{x}}(t_0) = [\ 0 \quad \boldsymbol{x}_0 \quad t_0\ ]^T. \tag{11.165}$$

In particular, $x_0(t)$ represents the cost of transferring the state from an initial $\boldsymbol{x}(t_0) = \boldsymbol{x}_0 \in \mathrm{X}_0$ by the control $\boldsymbol{u}(\cdot)$ and $x_{n+1}(t) = t$ is the time variable.

Assume that $(\boldsymbol{x}^*(\cdot), \boldsymbol{u}^*(\cdot))$ is an optimal pair. In particular, $\boldsymbol{u}^*(\cdot) \in \Theta$ steers $\boldsymbol{x}_0^* \in \mathrm{X}_0$ to $\boldsymbol{x}_1^* \in \mathrm{X}_1$ in time t_1^* and $\boldsymbol{x}^*(\cdot)$ is the corresponding optimal trajectory that satisfies the initial value problem (10.1) - (10.2) with $\boldsymbol{x}_0 = \boldsymbol{x}_0^* \in \mathrm{X}_0$. Define the $(n+2) \times (n+2)$ matrix $\tilde{A}(t)$ by

$$\begin{aligned}\tilde{A}(t) &= \left[\frac{\partial \tilde{f}_i(x_0, \boldsymbol{x}, x_{n+1}, \boldsymbol{u})}{\partial x_j}\right]_{|(\hat{\boldsymbol{x}}^*(t), \boldsymbol{u}^*(t))} \\ &= \left[\frac{\partial \tilde{f}_i(\tilde{\boldsymbol{x}}^*(t), \boldsymbol{u}^*(t))}{\partial x_j}\right], \quad i, j = 0, 1, 2, ..., n, n+1.\end{aligned} \tag{11.166}$$

Thus,

$$\tilde{A}(t) = \left[\frac{\partial \tilde{f}_i(\hat{\boldsymbol{x}}^*(t), \boldsymbol{u}^*(t))}{\partial x_j}\right]$$

$$= \begin{bmatrix} \frac{\partial \tilde{f}_0(\hat{\boldsymbol{x}}^*(t),\boldsymbol{u}^*(t))}{\partial x_0} & \frac{\partial \tilde{f}_0(\hat{\boldsymbol{x}}^*(t),\boldsymbol{u}^*(t))}{\partial x_1} & \cdots & \frac{\partial \tilde{f}_0(\hat{\boldsymbol{x}}^*(t),\boldsymbol{u}^*(t))}{\partial x_n} & \frac{\partial \tilde{f}_0(\hat{\boldsymbol{x}}^*(t),\boldsymbol{u}^*(t))}{\partial x_{n+1}} \\ \frac{\partial \tilde{f}_1(\hat{\boldsymbol{x}}^*(t),\boldsymbol{u}^*(t))}{\partial x_0} & \frac{\partial \tilde{f}_1(\hat{\boldsymbol{x}}^*(t),\boldsymbol{u}^*(t))}{\partial x_1} & \cdots & \frac{\partial \tilde{f}_1(\hat{\boldsymbol{x}}^*(t),\boldsymbol{u}^*(t))}{\partial x_n} & \frac{\partial \tilde{f}_1(\hat{\boldsymbol{x}}^*(t),\boldsymbol{u}^*(t))}{\partial x_{n+1}} \\ \frac{\partial \tilde{f}_2(\hat{\boldsymbol{x}}^*(t),\boldsymbol{u}^*(t))}{\partial x_0} & \frac{\partial \tilde{f}_2(\hat{\boldsymbol{x}}^*(t),\boldsymbol{u}^*(t))}{\partial x_1} & \cdots & \frac{\partial \tilde{f}_2(\hat{\boldsymbol{x}}^*(t),\boldsymbol{u}^*(t))}{\partial x_n} & \frac{\partial \tilde{f}_2(\hat{\boldsymbol{x}}^*(t),\boldsymbol{u}^*(t))}{\partial x_{n+1}} \\ \vdots & \vdots & \ddots & \vdots & \vdots \\ \frac{\partial \tilde{f}_n(\hat{\boldsymbol{x}}^*(t),\boldsymbol{u}^*(t))}{\partial x_0} & \frac{\partial \tilde{f}_n(\hat{\boldsymbol{x}}^*(t),\boldsymbol{u}^*(t))}{\partial x_1} & \cdots & \frac{\partial \tilde{f}_n(\hat{\boldsymbol{x}}^*(t),\boldsymbol{u}^*(t))}{\partial x_n} & \frac{\partial \tilde{f}_n(\hat{\boldsymbol{x}}^*(t),\boldsymbol{u}^*(t))}{\partial x_{n+1}} \\ \frac{\partial \tilde{f}_{n+1}(\hat{\boldsymbol{x}}^*(t),\boldsymbol{u}^*(t))}{\partial x_0} & \frac{\partial \tilde{f}_{n+1}(\hat{\boldsymbol{x}}^*(t),\boldsymbol{u}^*(t))}{\partial x_1} & \cdots & \frac{\partial \tilde{f}_{n+1}(\hat{\boldsymbol{x}}^*(t),\boldsymbol{u}^*(t))}{\partial x_n} & \frac{\partial \tilde{f}_{n+1}(\hat{\boldsymbol{x}}^*(t),\boldsymbol{u}^*(t))}{\partial x_{n+1}} \end{bmatrix}.$$

Since $\tilde{f}_i(\tilde{\boldsymbol{x}}, \boldsymbol{u}) = \tilde{f}_i(\hat{\boldsymbol{x}}, x_{n+1}, \boldsymbol{u}) = f_i(x_{n+1}, \boldsymbol{x}, \boldsymbol{u}) = f_i(t, x_1, x_2, ..., x_n, u_1, u_{2,}..., u_m)$ does not depend on x_0, it follows that

$$\frac{\partial f_i(\tilde{\boldsymbol{x}}^*(t), \boldsymbol{u}^*(t))}{\partial x_0} = 0$$

for all $i = 0, 1, 2, ..., n, n+1$ and since $\tilde{f}_{n+1}(\tilde{\boldsymbol{x}}, \boldsymbol{u}) = 1$ we have that

$$\frac{\partial \tilde{f}_{n+1}(\tilde{\boldsymbol{x}}^*(t), \boldsymbol{u}^*(t))}{\partial x_j} = 0$$

for all $i = 0, 1, 2, ..., n, n+1$. Consequently,

$$\tilde{A}(t) = \begin{bmatrix} 0 & \frac{\partial \tilde{f}_0(\hat{\boldsymbol{x}}^*(t),\boldsymbol{u}^*(t))}{\partial x_1} & \cdots & \frac{\partial \tilde{f}_0(\hat{\boldsymbol{x}}^*(t),\boldsymbol{u}^*(t))}{\partial x_n} & \frac{\partial \tilde{f}_0(\hat{\boldsymbol{x}}^*(t),\boldsymbol{u}^*(t))}{\partial x_{n+1}} \\ 0 & \frac{\partial \tilde{f}_1(\hat{\boldsymbol{x}}^*(t),\boldsymbol{u}^*(t))}{\partial x_1} & \cdots & \frac{\partial \tilde{f}_1(\hat{\boldsymbol{x}}^*(t),\boldsymbol{u}^*(t))}{\partial x_n} & \frac{\partial \tilde{f}_1(\hat{\boldsymbol{x}}^*(t),\boldsymbol{u}^*(t))}{\partial x_{n+1}} \\ 0 & \frac{\partial \tilde{f}_2(\hat{\boldsymbol{x}}^*(t),\boldsymbol{u}^*(t))}{\partial x_1} & \cdots & \frac{\partial \tilde{f}_2(\hat{\boldsymbol{x}}^*(t),\boldsymbol{u}^*(t))}{\partial x_n} & \frac{\partial \tilde{f}_2(\hat{\boldsymbol{x}}^*(t),\boldsymbol{u}^*(t))}{\partial x_{n+1}} \\ \vdots & \vdots & \ddots & \vdots & \vdots \\ 0 & \frac{\partial \tilde{f}_n(\hat{\boldsymbol{x}}^*(t),\boldsymbol{u}^*(t))}{\partial x_1} & \cdots & \frac{\partial \tilde{f}_n(\hat{\boldsymbol{x}}^*(t),\boldsymbol{u}^*(t))}{\partial x_n} & \frac{\partial \tilde{f}_n(\hat{\boldsymbol{x}}^*(t),\boldsymbol{u}^*(t))}{\partial x_{n+1}} \\ 0 & 0 & \cdots & 0 & 0 \end{bmatrix} \tag{11.167}$$

and using the definition of $\tilde{f}_i(\tilde{\boldsymbol{x}}, \boldsymbol{u}) = f_i(x_{n+1}, \boldsymbol{x}, \boldsymbol{u}) = f_i(t, \boldsymbol{x}^*(t), \boldsymbol{u}^*(t))$, it follows that

$$\frac{\partial \tilde{f}_i(\tilde{\boldsymbol{x}}^*(t), \boldsymbol{u}^*(t))}{\partial x_{n+1}} = \frac{\partial f_i(t, \boldsymbol{x}^*(t), \boldsymbol{u}^*(t))}{\partial t}.$$

Therefore, the matrix $\tilde{A}(t)$ is given by

$$\tilde{A}(t) = \begin{bmatrix} 0 & \frac{\partial f_0(t,\boldsymbol{x}^*(t),\boldsymbol{u}^*(t))}{\partial x_1} & \cdots & \frac{\partial f_0(t,\boldsymbol{x}^*(t),\boldsymbol{u}^*(t))}{\partial x_n} & \frac{\partial f_0(t,\boldsymbol{x}^*(t),\boldsymbol{u}^*(t))}{\partial t} \\ 0 & \frac{\partial f_1(t,\boldsymbol{x}^*(t),\boldsymbol{u}^*(t))}{\partial x_1} & \cdots & \frac{\partial f_1(\hat{\boldsymbol{x}}^*(t),\boldsymbol{u}^*(t))}{\partial x_n} & \frac{\partial f_1(t,\boldsymbol{x}^*(t),\boldsymbol{u}^*(t))}{\partial t} \\ 0 & \frac{\partial f_2(t,\boldsymbol{x}^*(t),\boldsymbol{u}^*(t))}{\partial x_1} & \cdots & \frac{\partial f_2(\hat{\boldsymbol{x}}^*(t),\boldsymbol{u}^*(t))}{\partial x_n} & \frac{\partial f_2(t,\boldsymbol{x}^*(t),\boldsymbol{u}^*(t))}{\partial t} \\ \vdots & \vdots & \ddots & \vdots & \vdots \\ 0 & \frac{\partial f_n(t,\boldsymbol{x}^*(t),\boldsymbol{u}^*(t))}{\partial x_1} & \cdots & \frac{\partial f_n(t,\boldsymbol{x}^*(t),\boldsymbol{u}^*(t))}{\partial x_n} & \frac{\partial f_n(t,\boldsymbol{x}^*(t),\boldsymbol{u}^*(t))}{\partial t} \\ 0 & 0 & \cdots & 0 & 0 \end{bmatrix},$$

which has the block form

$$\tilde{A}(t) = \begin{bmatrix} 0 & [\nabla_{\boldsymbol{x}} f_0(t,\boldsymbol{x}^*(t),\boldsymbol{u}^*(t))]^T & \frac{\partial f_0(t,\boldsymbol{x}^*(t),\boldsymbol{u}^*(t))}{\partial t} \\ \mathbf{0} & \left[\frac{\partial f_i(t,\boldsymbol{x}^*(t),\boldsymbol{u}^*(t))}{\partial x_j}\right]_{n\times n} & \left[\frac{\partial f_i(t,\boldsymbol{x}^*(t),\boldsymbol{u}^*(t))}{\partial t}\right]_{n\times 1} \\ 0 & \mathbf{0} & 0 \end{bmatrix}.$$

Observe that $\tilde{A}(t)$ can be written in block form

$$\tilde{A}(t) = \begin{bmatrix} \hat{A}(t) & \left[\frac{\partial \hat{f}(t,\boldsymbol{x}^*(t),\boldsymbol{u}^*(t))}{\partial t}\right]_{(n+1)\times 1} \\ \mathbf{0} & 0 \end{bmatrix}$$

where

$$\hat{A}(t) = \begin{bmatrix} 0 & [\nabla_{\boldsymbol{x}} f_0(t,\boldsymbol{x}^*(t),\boldsymbol{u}^*(t))]^T \\ \mathbf{0} & \left[\frac{\partial f_i(t,\boldsymbol{x}^*(t),\boldsymbol{u}^*(t))}{\partial x_j}\right]_{n\times n} \end{bmatrix} \tag{11.168}$$

is time dependent because of the explicit dependence of time in the problem as well as the inclusion of the optimal state and control.

The negative transpose of $\tilde{A}(t)$ is given by

$$-[\tilde{A}(t)]^T = \begin{bmatrix} 0 & [0]_{1\times n} & 0 \\ -[\nabla_{\boldsymbol{x}} f_0(t,\boldsymbol{x}^*(t),\boldsymbol{u}^*(t))]_{n\times 1} & -\left[\frac{\partial f_i(t,\boldsymbol{x}^*(t),\boldsymbol{u}^*(t))}{\partial x_j}\right]^T_{n\times n} & \mathbf{0} \\ -\frac{\partial f_0(t,\boldsymbol{x}^*(t),\boldsymbol{u}^*(t))}{\partial t} & -\left[\frac{\partial f(t,\boldsymbol{x}^*(t),\boldsymbol{u}^*(t))}{\partial t}\right]^T_{n\times 1} & 0 \end{bmatrix}$$

so that the augmented adjoint system

$$\frac{d}{dt}\begin{bmatrix} \eta_0(t) \\ \boldsymbol{\eta}(t) \\ \eta_{n+1}(t) \end{bmatrix} = -[\tilde{A}(t)]^T \begin{bmatrix} \eta_0(t) \\ \boldsymbol{\eta}(t) \\ \eta_{n+1}(t) \end{bmatrix}$$

is equivalent to

$$\frac{d}{dt}\eta_0(t) = 0, \tag{11.169}$$

$$\frac{d}{dt}\boldsymbol{\eta}(t) = -\eta_0(t)\left[\nabla_{\boldsymbol{x}} f_0(t, \boldsymbol{x}^*(t), \boldsymbol{u}^*(t))\right] - [\mathbb{J}_{\boldsymbol{x}} f(t, \boldsymbol{x}^*(t), \boldsymbol{u}^*(t)]^T \boldsymbol{\eta}(t), \tag{11.170}$$

$$\frac{d}{dt}\eta_{n+1}(t) = -\eta_0(t)\frac{\partial f_0(t, \boldsymbol{x}^*(t), \boldsymbol{u}^*(t))}{\partial t} - \sum_{i=1}^{n} \eta_i(t)\frac{\partial f_i(t, \boldsymbol{x}^*(t), \boldsymbol{u}^*(t))}{\partial t}. \tag{11.171}$$

Note that the first two equations have the form

$$\frac{d}{dt}\begin{bmatrix} \eta_0(t) \\ \boldsymbol{\eta}(t) \end{bmatrix} = -[\hat{A}(t)]^T \begin{bmatrix} \eta_0(t) \\ \boldsymbol{\eta}(t) \end{bmatrix},$$

where the matrix $\hat{A}(t)$ is given by (11.168).

The augmented Hamiltonian function is given by

$$\begin{aligned}\tilde{H}(\tilde{\boldsymbol{\eta}}, \tilde{\boldsymbol{x}}, \boldsymbol{u}) &= \tilde{H}(\eta_0, \eta_1, \eta_2, ..., \eta_n, \eta_{n+1}, x_0, x_1, x_2, ..., \\ &\qquad x_n, x_{n+1}, u_1, u_{2,}..., u_m) \qquad (11.172)\\ &= \eta_0 f_0(x_{n+1}, \boldsymbol{x}, \boldsymbol{u}) + \langle \eta, f(x_{n+1}, \boldsymbol{x}, \boldsymbol{u})\rangle + \eta_{n+1} \\ &= \eta_0 f_0(x_{n+1}, \boldsymbol{x}, \boldsymbol{u}) + \sum_{i=1}^{n} \eta_i f_i(x_{n+1}, \boldsymbol{x}, \boldsymbol{u}) + \eta_{n+1}.\end{aligned}$$

Since $\tilde{\boldsymbol{x}} = [\hat{\boldsymbol{x}} \ \ x_{n+1}]^T$ and $\tilde{\boldsymbol{\eta}} = [\hat{\boldsymbol{\eta}} \ \ \eta_{n+1}]^T$, we can write

$$\tilde{H}(\tilde{\boldsymbol{\eta}}, \tilde{\boldsymbol{x}}, \boldsymbol{u}) = \hat{H}(t, \hat{\boldsymbol{\eta}}, \hat{\boldsymbol{x}}, \boldsymbol{u}) + \eta_{n+1},$$

where

$$\hat{H}(t, \hat{\boldsymbol{\eta}}, \hat{\boldsymbol{x}}, \boldsymbol{u})$$

is called the *time dependent augmented Hamiltonian.* Note also that $\tilde{M} : \mathbb{R}^{n+2} \times \mathbb{R}^{n+2} \to \mathbb{R}$ defined by

$$\tilde{M}(\tilde{\boldsymbol{\eta}}, \tilde{\boldsymbol{x}}) = \max_{\boldsymbol{u}\in\Omega} \tilde{H}(\tilde{\boldsymbol{\eta}}, \tilde{\boldsymbol{x}}, u) \tag{11.173}$$

has the form

$$\tilde{M}(\tilde{\boldsymbol{\eta}}, \tilde{\boldsymbol{x}}) = \max_{\boldsymbol{u}\in\Omega} \tilde{H}(\tilde{\boldsymbol{\eta}}, \tilde{\boldsymbol{x}}, \boldsymbol{u}) = \max_{\boldsymbol{u}\in\Omega} \hat{H}(t, \hat{\boldsymbol{\eta}}, \hat{\boldsymbol{x}}, \boldsymbol{u}) + \eta_{n+1}$$
$$= \hat{M}(t, \hat{\boldsymbol{\eta}}, \hat{\boldsymbol{x}}) + \eta_{n+1}. \tag{11.174}$$

so that maximizing $\tilde{H}(\tilde{\boldsymbol{\eta}}, \tilde{\boldsymbol{x}}, \boldsymbol{u})$ over Ω is equivalent to maximizing $\hat{H}(t, \hat{\boldsymbol{\eta}}, \hat{\boldsymbol{x}}, \boldsymbol{u})$ over Ω.

We now state the (Pontryagin) Maximum Principle for the nonautonomous optimal control problem and use the basic Maximum Principle to prove this result.

Theorem 11.4 ***(Maximum Principle for Nonautonomous Systems)*** *Assume that* $f : \mathbb{R}^1 \times \mathbb{R}^n \times \mathbb{R}^m \to \mathbb{R}^n$, $f_0 : \mathbb{R}^1 \times \mathbb{R}^n \times \mathbb{R}^m \to \mathbb{R}$, $\mathrm{X}_0 \subseteq \mathbb{R}^n$, $t_0 \in \mathbb{R}$, $\mathrm{X}_1 \subseteq \mathbb{R}^n$ *and* $\Omega \subseteq \mathbb{R}^m$ *are given as above and consider the control system*

$$(\mathcal{S}) \qquad \dot{\boldsymbol{x}}(t) = f(t, \boldsymbol{x}(t), \boldsymbol{u}(t)), \quad t_0 < t \leq t_1, \tag{11.175}$$

with piecewise continuous controllers $\boldsymbol{u}(\cdot) \in PWC(t_0, +\infty; \mathbb{R}^m)$ *satisfying* $\boldsymbol{u}(t) \in \Omega \subseteq \mathbb{R}^m$ *e.f. If*

$$\boldsymbol{u}^*(\cdot) \in \Theta = \left\{ \begin{array}{c} \boldsymbol{u}(\cdot) \in PWC(t_0, t_1; \mathbb{R}^m) : \boldsymbol{u}(t) \in \Omega \ e.f. \text{ and} \\ \boldsymbol{u}(\cdot) \text{ steers } \mathrm{X}_0 \text{ to } \mathrm{X}_1 \text{ at a finite time } t_1. \end{array} \right\} \tag{11.176}$$

minimizes

$$J(\boldsymbol{u}(\cdot)) = \int_{t_0}^{t_1} f_0(s, \boldsymbol{x}(s), \boldsymbol{u}(s)) ds, \tag{11.177}$$

on the set of admissible controls Θ *with optimal response* $\boldsymbol{x}^*(\cdot)$ *satisfying* $\boldsymbol{x}^*(t_0) = \boldsymbol{x}_0^* \in \mathrm{X}_0$ *and* $\boldsymbol{x}^*(t_1^*) = \boldsymbol{x}_1^* \in \mathrm{X}_1$*, then there exists a non-trivial solution*

$$\tilde{\boldsymbol{\eta}}^*(t) = \begin{bmatrix} \eta_0^*(t) \\ \boldsymbol{\eta}^*(t) \\ \eta_{n+1}^*(t) \end{bmatrix} \tag{11.178}$$

to the augmented adjoint equation

$$\frac{d}{dt} \begin{bmatrix} \eta_0(t) \\ \boldsymbol{\eta}(t) \\ \eta_{n+1}(t) \end{bmatrix} = -[\tilde{A}(t)]^T \begin{bmatrix} \eta_0(t) \\ \boldsymbol{\eta}(t) \\ \eta_{n+1}(t) \end{bmatrix}, \tag{11.179}$$

such that

$$\tilde{H}(\tilde{\boldsymbol{\eta}}^*(t), \tilde{\boldsymbol{x}}^*(t), \boldsymbol{u}^*(t)) = \tilde{M}(\tilde{\boldsymbol{\eta}}^*(t), \tilde{\boldsymbol{x}}^*(t)) = \max_{\boldsymbol{u}\in\Omega} \tilde{H}(\tilde{\boldsymbol{\eta}}^*(t), \tilde{\boldsymbol{x}}^*(t), \boldsymbol{u}). \tag{11.180}$$

Moreover, there is a constant $\eta_0^* \leq 0$ *such that*

$$\eta_0^*(t) \equiv \eta_0^* \leq 0 \tag{11.181}$$

and for all $t \in [t_0, t_1^*]$,

$$\tilde{M}(\tilde{\boldsymbol{\eta}}^*(t), \tilde{\boldsymbol{x}}^*(t)) = \max_{\boldsymbol{u}\in\Omega} \tilde{H}(\tilde{\boldsymbol{\eta}}^*(t), \tilde{\boldsymbol{x}}^*(t), \boldsymbol{u}) \equiv 0. \tag{11.182}$$

The above maximal principle is equivalent to the maximization of the time augmented Hamiltonian

$$\hat{H}(t, \hat{\boldsymbol{\eta}}^*(t), \hat{\boldsymbol{x}}^*(t), \boldsymbol{u})$$

and

$$\max_{\boldsymbol{u}\in\Omega} \hat{H}(t, \hat{\boldsymbol{\eta}}^*(t), \hat{\boldsymbol{x}}^*(t), \boldsymbol{u}) = \int_{t_1^*}^{t} \sum_{i=0}^{n} \eta_i^*(s) \frac{\partial f_i(s, \boldsymbol{x}^*(s), \boldsymbol{u}^*(s))}{\partial t} ds. \tag{11.183}$$

The transversality conditions imply that

$$\eta_{n+1}^*(t_1^*) = 0 \tag{11.184}$$

and hence

$$\max_{\boldsymbol{u}\in\Omega} \hat{H}(t_1^*, \hat{\boldsymbol{\eta}}^*(t_1^*), \hat{\boldsymbol{x}}^*(t_1^*), \boldsymbol{u}) = \hat{M}(t_1^*, \hat{\boldsymbol{\eta}}^*(t_1^*), \hat{\boldsymbol{x}}^*(t_1^*)) = 0. \tag{11.185}$$

Also, if $\mathrm{X}_0 \subseteq \mathbb{R}^n$ *and* $\mathrm{X}_1 \subseteq \mathbb{R}^n$ *are manifolds with tangent spaces* $\mathbb{T}_0$ *and* $\mathbb{T}_1$ *at* $\boldsymbol{x}^*(t_0) = \boldsymbol{x}_0^* \in \mathrm{X}_0$ *and* $\boldsymbol{x}^*(t_1) = \boldsymbol{x}_1^* \in \mathrm{X}_1$*, respectively, then*

$$\hat{\boldsymbol{\eta}}^*(t) = [\eta_0^*(t) \;\; \eta_1^*(t) \;\; \eta_2^*(t) \;\; \ldots \;\; \eta_n^*(t)]^T = [\eta_0^*(t) \;\; \boldsymbol{\eta}^*(t)]^T$$

can be selected to satisfy the transversality conditions

$$\boldsymbol{\eta}^*(t_0) \perp \mathbb{T}_0 \tag{11.186}$$

and

$$\boldsymbol{\eta}^*(t_1^*) \perp \mathbb{T}_1. \tag{11.187}$$

The proof of this result comes from a direct application of the the Maximum Principle given in Theorem 11.4 to this case where $\boldsymbol{x}$ is replaced by $[\boldsymbol{x}\ \ x_{n+1}]^T$ (see pages 318 - 322 in Lee and Markus [119]).

Remark 11.5 *One can still derive a Maximum Principle where the functions* $f : \mathbb{R}^1 \times \mathbb{R}^n \times \mathbb{R}^m \to \mathbb{R}^n$ *and* $f_0 : \mathbb{R}^1 \times \mathbb{R}^n \times \mathbb{R}^m \to \mathbb{R}$ *are not differentiable in time. However, some information is lost since (11.171) and (11.183) do not make sense when the partial derivatives* $\frac{\partial}{\partial t} f_i(t, x_1, x_2, ..., x_n, u_1, u_{2,}..., u_m)$ *do not exist. Also, it is important to note that* $\tilde{A}(\cdot)$ *given by equation (11.167) is not well defined unless* $f_i(t, \boldsymbol{x}, \boldsymbol{u})$ *are differentiable in time. Finally, note that if the adjoint states* η_0^* *and* $\boldsymbol{\eta}^*(\cdot)$ *are both zero, then* $\eta_{n+1}^*(\cdot)$ *would be a constant and the transversality condition (11.184) implies that* $\eta_{n+1}^*(\cdot) = 0$. *Thus,* η_0^* *and* $\boldsymbol{\eta}^*(\cdot)$ *can not be both zero.*

We provide a statement of the Maximum Principle for the nonautonomous control problem that does not require smoothness in time. The result can be found in the books by Young [186] and Fleming [79].

Theorem 11.5 ***(Second Maximum Principle for Nonautonomous Systems)*** *Assume that* $f : \mathbb{R}^1 \times \mathbb{R}^n \times \mathbb{R}^m \to \mathbb{R}^n$, $f_0 : \mathbb{R}^1 \times \mathbb{R}^n \times \mathbb{R}^m \to \mathbb{R}$, $\mathrm{X}_0 \subseteq \mathbb{R}^n$, $t_0 \in \mathbb{R}$, $\mathrm{X}_1 \subseteq \mathbb{R}^n$ *and* $\Omega \subseteq \mathbb{R}^m$ *are given as above and consider the control system*

$$(\mathcal{S}) \qquad \dot{\boldsymbol{x}}(t) = f(t, \boldsymbol{x}(t), \boldsymbol{u}(t)), \quad t_0 < t \leq t_1, \tag{11.188}$$

with piecewise continuous controllers $\boldsymbol{u}(\cdot) \in PWC(t_0, +\infty; \mathbb{R}^m)$ *satisfying* $\boldsymbol{u}(t) \in \Omega \subseteq \mathbb{R}^m$ *e.f. If*

$$\boldsymbol{u}^*(\cdot) \in \Theta = \left\{ \begin{array}{c} \boldsymbol{u}(\cdot) \in PWC(t_0, t_1; \mathbb{R}^m) : \boldsymbol{u}(t) \in \Omega \ e.f. \text{ and} \\ \boldsymbol{u}(\cdot) \text{ steers } \mathrm{X}_0 \text{ to } \mathrm{X}_1 \text{ at a finite time } t_1. \end{array} \right\} \tag{11.189}$$

minimizes

$$J(\boldsymbol{u}(\cdot)) = \int_{t_0}^{t_1} f_0(s, \boldsymbol{x}(s), \boldsymbol{u}(s)) ds, \tag{11.190}$$

on the set of admissible controls Θ *with optimal response* $\boldsymbol{x}^*(\cdot)$ *satisfying* $\boldsymbol{x}^*(t_0) = \boldsymbol{x}_0^* \in \mathrm{X}_0$ *and* $\boldsymbol{x}^*(t_1^*) = \boldsymbol{x}_1^* \in \mathrm{X}_1$*, then there exists a non-trivial solution*

$$\hat{\boldsymbol{\eta}}^*(t) = \begin{bmatrix} \eta_0^*(t) \\ \boldsymbol{\eta}^*(t) \end{bmatrix} \tag{11.191}$$

to the time dependent augmented adjoint equation

$$\frac{d}{dt}\begin{bmatrix} \eta_0(t) \\ \boldsymbol{\eta}(t) \end{bmatrix} = -[\hat{A}(t)]^T \begin{bmatrix} \eta_0(t) \\ \boldsymbol{\eta}(t) \end{bmatrix}, \tag{11.192}$$

such that the time dependent augmented Hamiltonian is maximized

$$\begin{aligned} \hat{H}(t, \hat{\boldsymbol{\eta}}^*(t), \hat{\boldsymbol{x}}^*(t), \boldsymbol{u}^*(t)) &= \hat{M}(t, \hat{\boldsymbol{\eta}}^*(t), \hat{\boldsymbol{x}}^*(t)) \\ &= \max_{\boldsymbol{u}\in\Omega} \hat{H}(t, \hat{\boldsymbol{\eta}}^*(t), \hat{\boldsymbol{x}}^*(t), \boldsymbol{u}). \end{aligned} \tag{11.193}$$

Moreover, there is a constant $\eta_0^* \leq 0$ *such that*

$$\eta_0^*(t) \equiv \eta_0^* \leq 0. \tag{11.194}$$

The transversality conditions imply that

$$\max_{\boldsymbol{u}\in\Omega} \hat{H}(t_1^*, \hat{\boldsymbol{\eta}}^*(t_1^*), \hat{\boldsymbol{x}}^*(t_1^*), \boldsymbol{u}) = \hat{\boldsymbol{M}}(t_1^*, \hat{\boldsymbol{\eta}}^*(t_1^*), \hat{\boldsymbol{x}}^*(t_1^*)) = 0. \tag{11.195}$$

Also, if $\mathrm{X}_0 \subseteq \mathbb{R}^n$ *and* $\mathrm{X}_1 \subseteq \mathbb{R}^n$ *are manifolds with tangent spaces* $\mathbb{T}_0$ *and* $\mathbb{T}_1$ *at* $\boldsymbol{x}^*(t_0) = \boldsymbol{x}_0^* \in \mathrm{X}_0$ *and* $\boldsymbol{x}^*(t_1) = \boldsymbol{x}_1^* \in \mathrm{X}_1$*, respectively, then*

$$\hat{\boldsymbol{\eta}}^*(t) = [\eta_0^*(t) \;\; \eta_1^*(t) \;\; \eta_2^*(t) \;\; \ldots \;\; \eta_n^*(t)]^T = [\eta_0^*(t) \;\; \boldsymbol{\eta}^*(t)]^T$$

can be selected to satisfy the transversality conditions

$$\boldsymbol{\eta}^*(t_0) \perp \mathbb{T}_0 \tag{11.196}$$

and

$$\boldsymbol{\eta}^*(t_1^*) \perp \mathbb{T}_1. \tag{11.197}$$

If in addition, the functions $f_i(t, \boldsymbol{x}, \boldsymbol{u})$*,* $i = 0, 1, 2, \ldots, n$ *are* C^1 *in* t*, then*

$$\hat{M}(t, \hat{\boldsymbol{\eta}}^*(t), \hat{\boldsymbol{x}}^*(t)) = \int_{t_1^*}^{t} \sum_{i=0}^{n} \eta_i^*(s) \frac{\partial f_i(s, \boldsymbol{x}^*(s), \boldsymbol{u}^*(s))}{\partial t} ds. \tag{11.198}$$

11.8 Application to the Nonautonomous LQ Control Problem

The Maximum Principle **Theorem** 11.5 will be applied to the Nonautonomous Linear Quadratic (LQ) control problem. We use the notation

$$M : I \to \mathbb{R}^{p\times q}$$

to denote a $p \times q$ (real valued) matrix function defined on an interval I. In particular, if $M : I \to \mathbb{R}^{p\times q}$, then

$$M(t) = \begin{bmatrix} m_{1,1}(t) & m_{1,2}(t) & \cdots & m_{1,q}(t) \\ m_{2,1}(t) & m_{2,2}(t) & \cdots & m_{2,q}(t) \\ \vdots & \vdots & \ddots & \vdots \\ m_{p,1}(t) & m_{p,2}(t) & \cdots & m_{p,q}(t) \end{bmatrix}$$

where each entry is a real valued function defined on the interval I. Likewise we say that $M(\cdot)$ is piecewise continuous, continuous, etc. if each component is piecewise continuous, continuous, etc.

We assume that t_0, $\boldsymbol{x}_0 \in \mathbb{R}^n$ and $t_0 < t_1$ are given and that $A : [t_0, t_1] \to \mathbb{R}^{n\times n}$ and $B : [t_0, t_1] \to \mathbb{R}^{n\times m}$ are piecewise continuous matrix valued functions. Consider the nonautonomous linear control system

$$\dot{\boldsymbol{x}}(t) = A(t)\boldsymbol{x}(t) + B(t)\boldsymbol{u}(t) \tag{11.199}$$

with initial data

$$\boldsymbol{x}(0) = \boldsymbol{x}_0 \in \mathbb{R}^n. \tag{11.200}$$

Also, let $Q : [t_0, t_1] \to \mathbb{R}^{n\times n}$ and $R : [t_0, t_1] \to \mathbb{R}^{m\times m}$ be piecewise continuous and symmetric matrix valued functions such that

$$Q^T(t) = Q(t) \geq 0 \text{ and } R^T(t) = R(t) > 0,$$

for all $t \in [t_0, t_1]$. In addition we require that there is an $\alpha > 0$ such that for all $t \in [t_0, t_1]$ we have

$$0 < \alpha I_{m\times m} \leq R(t). \tag{11.201}$$

Condition (11.201) implies that for any $\boldsymbol{u} \in \mathbb{R}^m$,

$$\alpha \|\boldsymbol{u}\|^2 = \alpha \langle \boldsymbol{u}, \boldsymbol{u} \rangle \leq \langle R(t)\boldsymbol{u}, \boldsymbol{u} \rangle$$

and $[R(t)]^{-1}$ exists. Let $\boldsymbol{w} \in \mathbb{R}^m$ and define

$$\boldsymbol{u} = [R(t)]^{-1}\,\boldsymbol{w} \in \mathbb{R}^m$$

so that

$$\begin{aligned}\alpha \left\|[R(t)]^{-1}\boldsymbol{w}\right\|^2 &= \alpha \left\|\boldsymbol{u}\right\|^2 \le \langle R(t)\boldsymbol{u}, \boldsymbol{u}\rangle \le \left\|R(t)\boldsymbol{u}\right\| \left\|\boldsymbol{u}\right\| \\ &= \left\|\boldsymbol{w}\right\| \left\|[R(t)]^{-1}\boldsymbol{w}\right\|.\end{aligned}$$

If $\boldsymbol{w} \neq \boldsymbol{0}$, then $[R(t)]^{-1}\boldsymbol{w} \neq \boldsymbol{0}$ and dividing by $\left\|[R(t)]^{-1}\boldsymbol{w}\right\|$ it follows that

$$\alpha \left\|[R(t)]^{-1}\boldsymbol{w}\right\| \le \left\|\boldsymbol{w}\right\|$$

so that

$$\left\|[R(t)]^{-1}\boldsymbol{w}\right\| \le \frac{\left\|\boldsymbol{w}\right\|}{\alpha}$$

for all $\boldsymbol{w} \in \mathbb{R}^m$. Therefore,

$$\left\|[R(t)]^{-1}\right\| \le \frac{1}{\alpha}$$

and $[R(t)]^{-1}$ is bounded on $[t_0, t_1]$. Since $R : [t_0, t_1] \to \mathbb{R}^{m\times m}$ is piecewise continuous, it then follows that $R^{-1} : [t_0, t_1] \to \mathbb{R}^{m\times m}$ is also piecewise continuous. This is important since we need the mapping $\Lambda : [t_0, T] \to \mathbb{R}^{m\times n}$ defined by

$$\Lambda(t) = [B(t)][R(t)]^{-1}[B(t)]^T$$

to also be piecewise continuous.

We assume that the controls belong to the space $PWC(t_0, T; \mathbb{R}^m)$ and the quadratic cost functional is defined by

$$\begin{aligned}J(\boldsymbol{u}(\cdot)) &= \langle S\boldsymbol{x}(t_1), \boldsymbol{x}(t_1)\rangle \\ &\quad + \frac{1}{2}\int_{t_0}^{t_1} \{\langle Q(s)\boldsymbol{x}(s), \boldsymbol{x}(s)\rangle + \langle R(s)\boldsymbol{u}(s), \boldsymbol{u}(s)\rangle\}\, ds,\end{aligned} \tag{11.202}$$

where $S = S^T \ge 0$ is a constant symmetric matrix and $\boldsymbol{x}(t) = \boldsymbol{x}(t; \boldsymbol{u}(\cdot))$ is the solution to the system (11.199) - (11.200). The

Linear Quadratic (LQ) Optimal Control problem is to find $\boldsymbol{u}^*(\cdot) \in PWC(t_0, t_1; \mathbb{R}^m)$ so that

$$\begin{aligned} J(\boldsymbol{u}^*(\cdot)) &= \langle S\boldsymbol{x}^*(T), \boldsymbol{x}^*(T)\rangle \\ &\quad + \frac{1}{2}\int_{t_0}^{t_1} \{\langle Q(s)\boldsymbol{x}^*(s), \boldsymbol{x}^*(s)\rangle + \langle R(s)\boldsymbol{u}^*(s), \boldsymbol{u}^*(s)\rangle\}\, ds \\ &\le \langle S\boldsymbol{x}(T), \boldsymbol{x}(T)\rangle \\ &\quad + \frac{1}{2}\int_{t_0}^{t_1} \{\langle Q(s)\boldsymbol{x}(s), \boldsymbol{x}(s)\rangle + \langle R(s)\boldsymbol{u}(s), \boldsymbol{u}(s)\rangle\}\, ds, \end{aligned}$$

for all $\boldsymbol{u}(\cdot) \in PWC(0, t_1; \mathbb{R}^m)$.

Formulation as an Optimal Control Problem: In order to set up the nonautonomous linear quadratic optimal control problem as a fixed time optimal control problem and apply the Maximum Principle, we need to:

1. Identify the initial time t_0 and final time t_1;
2. Identify the initial set $X_0 \subseteq \mathbb{R}^n$, the terminal set $X_1 \subseteq \mathbb{R}^n$;
3. Identify the control constraint set $\Omega \subseteq \mathbb{R}^m$;
4. Identify the functions $f : \mathbb{R}^1 \times \mathbb{R}^n \times \mathbb{R}^m \longrightarrow \mathbb{R}^n$, $f_0 : \mathbb{R}^1 \times \mathbb{R}^n \times \mathbb{R}^m \longrightarrow \mathbb{R}^1$;
5. Define the time dependent augmented Hamiltonian $\hat{H}(t, \hat{\boldsymbol{\eta}}, \hat{\boldsymbol{x}}, \boldsymbol{u})$;
6. Form the time dependent augmented adjoint system matrix $\hat{A}(t)$.

The initial time is t_0 and the final time t_1 is given and fixed. The initial set is the single initial vector $X_0 = \{\boldsymbol{x}_0\}$ and since there is no terminal constraint, the terminal set is the whole state space $X_1 = \mathbb{R}^n$. Also, there is no constraint on the control $\boldsymbol{u}(\cdot) \in PWC(t_0, t_1; \mathbb{R}^m)$, hence the control constraint set is all of $\mathbb{R}^m$ and $\Omega = \mathbb{R}^m$. The functions $f : \mathbb{R}^1 \times \mathbb{R}^n \times \mathbb{R}^m \longrightarrow \mathbb{R}^n$ and $f_0 : \mathbb{R}^1 \times \mathbb{R}^n \times \mathbb{R}^m \longrightarrow \mathbb{R}^1$ are given by

$$f(t, \boldsymbol{x}, \boldsymbol{u}) = A(t)\boldsymbol{x} + B(t)\boldsymbol{u},$$

and

$$f_0(t, \boldsymbol{x}, \boldsymbol{u}) = \frac{1}{2}\langle Q(t)\boldsymbol{x}, \boldsymbol{x}\rangle + \frac{1}{2}\langle R(t)\boldsymbol{u}, \boldsymbol{u}\rangle \geq 0,$$

respectively.

The time dependent augmented Hamiltonian $\hat{H} : \mathbb{R}^1 \times \mathbb{R}^{n+1} \times \mathbb{R}^{n+1} \times \mathbb{R}^m \longrightarrow \mathbb{R}^1$ is defined by

$$\begin{aligned}\hat{H}(t, \hat{\boldsymbol{\eta}}, \hat{\boldsymbol{x}}, \boldsymbol{u}) &= \frac{\eta_0}{2}\{\langle Q(t)\boldsymbol{x}, \boldsymbol{x}\rangle + \langle R(t)\boldsymbol{u}, \boldsymbol{u}\rangle\} \\ &\quad + \langle \boldsymbol{\eta},\ A(t)\boldsymbol{x} + B(t)\boldsymbol{u}\rangle \qquad (11.203) \\ &= \frac{\eta_0}{2}\boldsymbol{x}^T Q(t)\,\boldsymbol{x} + \frac{\eta_0}{2}\boldsymbol{u}^T R(t)\boldsymbol{u} + [A(t)\boldsymbol{x}]^T\boldsymbol{\eta} \\ &\quad + [B(t)\boldsymbol{u}]^T\boldsymbol{\eta} \\ &= \frac{\eta_0}{2}\boldsymbol{x}^T Q(t)\,\boldsymbol{x} + \frac{\eta_0}{2}\boldsymbol{u}^T R(t)\boldsymbol{u} + \boldsymbol{x}^T[A(t)]^T\boldsymbol{\eta} \\ &\quad + \boldsymbol{u}^T\,[B(t)]^T\boldsymbol{\eta},\end{aligned}$$

where $\hat{\boldsymbol{x}} = \begin{bmatrix} x_0 & \boldsymbol{x} \end{bmatrix}^T \in \mathbb{R}^{n+1}$, $\hat{\boldsymbol{\eta}} = \begin{bmatrix} \eta_0 & \boldsymbol{\eta} \end{bmatrix}^T \in \mathbb{R}^{n+1}$ and $\boldsymbol{u} \in \mathbb{R}^m$. The time dependent augmented function $f : \mathbb{R}^1 \times \mathbb{R}^{n+1} \times \mathbb{R}^m \longrightarrow \mathbb{R}^{n+1}$ is defined by

$$\begin{aligned}\hat{f}(t, \hat{\boldsymbol{x}}, \boldsymbol{u}) &= \begin{bmatrix} \frac{1}{2}\langle Q(t)\boldsymbol{x}, \boldsymbol{x}\rangle + \frac{1}{2}\langle R(t)\boldsymbol{u}, \boldsymbol{u}\rangle \\ A(t)\boldsymbol{x} + B(t)\boldsymbol{u}, \end{bmatrix} \\ &= \begin{bmatrix} \frac{1}{2}\boldsymbol{x}^T Q(t)\boldsymbol{x} + \frac{1}{2}\boldsymbol{u}^T R(t)\boldsymbol{u} \\ A(t)\boldsymbol{x} + B(t)\boldsymbol{u}, \end{bmatrix}, \qquad (11.204)\end{aligned}$$

so that the time dependent Jacobian is given by

$$\mathbb{J}_{\hat{\boldsymbol{x}}}\hat{f}(t, \hat{\boldsymbol{x}}, \boldsymbol{u}) = \mathbb{J}_{\hat{\boldsymbol{x}}}\hat{f}(t, \boldsymbol{x}, u) = \begin{bmatrix} 0 & [Q(t)\boldsymbol{x}]^T \\ \boldsymbol{0} & A(t) \end{bmatrix}. \qquad (11.205)$$

Now assume that $(\boldsymbol{x}^*(\cdot), \boldsymbol{u}^*(\cdot))$ is an optimal pair so that the matrix $\hat{A}(t)$ is given by

$$\begin{aligned}\hat{A}(t) &= \mathbb{J}_{\hat{\boldsymbol{x}}}\hat{f}(t, \hat{\boldsymbol{x}}, \boldsymbol{u})_{|(\boldsymbol{x}^*(t), u^*(t))} = \mathbb{J}_{\hat{\boldsymbol{x}}}\hat{f}(t, \boldsymbol{x}^*(t), \boldsymbol{u}^*(t)) \\ &= \begin{bmatrix} 0 & [Q(t)\boldsymbol{x}^*(t)]^T \\ \boldsymbol{0} & A(t) \end{bmatrix} \qquad (11.206)\end{aligned}$$

and

$$-[\hat{A}(t)]^T = -\begin{bmatrix} 0 & \mathbf{0} \\ [Q(t)\boldsymbol{x}^*(t)] & [A(t)]^T \end{bmatrix}$$
$$= \begin{bmatrix} 0 & \mathbf{0} \\ -[Q(t)\boldsymbol{x}^*(t)] & -[A(t)]^T \end{bmatrix}.$$

The time dependent augmented adjoint equation has the form

$$\frac{d}{dt}\begin{bmatrix} \eta_0(t) \\ \boldsymbol{\eta}(t) \end{bmatrix} = -[\hat{A}(t)]^T \begin{bmatrix} \eta_0(t) \\ \boldsymbol{\eta}(t) \end{bmatrix}$$
$$= \begin{bmatrix} 0 & \mathbf{0} \\ -[Q(t)\boldsymbol{x}^*(t)] & -[A(t)]^T \end{bmatrix} \begin{bmatrix} \eta_0(t) \\ \boldsymbol{\eta}(t) \end{bmatrix} \tag{11.207}$$

which is equivalent to the system

$$\frac{d}{dt}\eta_0(t) = 0$$
$$\frac{d}{dt}\boldsymbol{\eta}(t) = -\eta_0(t)Q(t)x^*(t) - [A(t)]^T\eta(t),$$

where again $\boldsymbol{x}^*(\cdot)$ is the optimal trajectory. The first equation above implies (as always) that the zero adjoint state is a constant $\eta_0(t) \equiv \eta_0$. The second equation is coupled to the state equation by the term $-\eta_0(t)Q(t)\boldsymbol{x}^*(t)$.

If $(\boldsymbol{x}^*(\cdot), \boldsymbol{u}^*(\cdot))$ is an optimal pair for the LQ optimal control problem, then the Maximum Principle implies that there is a nontrivial solution

$$\hat{\boldsymbol{\eta}}^*(t) = \begin{bmatrix} \eta_0^*(t) \\ \boldsymbol{\eta}^*(t) \end{bmatrix} \in \mathbb{R}^{n+1}$$

to the augmented adjoint equation (11.81) such that $\eta_0^*(t) \equiv \eta_0^* \leq 0$,

$$\frac{d}{dt}\boldsymbol{\eta}^*(t) = -\eta_0^* Q(t)\boldsymbol{x}^*(t) - [A(t)]^T\boldsymbol{\eta}^*(t) \tag{11.208}$$

and

$$\hat{H}(t, \hat{\boldsymbol{\eta}}^*(t), \hat{\boldsymbol{x}}^*(t), \boldsymbol{u}^*(t)) = \max_{\boldsymbol{u} \in \mathbb{R}^m} \hat{H}(t, \hat{\boldsymbol{\eta}}^*(t), \hat{\boldsymbol{x}}^*(t), \boldsymbol{u}) \equiv c(t).$$

Since $\Omega = \mathbb{R}^m$ is open, then

$$D_{\boldsymbol{u}}\hat{H}(t, \hat{\boldsymbol{\eta}}^*(t), \hat{\boldsymbol{x}}^*(t), \boldsymbol{u})_{|\boldsymbol{u}=\boldsymbol{u}^*(t)} = D_{\boldsymbol{u}}\hat{H}(t, \hat{\boldsymbol{\eta}}^*(t), \hat{\boldsymbol{x}}^*(t), \boldsymbol{u}^*(t)) = \mathbf{0},$$

where

$$\begin{aligned}
D_{\boldsymbol{u}}\hat{H}(t,&\hat{\boldsymbol{\eta}}^*(t), \hat{\boldsymbol{x}}^*(t), \boldsymbol{u}) \\
&= \frac{\partial}{\partial \boldsymbol{u}}\left\{\frac{\eta_0^*}{2}[\boldsymbol{x}^*(t)]^T Q(t)\boldsymbol{x}^*(t) + \frac{\eta_0^*}{2}\boldsymbol{u}^T R(t)\boldsymbol{u}\right\} \\
&\quad + \frac{\partial}{\partial \boldsymbol{u}}\left\{[\boldsymbol{x}^*(t)]^T[A(t)]^T\boldsymbol{\eta}^*(t) + \boldsymbol{u}^T[B(t)]^T\boldsymbol{\eta}^*(t)\right\} \\
&= \frac{\partial}{\partial \boldsymbol{u}}\left\{\frac{\eta_0^*}{2}\boldsymbol{u}^T R(t)\boldsymbol{u} + \boldsymbol{u}^T[B(t)]^T\boldsymbol{\eta}^*(t)\right\} \\
&= \eta_0^* R(t)\boldsymbol{u} + [B(t)]^T\boldsymbol{\eta}^*(t).
\end{aligned}$$

Hence, when $\boldsymbol{u} = \boldsymbol{u}^*(t)$ it follows that

$$\begin{aligned}
D_{\boldsymbol{u}}\hat{H}(t, \hat{\boldsymbol{\eta}}^*(t), \hat{\boldsymbol{x}}^*(t), \boldsymbol{u})_{|\boldsymbol{u}=\boldsymbol{u}^*(t)} &= \eta_0^* R(t)\boldsymbol{u} + [B(t)]^T\boldsymbol{\eta}^*(t)\ |_{u=u^*(t)} \qquad (11.209) \\
&= \eta_0^* R(t)\boldsymbol{u}^*(t) + [B(t)]^T\boldsymbol{\eta}^*(t) = 0.
\end{aligned}$$

Applying the transversality condition at $\boldsymbol{x}^*(0) = \boldsymbol{x}_0 \in \mathrm{X}_0 = \{\boldsymbol{x}_0\}$ yields that $\eta^*(0)$ can be any vector since $\mathrm{T}_0 = \{\boldsymbol{x}_0\}$. However, at t_1 we have the transversality condition

$$\boldsymbol{\eta}^*(t_1) = \eta_0 S\boldsymbol{x}^*(t_1). \tag{11.210}$$

This boundary condition in turn implies that $\eta_0^* < 0$. To see this assume that $\eta_0^* = 0$ and observe that the adjoint equation (11.82) reduces to the linear system

$$\begin{aligned}
\frac{d}{dt}\boldsymbol{\eta}^*(t) &= -\eta_0^* Q\boldsymbol{x}^*(t) - A^T\boldsymbol{\eta}^*(t) \\
&= -A^T\boldsymbol{\eta}^*(t).
\end{aligned}$$

Therefore, $\boldsymbol{\eta}^*(\cdot)$ would be a solution of the homogenous linear initial value problem

$$\frac{d}{dt}\boldsymbol{\eta}^*(t) = -A^T\boldsymbol{\eta}^*(t), \quad \boldsymbol{\eta}^*(t_1^*) = \eta_0 S\boldsymbol{x}^*(t_1) = \mathbf{0}$$

and hence it follows that $\boldsymbol{\eta}^*(t) \equiv \mathbf{0}$. Consequently, we have shown that if $\eta_0^* = 0$, then $\boldsymbol{\eta}^*(t) \equiv \mathbf{0}$ and hence

$$\hat{\boldsymbol{\eta}}^*(t) = \begin{bmatrix} \eta_0^*(t) \\ \boldsymbol{\eta}^*(t) \end{bmatrix} \equiv \mathbf{0}$$

which contradicts the statement that $\hat{\boldsymbol{\eta}}^*(t)$ is a nontrivial solution of the time augmented adjoint equation (11.207).

Since $\eta_0^* < 0$ we can solve (11.209) for the optimal control. We have

$$\eta_0^* R(t)\boldsymbol{u}^*(t) + [B(t)]^T \boldsymbol{\eta}^*(t) = 0$$

which yields

$$\eta_0^* R(t)\boldsymbol{u}^*(t) = -[B(t)]^T \boldsymbol{\eta}^*(t)$$

and

$$\boldsymbol{u}^*(t) = [R(t)]^{-1}[B(t)]^T \left[\frac{\boldsymbol{\eta}^*(t)}{-\eta_0^*}\right].$$

Since the matrix $R(t) = [R(t)]^T > 0$ is nonsingular and $[R(t)]^{-1}$ is piecewise smooth we have the following expression for the (possibly piecewise continuous) optimal control

$$\boldsymbol{u}^*(t) = [R(t)]^{-1}[B(t)]^T \left[\frac{\boldsymbol{\eta}^*(t)}{-\eta_0^*}\right]. \tag{11.211}$$

Summarizing, it follows that the optimal trajectory can be obtained by solving the two point boundary value problem defined by the coupled state and adjoint equations

$$\begin{aligned}
&\tfrac{d}{dt}\boldsymbol{x}^*(t) = A(t)\boldsymbol{x}^*(t) - \tfrac{1}{\eta_0^*}[B(t)][R(t)]^{-1}[B(t)]^T\boldsymbol{\eta}^*(t), \quad \boldsymbol{x}^*(0) = \boldsymbol{x}_0,\\
&\tfrac{d}{dt}\boldsymbol{\eta}^*(t) = -\eta_0^* Q(t)\boldsymbol{x}^*(t) - [A(t)]^T\boldsymbol{\eta}^*(t), \qquad \boldsymbol{\eta}^*(t_1) = \eta_0 S\boldsymbol{x}^*(t_1),
\end{aligned} \tag{11.212}$$

and setting

$$\boldsymbol{u}^*(t) = [R(t)]^{-1}[B(t)]^T \left[\frac{\boldsymbol{\eta}^*(t)}{-\eta_0^*}\right].$$

To eliminate the η_0^* term, we divide the adjoint equation above by $-\eta_0^*$ which yields

$$\frac{d}{dt}\left[\frac{\boldsymbol{\eta}^*(t)}{-\eta_0^*}\right] = Q(t)\boldsymbol{x}^*(t) - [A(t)]^T\left[\frac{\boldsymbol{\eta}^*(t)}{-\eta_0^*}\right], \quad \left[\frac{\boldsymbol{\eta}^*(t_1)}{-\eta_0^*}\right] = -S\boldsymbol{x}^*(t_1).$$

Defining the normalized adjoint state $\boldsymbol{\lambda}^*(t)$ by

$$\boldsymbol{\lambda}^*(t) \triangleq \frac{\boldsymbol{\eta}^*(t)}{-\eta_0^*}$$

produces the optimality conditions

$$\begin{aligned} \tfrac{d}{dt}\boldsymbol{x}^*(t) &= A(t)\boldsymbol{x}^*(t) + [B(t)][R(t)]^{-1}[B(t)]^T\boldsymbol{\lambda}^*(t), & \boldsymbol{x}^*(0) &= \boldsymbol{x}_0, \\ \tfrac{d}{dt}\boldsymbol{\lambda}^*(t) &= Q(t)\boldsymbol{x}^*(t) - [A(t)]^T\boldsymbol{\lambda}^*(t), & \boldsymbol{\lambda}^*(t_1) &= -S\boldsymbol{x}^*(t_1), \end{aligned} \tag{11.213}$$

where the optimal control is defined by

$$\boldsymbol{u}^*(t) = R^{-1}(t)B^T(t)\boldsymbol{\lambda}^*(t). \tag{11.214}$$

We can write the optimality system as

$$\frac{d}{dt}\begin{bmatrix} \boldsymbol{x}^*(t) \\ \boldsymbol{\lambda}^*(t) \end{bmatrix} = \begin{bmatrix} A(t) & [B(t)][R(t)]^{-1}[B(t)]^T \\ Q(t) & -[A(t)]^T \end{bmatrix}\begin{bmatrix} \boldsymbol{x}^*(t) \\ \boldsymbol{\lambda}^*(t) \end{bmatrix}, \tag{11.215}$$

with boundary conditions

$$\boldsymbol{x}^*(0) = \begin{bmatrix} I_{n\times n} & 0_{n\times n} \end{bmatrix}\begin{bmatrix} \boldsymbol{x}^*(0) \\ \boldsymbol{\lambda}^*(0) \end{bmatrix} = \boldsymbol{x}_0, \tag{11.216}$$

and

$$\boldsymbol{\lambda}^*(t_1) = \begin{bmatrix} -S & 0_{n\times n} \end{bmatrix}\begin{bmatrix} \boldsymbol{x}^*(t_1) \\ \boldsymbol{\lambda}^*(t_1) \end{bmatrix}. \tag{11.217}$$

Thus, if one solves the two point boundary value problem (11.215) - (11.217), the optimal control is defined by (11.214).

11.9 Problem Set for Chapter 11

Problem 11.1 *Consider the control system*

$$\begin{aligned} \dot{x}_1(t) &= x_2(t) + u_1(t), \\ \dot{x}_2(t) &= u_1(t), \end{aligned}$$

with initial condition

$$x_1(0) = 1, \quad x_2(0) = 1,$$

and terminal condition

$$x_2(2) = 0.$$

Let the control constraint set be $\Omega = \mathbb{R}^2$. *Investigate the optimal control problem for the cost functional*

$$J(u(\cdot)) = \frac{1}{2}\int_0^2 \sqrt{[u_1(s)]^2 + [u_2(s)]^2} ds.$$

Problem 11.2 *Consider the control system*

$$\dot{x}_1(t) = x_2(t),$$
$$\dot{x}_2(t) = u(t),$$

with initial condition

$$x_1(0) = 1, \quad x_2(0) = 1$$

and terminal condition

$$x_1(1) = 0, \quad x_2(1) \text{ is free.}$$

Let the control constraint set be $\Omega = \mathbb{R}^1$. *Investigate the optimal control problem for the cost functional*

$$J(u(\cdot)) = \frac{1}{2}\int_0^1 [u(s)]^2 ds.$$

Problem 11.3 *Consider the control system*

$$\dot{x}_1(t) = x_2(t),$$
$$\dot{x}_2(t) = u(t),$$

with initial condition

$$x_1(0) = 1, \quad x_2(0) = 1$$

and terminal condition

$$x_1(1) = 0, \quad x_2(1) = 0.$$

Let the control constraint set be $\Omega = \mathbb{R}^1$. *Investigate the optimal control problem for the cost functional*

$$J(u(\cdot)) = \frac{1}{2}\int_0^1 [u(s)]^2 ds.$$

Problem 11.4 *Consider the control system*

$$\dot{x}_1(t) = x_2(t),$$
$$\dot{x}_2(t) = u(t),$$

with initial condition

$$x_1(0) = 1, \quad x_2(0) = 1,$$

and terminal condition

$$x_1(t_1) = 0, \quad x_2(1) = 0.$$

Let the control constraint set be $\Omega = [-1, +1]$. *Investigate the optimal control problem for the cost functional*

$$J(u(\cdot)) = \int_0^1 \sqrt{1 + [u(s)]^2} ds.$$

Problem 11.5 *Consider the control system*

$$\dot{x}(t) = x(t) + u(t),$$

with initial condition

$$x(0) = 1,$$

and terminal condition

$$x(2) \text{ is free.}$$

Let the control constraint set be $\Omega = \mathbb{R}^1$. *Investigate the optimal control problem for the cost functional*

$$J(u(\cdot)) = \frac{1}{2}[x(2)]^2 + \frac{1}{2}\int_0^2 \{[x(s)]^2 + [u(s)]^2\} ds.$$

Problem 11.6 *Consider the control system*

$$\dot{x}(t) = x(t) + u(t),$$

with initial condition

$$x(0) = 1,$$

and terminal condition

$$x(1) \textit{ is free.}$$

Let the control constraint set be $\Omega = \mathbb{R}^1$. *Investigate the optimal control problem for the cost functional*

$$J(u(\cdot)) = \frac{1}{2}\int_0^1 \left\{[x(s) - e^{-s}]^2 + [u(s)]^2\right\} ds.$$

Problem 11.7 *Consider the control system*

$$\dot{x}(t) = x(t) + u(t),$$

with initial condition

$$x(0) = 1,$$

and terminal condition

$$x(1) \textit{ is free.}$$

Let the control constraint set be $\Omega = [-1, +1]$. *Investigate the optimal control problem for the cost functional*

$$J(u(\cdot)) = \frac{1}{2}\int_0^2 [x(s) - e^{-s}]^2 ds.$$

Problem 11.8 *Consider the control system*

$$\dot{x}_1(t) = x_2(t) + \cos(u(t)),$$
$$\dot{x}_2(t) = \sin(u(t)),$$

with initial condition

$$x_1(0) = 0, \quad x_2(0) = 0,$$

and terminal condition

$$x_1(1) \text{ and } x_2(1) \text{ are free.}$$

Let the control constraint set be $\Omega = [-\pi, +\pi]$. *Investigate the optimal control problem for the cost functional*

$$J(u(\cdot)) = -\int_0^1 \{x_2(s) + \cos(u(s))\}\, ds.$$

Problem 11.9 *Consider the control system*

$$\dot{x}(t) = -x(t) + u(t),$$

with initial condition

$$x(0) = 1,$$

and terminal condition

$$x(2) \text{ is free.}$$

Let the control constraint set be $\Omega = \mathbb{R}^1$. *Investigate the optimal control problem for the cost functional*

$$J(u(\cdot)) = \int_0^2 \left\{[x(s)]^2 + [u(s)]^2\right\} ds.$$

Derive and solve the Riccati differential equation that defines the optimal feedback controller.

Problem 11.10 *Consider the control system*

$$\dot{x}(t) = x(t) + bu(t),$$

with initial condition

$$x(0) = 1,$$

and terminal condition

$$x(1) \text{ is free.}$$

Here, $b \neq 0$ and the control constraint set is $\Omega = \mathbb{R}^1$. Investigate the optimal control problem for the cost functional

$$J(u(\cdot)) = \int_0^1 \left\{q[x(s)]^2 + [u(s)]^2\right\} ds,$$

where $q \geq 0$. Derive and solve the Riccati differential equation that defines the optimal feedback controller. What happens when $q \longrightarrow 0$?

Chapter 12

Linear Control Systems

Although we have focused on control and optimal control of nonlinear control systems, linear control systems play a key role in practical applications of modern control. Linear systems arise naturally when the dynamic model is linear and when linearization is used to produce an approximate system. We discuss linearization later, but first we present a short introduction to the basic ideas and control system properties. Good references for this section include [7], [45], [110] and [115].

12.1 Introduction to Linear Control Systems

We focus on linear control systems defined by a linear differential equation

$$\dot{\boldsymbol{x}}(t) = A(t)\boldsymbol{x}(t) + B(t)\boldsymbol{u}(t) + G(t)\boldsymbol{w}(t) \tag{12.1}$$

with initial data

$$\boldsymbol{x}(t_0) = \boldsymbol{x}_0 \in \mathbb{R}^n. \tag{12.2}$$

The inputs to the system are defined by the control $\boldsymbol{u}(\cdot)$ and a "disturbance" $\boldsymbol{w}(\cdot)$. As before, we assume that $\boldsymbol{u}(\cdot)$ is piecewise continuous and in this section we assume that the disturbance $\boldsymbol{w}(\cdot)$ is also piecewise continuous. The sensed output is defined by

$$\boldsymbol{y}(t) = C(t)\boldsymbol{x}(t) + H(t)\boldsymbol{v}(t) \in \mathbb{R}^p \tag{12.3}$$

where $\boldsymbol{y}(\cdot)$ is called the *sensed (or measured) output* and $\boldsymbol{v}(\cdot)$ represents *sensor noise*. Again, in this section it is useful to think of $\boldsymbol{v}(\cdot)$ as a piecewise continuous function. However, later it will be important to view both $\boldsymbol{w}(\cdot)$ and $\boldsymbol{v}(\cdot)$ as white noise. In the case with no disturbances, the control system is defined by

$$\begin{cases} \dot{\boldsymbol{x}}(t) = A(t)\boldsymbol{x}(t) + B(t)\boldsymbol{u}(t), \\ \boldsymbol{y}(t) = C(t)\boldsymbol{x}(t). \end{cases} \tag{12.4}$$

Recall that many of the previous optimal control problems are formulated in terms of finding a control $\boldsymbol{u}(\cdot)$ that steers an initial state $\boldsymbol{x}_0$ to a terminal state $\boldsymbol{x}_1$ in a finite time $t_1 > t_0$. Thus, we must consider the following question.

> Given an initial state $\boldsymbol{x}_0 \in \mathbb{R}^n$ and a final state $\boldsymbol{x}_1 \in \mathbb{R}^n$ is there a time $t_1 > t_0$ and a corresponding control $\boldsymbol{u}(\cdot) \in PWC(t_0, t_1; \mathbb{R}^m)$ such that $\boldsymbol{u}(\cdot)$ steers $\boldsymbol{x}_0 \in \mathbb{R}^n$ to $\boldsymbol{x}_1 \in \mathbb{R}^n$ at time t_1? This question motivates the following definition.

Definition 12.1 (Controllability) *The system*

$$\dot{\boldsymbol{x}}(t) = A(t)\boldsymbol{x}(t) + B(t)\boldsymbol{u}(t) \tag{12.5}$$

is said to be (completely) controllable if for any initial time t_0 and states $\boldsymbol{x}_0 \in \mathbb{R}^n$ and $\boldsymbol{x}_1 \in \mathbb{R}^n$, there exists a finite time $t_1 > t_0$ and a piecewise continuous control $\boldsymbol{u}(\cdot) \in PWC(t_0, t_1; \mathbb{R}^m)$ such that the solution to (12.5) satisfying the initial condition

$$\boldsymbol{x}(t_0) = \boldsymbol{x}_0$$

also satisfies

$$\boldsymbol{x}(t_1) = \boldsymbol{x}_1.$$

In particular, (12.5) is controllable if and only if $\boldsymbol{x}_0$ can be steered to $\boldsymbol{x}_1$ in finite time $t_1 - t_0$.

We shall also be interested in the concept of observability. Loosely speaking, observability implies that one can reconstruct the initial state of the system from given input - output pairs.

Definition 12.2 (Observability) *The system*

$$\begin{cases} \dot{\boldsymbol{x}}(t) = A(t)\boldsymbol{x}(t) + B(t)\boldsymbol{u}(t) \\ \boldsymbol{y}(t) = C(t)\boldsymbol{x}(t) \end{cases}$$

is (completely) observable if for all t_0, there exists a $t_1 > t_0$ such that if

$$\begin{aligned} \boldsymbol{y}(t; t_0, \boldsymbol{x}_0, \boldsymbol{u}(\cdot)) = C(t)\boldsymbol{x}(t; t_0, \boldsymbol{x}_0, \boldsymbol{u}(\cdot)) &= \boldsymbol{y}(t; t_0, \tilde{\boldsymbol{x}}_0, \boldsymbol{u}(\cdot)) \\ &= C(t)\boldsymbol{x}(t; t_0, \tilde{\boldsymbol{x}}_0, \boldsymbol{u}(\cdot)) \end{aligned}$$

for all $t \in [t_0, t_1]$ and all controls $\boldsymbol{u}(\cdot) \in PWC(t_0, t_1; \mathbb{R}^m)$, then

$$\boldsymbol{x}_0 = \tilde{\boldsymbol{x}}_0.$$

Here, $\boldsymbol{x}(t; t_0, \boldsymbol{x}_0, \boldsymbol{u}(\cdot))$ denotes the solution to (12.5) with initial condition $\boldsymbol{x}(t_0) = \boldsymbol{x}_0$. In particular, the initial state $\boldsymbol{x}_0$ at time t_0 is uniquely determined by the output $\boldsymbol{y}(t; t_0, \boldsymbol{x}_0, \boldsymbol{u}(\cdot))$ on the interval $[t_0, t_1]$.

Remark 12.1 *In the special case where the matrix $C(t)$ is an $n \times n$ nonsingular matrix, then*

$$\boldsymbol{y}(t_0) = C(t_0)\boldsymbol{x}(t_0)$$

implies $\boldsymbol{x}_0 = \boldsymbol{x}(t_0) = [C(t_0)]^{-1}\boldsymbol{y}(t_0)$ so that the system is trivially observable. Note that controllability is determined by the matrices $A(\cdot)$ and $B(\cdot)$ and observability is determined by the matrices $A(\cdot)$ and $C(\cdot)$. Thus, we shall simply say that the pair $(A(\cdot), B(\cdot))$ is controllable or that $(A(\cdot), C(\cdot))$ is observable.

In the previous section we saw where linear control systems occur either naturally or from linearization. In this section we consider the LQ control problem with time varying matrix coefficients. Most of the following material can be found in the standard books [6], [5], [7], [16], [45], [71] and [73], [81], [115] and [187].

Consider the linear system on the time interval $[t_0, t_1]$ given by

$$\dot{\boldsymbol{x}}(t) = A(t)\boldsymbol{x}(t) + B(t)\boldsymbol{u}(t), \;\; \boldsymbol{x}(t_0) = \boldsymbol{x}_0 \in \mathbb{R}^n. \tag{12.6}$$

Recall that the state transition matrix is defined by

$$\frac{d}{dt}\Phi(t,s) = A(t)\Phi(t,s), \qquad \Phi(\tau,\tau) = I_{n\times n} \tag{12.7}$$

and the solution to (12.6) is given by the Variation of Parameters Formula

$$\boldsymbol{x}(t) = \Phi(t,t_0)\boldsymbol{x}_0 + \int_{t_0}^{t} \Phi(t,s)B(s)\boldsymbol{u}(s)ds. \tag{12.8}$$

Let

$$W_c(t_0,t_1) \triangleq \int_{t_0}^{t_1} \Phi(t_0,s)B(s)B^T(s)\Phi^T(t_0,s)ds$$

and

$$W_r(t_0,t_1) \triangleq \int_{t_0}^{t_1} \Phi(t_1,s)B(s)B^T(s)\Phi^T(t_1,s)ds.$$

The matrices $W_c(t_0,t_1)$ and $W_r(t_0,t_1)$ are called the *controllability Gramian* and *reachability Gramian*, respectively (see Chapter 3 in [7]). Observe that

$$\begin{aligned}
&\Phi(t_1,t_0)W_c(t_0,t_1)\Phi^T(t_1,t_0)\\
&\quad= \Phi(t_1,t_0)\left[\int_{t_0}^{t_1} \Phi(t_0,s)B(s)B^T(s)\Phi^T(t_0,s)ds\right]\Phi^T(t_1,t_0)\\
&\quad= \left[\int_{t_0}^{t_1} \Phi(t_1,t_0)\Phi(t_0,s)B(s)B^T(s)\Phi^T(t_0,s)\Phi^T(t_1,t_0)ds\right]\\
&\quad= \left[\int_{t_0}^{t_1} [\Phi(t_1,t_0)\Phi(t_0,s)]B(s)B^T(s)[\Phi(t_1,t_0)\Phi(t_0,s)]^T ds\right]\\
&\quad= \left[\int_{t_0}^{t_1} [\Phi(t_1,s)]B(s)B^T(s)[\Phi(t_1,s)]^T ds\right]\\
&\quad= W_r(t_0,t_1),
\end{aligned}$$

so that

$$W_r(t_0,t_1) = \Phi(t_1,t_0)W_c(t_0,t_1)\Phi^T(t_1,t_0).$$

Moreover, since $\Phi(t_1,t_0)$ is non-singular it follows that $W_r(t_0,t_1)$ and $W_c(t_0,t_1)$ have the same range.

Theorem 12.1 (Minimum Energy Theorem) *The state $\boldsymbol{x}_0$ at time t_0 can be transferred to the state $\boldsymbol{x}_1$ at time t_1 if and only if*

$$\boldsymbol{w} = \boldsymbol{x}_1 - \Phi(t_1, t_0)\boldsymbol{x}_0 \in Range(W_r(t_0, t_1)).$$

If $\boldsymbol{w} = \boldsymbol{x}_1 - \Phi(t_1, t_0)\boldsymbol{x}_0 \in Range(W_r(t_0, t_1))$, then there is a vector $\boldsymbol{v} \in \mathbb{R}^n$ such that $\boldsymbol{w} = \boldsymbol{x}_1 - \Phi(t_1, t_0)\boldsymbol{x}_0 = W_r(t_0, t_1)\boldsymbol{v}$ and a control that transfers $\boldsymbol{x}_0$ at time t_0 to $\boldsymbol{x}_1$ at time t_1 is given by

$$\hat{\boldsymbol{u}}(t) = B^T(t)\Phi^T(t_1, t)\boldsymbol{v}. \tag{12.9}$$

If $\boldsymbol{u}_o(\cdot)$ is any other controller that steers $\boldsymbol{x}_0$ at time t_0 to the state $\boldsymbol{x}_1$ at time t_1, then

$$\frac{1}{2}\int_{t_0}^{t_1} \hat{\boldsymbol{u}}^T(s)\hat{\boldsymbol{u}}(s)ds \leq \frac{1}{2}\int_{t_0}^{t_1} \boldsymbol{u}_o^T(s)\boldsymbol{u}_o(s)ds \tag{12.10}$$

and $\hat{\boldsymbol{u}}(t) = B^T(t)\Phi^T(t_1, t)\boldsymbol{v}$ minimizes the "energy" functional defined by

$$E = J(\boldsymbol{u}(\cdot)) = \frac{1}{2}\int_{t_0}^{t} \langle \boldsymbol{u}(s), \boldsymbol{u}(s)\rangle\, ds = \frac{1}{2}\int_{t_0}^{t} \boldsymbol{u}(s)^T\boldsymbol{u}(s)ds.$$

Moreover, if $[W_r(t_0, t_1)]^{-1}$ exists, then the energy required to make this transfer is given by

$$\begin{aligned} E &= \frac{1}{2}\int_{t_0}^{t_1} \hat{\boldsymbol{u}}^T(s)\hat{\boldsymbol{u}}(s)ds = \frac{1}{2}\boldsymbol{w}^T[W_r(t_0, t_1)]^{-1}\boldsymbol{w} \\ &= \frac{1}{2}\left\langle [W_r(t_0, t_1)]^{-1}\boldsymbol{w}, \boldsymbol{w}\right\rangle. \end{aligned} \tag{12.11}$$

Proof: Note that if $\boldsymbol{x}_1 - \Phi(t_1, t_0)\boldsymbol{x}_0 = W_r(t_0, t_1)\boldsymbol{v}$ and $\hat{\boldsymbol{u}}(t) = B^T(s)\Phi^T(t_1, t)\boldsymbol{v}$, the Variation of Parameters formula implies

$$\begin{aligned} \boldsymbol{x}(t) &= \Phi(t, t_0)\boldsymbol{x}_0 + \int_{t_0}^{t} \Phi(t, s)B(s)\hat{\boldsymbol{u}}(s)ds. \\ &= \Phi(t, t_0)\boldsymbol{x}_0 + \int_{t_0}^{t} \Phi(t, s)B(s)B^T(s)\Phi^T(t_1, s)\boldsymbol{v}ds \end{aligned}$$

so that

$$\begin{aligned}\boldsymbol{x}(t_1) &= \Phi(t_1,t_0)\boldsymbol{x}_0 + \left[\int_{t_0}^{t_1} \Phi(t_1,s)B(s)B^T(s)\Phi^T(t_1,s)ds\right]\boldsymbol{v}\\ &= \Phi(t_1,t_0)\boldsymbol{x}_0 + W_r(t_0,t_1)\boldsymbol{v}.\end{aligned}$$

Hence, we have

$$\boldsymbol{x}(t_1) = \Phi(t_1,t_0)\boldsymbol{x}_0 + W_r(t_0,t_1)\boldsymbol{v} = \boldsymbol{x}_1$$

so that $\hat{\boldsymbol{u}}(t) = B^T(s)\Phi^T(t_1,t)\boldsymbol{v}$ steers $\boldsymbol{x}_0$ at time t_0 to $\boldsymbol{x}_1$ at time t_1.

Assume now that $\boldsymbol{u}_o(\cdot)$ is any other controller that steers $\boldsymbol{x}_0$ at time t_0 to the state $\boldsymbol{x}_1$ at time t_1. The Variation of Parameters formula implies

$$\boldsymbol{x}_1 = \Phi(t_1,t_0)\boldsymbol{x}_0 + \int_{t_0}^{t_1} \Phi(t_1,s)B(s)\boldsymbol{u}_o(s)ds.$$

Likewise,

$$\boldsymbol{x}_1 = \Phi(t_1,t_0)\boldsymbol{x}_0 + \int_{t_0}^{t_1} \Phi(t_1,s)B(s)\hat{\boldsymbol{u}}(s)ds$$

and subtracting the two we obtain

$$\int_{t_0}^{t_1} \Phi(t_1,s)B(s)[\boldsymbol{u}_o(s) - \hat{\boldsymbol{u}}(s)]ds = 0.$$

Pre-multiplying by $\boldsymbol{v}^T$ yields

$$\begin{aligned}0 &= \boldsymbol{v}^T \int_{t_0}^{t_1} \Phi(t_0,s)B(s)[\boldsymbol{u}_o(s) - \hat{\boldsymbol{u}}(s)]ds\\ &= \int_{t_0}^{t_1} \boldsymbol{v}^T\Phi(t_0,s)B(s)[\boldsymbol{u}_o(s) - \hat{\boldsymbol{u}}(s)]ds\\ &= \int_{t_0}^{t_1} \left[B^T(s)\Phi^T(t_1,t)\boldsymbol{v}\right]^T[\boldsymbol{u}_o(s) - \hat{\boldsymbol{u}}(s)]ds\end{aligned}$$

and since

$$\hat{\boldsymbol{u}}(t) = B^T(t)\Phi^T(t_1,t)\boldsymbol{v},$$

it follows that

$$0 = \int_{t_0}^{t_1} [\hat{\boldsymbol{u}}(t)]^T [\boldsymbol{u}_o(s) - \hat{\boldsymbol{u}}(s)] ds.$$

Therefore, we have

$$\begin{aligned}
0 &\leq \frac{1}{2}\int_{t_0}^{t_1} [\boldsymbol{u}_o(s) - \hat{\boldsymbol{u}}(s)]^T [\boldsymbol{u}_o(s) - \hat{\boldsymbol{u}}(s)] ds \\
&= \frac{1}{2}\int_{t_0}^{t_1} \left[\boldsymbol{u}_o^T(s)\boldsymbol{u}_o(s) - 2\hat{\boldsymbol{u}}^T(s)\boldsymbol{u}_o(s) + \hat{\boldsymbol{u}}^T(s)\hat{\boldsymbol{u}}(s)\right] ds \\
&= \frac{1}{2}\int_{t_0}^{t_1} \left[\boldsymbol{u}_o^T(s)\boldsymbol{u}_o(s) - 2\hat{\boldsymbol{u}}^T(s)\boldsymbol{u}_o(s) + 2\hat{\boldsymbol{u}}^T(s)\hat{\boldsymbol{u}}(s) - \hat{\boldsymbol{u}}^T(s)\hat{\boldsymbol{u}}(s)\right] ds \\
&= \frac{1}{2}\int_{t_0}^{t_1} \boldsymbol{u}_o^T(s)\boldsymbol{u}_o(s) ds - \int_{t_0}^{t_1} \left[\hat{\boldsymbol{u}}^T(s)\boldsymbol{u}_o(s) - \hat{\boldsymbol{u}}^T(s)\hat{\boldsymbol{u}}(s)\right] ds \\
&\quad - \frac{1}{2}\int_{t_0}^{t_1} \hat{\boldsymbol{u}}^T(s)\hat{\boldsymbol{u}}(s) ds \\
&= \frac{1}{2}\int_{t_0}^{t_1} \boldsymbol{u}_o^T(s)\boldsymbol{u}_o(s) ds - \int_{t_0}^{t_1} \hat{\boldsymbol{u}}^T(s)\left[\boldsymbol{u}_o(s) - \hat{\boldsymbol{u}}(s)\right] ds \\
&\quad - \frac{1}{2}\int_{t_0}^{t_1} \hat{\boldsymbol{u}}^T(s)\hat{\boldsymbol{u}}(s) ds \\
&= \frac{1}{2}\int_{t_0}^{t_1} \boldsymbol{u}_o^T(s)\boldsymbol{u}_o(s) ds - \frac{1}{2}\int_{t_0}^{t_1} \hat{\boldsymbol{u}}^T(s)\hat{\boldsymbol{u}}(s) ds
\end{aligned}$$

and hence

$$\frac{1}{2}\int_{t_0}^{t_1} \hat{\boldsymbol{u}}^T(s)\hat{\boldsymbol{u}}(s) ds \leq \frac{1}{2}\int_{t_0}^{t_1} \boldsymbol{u}_o^T(s)\boldsymbol{u}_o(s) ds,$$

so that $\hat{\boldsymbol{u}}(t) = B^T(t)\Phi^T(t_1, t)\boldsymbol{v}$ minimizes the energy needed to transfer $\boldsymbol{x}_0$ at time t_0 to the state $\boldsymbol{x}_1$ at time t_1.

Finally, if $[W_r(t_0, t_1)]^{-1}$ exists then

$$\boldsymbol{v} = [W_r(t_0, t_1)]^{-1}[\boldsymbol{x}_1 - \Phi(t_1, t_0)\boldsymbol{x}_0] = [W_r(t_0, t_1)]^{-1}\boldsymbol{w}]$$

and the control is given by

$$\hat{\boldsymbol{u}}(t) = B^T(t)\Phi^T(t_1, t)\,\boldsymbol{v} = B^T(t)\Phi^T(t_1, t)[W_r(t_0, t_1)]^{-1}\boldsymbol{w}.$$

Thus, the minimum energy required to transfer $\boldsymbol{x}_0$ at time t_0 to the state $\boldsymbol{x}_1$ at time t_1 is given by

$$\begin{aligned}
E &= \frac{1}{2}\int_{t_0}^{t} \langle \boldsymbol{u}(s), \boldsymbol{u}(s)\rangle\, ds = \frac{1}{2}\int_{t_0}^{t} \boldsymbol{u}(s)^T \boldsymbol{u}(s) ds \\
&= \frac{1}{2}\int_{t_0}^{t} [B^T(s)\Phi^T(t_1, s)[W_r(t_0, t_1)]^{-1}\boldsymbol{w}]^T \\
&\qquad \times [B^T(s)\Phi^T(t_1, s)[W_r(t_0, t_1)]^{-1}\boldsymbol{w}] ds \\
&= \frac{1}{2}\int_{t_0}^{t} \boldsymbol{w}^T[[W_r(t_0, t_1)]^{-1}]^T\Phi(t_1, s)B^T(s) \\
&\qquad \times B^T(s)\Phi^T(t_1, s)[W_r(t_0, t_1)]^{-1}\boldsymbol{w}] ds \\
&= \frac{1}{2}\boldsymbol{w}^T[[W_r(t_0, t_1)]^{-1}]^T \left[\int_{t_0}^{t} \Phi(t_1, s)B^T(s)B^T(s)\Phi^T(t_1, s) ds\right] \\
&\quad \times [W_r(t_0, t_1)]^{-1}\boldsymbol{w}] \\
&= \frac{1}{2}\boldsymbol{w}^T[[W_r(t_0, t_1)]^{-1}]^T\, [W_r(t_0, t_1)]\, [W_r(t_0, t_1)]^{-1}\boldsymbol{w}] \\
&= \frac{1}{2}\boldsymbol{w}^T[[W_r(t_0, t_1)]^{-1}]^T\boldsymbol{w}].
\end{aligned}$$

Since $[W_r(t_0, t_1)]^T = W_r(t_0, t_1)$, it follows that $[[W_r(t_0, t_1)]^{-1}]^T = [W_r(t_0, t_1)]^{-1}$ and hence

$$E = \frac{1}{2}\boldsymbol{w}^T[[W_r(t_0, t_1)]^{-1}]^T\boldsymbol{w} = \frac{1}{2}\boldsymbol{w}^T[W_r(t_0, t_1)]^{-1}\boldsymbol{w} \qquad (12.12)$$

and this completes the proof. □

Observe that the previous theorem provides a characterization of the (complete) controllability of the system (12.6). In particular, we have the following result.

Theorem 12.2 *The system (12.6) is completely controllable at time t_0 if and only if there is a time $t_1 > t_0$ such that $W_r(t_0, t_1)$ has maximal rank n. Since $rank[W_r(t_0, t_1)] = rank[W_c(t_0, t_1)]$, it also follows that (12.6) is completely controllable at time t_0 if and only if there is a time $t_1 > t_0$ such that $W_c(t_0, t_1)$ has maximal rank n.*

For time invariant systems

$$\dot{\boldsymbol{x}}(t) = A\boldsymbol{x}(t) + B\boldsymbol{u}(t) \tag{12.13}$$

one has the following result (see [7], [45], [110] and [115]).

Theorem 12.3 *For the time invariant system (12.13),*

$$\Phi(t, s) = e^{A(t-s)}$$

and $Range(\mathcal{CM}) = Range(W_r(0,t))$ *for all time* $t > 0$, *where*

$$\mathcal{CM} = [B, AB, A^2B, \ldots, A^{n-1}B].$$

Remark 12.2 *It is important to note that for time invariant systems where*

$$\Phi(t, s) = e^{A(t-s)}$$

the transpose of $\Phi(t, s)$ *is given by*

$$[\Phi(t, s)]^T = e^{A^T(t-s)}$$

so that

$$\frac{d}{dt}[\Phi(t, s)]^T = A^T e^{A^T(t-s)} = A^T[\Phi(t, s)]^T.$$

However, in general one has

$$\frac{d}{dt}[\Phi(t, s)]^T = \left[\frac{d}{dt}[\Phi(t, s)]\right]^T = [A(t)\Phi(t, s)]^T = [\Phi(t, s)]^T A(t)^T.$$

Moreover, (in general) $\Phi(t, s)$ *and* $A(t)$ *do not commute so that*

$$[\Phi(t, s)]^T A(t)^T \neq A(t)^T[\Phi(t, s)]^T$$

and one cannot imply that $\frac{d}{dt}[\Phi(t, s)]^T = A(t)^T[\Phi(t, s)]^T$. *For example, consider the problem with*

$$A(t) = \begin{bmatrix} 0 & 1 \\ \frac{-2}{t^2} & \frac{2}{t} \end{bmatrix}$$

for $t > 1$. *The state transition matrix is given by*

$$\Phi(t, s) = \begin{bmatrix} (\frac{2t}{s} - \frac{t^2}{s^2}) & (\frac{t^2}{s} - t) \\ (\frac{2}{s} - \frac{2t^2}{s^2}) & (\frac{2t}{s} - 1) \end{bmatrix}$$

and it is easy to show that

$$[\Phi(t, s)]A(t) \neq A(t)[\Phi(t, s)].$$

Example 12.1 *Consider the model of the rocket sled*

$$\frac{d}{dt}\begin{bmatrix} x_1(t) \\ x_2(t) \end{bmatrix} = \begin{bmatrix} 0 & 1 \\ 0 & 0 \end{bmatrix}\begin{bmatrix} x_1(t) \\ x_2(t) \end{bmatrix} + \begin{bmatrix} 0 \\ 1/m \end{bmatrix} u(t).$$

A direct calculation yields

$$\Phi(t,s) = e^{A(t-s)} = \begin{bmatrix} 1 & (t-s) \\ 0 & 1 \end{bmatrix}$$

and

$$\Phi(t,s)BB^T[\Phi(t,s)]^T = \frac{1}{m^2}\begin{bmatrix} (t-s)^2 & (t-s) \\ (t-s) & 1 \end{bmatrix}.$$

Therefore,

$$W_r(0,t) = \int_0^t \Phi(t_1,s)B(s)B^T(s)\Phi^T(t_1,s)ds = \frac{1}{m^2}\begin{bmatrix} \frac{t3}{3} & \frac{t2}{2} \\ \frac{t2}{2} & t \end{bmatrix}$$

is non-singular.

Again, in the special case where the system is autonomous (i.e. time-invariant) and has the form

$$\begin{cases} \dot{\boldsymbol{x}}(t) = A\boldsymbol{x}(t) + B\boldsymbol{u}(t) \in \mathbb{R}^n \\ \boldsymbol{y}(t) = C\boldsymbol{x}(t) \in \mathbb{R}^p \end{cases}, \tag{12.14}$$

one has the following results which can be found in [7], [45], [110], [115] and [119].

Theorem 12.4 *Consider the linear autonomous system (12.14).*
(i) The autonomous linear system (12.14) is controllable if and only if the $n \times nm$ controllability matrix

$$\mathcal{CM} = [B, AB, A^2B, \ldots, A^{n-1}B] \tag{12.15}$$

has rank n.

(ii) The linear autonomous system (12.14) is observable if and only if the $n \times np$ observability matrix

$$\mathcal{OM} = [C^T, A^TC^T, [A^T]^2C^T, \ldots, [A^T]^{n-1}C^T] \tag{12.16}$$

has rank n.

The previous results point to an important dual relationship between controllability and observability. Given the system (12.14) we define the *dual system* by

$$\begin{cases} \dot{\boldsymbol{z}}(t) = A^T \boldsymbol{z}(t) + C^T \boldsymbol{v}(t) \in \mathbb{R}^n \\ \boldsymbol{\varsigma}(t) = B^T \boldsymbol{z}(t) \in \mathbb{R}^m \end{cases} \tag{12.17}$$

where C^T is now an $n \times p$ matrix and B^T is an $m \times n$ matrix. Note that the dual system has inputs from the output space of the system (12.14) and outputs from the input space of (12.14). Thus, we have the following duality result.

Theorem 12.5 *The system (12.14) is controllable if and only if the dual system (12.17) is observable and (12.14) is observable if and only if the dual system (12.17) is controllable.*

Example 12.2 *Consider the rocket control problem defined by the system*

$$\frac{d}{dt}\begin{bmatrix} x_1(t) \\ x_2(t) \end{bmatrix} = \begin{bmatrix} 0 & 1 \\ 0 & 0 \end{bmatrix}\begin{bmatrix} x_1(t) \\ x_2(t) \end{bmatrix} + \begin{bmatrix} 0 \\ 1/m \end{bmatrix} u(t), \tag{12.18}$$

or

$$\dot{\boldsymbol{x}}(t) = A\boldsymbol{x}(t) + Bu(t), \tag{12.19}$$

where the matrices A and B are defined by

$$A = \begin{bmatrix} 0 & 1 \\ 0 & 0 \end{bmatrix} \quad \textit{and } B = \begin{bmatrix} 0 \\ 1/m \end{bmatrix},$$

respectively. Here, $\boldsymbol{x}(t) = \begin{bmatrix} x_1(t) & x_2(t) \end{bmatrix}^T$, where $x_1(t)$ is the position of the sled and $x_1(t)$ is the velocity. Since $n = 2$, the controllability matrix is given by

$$\mathcal{CM} = [B, AB] = \begin{bmatrix} 0 & 1/m \\ 1/m & 0 \end{bmatrix}$$

and clearly $rank(\mathcal{CM}) = 2$ so that the system is controllable. Consider the output given by

$$y(t) = C\boldsymbol{x}(t) = \begin{bmatrix} 0 & 1 \end{bmatrix}\begin{bmatrix} x_1(t) \\ x_2(t) \end{bmatrix} = x_2(t).$$

Here $C = \begin{bmatrix} 0 & 1 \end{bmatrix}$ *and hence*

$$C^T = \begin{bmatrix} 0 \\ 1 \end{bmatrix}.$$

Moreover,

$$A^T = \begin{bmatrix} 0 & 0 \\ 1 & 0 \end{bmatrix}$$

so that the observability matrix is given by

$$\mathcal{OM} = [C^T, A^T C^T] = \begin{bmatrix} 0 & 0 \\ 1 & 1 \end{bmatrix}.$$

Since $\text{rank}(\mathcal{OM}) = 1$ *the system is not observable. On the other hand, consider the case where the the output is given by*

$$y(t) = C\boldsymbol{x}(t) = \begin{bmatrix} 1 & 0 \end{bmatrix} \begin{bmatrix} x_1(t) \\ x_2(t) \end{bmatrix} = x_1(t).$$

Here $C = \begin{bmatrix} 1 & 0 \end{bmatrix}$ *and hence*

$$C^T = \begin{bmatrix} 1 \\ 0 \end{bmatrix}$$

so that the observability matrix is given by

$$\mathcal{OM} = [C^T, A^T C^T] = \begin{bmatrix} 1 & 0 \\ 0 & 1 \end{bmatrix}$$

and $\text{rank}(\mathcal{OM}) = 2$. *In this case the system is observable.*

We see that sensing (measuring) the velocity leads to an unobservable system while sensing the position produces an observable system.

Example 12.3 *Consider the system*

$$\frac{d}{dt} \begin{bmatrix} x_1(t) \\ x_2(t) \end{bmatrix} = \begin{bmatrix} -1 & 0 \\ 0 & 1 \end{bmatrix} \begin{bmatrix} x_1(t) \\ x_2(t) \end{bmatrix} + \begin{bmatrix} 0 \\ 1 \end{bmatrix} u(t), \qquad (12.20)$$

with output

$$y(t) = y(t) = C\boldsymbol{x}(t) = \begin{bmatrix} 1 & 1 \end{bmatrix} \begin{bmatrix} x_1(t) \\ x_2(t) \end{bmatrix} = x_1(t) + x_2(t).$$

Here,

$$A = \begin{bmatrix} -1 & 0 \\ 0 & 1 \end{bmatrix}, \; B = \begin{bmatrix} 0 \\ 1 \end{bmatrix} \; \textit{and} \; C = \begin{bmatrix} 1 & 1 \end{bmatrix}$$

so that

$$\mathcal{CM} = [B, AB] = \begin{bmatrix} 0 & 0 \\ 1 & 1 \end{bmatrix}$$

and

$$\mathcal{OM} = [C^T, A^T C^T] = \begin{bmatrix} 1 & -1 \\ 1 & 1 \end{bmatrix}.$$

Since $rank(\mathcal{CM}) = 1$ *the system is not controllable, but since* $rank(\mathcal{OM}) = 2$ *the system is observable.*

Stability is of primary importance in many control systems and ensuring stability often is a major goal in feedback design. We recall the basic results for stability of linear autonomous systems.

Theorem 12.6 *Let* A *be a constant* $n \times n$ *matrix and suppose that all the eigenvalues* $\lambda_i, i = 1, 2, \ldots, n$ *of* A *have negative real part. Then every solution of* $\dot{\boldsymbol{x}}(t) = A\boldsymbol{x}(t)$ *is exponentially asymptotically stable. In particular, there exists constants* $M \geq 1$ *and* $\gamma > 0$ *such that*

$$\|\boldsymbol{x}(t)\| \leq M e^{-\gamma t} \|\boldsymbol{x}(0)\| \tag{12.21}$$

for all $t \geq 0$*. On the other hand, if* A *has one eigenvalue with positive real part, then every solution of* $\dot{\boldsymbol{x}}(t) = A\boldsymbol{x}(t)$ *is unstable.*

Since the stability of the system

$$\dot{\boldsymbol{x}}(t) = A\boldsymbol{x}(t)$$

is determined by the eigenvalues of A, we say that A is a *stable matrix* if and only if

$$\Re(\lambda_i) < 0 \tag{12.22}$$

for all eigenvalues $\lambda_i, i = 1, 2, \ldots, n$ of A.

As observed in the previous chapter (see Section 11.5), linear quadratic optimal control problems yield optimal controllers that are of the form of linear state feedback

$$\boldsymbol{u}^*(t) = -K(t)\boldsymbol{x}^*(t).$$

We shall see later that in some cases the gain matrix $K(t)$ is constant so that the feedback controller has the form

$$\boldsymbol{u}^*(t) = -K\boldsymbol{x}^*(t). \tag{12.23}$$

If we apply a feedback control law of the form (12.23) to the control system

$$\dot{\boldsymbol{x}}(t) = A\boldsymbol{x}(t) + B\boldsymbol{u}(t),$$

then the closed-loop system has the form

$$\dot{\boldsymbol{x}}(t) = A\boldsymbol{x}(t) - BK\boldsymbol{x}(t) = [A - BK]\boldsymbol{x}(t). \tag{12.24}$$

Consequently, the closed-loop system is stable if and only if the matrix $[A - BK]$ is a stable matrix. Thus, we are interested in the issue of finding a gain matrix such that $[A - BK]$ is a stable matrix and this motivates the following definitions.

Definition 12.3 *The control system*

$$\begin{cases} \dot{\boldsymbol{x}}(t) = A\boldsymbol{x}(t) + B\boldsymbol{u}(t) \in \mathbb{R}^n \\ \boldsymbol{y}(t) = C\boldsymbol{x}(t) \in \mathbb{R}^p \end{cases} \tag{12.25}$$

is stabilizable if there exists a $m \times n$ matrix K such that $[A - BK]$ is a stable matrix.

Definition 12.4 *The control system*

$$\begin{cases} \dot{\boldsymbol{x}}(t) = A\boldsymbol{x}(t) + B\boldsymbol{u}(t) \in \mathbb{R}^n \\ \boldsymbol{y}(t) = C\boldsymbol{x}(t) \in \mathbb{R}^p \end{cases} \tag{12.26}$$

is detectable if there exists a $n \times p$ matrix F such that $[A - FC]$ is a stable matrix.

Observe that if one considers the dual system (12.17)

$$\begin{cases} \dot{\boldsymbol{z}}(t) = A^T\boldsymbol{z}(t) + C^T\boldsymbol{v}(t) \in \mathbb{R}^n \\ \boldsymbol{\zeta}(t) = B^T\boldsymbol{z}(t) \in \mathbb{R}^m \end{cases}, \tag{12.27}$$

then the dual system is stabilizable if there is a matrix $\tilde{K}$ such that $[A^T - C^T\tilde{K}]$ is a stable matrix. However, since a matrix and its transpose have the same eigenvalues it follows that

$[A^T - C^T\tilde{K}]^T = [A - \tilde{K}^T C]$ is a stable matrix. Thus, if we set $F = \tilde{K}^T$, then $[A - FC]$ is a stable matrix and the system (12.26) is detectable. Likewise, we see that if the dual system (12.27) is detectable, then the system (12.26) is stabilizable. Consequently, like controllability and observability, stabilizability and detectability are dual concepts for autonomous systems.

Consider the example above defined by the system

$$\frac{d}{dt}\begin{bmatrix} x_1(t) \\ x_2(t) \end{bmatrix} = \begin{bmatrix} -1 & 0 \\ 0 & 1 \end{bmatrix}\begin{bmatrix} x_1(t) \\ x_2(t) \end{bmatrix} + \begin{bmatrix} 0 \\ 1 \end{bmatrix} u(t). \tag{12.28}$$

Recall that (12.28) is not controllable. However, let $K = \begin{bmatrix} 0 & 4 \end{bmatrix}$ and compute

$$\begin{aligned} [A - BK] &= \begin{bmatrix} -1 & 0 \\ 0 & 1 \end{bmatrix} - \begin{bmatrix} 0 \\ 1 \end{bmatrix}\begin{bmatrix} 0 & 4 \end{bmatrix} \\ &= \begin{bmatrix} -1 & 0 \\ 0 & 1 \end{bmatrix} - \begin{bmatrix} 0 & 0 \\ 0 & 4 \end{bmatrix} = \begin{bmatrix} -1 & 0 \\ 0 & -3 \end{bmatrix}. \end{aligned}$$

Since the eigenvalues of

$$[A - BK] = \begin{bmatrix} -1 & 0 \\ 0 & -3 \end{bmatrix}$$

are $\lambda_1 = -1$ and $\lambda_2 = -3$, $[A - BK]$ is a stable matrix and hence (12.28) is stabilizable. Thus, uncontrollable systems can be stabilizable. Likewise, there are detectable systems that are not observable so that stabilizability and detectability are weaker conditions than controllability and observability.

12.2 Linear Control Systems Arising from Nonlinear Problems

Although linear control systems occur naturally in many applications, linear systems also play a central role in control problems governed by nonlinear systems. In some cases, linearization offers the only practical approach to the analysis and design of control systems. Also, of immense importance is the issue of robustness and sensitivity of optimal controllers with respect to parameters and disturbances. We briefly discuss these processes below.

12.2.1 Linearized Systems

Consider the case where $\bar{\boldsymbol{u}}(\cdot) \in PWC(t_0, t_1)$ and $\bar{\boldsymbol{x}}(\cdot) \in PWS(t_0, t_1)$ is the response to the nonlinear control system

$$(\mathcal{NS}) \qquad \frac{d}{dt}\bar{\boldsymbol{x}}(t) = f(\bar{\boldsymbol{x}}(t), \bar{\boldsymbol{u}}(t)), \tag{12.29}$$

with initial condition

$$\bar{\boldsymbol{x}}(t_0) = \bar{\boldsymbol{x}}_0.$$

We call $\bar{\boldsymbol{u}}(\cdot)$ the *nominal control* and $\bar{\boldsymbol{x}}(\cdot)$ the *nominal trajectory.* In many cases the nominal control could be an optimal controller for a given problem, but for the moment we simply assume it is a given control. Let

$$\boldsymbol{u}(t) = \bar{\boldsymbol{u}}(t) + \boldsymbol{v}(t)$$

where we assume that $\boldsymbol{v}(t)$ is "small" and let

$$\boldsymbol{x}(t) = \bar{\boldsymbol{x}}(t) + \boldsymbol{z}(t)$$

be the response to the control $\boldsymbol{u}(t) = \bar{\boldsymbol{u}}(t) + \boldsymbol{v}(t)$ with initial condition

$$\boldsymbol{x}(t_0) = \bar{\boldsymbol{x}}(t_0) + \boldsymbol{z}(t_0) = \bar{\boldsymbol{x}}_0 + \boldsymbol{z}_0.$$

Assuming that both $\boldsymbol{v}(t)$ and $\boldsymbol{z}_0$ are small, one would expect that $\boldsymbol{z}(t)$ is small. Applying Taylor's theorem we have

$$\begin{aligned}\frac{d}{dt}\boldsymbol{x}(t) &= \frac{d}{dt}\bar{\boldsymbol{x}}(t) + \frac{d}{dt}\boldsymbol{z}(t) = f(\bar{\boldsymbol{x}}(t) + \boldsymbol{z}(t), \bar{\boldsymbol{u}}(t) + \boldsymbol{v}(t))\\ &= f(\bar{\boldsymbol{x}}(t), \bar{\boldsymbol{u}}(t)) + [\mathbb{J}_{\boldsymbol{x}} f(\bar{\boldsymbol{x}}(t), \bar{\boldsymbol{u}}(t))]\boldsymbol{z}(t)\\ &\quad + [\mathbb{J}_{\boldsymbol{u}} f(\bar{\boldsymbol{x}}(t), \bar{\boldsymbol{u}}(t))]\boldsymbol{v}(t) + HOT(t).\end{aligned}$$

Here, $[\mathbb{J}_{\boldsymbol{x}} f(\bar{\boldsymbol{x}}(t), \bar{\boldsymbol{u}}(t))]$ and $[\mathbb{J}_{\boldsymbol{u}} f(\bar{\boldsymbol{x}}(t), \bar{\boldsymbol{u}}(t))]$ are the Jacobian matrices and the higher order terms $HOT(t)$ are "small" with respect to the control $\boldsymbol{v}(t)$ and response $\boldsymbol{z}(t)$. Hence,

$$\begin{aligned}\frac{d}{dt}\bar{\boldsymbol{x}}(t) + \frac{d}{dt}\boldsymbol{z}(t) &= f(\bar{\boldsymbol{x}}(t), \bar{\boldsymbol{u}}(t)) + [\mathbb{J}_{\boldsymbol{x}} f(\bar{\boldsymbol{x}}(t), \bar{\boldsymbol{u}}(t))]\boldsymbol{z}(t)\\ &\quad + [\mathbb{J}_{\boldsymbol{u}} f(\bar{\boldsymbol{x}}(t), \bar{\boldsymbol{u}}(t))]\boldsymbol{v}(t) + HOT(t),\end{aligned}$$

and since

$$\frac{d}{dt}\bar{\boldsymbol{x}}(t) = f(\bar{\boldsymbol{x}}(t), \bar{\boldsymbol{u}}(t)),$$

it follows that

$$\frac{d}{dt}\boldsymbol{z}(t) = [\mathbb{J}_{\boldsymbol{x}} f(\bar{\boldsymbol{x}}(t), \bar{\boldsymbol{u}}(t))]\boldsymbol{z}(t) + [\mathbb{J}_{\boldsymbol{u}} f(\bar{\boldsymbol{x}}(t), \bar{\boldsymbol{u}}(t))]\boldsymbol{v}(t) + HOT(t).$$

Therefore, the "variation" $\boldsymbol{z}(t)$ is an approximate solution to the linear system

$$\frac{d}{dt}\boldsymbol{z}(t) = [\mathbb{J}_{\boldsymbol{x}} f(\bar{\boldsymbol{x}}(t), \bar{\boldsymbol{u}}(t))]\boldsymbol{z}(t) + [\mathbb{J}_{\boldsymbol{u}} f(\bar{\boldsymbol{x}}(t), \bar{\boldsymbol{u}}(t))]\boldsymbol{v}(t)$$

with input $\boldsymbol{v}(t)$. Defining

$$A(t) \triangleq [\mathbb{J}_{\boldsymbol{x}} f(\bar{\boldsymbol{x}}(t), \bar{\boldsymbol{u}}(t))] \quad \text{and} \quad B(t) \triangleq [\mathbb{J}_{\boldsymbol{u}} f(\bar{\boldsymbol{x}}(t), \bar{\boldsymbol{u}}(t))]$$

leads to the linearized system

$$\dot{\boldsymbol{z}}(t) = A(t)\boldsymbol{z}(t) + B(t)\boldsymbol{v}(t). \tag{12.30}$$

The theory of ordinary differential equations can be used to provide a rigorous basis of the linearization (see [58], [138] and [153]).

In the special case where $\bar{\boldsymbol{u}}(t) \equiv \bar{\boldsymbol{u}}$ and $\bar{\boldsymbol{x}}(t) \equiv \bar{\boldsymbol{x}}$ are constants, then

$$A \triangleq [\mathbb{J}_{\boldsymbol{x}} f(\bar{\boldsymbol{x}}, \bar{\boldsymbol{u}})] \quad \text{and} \quad B \triangleq [\mathbb{J}_{\boldsymbol{u}} f(\bar{\boldsymbol{x}}, \bar{\boldsymbol{u}})]$$

are time independent and the linearized system (12.30) is an *autonomous* system

$$\dot{\boldsymbol{z}}(t) = A\boldsymbol{z}(t) + B\boldsymbol{v}(t). \tag{12.31}$$

12.2.2 Sensitivity Systems

Assume the control is parameterized, say $\boldsymbol{u}(t) = \boldsymbol{u}(t, \boldsymbol{q})$ where $\boldsymbol{q} \in \mathbb{R}^p$. For a fixed parameter $\bar{\boldsymbol{q}}$ let the nominal control be defined by $\bar{\boldsymbol{u}}(t) = \boldsymbol{u}(t, \bar{\boldsymbol{q}})$ with nominal trajectory $\bar{\boldsymbol{x}}(t) = \boldsymbol{x}(t, \bar{\boldsymbol{q}})$. The **raw sensitivity** of $\boldsymbol{x}(t, \boldsymbol{q})$ at $\boldsymbol{q} = \bar{\boldsymbol{q}}$ is defined by the partial derivative

$$\frac{\partial \boldsymbol{x}(t, \bar{\boldsymbol{q}})}{\partial \boldsymbol{q}} = \partial_{\boldsymbol{q}} \boldsymbol{x}(t, \bar{\boldsymbol{q}}) = \boldsymbol{x}_q(t, \boldsymbol{q})|_{\boldsymbol{q}=\bar{\boldsymbol{q}}} = \boldsymbol{x}_q(t, \bar{\boldsymbol{q}}).$$

Observe that for a fixed $\hat{t}$, the raw sensitivity is nothing more than the derivative of the nonlinear mapping $G(\hat{t}) : \mathbb{R}^p \longrightarrow \mathbb{R}^n$ defined by

$$G(\hat{t})\boldsymbol{q} \triangleq \boldsymbol{x}(\hat{t}, \boldsymbol{q}).$$

Hence, for each t and $\bar{\mathbf{q}}$ the derivative $\partial_{\boldsymbol{q}}\boldsymbol{x}(t, \bar{\boldsymbol{q}})$ is a linear operator $\partial_{\boldsymbol{q}}\boldsymbol{x}(t, \bar{\boldsymbol{q}}) : \mathbb{R}^p \longrightarrow \mathbb{R}^n$ which has the matrix representation

$$\mathbb{J}_{\boldsymbol{q}}\boldsymbol{x}(t, \bar{\boldsymbol{q}}) = \begin{bmatrix} \frac{\partial x_1(t,\bar{\boldsymbol{q}})}{\partial q_1} & \frac{\partial x_1(t,\bar{\boldsymbol{q}})}{\partial q_2} & \cdots & \frac{\partial x_1(t,\bar{\boldsymbol{q}})}{\partial q_p} \\ \frac{\partial x_2(t,\bar{\boldsymbol{q}})}{\partial q_1} & \frac{\partial x_2(t,\bar{\boldsymbol{q}})}{\partial q_2} & \cdots & \frac{\partial x_2(t,\bar{\boldsymbol{q}})}{\partial q_p} \\ \vdots & \vdots & \ddots & \vdots \\ \frac{\partial x_n(t,\bar{\boldsymbol{q}})}{\partial q_1} & \frac{\partial x_n(t,\bar{\boldsymbol{q}})}{\partial q_2} & \cdots & \frac{\partial x_n(t,\bar{\boldsymbol{q}})}{\partial q_p} \end{bmatrix}. \tag{12.32}$$

We are interested in computing the raw sensitivity as a function of time.

Given a parameter $\boldsymbol{q} \in \mathbb{R}^p$ and the control $\boldsymbol{u}(t) = \boldsymbol{u}(t, \boldsymbol{q})$, then the corresponding trajectory $\boldsymbol{x}(t) = \boldsymbol{x}(t, \boldsymbol{q})$ satisfies the nonlinear system (12.29) which explicitly has the form

$$\dot{\boldsymbol{x}}(t, \boldsymbol{q}) = \frac{d}{dt}\boldsymbol{x}(t, \boldsymbol{q}) = f(\boldsymbol{x}(t, \boldsymbol{q}), \boldsymbol{u}(t, \boldsymbol{q})).$$

It is important to note that $\boldsymbol{x}(t, \boldsymbol{q})$ is a function of two variables t and $\boldsymbol{q}$ so that when we write $\frac{d}{dt}\boldsymbol{x}(t, \boldsymbol{q})$ or $\dot{\boldsymbol{x}}(t, \boldsymbol{q})$ we mean

$$\frac{d}{dt}\boldsymbol{x}(t, \boldsymbol{q}) = \frac{\partial}{\partial t}\boldsymbol{x}(t, \boldsymbol{q})$$

so that

$$\frac{\partial}{\partial t}\boldsymbol{x}(t, \boldsymbol{q}) = f(\boldsymbol{x}(t, \boldsymbol{q}), \boldsymbol{u}(t, \boldsymbol{q})). \tag{12.33}$$

Consequently, taking the partial derivatives of both sides of (12.33) with respect to $\boldsymbol{q}$ yields

$$\frac{\partial}{\partial \boldsymbol{q}}\left[\frac{\partial}{\partial t}\boldsymbol{x}(t, \boldsymbol{q})\right] = \frac{\partial}{\partial \boldsymbol{q}}\left[f(\boldsymbol{x}(t, \boldsymbol{q}), \boldsymbol{u}(t, \boldsymbol{q}))\right]$$

and applying the chain rule we obtain

$$\begin{aligned}\frac{\partial}{\partial \boldsymbol{q}}\left[\frac{\partial}{\partial t}\boldsymbol{x}(t,\boldsymbol{q})\right] &= \frac{\partial}{\partial \boldsymbol{q}}\left[f(\boldsymbol{x}(t,\boldsymbol{q}),\boldsymbol{u}(t,\boldsymbol{q}))\right] \\ &= \frac{\partial}{\partial \boldsymbol{x}}f(\boldsymbol{x}(t,\boldsymbol{q}),\boldsymbol{u}(t,\boldsymbol{q}))\circ\frac{\partial}{\partial \boldsymbol{q}}\boldsymbol{x}(t,\boldsymbol{q}) \\ &\quad +\frac{\partial}{\partial \boldsymbol{u}}f(\boldsymbol{x}(t,\boldsymbol{q}),\boldsymbol{u}(t,\boldsymbol{q}))\circ\frac{\partial}{\partial \boldsymbol{q}}\boldsymbol{u}(t,\boldsymbol{q}),\end{aligned}$$

where "∘" denotes function composition. Assuming that $\boldsymbol{x}(t,\boldsymbol{q})$ is smooth in both parameters, we can interchange the order of integration to produce

$$\begin{aligned}\frac{\partial}{\partial t}\left[\frac{\partial}{\partial \boldsymbol{q}}\boldsymbol{x}(t,\boldsymbol{q})\right] &= \frac{\partial}{\partial \boldsymbol{q}}\left[f(\boldsymbol{x}(t,\boldsymbol{q}),\boldsymbol{u}(t,\boldsymbol{q}))\right] \\ &= \frac{\partial}{\partial \boldsymbol{x}}f(\boldsymbol{x}(t,\boldsymbol{q}),\boldsymbol{u}(t,\boldsymbol{q}))\circ\frac{\partial}{\partial \boldsymbol{q}}\boldsymbol{x}(t,\boldsymbol{q}) \\ &\quad +\frac{\partial}{\partial \boldsymbol{u}}f(\boldsymbol{x}(t,\boldsymbol{q}),\boldsymbol{u}(t,\boldsymbol{q}))\circ\frac{\partial}{\partial \boldsymbol{q}}\boldsymbol{u}(t,\boldsymbol{q}).\end{aligned}$$

In terms of the matrix representations of the operators, we have that

$$\begin{aligned}\frac{\partial}{\partial t}\left[\mathbb{J}_{\boldsymbol{q}}\boldsymbol{x}(t,\boldsymbol{q})\right] &= \left[\mathbb{J}_{\boldsymbol{x}}f(\boldsymbol{x}(t,\boldsymbol{q}),\boldsymbol{u}(t,\boldsymbol{q}))\right]\left[\mathbb{J}_{\boldsymbol{q}}\boldsymbol{x}(t,\boldsymbol{q})\right] \\ &\quad +\left[\mathbb{J}_{\boldsymbol{u}}f(\boldsymbol{x}(t,\boldsymbol{q}),\boldsymbol{u}(t,\boldsymbol{q}))\right]\left[\mathbb{J}_{\boldsymbol{q}}\boldsymbol{u}(t,\boldsymbol{q})\right].\end{aligned}$$

Again, we define

$$\bar{A}(t) \triangleq \mathbb{J}_{\boldsymbol{x}}f(\boldsymbol{x}(t,\bar{\boldsymbol{q}}),\boldsymbol{u}(t,\bar{\boldsymbol{q}})) \quad \text{and} \quad \bar{B}(t) \triangleq \mathbb{J}_{\boldsymbol{u}}f(\boldsymbol{x}(t,\bar{\boldsymbol{q}}),\boldsymbol{u}(t,\bar{\boldsymbol{q}}))$$

and let

$$\boldsymbol{s}(t,\bar{\boldsymbol{q}}) \triangleq \mathbb{J}_{\boldsymbol{q}}\boldsymbol{x}(t,\bar{\boldsymbol{q}}).$$

It now follows that the sensitivity (matrix) $\boldsymbol{s}(t,\bar{\boldsymbol{q}})$ satisfies the linear matrix equation

$$\frac{\partial}{\partial t}\left[\boldsymbol{s}(t,\bar{\boldsymbol{q}})\right] = \bar{A}(t)\boldsymbol{s}(t,\bar{\boldsymbol{q}}) + \bar{B}(t)\left[\mathbb{J}_{\boldsymbol{q}}\boldsymbol{u}(t,\bar{\boldsymbol{q}})\right], \tag{12.34}$$

where $\mathbb{J}_{\boldsymbol{q}}\boldsymbol{u}(t,\boldsymbol{q})$ is the Jacobian matrix for $\boldsymbol{u}(t,\boldsymbol{q})$.

Therefore, the linear matrix differential equation (12.34) defines the dynamics of the time evolution of the sensitivities $\boldsymbol{s}(t,\boldsymbol{q}) \triangleq \mathbb{J}_{\boldsymbol{q}}\boldsymbol{x}(t,\boldsymbol{q})$ at $\boldsymbol{q} = \bar{\boldsymbol{q}}$. Equation (12.34) is called the **Continuous Sensitivity Equation.** If we introduce the additional notation

$$\boldsymbol{S}(t) \triangleq \boldsymbol{s}(t,\bar{\boldsymbol{q}}) = \mathbb{J}_{\boldsymbol{q}}\boldsymbol{x}(t,\bar{\boldsymbol{q}})$$

and

$$\boldsymbol{V}(t) \triangleq \mathbb{J}_{\boldsymbol{q}}\boldsymbol{u}(t,\bar{\boldsymbol{q}}),$$

then the continuous sensitivity equation has the form

$$\dot{\boldsymbol{S}}(t) = \bar{A}(t)\boldsymbol{S}(t) + \bar{B}(t)\boldsymbol{V}(t). \tag{12.35}$$

Observe that if the initial state $\bar{\boldsymbol{x}}(t_0) = \bar{\boldsymbol{x}}_0$ does not depend on $\boldsymbol{q}$, then the initial condition for the continuous sensitivity equation (12.35) is given by

$$\boldsymbol{S}(t_0) = \boldsymbol{0}.$$

However, if $\bar{\boldsymbol{x}}(t_0) = \bar{\boldsymbol{x}}_0(\boldsymbol{q})$ also depends on the parameter $\boldsymbol{q}$, then the appropriate initial condition for the matrix sensitivity equation (12.35) is given by

$$\boldsymbol{S}(t_0) = \mathbb{J}_{\boldsymbol{q}}\boldsymbol{x}_0(\bar{\boldsymbol{q}}). \tag{12.36}$$

Even in the special case where $x(t) = x(t,\boldsymbol{q})$ is a scalar, the sensitivity $\boldsymbol{s}(t,\bar{\boldsymbol{q}})$ is a vector,

$$\boldsymbol{s}(t,\bar{\boldsymbol{q}}) \triangleq \mathbb{J}_{\boldsymbol{q}}x(t,\bar{\boldsymbol{q}}) = \nabla_{\boldsymbol{q}}x(t,\bar{\boldsymbol{q}})$$

and the sensitivity system has the form

$$\frac{\partial}{\partial t}\left[\nabla_{\boldsymbol{q}}x(t,\bar{\boldsymbol{q}})\right] = \bar{A}(t)[\nabla_{\boldsymbol{q}}x(t,\bar{\boldsymbol{q}})] + \bar{B}(t)\left[\nabla_{\boldsymbol{q}}u(t,\bar{\boldsymbol{q}})\right].$$

12.3 Linear Quadratic Optimal Control

In the previous sections we saw where linear control systems occur naturally as the basic model, from linearization of a nonlinear system and as the sensitivity equations for a nonlinear problem. In many cases these systems can be time dependent (nonautonomous) even if the nonlinear system is autonomous. In this section we focus on the LQ optimal control problem with time varying matrix coefficients and a terminal cost. Thus, we consider nonautonomous problems of Bolza type. Most of the following material can be found in the standard books [6], [5], [16], [45], [71] and [73], [81], [115] and [187].

We begin with the LQ optimal control problem defined on a fixed time interval $[t_0, t_1]$ which is governed by the system

$$\dot{\boldsymbol{x}}(t) = A(t)\boldsymbol{x}(t) + B(t)\boldsymbol{u}(t), \ \boldsymbol{x}(t_0) = \boldsymbol{x}_0 \tag{12.37}$$

with cost function of Bolza type

$$J(\boldsymbol{u}(\cdot), \boldsymbol{x}_0) = \langle S\boldsymbol{x}(t_1), \boldsymbol{x}(t_1)\rangle + \int_{t_0}^{t_1} [\langle Q(s)\boldsymbol{x}(s), \boldsymbol{x}(s)\rangle + \langle R(s)\boldsymbol{u}(s), \boldsymbol{u}(s)\rangle] ds. \tag{12.38}$$

Here, we assume that $S = S^T \geq 0$, $Q(t) = Q(t)^T \geq 0$ and $R(t) = R(t)^T \geq \alpha I_{m\times m} > 0$ for some $\alpha > 0$.

The results in Section 11.8 can be extended to this Bolza problem. In particular, a direct combination of the Maximum Principle for the Problem of Bolza (**Theorem** 11.3) and the Maximum Principle for nonautonomous systems (**Theorem** 11.4) yields the following result.

Theorem 12.7 *If an optimal control exists, then the optimality conditions are given by*

$$\begin{aligned} \tfrac{d}{dt}\boldsymbol{x}^*(t) &= A(t)\boldsymbol{x}^*(t) + [B(t)][R(t)]^{-1}[B(t)]^T\boldsymbol{\lambda}^*(t), & \boldsymbol{x}^*(t_0) &= \boldsymbol{x}_0, \\ \tfrac{d}{dt}\boldsymbol{\lambda}^*(t) &= Q(t)\boldsymbol{x}^*(t) - [A(t)]^T\boldsymbol{\lambda}^*(t), & \boldsymbol{\lambda}^*(t_1) &= -S\boldsymbol{x}^*(t_1), \end{aligned} \tag{12.39}$$

where the optimal control is defined by

$$\boldsymbol{u}^*(t) = R^{-1}(t)B^T(t)\boldsymbol{\lambda}^*(t). \tag{12.40}$$

We can write the optimality system as

$$\frac{d}{dt}\begin{bmatrix} \boldsymbol{x}^*(t) \\ \boldsymbol{\lambda}^*(t) \end{bmatrix} = \begin{bmatrix} A(t) & [B(t)][R(t)]^{-1}[B(t)]^T \\ Q(t) & -[A(t)]^T \end{bmatrix} \begin{bmatrix} \boldsymbol{x}^*(t) \\ \boldsymbol{\lambda}^*(t) \end{bmatrix}, \tag{12.41}$$

with boundary conditions

$$\boldsymbol{x}^*(t_0) = \begin{bmatrix} I_{n\times n} & 0_{n\times n} \end{bmatrix} \begin{bmatrix} \boldsymbol{x}^*(t_0) \\ \boldsymbol{\lambda}^*(t_0) \end{bmatrix} = \boldsymbol{x}_0, \tag{12.42}$$

and

$$\boldsymbol{\lambda}^*(t_1) = \begin{bmatrix} -S & 0_{n\times n} \end{bmatrix} \begin{bmatrix} \boldsymbol{x}^*(t_1) \\ \boldsymbol{\lambda}^*(t_1) \end{bmatrix}. \tag{12.43}$$

Thus, if one solves the two point boundary value problem (12.41) - (12.43), the optimal control is defined by (12.40).

We will show that the optimal control has the form

$$\boldsymbol{u}^*(t) = -K(t)\boldsymbol{x}^*(t), \tag{12.44}$$

where the gain matrix $K(t)$ is given by

$$K(t) = R(t)^{-1}B(t)^T P(t) \tag{12.45}$$

and $P(t)$ is a non-negative definite solution to the Riccati differential equation (RDE)

$$-\dot{P}(t) = P(t)A(t) + A(t)^T P(t) - P(t)B(t)R(t)^{-1}B(t)^T P(t) + Q(t) \tag{12.46}$$

with final data

$$P(t_1) = S. \tag{12.47}$$

In addition, the optimal cost is given by

$$J(\boldsymbol{u}(\cdot), \boldsymbol{x}_0) = \langle P(t_0)\boldsymbol{x}(t_0), \boldsymbol{x}(t_0)\rangle . \tag{12.48}$$

In order to establish this result, we follow the basic approach first used in Sections 11.5 and 11.8.

12.4 The Riccati Differential Equation for a Problem of Bolza

Here we consider the Bolza type LQ optimal control problem for the general nonautonomous system discussed in the previous section. We extend the results on Riccati equations presented in Section 11.5.2 to this case. Again, the starting point is the optimality system defined by the two point boundary value problem (12.41) - (12.43) and the goal is to show that one can transform the "state variable" to the "adjoint variable" by a matrix that satisfies a Riccati differential equation.

First write (12.41) as the linear system

$$\frac{d}{dt}\begin{bmatrix} \boldsymbol{x}(t) \\ \boldsymbol{\lambda}(t) \end{bmatrix} = \begin{bmatrix} A(t) & B(t)[R(t)]^{-1}[B(t)]^T \\ Q(t) & -[A(t)]^T \end{bmatrix} \begin{bmatrix} \boldsymbol{x}(t) \\ \boldsymbol{\lambda}(t) \end{bmatrix}$$
$$\triangleq H(t) \begin{bmatrix} \boldsymbol{x}(t) \\ \boldsymbol{\lambda}(t) \end{bmatrix} \tag{12.49}$$

where $H(t)$ is the $2n \times 2n$ matrix

$$H(t) = \begin{bmatrix} A(t) & B(t)[R(t)]^{-1}[B(t)]^T \\ Q(t) & -[A(t)]^T \end{bmatrix}. \tag{12.50}$$

Let $\Psi(t, \tau)$ be the fundamental matrix for $H(t)$. In particular, $\Psi(t, \tau)$ satisfies the matrix differential equation

$$\frac{d}{dt}\Psi(t, \tau) = H(t)\Psi(t, \tau), \tag{12.51}$$

with initial conditions

$$\Psi(\tau, \tau) = I_{2n \times 2n}. \tag{12.52}$$

The solution to (12.41) - (12.43) has the form

$$\begin{bmatrix} \boldsymbol{x}(t) \\ \boldsymbol{\lambda}(t) \end{bmatrix} = \Psi(t, t_1) \begin{bmatrix} \boldsymbol{x}(t_1) \\ \boldsymbol{\lambda}(t_1) \end{bmatrix} = \Psi(t, t_1) \begin{bmatrix} \boldsymbol{x}(t_1) \\ -S\boldsymbol{x}(t_1) \end{bmatrix},$$

since $\boldsymbol{\lambda}(t_1) = -S\boldsymbol{x}(t_1)$. Setting

$$\Psi(t, \tau) = \begin{bmatrix} \psi_{11}(t, \tau) & \psi_{12}(t, \tau) \\ \psi_{21}(t, \tau) & \psi_{22}(t, \tau) \end{bmatrix},$$

it follows that

$$\begin{bmatrix} \boldsymbol{x}(t) \\ \boldsymbol{\lambda}(t) \end{bmatrix} = \Psi(t, t_1) \begin{bmatrix} \boldsymbol{x}(t_1) \\ -S\boldsymbol{x}(t_1) \end{bmatrix}$$
$$= \begin{bmatrix} \psi_{11}(t, t_1) & \psi_{12}(t, t_1) \\ \psi_{21}(t, t_1) & \psi_{22}(t, t_1) \end{bmatrix} \begin{bmatrix} \boldsymbol{x}(t_1) \\ -S\boldsymbol{x}(t_1) \end{bmatrix}$$

so that

$$\begin{aligned}\boldsymbol{x}(t) &= \psi_{11}(t,t_1)\boldsymbol{x}(t_1) - \psi_{12}(t,t_1)S\boldsymbol{x}(t_1)\\ &= [\psi_{11}(t,t_1) - \psi_{12}(t,t_1)S]\boldsymbol{x}(t_1),\end{aligned} \tag{12.53}$$

and

$$\begin{aligned}\boldsymbol{\lambda}(t) &= \psi_{21}(t-t_1)\boldsymbol{x}(t_1) - \psi_{22}(t,t_1)S\boldsymbol{x}(t_1)\\ &= [\psi_{21}(t-t_1) - \psi_{22}(t,t_1)S]\boldsymbol{x}(t_1).\end{aligned} \tag{12.54}$$

If $[\psi_{11}(t,t_1) - \psi_{12}(t,t_1)S]$ is non-singular for $0 \le t \le t_1$, then we can solve (12.53) for $\boldsymbol{x}(t_1)$. In particular,

$$\boldsymbol{x}(t_1) = [\psi_{11}(t,t_1) - \psi_{12}(t,t_1)S]^{-1}\boldsymbol{x}(t)$$

which, when substituted into (12.54), yields

$$\boldsymbol{\lambda}(t) = [\psi_{21}(t-t_1) - \psi_{22}(t,t_1)S][\psi_{11}(t,t_1) - \psi_{12}(t,t_1)S]^{-1}\boldsymbol{x}(t).$$

If $P(t)$ is the $n \times n$ matrix defined by

$$P(t) \triangleq -[\psi_{21}(t-t_1) - \psi_{22}(t,t_1)S][\psi_{11}(t,t_1) - \psi_{12}(t,t_1)S]^{-1}, \tag{12.55}$$

then we have that $\boldsymbol{\lambda}(t)$ and $\boldsymbol{x}(t)$ are linearly related by the matrix $P(t)$ and the relationship is given by

$$\boldsymbol{\lambda}(t) = -P(t)\boldsymbol{x}(t).$$

The choice of the negative sign in defining $P(\cdot)$ is made to be consistent with much of the existing literature. In order to make this step rigorous, one needs to prove that $[\psi_{11}(t,t_1) - \psi_{12}(t,t_1)S]$ is non-singular for $t_0 \le t \le t_1$. On the other hand, we could simply ask the question:

Is there a matrix $P(t)$ so that $\boldsymbol{\lambda}(t) = -P(t)\boldsymbol{x}(t)$ and how can $P(t)$ be computed?

We will address the issue of the existence of $P(t)$ later. However, assume for the moment that $\boldsymbol{x}(\cdot)$ and $\boldsymbol{\lambda}(t)$ satisfying (12.41) - (12.43) and

$$\boldsymbol{\lambda}(t) = -P(t)\boldsymbol{x}(t), \tag{12.56}$$

with $P(t)$ differentiable. Differentiating the equation (12.56) one obtains

$$\begin{aligned}\frac{d}{dt}\boldsymbol{\lambda}(t) &= -\left[\frac{d}{dt}P(t)\right]\boldsymbol{x}(t) - P(t)\left[\frac{d}{dt}\boldsymbol{x}(t)\right]\\ &= -\left[\frac{d}{dt}P(t)\right]\boldsymbol{x}(t)\\ &\quad - P(t)\left[A(t)\boldsymbol{x}(t) + B(t)[R(t)]^{-1}[B(t)]^T\boldsymbol{\lambda}(t)\right]\\ &= -\left[\frac{d}{dt}P(t)\right]\boldsymbol{x}(t)\\ &\quad - P(t)\left[A(t)\boldsymbol{x}(t) - B(t)[R(t)]^{-1}[B(t)]^T P(t)\boldsymbol{x}(t)\right]\\ &= -\left[\frac{d}{dt}P(t)\right]\boldsymbol{x}(t)\\ &\quad - P(t)A(t)\boldsymbol{x}(t) + P(t)B(t)[R(t)]^{-1}[B(t)]^T P(t)\boldsymbol{x}(t).\end{aligned}$$

However, from (12.49) we have

$$\begin{aligned}\frac{d}{dt}\boldsymbol{\lambda}(t) &= Q(t)\boldsymbol{x}(t) - [A(t)]^T\boldsymbol{\lambda}(t)\\ &= Q(t)\boldsymbol{x}(t) + [A(t)]^T P(t)\boldsymbol{x}(t)\end{aligned}$$

so that

$$\begin{aligned}Q(t)\boldsymbol{x}(t) + [A(t)]^T P(t)\boldsymbol{x}(t) &= -\left[\frac{d}{dt}P(t)\right]\boldsymbol{x}(t) - P(t)A(t)\boldsymbol{x}(t)\\ &\quad + P(t)B(t)[R(t)]^{-1}[B(t)]^T P(t)\boldsymbol{x}(t).\end{aligned}$$

Rearranging the terms we have

$$\begin{aligned}-\left[\frac{d}{dt}P(t)\right]\boldsymbol{x}(t) &= [A(t)]^T P(t)\boldsymbol{x}(t) + P(t)A(t)\boldsymbol{x}(t) \qquad (12.57)\\ &\quad - P(t)B(t)[R(t)]^{-1}[B(t)]^T P(t)\boldsymbol{x}(t) + Q(t)\boldsymbol{x}(t).\end{aligned}$$

Consequently, $P(t)$ satisfies (12.57) along the trajectory $\boldsymbol{x}(t)$. Observe that (12.57) is satisfied for any solution of the system (12.49) with $\boldsymbol{\lambda}(t_1) = -S\boldsymbol{x}(t_1)$ and all values of $\boldsymbol{x}(t_1)$. Therefore, if

$$\boldsymbol{\lambda}(t) = -P(t)\boldsymbol{x}(t),$$

then $P(t)$ satisfies the matrix Riccati differential equation

$$-\dot{P}(t) = A^T P(t) + P(t)A - P(t)BR^{-1}B^T P(t) + Q, \quad t_0 \le t < t_1, \tag{12.58}$$

with terminal condition

$$P(t_1) = S, \tag{12.59}$$

since

$$-P(t_1)\boldsymbol{x}(t_1) = \boldsymbol{\lambda}(t_1) = -S\boldsymbol{x}(t_1)$$

and $\boldsymbol{x}(t_1)$ can be any vector in $\mathbb{R}^n$.

We shall show below that under the assumption that there is a solution $P(t)$ to the Riccati differential equation (12.58) satisfying (12.59), then the LQ optimal control problem has a solution and the optimal control is given by

$$\boldsymbol{u}^*(t) = -[R(t)]^{-1}[B(t)]^T P(t)\boldsymbol{x}^*(t). \tag{12.60}$$

In order to provide a rigorous treatment of this problem, we present two lemmas. These results relate the existence of a solution to the Riccati equation (12.58) to the existence of an optimal control for the LQ optimal control problem. First we note that any solution to the Riccati differential equation must be symmetric. In particular, $P(t) = [P(t)]^T$ for all t.

Lemma 12.1 *Suppose that $P(t) = [P(t)]^T$ is any $n \times n$ matrix function with $P(t)$ differentiable on the interval $[t_0, t_1]$. If $\boldsymbol{u}(\cdot) \in PWC(t_0, t_1; \mathbb{R}^m)$ and*

$$\dot{\boldsymbol{x}}(t) = A(t)\boldsymbol{x}(t) + B(t)\boldsymbol{u}(t), \quad t_0 \le t \le t_1,$$

then

$$\langle P(s)\boldsymbol{x}(s), \boldsymbol{x}(s)\rangle \,\Big|_{t_0}^{t_1}$$

$$= \int_{t_0}^{t_1} \left\langle \left[\dot{P}(s) + P(s)A(s) + [A(s)]^T P(s)\right] \boldsymbol{x}(s), \boldsymbol{x}(s)\right\rangle ds \tag{12.61}$$

$$+ \int_{t_0}^{t_1} \langle P(s)B(s)\boldsymbol{u}(s), \boldsymbol{x}(s)\rangle \, ds \tag{12.62}$$

$$+ \int_{t_0}^{t_1} \left\langle [B(s)]^T P(s)\boldsymbol{x}(s), \boldsymbol{u}(s)\right\rangle ds$$

Proof: Observe that

$$\begin{aligned}\langle P(s)\boldsymbol{x}(s), \boldsymbol{x}(s)\rangle\,|_{t_0}^{t_1} &= \int_{t_0}^{t_1} \frac{d}{ds}\langle P(s)\boldsymbol{x}(s), \boldsymbol{x}(s)\rangle\, ds \\ &= \int_{t_0}^{t_1} \left\langle \dot{P}(s)\boldsymbol{x}(s), \boldsymbol{x}(s)\right\rangle ds \\ &\quad + \int_{t_0}^{t_1} \langle P(s)\dot{\boldsymbol{x}}(s), \boldsymbol{x}(s)\rangle\, ds \\ &\quad + \int_{t_0}^{t_1} \langle P(s)\boldsymbol{x}(s), \dot{\boldsymbol{x}}(s)\rangle\, ds\end{aligned}$$

and by substituting $A\boldsymbol{x}(s) + B\boldsymbol{u}(s)$ for $\dot{\boldsymbol{x}}(s)$ we obtain

$$\begin{aligned}\langle P(s)\boldsymbol{x}(s), \boldsymbol{x}(s)\rangle\,|_{t_0}^{t_1} &= \int_{t_0}^{t_1} \left\langle \dot{P}(s)\boldsymbol{x}(s), \boldsymbol{x}(s)\right\rangle ds \\ &\quad + \int_{t_0}^{t_1} \langle P(s)\left[A\boldsymbol{x}(s) + B\boldsymbol{u}(s)\right], \boldsymbol{x}(s)\rangle\, ds \\ &\quad + \int_{t_0}^{t_1} \langle P(s)\boldsymbol{x}(s), \left[A\boldsymbol{x}(s) + B\boldsymbol{u}(s)\right]\rangle\, ds.\end{aligned}$$

Simplifying this expression we obtain

$$\begin{aligned}\langle P(s)\boldsymbol{x}(s), \boldsymbol{x}(s)\rangle\,|_{t_0}^{t_1} &= \int_{t_0}^{t_1} \left\langle \dot{P}(s)\boldsymbol{x}(s), \boldsymbol{x}(s)\right\rangle ds \\ &\quad + \int_{t_0}^{t_1} \langle P(s)\left[A(s)\boldsymbol{x}(s)\right], \boldsymbol{x}(s)\rangle\, ds \\ &\quad + \int_{t_0}^{t_1} \langle P(s)\left[B(s)\boldsymbol{u}(s)\right], \boldsymbol{x}(s)\rangle\, ds \\ &\quad + \int_{t_0}^{t_1} \langle P(s)\boldsymbol{x}(s), \left[A(s)\boldsymbol{x}(s)\right]\rangle\, ds \\ &\quad + \int_{t_0}^{t_1} \langle P(s)\boldsymbol{x}(s), \left[B(s)\boldsymbol{u}(s)\right]\rangle\, ds\end{aligned}$$

and collecting terms yields

$$\begin{aligned}
&\langle P(s)\boldsymbol{x}(s), \boldsymbol{x}(s)\rangle \big|_{t_0}^{t_1} \\
&\quad = \int_{t_0}^{t_1} \left\langle [\dot{P}(s) + P(s)A(s) + [A(s)]^T P(s)]\boldsymbol{x}(s), \boldsymbol{x}(s) \right\rangle ds \\
&\qquad + \int_{t_0}^{t_1} \langle P(s)B(s)\boldsymbol{u}(s), \boldsymbol{x}(s)\rangle \, ds \\
&\qquad + \int_{t_0}^{t_1} \langle [B(s)]^T P(s)\boldsymbol{x}(s), \boldsymbol{u}(s)\rangle \, ds
\end{aligned}$$

which establishes (12.62). □

Lemma 12.2 *Assume that the Riccati differential equation (12.58) has a solution* $P(t) = [P(t)]^T$ *for* $t_0 \leq t < t_1$ *and* $P(t_1) = 0_{n\times n}$. *If* $\boldsymbol{u}(\cdot) \in PWC(t_0, t_1; \mathbb{R}^m)$ *and*

$$\dot{\boldsymbol{x}}(t) = A(t)\boldsymbol{x}(t) + B(t)\boldsymbol{u}(t), \quad t_0 \leq t \leq t_1,$$

then the cost function $J(\boldsymbol{u}(\cdot)) = \langle S\boldsymbol{x}(t_1), \boldsymbol{x}(t_1)\rangle + \int_{t_0}^{t_1} [\langle Q(s)\boldsymbol{x}(s), \boldsymbol{x}(s)\rangle + \langle R(s)\boldsymbol{u}(s), \boldsymbol{u}(s)\rangle] ds$ *has the representation*

$$\begin{aligned}
J(\boldsymbol{u}(\cdot)) = &\int_{t_0}^{t_1} \left\| [R(s)]^{1/2}\boldsymbol{u}(s) + [R(s)]^{-1/2}[B(s)]^T P(s)\boldsymbol{x}(s) \right\|^2 ds \\
&+ \langle P(t_0)\boldsymbol{x}(t_0), \boldsymbol{x}(t_0)\rangle .
\end{aligned}$$

Proof: Let

$$N(\boldsymbol{x}(\cdot), \boldsymbol{u}(\cdot)) = \int_{t_0}^{t_1} \left\| [R(s)]^{1/2}\boldsymbol{u}(s) + [R(s)]^{-1/2}[B(s)]^T P(s)\boldsymbol{x}(s) \right\|^2 ds$$

and expanding $N(\boldsymbol{x}(\cdot), \boldsymbol{u}(\cdot))$ we obtain

$$\begin{aligned}
N(\boldsymbol{x}(\cdot), \boldsymbol{u}(\cdot)) = &\int_{t_0}^{t_1} \langle [R(s)]^{1/2}\boldsymbol{u}(s), [R(s)]^{1/2}\boldsymbol{u}(s)\rangle \, ds \\
&+ \int_{t_0}^{t_1} \langle [R(s)]^{1/2}\boldsymbol{u}(s), [R(s)]^{-1/2}[B(s)]^T P(s)\boldsymbol{x}(s)\rangle \, ds \\
&+ \int_{t_0}^{t_1} \langle [R(s)]^{-1/2}[B(s)]^T P(s)\boldsymbol{x}(s), [R(s)]^{1/2}\boldsymbol{u}(s)\rangle \, ds \\
&+ \int_{t_0}^{t_1} \langle [R(s)]^{-1/2}[B(s)]^T P(s)\boldsymbol{x}(s), [R(s)]^{-1/2} \\
&\quad \times [B(s)]^T P(s)\boldsymbol{x}(s)\rangle ds.
\end{aligned}$$

Simplifying each term we have

$$
\begin{aligned}
N(\boldsymbol{x}(\cdot), \boldsymbol{u}(\cdot)) = & \\
& \int_{t_0}^{t_1} \big\langle [R(s)]^{1/2}[R(s)]^{1/2}\boldsymbol{u}(s), \boldsymbol{u}(s)\big\rangle \, ds \\
& + \int_{t_0}^{t_1} \big\langle \boldsymbol{u}(s), [R(s)]^{1/2}[R(s)]^{-1/2}[B(s)]^T P(s)\boldsymbol{x}(s)\big\rangle \, ds \\
& + \int_{t_0}^{t_1} \big\langle [R(s)]^{1/2}[R(s)]^{-1/2}[B(s)]^T P(s)\boldsymbol{x}(s), \boldsymbol{u}(s)\big\rangle \, ds \\
& + \int_{t_0}^{t_1} \big\langle [R(s)]^{-1/2}[R(s)]^{-1/2}[B(s)]^T P(s)\boldsymbol{x}(s), [B(s)]^T P(s)\boldsymbol{x}(s)\big\rangle ds,
\end{aligned}
$$

which implies

$$
\begin{aligned}
N(\boldsymbol{x}(\cdot), \boldsymbol{u}(\cdot)) = & \int_{t_0}^{t_1} \langle R(s)\boldsymbol{u}(s), \boldsymbol{u}(s)\rangle \, ds \\
& + \int_{t_0}^{t_1} \big\langle \boldsymbol{u}(s), [B(s)]^T P(s)\boldsymbol{x}(s)\big\rangle \, ds \\
& + \int_{t_0}^{t_1} \big\langle [B(s)]^T P(s)\boldsymbol{x}(s), \boldsymbol{u}(s)\big\rangle \, ds \\
& + \int_{t_0}^{t_1} \big\langle [R(s)]^{-1}[B(s)]^T P(s)\boldsymbol{x}(s), [B(s)]^T P(s)\boldsymbol{x}(s)\big\rangle \, ds,
\end{aligned}
$$

or equivalently

$$
\begin{aligned}
N(\boldsymbol{x}(\cdot), \boldsymbol{u}(\cdot)) = & \int_{t_0}^{t_1} \langle R(s)\boldsymbol{u}(s), \boldsymbol{u}(s)\rangle \\
& + \int_{t_0}^{t_1} \big\langle \boldsymbol{u}(s), [B(s)]^T P(s)\boldsymbol{x}(s)\big\rangle \, ds \qquad (12.63) \\
& + \int_{t_0}^{t_1} \big\langle [B(s)]^T P(s)\boldsymbol{x}(s), \boldsymbol{u}(s)\big\rangle \, ds \\
& + \int_{t_0}^{t_1} \big\langle P(s)B(s)[R(s)]^{-1}[B(s)]^T P(s)\boldsymbol{x}(s), \boldsymbol{x}(s)\big\rangle \, ds.
\end{aligned}
$$

Since the matrix $P(s)$ satisfies the Riccati equation (12.58), it follows that

$$\begin{aligned} &P(s)B(s)[R(s)]^{-1}[B(s)]^T P(s)\boldsymbol{x}(s) \\ &\quad = \left[\dot{P}(s) + [A(s)]^T P(s) + P(s)A(s) + Q(s)\right] \boldsymbol{x}(s) \end{aligned}$$

and the last term above becomes

$$\int_{t_0}^{t_1} \left\langle \left[\dot{P}(s) + [A(s)]^T P(s) + P(s)A(s) + Q(s)\right] \boldsymbol{x}(s), \boldsymbol{x}(s) \right\rangle ds.$$

Substituting this expression into (11.131) and rearranging yields

$$\begin{aligned} &N(\boldsymbol{x}(\cdot), \boldsymbol{u}(\cdot)) = \\ &\int_{t_0}^{t_1} \langle R(s)\boldsymbol{u}(s), \boldsymbol{u}(s)\rangle \, ds + \int_{t_0}^{t_1} \langle Q(s)\boldsymbol{x}(s), \boldsymbol{x}(s)\rangle \, ds \\ &\quad + \int_{t_0}^{t_1} \left\langle \left[\dot{P}(s) + [A(s)]^T P(s) + P(s)A(s)\right] \boldsymbol{x}(s), \boldsymbol{x}(s) \right\rangle ds \\ &\quad + \int_{t_0}^{t_1} \left\langle \boldsymbol{u}(s), [B(s)]^T P(s)\boldsymbol{x}(s) \right\rangle ds \\ &\quad + \int_{t_0}^{t_1} \left\langle [B(s)]^T P(s)\boldsymbol{x}(s), \boldsymbol{u}(s) \right\rangle ds, \end{aligned}$$

which implies

$$\begin{aligned} &N(\boldsymbol{x}(\cdot), \boldsymbol{u}(\cdot)) = \\ &\quad J(\boldsymbol{u}(\cdot)) - \langle S\boldsymbol{x}(t_1), \boldsymbol{x}(t_1)\rangle \\ &\quad + \int_{t_0}^{t_1} \left\langle \left[\dot{P}(s) + [A(s)]^T P(s) + P(s)A(s)\right] \boldsymbol{x}(s), \boldsymbol{x}(s) \right\rangle ds \\ &\quad + \int_{t_0}^{t_1} \left\langle \boldsymbol{u}(s), [B(s)]^T P(s)\boldsymbol{x}(s) \right\rangle ds \\ &\quad + \int_{t_0}^{t_1} \left\langle [B(s)]^T P(s)\boldsymbol{x}(s), \boldsymbol{u}(s) \right\rangle ds. \end{aligned}$$

Applying (12.62) from the previous Lemma yields

$$\begin{aligned} N(\boldsymbol{x}(\cdot), \boldsymbol{u}(\cdot)) &= J(\boldsymbol{u}(\cdot)) - \langle S\boldsymbol{x}(t_1), \boldsymbol{x}(t_1)\rangle + \langle P(s)\boldsymbol{x}(s), \boldsymbol{x}(s)\rangle \big|_{t_0}^{t_1} \\ &= J(\boldsymbol{u}(\cdot)) - \langle S\boldsymbol{x}(t_1), \boldsymbol{x}(t_1)\rangle + \langle P(t_1)\boldsymbol{x}(t_1), \boldsymbol{x}(t_1)\rangle \\ &\quad - \langle P(t_0)\boldsymbol{x}(t_0), \boldsymbol{x}(t_0)\rangle, \end{aligned}$$

or equivalently

$$J(\boldsymbol{u}(\cdot)) = N(\boldsymbol{x}(\cdot), \boldsymbol{u}(\cdot)) + \langle S\boldsymbol{x}(t_1), \boldsymbol{x}(t_1)\rangle - \langle P(t_1)\boldsymbol{x}(t_1), \boldsymbol{x}(t_1)\rangle + \langle P(t_0)\boldsymbol{x}(t_0), \boldsymbol{x}(t_0)\rangle .$$

Since (12.59) holds, we have that $P(t_1) = S$ and hence

$$\begin{aligned} J(\boldsymbol{u}(\cdot)) &= N(\boldsymbol{x}(\cdot), \boldsymbol{u}(\cdot)) + \langle S\boldsymbol{x}(t_1), \boldsymbol{x}(t_1)\rangle - \langle S\boldsymbol{x}(t_1), \boldsymbol{x}(t_1)\rangle \\ &\quad + \langle P(t_0)\boldsymbol{x}(t_0), \boldsymbol{x}(t_0)\rangle \\ &= N(\boldsymbol{x}(\cdot), \boldsymbol{u}(\cdot)) + \langle P(t_0)\boldsymbol{x}(t_0), \boldsymbol{x}(t_0)\rangle . \end{aligned}$$

We conclude that

$$J(\boldsymbol{u}(\cdot)) = \int_{t_0}^{t_1} \left\| R^{1/2}\boldsymbol{u}(s) + R^{-1/2}B^T P(s)\boldsymbol{x}(s)\right\|^2 ds + \langle P(t_0)\boldsymbol{x}(t_0), \boldsymbol{x}(t_0)\rangle \tag{12.64}$$

which completes the proof. □

We now have the fundamental result on the relationship between solutions to the Riccati equation and the existence of an optimal control for the LQ optimal problem.

Theorem 12.8 (Existence of LQ Optimal Control) *If the Riccati differential equation (12.58) has a solution $P(t) = [P(t)]^T$ for $t_0 \le t < t_1$ and $P(t_1) = S$, then there is a control $\boldsymbol{u}^*(\cdot) \in PWC(t_0, t_1; \mathbb{R}^m)$ such that $\boldsymbol{u}^*(\cdot)$ minimizes*

$$J(\boldsymbol{u}(\cdot)) = \langle S\boldsymbol{x}(t_1), \boldsymbol{x}(t_1)\rangle + \int_{t_0}^{t_1} \{\langle Q(s)\boldsymbol{x}(s), \boldsymbol{x}(s)\rangle + \langle R(s)\boldsymbol{u}(s), \boldsymbol{u}(s)\rangle\} ds$$

on the set $PWC(t_0, t_1; \mathbb{R}^m)$. In addition, the optimal control is a linear feedback law

$$\boldsymbol{u}^*(t) = -[R(t)]^{-1}[B(t)]^T P(t)\boldsymbol{x}^*(t) \tag{12.65}$$

and the minimum value of $J(\cdot)$ is

$$J(\boldsymbol{u}^*(\cdot)) = \langle P(t_0)\boldsymbol{x}(t_0), \boldsymbol{x}(t_0)\rangle . \tag{12.66}$$

Proof: The identity (12.64) above implies that $J(\cdot)$ is minimized when the quadratic term

$$\int_{t_0}^{t_1} \left\|[R(s)]^{1/2}\boldsymbol{u}(s) + [R(s)]^{-1/2}[B(s)]^T P(s)\boldsymbol{x}(s)\right\|^2 ds \geq 0$$

is minimized. If $\boldsymbol{u}^*(s) = -[R(s)]^{-1}[B(s)]^T P(s)\boldsymbol{x}^*(s)$, then

$$[R(s)]^{1/2}\boldsymbol{u}^*(s) + [R(s)]^{-1/2}[B(s)]^T P(s)\boldsymbol{x}^*(s) = 0$$

and

$$\begin{aligned} J(\boldsymbol{u}^*(\cdot)) = & \int_{t_0}^{t_1} \left\|[R(s)]^{1/2}\boldsymbol{u}^*(s) + [R(s)]^{-1/2}[B(s)]^T P(s)\boldsymbol{x}^*(s)\right\|^2 ds \\ & + \langle P(t_0)\boldsymbol{x}^*(t_0), \boldsymbol{x}^*(t_0)\rangle \\ = & \langle P(t_0)\boldsymbol{x}^*(t_0), \boldsymbol{x}^*(t_0)\rangle . \end{aligned}$$

Consequently, if $\boldsymbol{u}(\cdot) \in PWC(t_0, t_1; \mathbb{R}^m)$, then

$$\begin{aligned} J(\boldsymbol{u}^*(\cdot)) = \langle P(t_0)\boldsymbol{x}_0, \boldsymbol{x}_0\rangle \leq & \langle P(t_0)\boldsymbol{x}_0, \boldsymbol{x}_0\rangle \\ & + \int_{t_0}^{t_1} \left\|[R(s)]^{1/2}\boldsymbol{u}(s) + [R(s)]^{-1/2}[B(s)]^T P(s)\boldsymbol{x}(s)\right\|^2 ds \\ = & J(\boldsymbol{u}(\cdot)), \end{aligned}$$

which completes the proof. □

The important observation here is that the optimal control is linear state feedback with gain operator $K(t)$ defined by (12.45). Thus, to implement the control law one needs the full state $\boldsymbol{x}^*(t)$. However, when the full state is not available one must develop a method to estimate those states that can not be sensed. This leads to the construction of state estimators or so-called observers.

12.5 Estimation and Observers

Assume that the system is described by the differential equation

$$\dot{\boldsymbol{x}}(t) = A(t)\boldsymbol{x}(t) + B(t)\boldsymbol{u}(t), \qquad \boldsymbol{x}(t_0) = \boldsymbol{x}_0 \in \mathbb{R}^n, \quad t \in [t_0, t_1] \tag{12.67}$$

with sensed or measured output

$$\boldsymbol{y}(t) = C(t)\boldsymbol{x}(t) \in \mathbb{R}^p. \tag{12.68}$$

If one has a feedback control law of the form

$$\boldsymbol{u}(t) = -K(t)\boldsymbol{x}(t) \tag{12.69}$$

for some *control gain matrix* $K(t)$, then in order to compute the control $\boldsymbol{u}(t)$ at a time t one must have the full state $\boldsymbol{x}(t)$ at time t. Since we can only sense $\boldsymbol{y}(t)$ and the control law requires $\boldsymbol{x}(t)$, we are led to the problem of finding a method to estimate $\boldsymbol{x}(t)$ from the data $\boldsymbol{y}(t)$. In particular, we look for a system of the form

$$\begin{aligned} &\dot{\boldsymbol{x}}_e(t) = A(t)\boldsymbol{x}_e(t) + F(t)\left[\boldsymbol{y}(t) - C(t)\boldsymbol{x}_e(t)\right] + B(t)\boldsymbol{u}(t), \\ &\boldsymbol{x}_e(t_0) = \boldsymbol{x}_{e,0} \in \mathbb{R}^n, \end{aligned} \tag{12.70}$$

such that the solution $\boldsymbol{x}_e(t)$ provides an estimate of the state $\boldsymbol{x}(t)$ on the interval of interest. Thus, we seek an **observer gain matrix** $F(t)$ so that the system (12.70) which is driven by the measured (sensed) output $\boldsymbol{y}(t)$ produces values $\boldsymbol{x}_e(t)$ that are as close to $\boldsymbol{x}(t)$ as possible. In this case, one replaces the full state feedback law (12.69) with

$$\boldsymbol{u}(t) = \boldsymbol{u}_e(t) = -K(t)\boldsymbol{x}_e(t). \tag{12.71}$$

In order for this to work and to be practical, one needs to know that $\boldsymbol{x}_e(t) \simeq \boldsymbol{x}(t)$ and that the equation (12.70) can be solved in real time. Under suitable conditions we shall say that the system (12.70) is an **observer** (or **state estimator**) for the system (12.67)-(12.68).

The selection of the system (12.70) as the form of an observer is not arbitrary, but follows naturally. A "good" observer should have the same structure as the system and be driven by the measured output $\boldsymbol{y}(t) = C(t)\boldsymbol{x}(t)$. Moreover, we require that if the initial data for the estimator is exactly the initial data for the system, then the estimated state should equal the actual state. Therefore, if $\boldsymbol{x}_e(t_0) = \boldsymbol{x}_{e,0} = \boldsymbol{x}_0 = \boldsymbol{x}(t_0)$, then $\boldsymbol{x}_e(t) = \boldsymbol{x}(t)$ for all $t \geq t_0$. Since the system (12.67)-(12.68) is linear, it is natural to think that the observer equation should also be linear so that an observer would have the form

$$\dot{\boldsymbol{x}}_e(t) = A_e(t)\boldsymbol{x}_e(t) + F(t)\boldsymbol{y}(t) + G(t)\boldsymbol{u}(t), \qquad \boldsymbol{x}_e(t_0) = \boldsymbol{x}_{e,0} \in \mathbb{R}^n, \tag{12.72}$$

for some matrices $A_e(t)$, $F(t)$ and $G(t)$. The corresponding error $\boldsymbol{e}(t) \triangleq \boldsymbol{x}_e(t) - \boldsymbol{x}(t)$ would satisfy the non-homogenous linear differential equation

$$\begin{aligned}
\dot{\boldsymbol{e}}(t) &= \dot{\boldsymbol{x}}_e(t) - \dot{\boldsymbol{x}}(t) \\
&= A_e(t)\boldsymbol{x}_e(t) + F(t)\boldsymbol{y}(t) + G(t)\boldsymbol{u}(t) - A(t)\boldsymbol{x}(t) - B(t)\boldsymbol{u}(t) \\
&= A_e(t)\boldsymbol{x}_e(t) - A_e(t)\boldsymbol{x}(t) + A_e(t)\boldsymbol{x}(t) \\
&\quad + F(t)\boldsymbol{y}(t) - A(t)\boldsymbol{x}(t) + [G(t) - B(t)]\boldsymbol{u}(t) \\
&= A_e(t)[\boldsymbol{x}_e(t) - \boldsymbol{x}(t)] \\
&\quad + A_e(t)\boldsymbol{x}(t) + F(t)\boldsymbol{y}(t) - A(t)\boldsymbol{x}(t) + [G(t) - B(t)]\boldsymbol{u}(t) \\
&= A_e(t)\boldsymbol{e}(t) + A_e(t)\boldsymbol{x}(t) + F(t)C(t)\boldsymbol{x}(t) - A(t)\boldsymbol{x}(t) \\
&\quad + [G(t) - B(t)]\boldsymbol{u}(t) \\
&= A_e(t)\boldsymbol{e}(t) + [A_e(t) + F(t)C(t) - A(t)]\boldsymbol{x}(t) \\
&\quad + [G(t) - B(t)]\boldsymbol{u}(t)
\end{aligned}$$

with initial data given by the error vector

$$\boldsymbol{e}(t_0) = \boldsymbol{x}_e(t_0) - \boldsymbol{x}(t_0).$$

The requirement that if $\boldsymbol{e}(t_0) = \boldsymbol{x}_e(t_0) - \boldsymbol{x}(t_0) = 0$, then $\boldsymbol{e}(t) = \boldsymbol{x}_e(t) - \boldsymbol{x}(t) \equiv 0$ for all states $\boldsymbol{x}(t)$ and controls $\boldsymbol{u}(t)$ implies that the matrices $[A_e(t) + F(t)C(t) - A(t)]$ and $[G(t) - B(t)]$ should be zero. Thus,

$$[A_e(t) + F(t)C(t) - A(t)] = 0$$

and

$$[G(t) - B(t)] = 0,$$

which implies that

$$A_e(t) = A(t) - F(t)C(t)$$

and

$$G(t) = B(t),$$

respectively. Therefore, the observer (12.72) must have the form

$$\begin{aligned}
\dot{\boldsymbol{x}}_e(t) &= A_e(t)\boldsymbol{x}_e(t) + F(t)\boldsymbol{y}(t) + G(t)\boldsymbol{u}(t) \\
&= [A(t) - F(t)C(t)]\boldsymbol{x}_e(t) + F(t)\boldsymbol{y}(t) + B(t)\boldsymbol{u}(t) \\
&= A(t)\boldsymbol{x}_e(t) + F(t)[\boldsymbol{y}(t) - C(t)\boldsymbol{x}_e(t)] + B(t)\boldsymbol{u}(t)
\end{aligned}$$

which is the form of the observer (12.70).

Let $\boldsymbol{e}(t) \triangleq \boldsymbol{x}_e(t) - \boldsymbol{x}(t)$ denote the error vector and note that

$$\begin{aligned}
\dot{\boldsymbol{e}}(t) &= \dot{\boldsymbol{x}}_e(t) - \dot{\boldsymbol{x}}(t) \\
&= [A(t)\boldsymbol{x}_e(t) + F(t)\,[\boldsymbol{y}(t) - C(t)\boldsymbol{x}_e(t)] + B(t)\boldsymbol{u}(t)] \\
&\quad -[A(t)\boldsymbol{x}(t) + B(t)\boldsymbol{u}(t)] \\
&= A(t)[\boldsymbol{x}_e(t) - \boldsymbol{x}(t)] + F(t)C(t)\boldsymbol{x}(t) - F(t)C(t)\boldsymbol{x}_e(t) \\
&= [A(t) - F(t)C(t)][\boldsymbol{x}_e(t) - \boldsymbol{x}(t)] \\
&= [A(t) - F(t)C(t)]\boldsymbol{e}(t).
\end{aligned}$$

Hence, the error satisfies the linear differential equation

$$\dot{\boldsymbol{e}}(t) = [A(t) - F(t)C(t)]\boldsymbol{e}(t) \tag{12.73}$$

with initial data

$$\boldsymbol{e}(t_0) = \boldsymbol{x}_e(t_0) - \boldsymbol{x}(t_0) = \boldsymbol{x}_{e,0} - \boldsymbol{x}_0. \tag{12.74}$$

Since the error equation (12.73) is linear and homogenous, if the error in the initial estimate $\boldsymbol{x}_{e,0}$ of $\boldsymbol{x}_0$ is zero (i.e. $\boldsymbol{x}_{e,0} = \boldsymbol{x}_0$), then

$$\boldsymbol{e}(t) = \boldsymbol{x}_e(t) - \boldsymbol{x}(t) \equiv 0$$

and the estimate of $\boldsymbol{x}(t)$ is exact as desired. Moreover, it follows that

$$\boldsymbol{e}(t) = \Psi_F(t, t_0)\boldsymbol{e}(t_0),$$

where $\Psi(t, \tau)$ is the fundamental matrix operator for (12.73). In particular, $\Psi_F(t, \tau)$ satisfies the matrix differential equation

$$\frac{d}{dt}\Psi_F(t, \tau) = [A(t) - F(t)C(t)]\Psi_F(t, \tau)$$

with initial condition

$$\Psi_F(\tau,\tau) = I_{n\times n}.$$

This leads to a bound on the error of the form

$$\|\boldsymbol{e}(t)\| = \|\Psi_F(t,t_0)\boldsymbol{e}(t_0)\| \leq \|\Psi_F(t,t_0)\| \, \|\boldsymbol{e}(t_0)\| \tag{12.75}$$

for $t_0 \leq t \leq t_1$. The selection of the observer gain $F(t)$ determines this error and hence one approach might be to select $F(t)$ so that

$$\sup_{t_0 \leq t \leq t_1} \|\Psi_F(t,t_0)\|$$

is minimized. We shall return to this issue later.

There are many approaches to "designing" a good observer, but we focus on the Luenberger Observer and the Kalman Filter.

12.5.1 The Luenberger Observer

As noted above, many optimal controllers require full state feedback which could be implemented if all of $\boldsymbol{x}(t)$ were available. However, in most complex systems the entire state is not available through sensing and hence such control laws cannot be implemented in practice. Thus, one must either consider an alternate form of the control or else devise a method for approximating the state $\boldsymbol{x}(t)$ using only knowledge about the system dynamics and sensed measurements. In 1964 Luenberger first considered this problem from a deterministic point of view (see [130], [132] and [129]). Although Luenberger considered only time invariant systems, the main results can be modified and extended to nonautonomous and even non-linear systems.

One of the basic results concerns the question of observing an output to the uncontrolled system given by

$$\dot{\boldsymbol{x}}(t) = A(t)\boldsymbol{x}(t), \qquad \boldsymbol{x}(t_0) = \boldsymbol{x}_0 \tag{12.76}$$

using only sensed measurements

$$\boldsymbol{y}(t) = C(t)\boldsymbol{x}(t), \tag{12.77}$$

where $A(t)$ and $C(t)$ are given $n \times n$ and $p \times n$ matrix functions, respectively. The idea is to use the output $\boldsymbol{y}(t)$ to drive another linear system

$$\dot{\boldsymbol{z}}(t) = A_e(t)\boldsymbol{z}(t) + F(t)\boldsymbol{y}(t), \qquad \boldsymbol{z}(t_0) = \boldsymbol{z}_0, \tag{12.78}$$

so that one matches a desired output

$$\boldsymbol{r}(t) = T(t)\boldsymbol{x}(t), \tag{12.79}$$

where $T(t)$ is a given $q \times n$ matrix function. Observe that if $A_e(t)$ is a $l \times l$ matrix, then $F(t)$ must be a $l \times p$ matrix since there are p measured outputs. Moreover, since

$$F(t)\boldsymbol{y}(t) = F(t)C(t)\boldsymbol{x}(t)$$

then $H(t) \triangleq F(t)C(t)$ must be a $l \times n$ matrix. The goal is to find $A_e(t)$ and $F(t)$ so that if $\boldsymbol{z}(t_0) = \boldsymbol{r}(t_0) = T(t_0)\boldsymbol{x}(t_0)$, then $\boldsymbol{z}(t) = \boldsymbol{r}(t)$ for all $t > t_0$. Clearly this implies that $l = q$ so that the size of system (12.78) is q. The following result is due to Luenberger and may be found in [132].

Theorem 12.9 *Let $\boldsymbol{x}(t)$ be a solution to the system (12.76) which drives the system (12.78). Suppose the matrix function $T(t)$ is differentiable and there exist $A_e(t)$ and $F(t)$ such that*

$$\dot{T}(t) + [T(t)A(t) - A_e(t)T(t)] = H(t) = F(t)C(t) \tag{12.80}$$

on the interval $t_0 < t < t_1$. If $\boldsymbol{z}(t_0) = \boldsymbol{r}(t_0) = T(t_0)\boldsymbol{x}(t_0)$, then $\boldsymbol{z}(t) = \boldsymbol{r}(t) = T(t)\boldsymbol{x}(t)$ for all $t_0 < t < t_1$.

Proof: Let the error vector be defined by $\boldsymbol{e}(t) = \boldsymbol{z}(t) - \boldsymbol{r}(t) = \boldsymbol{z}(t) - T(t)\boldsymbol{x}(t)$ and note that

$$\begin{aligned}
\dot{\boldsymbol{e}}(t) &= \dot{\boldsymbol{z}}(t) - \frac{d}{dt}[T(t)\boldsymbol{x}(t)] = \dot{\boldsymbol{z}}(t) - \dot{T}(t)\boldsymbol{x}(t) - T(t)\dot{\boldsymbol{x}}(t) \\
&= A_e(t)\boldsymbol{z}(t) + F(t)\boldsymbol{y}(t) - \dot{T}(t)\boldsymbol{x}(t) - T(t)[A(t)\boldsymbol{x}(t)] \\
&= A_e(t)\boldsymbol{z}(t) + F(t)C(t)\boldsymbol{x}(t) - \dot{T}(t)\boldsymbol{x}(t) - T(t)A(t)\boldsymbol{x}(t) \\
&= A_e(t)\boldsymbol{z}(t) + H(t)\boldsymbol{x}(t) - \dot{T}(t)\boldsymbol{x}(t) - T(t)A(t)\boldsymbol{x}(t).
\end{aligned}$$

However, since $H(t) = \dot{T}(t) + T(t)A(t) - A_e(t)T(t)$ it follows that

$$\begin{aligned}\dot{\boldsymbol{e}}(t) &= A_e(t)\boldsymbol{z}(t) + [\dot{T}(t) + T(t)A(t) - A_e(t)T(t)]\boldsymbol{x}(t)\\ &\qquad - \dot{T}(t)\boldsymbol{x}(t) - T(t)A(t)\boldsymbol{x}(t)\\ &= A_e(t)\boldsymbol{z}(t) - A_e(t)T(t)\boldsymbol{x}(t)\\ &= A_e(t)[\boldsymbol{z}(t) - T(t)\boldsymbol{x}(t)]\\ &= A_e(t)\boldsymbol{e}(t).\end{aligned}$$

Therefore, $\boldsymbol{e}(t)$ is given by

$$\boldsymbol{e}(t) = \Psi_e(t, t_0)\boldsymbol{e}(t_0)$$

where $\Psi_e(t, s)$ is the state transition matrix satisfying

$$\frac{\partial}{\partial t}\Psi_e(t, s) = A_e(t)\Psi_e(t, s), \qquad \Psi_e(s, s) = I_{q\times q}.$$

Hence we have

$$\boldsymbol{z}(t) - T(t)\boldsymbol{x}(t) = \Psi_e(t, t_0)[\boldsymbol{z}(t_0) - T(t_0)\boldsymbol{x}(t_0)],$$

so that

$$\boldsymbol{z}(t) = T(t)\boldsymbol{x}(t) + \Psi_e(t, t_0)[\boldsymbol{z}_0 - \boldsymbol{r}_0]. \tag{12.81}$$

Consequently, if $\boldsymbol{z}(t_0) = \boldsymbol{r}(t_0)$ then $\boldsymbol{z}(t) = \boldsymbol{r}(t)$ and this completes the proof. □

Observe that equation (12.81) implies that the error $\boldsymbol{e}(t) = \boldsymbol{z}(t) - \boldsymbol{r}(t) = \boldsymbol{z}(t) - T(t)\boldsymbol{x}(t)$ is equal to $\Psi_e(t, t_0)[\boldsymbol{z}_0 - \boldsymbol{r}_0]$ so that

$$\begin{aligned}\|\boldsymbol{e}(t)\| = \|\boldsymbol{z}(t) - T(t)\boldsymbol{x}(t)\| &= \|\Psi_e(t, t_0)[\boldsymbol{z}_0 - \boldsymbol{r}_0]\|\\ &\le \|\Psi_e(t, t_0)\|\,\|\boldsymbol{z}_0 - \boldsymbol{r}_0\|\end{aligned}$$

is bounded by the norm of the transition matrix. Consequently,

$$\|\boldsymbol{z}(t) - T(t)\boldsymbol{x}(t)\| \le \left[\sup_{t_0\le t\le t_1} \|\Psi_e(t, t_0)\|\right] \|\boldsymbol{z}_0 - \boldsymbol{r}_0\|\,.$$

Since $\Psi_e(t, t_0)$ depends on $A_e(t)$, and hence implicitly on $F(t)$ through the equation (12.80), the error can be made "small" if one can choose $A_e(t)$ and $F(t)$ satisfying (12.80) such that

$$\sup_{t_0\le t\le t_1} \|\Psi_e(t, t_0)\|$$

is small. Note that in the special autonomous case one has the following result which was first given in [130].

Corollary 12.1 *Let $\boldsymbol{x}(t)$ be a solution to the system $\dot{\boldsymbol{x}}(t) = A\boldsymbol{x}(t)$ which drives the system $\dot{\boldsymbol{z}}(t) = A_e\boldsymbol{z}(t)+F\boldsymbol{y}(t)$ where $\boldsymbol{y}(t) = C\boldsymbol{x}(t)$. Suppose there exist A_e and F such that*

$$[TA - A_eT] = H = FC. \tag{12.82}$$

If $\boldsymbol{z}(t_0) = \boldsymbol{r}(t_0) = T\boldsymbol{x}(t_0)$, then $\boldsymbol{z}(t) = \boldsymbol{r}(t) = T\boldsymbol{x}(t)$ for all $t_0 < t < t_1$.

Remark 12.3 *It is important to note that for a given $T(t)$ (or T in the time invariant case) it may not be possible to find $A_e(t)$ and $F(t)$ satisfying (12.80) (or (12.82)). See [130], [132] and [129] for more details. Luenberger actually posed the following question:*

> *Given $H = FC$ and A_e, then what conditions on T will insure that $\boldsymbol{z}(t_0) = \boldsymbol{r}(t_0) = T\boldsymbol{x}(t_0)$ implies $\boldsymbol{z}(t) = \boldsymbol{r}(t) = T\boldsymbol{x}(t)$ for all $t_0 < t < t_1$?*

Clearly, the answer is that T must satisfy $[TA - A_eT] = H$.

Remark 12.4 *Note that if T is an $n \times n$ nonsingular matrix so that T^{-1} exists, then given any F one can solve (12.82) for A_e. In particular,*

$$A_eT = TA - H = TA - FC$$

and hence

$$A_e = [TA - H]T^{-1} = TAT^{-1} - FCT^{-1}$$

satisfies (12.82). If $T = I_{n\times n}$, then $A_e = A - FC$ and

$$\dot{\boldsymbol{z}}(t) = A_e\boldsymbol{z}(t) + F\boldsymbol{y}(t)$$

is called the identity observer.

We will return to these issues later and consider related problems of asymptotic tracking. However, we turn now to a stochastic version of this problem which produces an "optimal observer" called the Kalman Filter.

12.5.2 An Optimal Observer: The Kalman Filter

In order to fully develop this topic requires a background in stochastic systems and control and is outside the scope of these notes. However, Section 1.10 in the book [115] by Kwakernaak and Sivan contains a very nice and brief discussion of the basic material. However, it is essential that we discuss the definition of a stochastic process and present a few fundamental definitions and results.

A Short Review of Random Variables and Stochastic Processes

We review the concepts needed to precisely define a stochastic process. This requires we discuss measures and some measure theory concepts. We begin with the concept of a σ-algebra.

Definition 12.5 *Given a set Ω, $\mathfrak{F}$ is called a **σ-algebra** of subsets of Ω if $\mathfrak{F}$ is a collection of subsets of Ω and:*
(1) $\mathfrak{F}$ is non-empty: i.e. there is at least one set $E \subseteq \Omega$ in $\mathfrak{F}$,
(2) $\mathfrak{F}$ is closed under complementation: i.e. if E is in Ω, then so is its complement $\Omega \backslash E$, where $\Omega \backslash E$ is defined by

$$\Omega \backslash E \triangleq \{x \in \Omega : x \notin E\},$$

(3) $\mathfrak{F}$ is closed under countable unions: i.e. if E_1, E_2, E_3, ... are in $\mathfrak{F}$, then $\bigcup_{i=1}^{+\infty} E_i$ is in $\mathfrak{F}$.

Definition 12.6 *A pair $(\Omega, \mathfrak{F})$, where Ω is a set and $\mathfrak{F}$ is a σ-algebra of subsets of Ω, is called a **measurable space**. A function $\mu : \mathfrak{F} \to [0, +\infty]$ is a (countably additive) measure on $(\Omega, \mathfrak{F})$ provided that*
(1) $\mu(\varnothing) = 0$, where $\varnothing$ is the empty subset of Ω;
(2) if $E_i \in \mathfrak{F}$ for each $i = 1, 2, 3, \ldots$ are disjoint sets, then

$$\mu\left(\bigcup_{i=1}^{+\infty} E_i\right) = \sum_{i=1}^{+\infty} \mu(E_i).$$

If $\mu : \mathfrak{F} \to [0, +\infty)$, then μ is called a ***finite measure****. If Ω can be decomposed into a countable union of measurable sets of finite measure, then μ is called a σ-finite measure. The triple $(\Omega, \mathfrak{F}, \mu)$ is called a* ***measure space*** *and sets in $\mathfrak{F}$ are called* ***measurable sets****.*

Although these definitions may seem rather abstract, consider the simple example where $\Omega = I$ is a subinterval of the real line (I could be all of $\mathbb{R}^1$) and let $\mathfrak{F}_{\mathfrak{B}}$ be the "smallest" collection of sets containing all open and closed subintervals and their countable unions and intersections. The σ-algebra $\mathfrak{F}_{\mathfrak{B}}$ is called the Borel σ-algebra. If $m : \mathfrak{F}_{\mathfrak{B}} \to [0, +\infty)$ is the usual Lebesgue measure (see [176]), then $(I, \mathfrak{F}_{\mathfrak{B}}, m)$ is a measure space. If I is a finite subinterval, then μ is a finite measure. If I is an infinite interval, then μ is a σ-finite measure. This measure space plays the central role in the development of modern integration theory. We will make use of the standard measure space $(\mathbb{R}^n, \mathfrak{F}_{\mathfrak{B}}, m)$ where again $\mathfrak{F}_{\mathfrak{B}}$ is the Borel σ-algebra on $\mathbb{R}^n$ and m is the usual Lebesgue measure.

Definition 12.7 *A* ***probability space*** *is a measure space $(\Omega, \mathfrak{F}, p)$ where the probability measure p satisfies*
(1) $0 \le p(E) \le 1$;
(2) $p(\Omega) = 1$.
The set Ω is called the sample space, $\mathfrak{F}$ is called the event space and p is called the probability measure.

Definition 12.8 *Let $(\Omega_1, \mathfrak{F}_1, \mu_1)$ and $(\Omega_2, \mathfrak{F}_2, \mu_2)$ be measure spaces. A function $X(\cdot) : \Omega_1 \to \Omega_2$ is said to be a* ***measurable function*** *if for every measurable set $S \in \mathfrak{F}_2$, the inverse image*

$$X^{-1}(S) = \{x \in \Omega_1 : X(x) \in S\}$$

is measurable, i.e. $X^{-1}(S) \in \mathfrak{F}_1$. If $(\Omega_1, \mathfrak{F}_1, \mu_1) = (\Omega, \mathfrak{F}, p)$ is a probability space, then a measurable function $X(\cdot) : \Omega \to \Omega_2$ is called a ***random variable****. In particular,* ***random variables are measurable functions*** *from a probability space into a measure space.*

Remark 12.5 *Note that we have used capital letters to denote random variables. This is done to distinguish between the random variable $X(\cdot)$ and the value of this function at a particular $x \in \Omega$. Thus, $X(x)$ is not the same as $X(\cdot)$. This is the same notation we have consistently used before. For example, recall that we write $u(\cdot) \in PWS(t_0, t_1)$ to mean that $u(\cdot)$ is a function and for a given fixed $t \in [t_0, t_1]$, $u(t)$ is the value of $u(\cdot)$ at t. However, this precise notation is often abused so that one sees things like "consider the function $u = t$". Of course we know that this means that $u(\cdot)$ is a function such that $u(\cdot)$ evaluated at t is just t or, $u(t) = t$. This same abuse of notation occurs in probability and statistics where you will see statements like "consider the random variable x". What this really means is that one is talking about the random variable $X(\cdot)$ where $X(x) = x$.*

Definition 12.9 *Let $(\Omega, \mathfrak{F}, p)$ be a probability space and $(\mathbb{R}^n, \mathfrak{F}_{\mathfrak{B}}, m)$ be the standard Lebesgue measure space. Let $\mathfrak{X}$ denote the set of all random variables $X(\cdot) : \Omega \to \mathbb{R}^n$. In particular,*

$$\mathfrak{X} = \{X(\cdot) : \Omega \to \mathbb{R}^n : X(\cdot) \quad \text{is measurable}\}.$$

Let I be a subinterval of the real line. A (vector-valued) ***stochastic process*** *on I is a family of $\mathbb{R}^n$-valued random variables*

$$\{[X_t](\cdot) \in \mathfrak{X} : t \in I\},$$

indexed by $t \in I$. In particular, for each $t \in I$, $X_t(\cdot)$ is a random variable (i.e. ***a measurable function****). Thus, a stochastic process is a function from I to $\mathfrak{X}$. We use the notation*

$$X(t, \cdot) = X_t(\cdot)$$

to indicate that for each fixed $t \in I$, $X(t, \cdot)$ is the random variable with range in $\mathbb{R}^n$. In this setting, we may consider a stochastic process as a function of two variables

$$X(\cdot, \cdot) : I \times \Omega \to \mathbb{R}^n$$

and if $\omega \in \Omega$, we write

$$X(t, \omega) = X_t(\omega).$$

If $X(t,\cdot) = X_t(\cdot)$ is a vector-valued stochastic process on I, then

$$X_t(\cdot) = \left[\ [X_t]_1(\cdot)\ \ [X_t]_2(\cdot)\ \ [X_t]_3(\cdot)\ \ \cdots\ \ [X_t]_n(\cdot)\ \right]^T,$$

where each random variable $[X_t]_i(\cdot)$ is a real-valued random variable. Thus, we write

$$X_t(\cdot) = X(t,\cdot) = \left[\ X_1(t,\cdot)\ \ X_2(t,\cdot)\ \ X_3(t,\cdot)\ \ \cdots\ \ X_n(t,\cdot)\ \right]^T$$

so that

$$\begin{aligned} X_t(\omega) &= X(t,\omega) \\ &= \left[\ [X_1(t)](\omega)\ \ [X_2(t)](\omega)\ \ [X_3(t)](\omega)\ \ \cdots\ \ [X_n(t)](\omega)\ \right]^T, \end{aligned}$$

or equivalently,

$$X(t,\omega) = \left[\ X_1(t,\omega)\ \ X_2(t,\omega)\ \ X_3(t,\omega)\ \ \cdots\ \ X_n(t,\omega)\ \right]^T.$$

At this point, one really starts to abuse notation. In particular, the statement that

$$V_t(\cdot) = \left[\ [V_t]_1(\cdot)\ \ [V_t]_2(\cdot)\ \ [V_t]_3(\cdot)\ \ \cdots\ \ [V_t]_n(\cdot)\ \right]^T$$

is a $\mathbb{R}^n$-valued stochastic process on I, is often written by stating that

$$\boldsymbol{v}(t) = \left[\ v_1(t)\ \ v_2(t)\ \ v_3(t)\ \ \cdots\ \ v_n(t)\ \right]^T$$

is a $\mathbb{R}^n$-valued stochastic process on I. It is important that one fully appreciates what is meant by this simplification of notation. In particular, always keep in mind that $v_i(t)$ is "representing" a random variable (i.e. a function) $[V_t]_i(\cdot)$. If this is well understood, then one has the background required to read Section 1.10 in the book [115].

The Kalman Filter and an Optimal Observer

We return now to the problem of constructing an observer that is optimal in some sense. As noted above, one use of an observer is to provide state estimates from measured (sensed) outputs that can be used in a practical feedback law. In particular,

$$\boldsymbol{u}(t) = -K(t)\boldsymbol{x}(t)$$

is replaced by

$$\boldsymbol{u}_e(t) = -K(t)\boldsymbol{x}_e(t)$$

and if $\boldsymbol{x}_e(t) \simeq \boldsymbol{x}(t)$, then one would expect that under suitable conditions the controllers satisfy $\boldsymbol{u}_e(t) \simeq \boldsymbol{u}(t)$. In this application, the selection of an observer for a given system depends on the choice of the gain matrix $K(t)$. However, we can formulate an "optimal observer" problem independent of the control problem and focus on the problem of finding an optimal observer. In addition, this approach allows us to deal with some practical issues where the control system is subject to disturbances and the sensors produce noisy signals. This is done by making specific assumptions concerning these disturbances and measurement errors that occur in the system that is to be observed. We follow the approach in [115].

It is assumed that the system equations are described by the stochastic differential equation on $t_0 \le t \le T$ given by

$$\dot{\boldsymbol{x}}(t) = A(t)\boldsymbol{x}(t) + B(t)\boldsymbol{u}(t) + G(t)\boldsymbol{w}(t), \qquad \boldsymbol{x}(t_0) = \boldsymbol{x}_0 \in \mathbb{R}^n, \tag{12.83}$$

with sensed output

$$\boldsymbol{y}(t) = C(t)\boldsymbol{x}(t) + E(t)\boldsymbol{v}(t) \in \mathbb{R}^p, \tag{12.84}$$

where $\boldsymbol{w}(t)$ and $\boldsymbol{v}(t)$ represent system disturbances and sensor noise, respectively.

A Comment on Notation: If $\boldsymbol{z}$ is a random variable, then we use $\mathcal{E}$ for the expectation operator so that $\mathcal{E}\left[\boldsymbol{z}\right]$ denotes the expected value of $\boldsymbol{z}$ and $\mathcal{E}\left[\boldsymbol{z}\boldsymbol{z}^T\right]$ is the covariance matrix.

We assume that the disturbance $\boldsymbol{w}(t)$ and sensor noise $\boldsymbol{v}(t)$ are white, Gaussian, have zero mean and are independent. Mathematically, this implies that for all t, $\tau \in [t_0, T]$,

$$\mathcal{E}\left[\boldsymbol{w}(t)\boldsymbol{w}(\tau)^T\right] = W(t)\delta(t-\tau), \qquad \mathcal{E}\left[\boldsymbol{v}(t)\right] \equiv 0, \tag{12.85}$$

$$\mathcal{E}\left[\boldsymbol{v}(t)\boldsymbol{v}(\tau)^T\right] = V(t)\delta(t-\tau), \qquad \mathcal{E}\left[\boldsymbol{w}(t)\right] \equiv 0, \tag{12.86}$$

and

$$\mathcal{E}\left[\boldsymbol{v}(t)\boldsymbol{w}(\tau)^T\right] = 0, \tag{12.87}$$

where $W(t) = W(t)^T \geq 0$ and $V(t) = V(t)^T \geq 0$ are the so-called intensity matrices. We define the $n \times n$ and matrices $M(t)$ and $N(t)$ by

$$M(t) = G(t)W(t)[G(t)]^T \text{ and } N(t) = E(t)V(t)[E(t)]^T, \quad (12.88)$$

respectively. In addition, we make the standing assumption that there is an $\alpha > 0$ such that

$$N(t) = E(t)V(t)[E(t)]^T = N(t)^T \geq \alpha I > 0. \quad (12.89)$$

The initial condition $\boldsymbol{x}_0$ is assumed to be a Gaussian random variable with mean $\boldsymbol{m}_0$ and covariance Π_0 so that

$$\mathcal{E}\left\{[\boldsymbol{x}_0 - \boldsymbol{m}_0][\boldsymbol{x}_0 - \boldsymbol{m}_0]^T\right\} = \Pi_0 \text{ and } \mathcal{E}[\boldsymbol{x}_0] = \boldsymbol{m}_0. \quad (12.90)$$

Finally, we assume that $\boldsymbol{x}_0$ is independent of the noise terms so that

$$\mathcal{E}[\boldsymbol{x}_0 \boldsymbol{v}(t)^T] = \mathcal{E}[\boldsymbol{x}_0 \boldsymbol{w}(t)^T] = 0 \qquad \text{for all } t \in [t_0, T]. \quad (12.91)$$

Suppose we are seeking an observer of the form

$$\begin{aligned} \dot{\boldsymbol{x}}_e(t) &= A(t)\boldsymbol{x}_e(t) + B(t)\boldsymbol{u}(t) + F(t)\left[y(t) - C(t)\boldsymbol{x}_e(t)\right], \\ \boldsymbol{x}_e(t_0) &= \boldsymbol{x}_{e,0} \in \mathbb{R}^n, \end{aligned} \quad (12.92)$$

to minimize the expected value of the error between the state and the estimate of the state. In particular, let

$$\boldsymbol{e}(t) = \boldsymbol{x}_e(t) - \boldsymbol{x}(t) \quad (12.93)$$

denote the error.
The **optimal estimation problem** is to find the observer gain $F(t)$ and the initial estimate $\boldsymbol{x}_{e,0}$ to minimize the weighted error

$$\begin{aligned} \mathcal{E}[\langle \Sigma(t)\boldsymbol{e}(t), \boldsymbol{e}(t)\rangle] &= \mathcal{E}[\boldsymbol{e}(t)^T \Sigma(t) \boldsymbol{e}(t)] \\ &= \mathcal{E}\left\{[\boldsymbol{x}_e(t) - \boldsymbol{x}(t)]^T \Sigma(t) [\boldsymbol{x}_e(t) - \boldsymbol{x}(t)]\right\}, \end{aligned} \quad (12.94)$$

where $\Sigma(t) = \Sigma(t)^T \geq 0$. The following theorem provides the answer to the optimal estimation problem and is known as the *Kalman Filter.*

Theorem 12.10 (Kalman Filter) *The optimal observer is obtained by choosing the observer gain matrix to be*

$$F^{opt}(t) = \Pi(t)C^T(t)N(t)^{-1} \tag{12.95}$$

where $\Pi(t) = [\Pi(t)]^T$ *is the solution to the matrix Riccati differential equation*

$$\dot{\Pi}(t) = A(t)\Pi(t) + \Pi(t)A(t)^T - \Pi(t)C(t)^T N(t)^{-1}C(t)\Pi(t) + M(t), \tag{12.96}$$

with initial condition

$$\Pi(t_0) = \Pi_0. \tag{12.97}$$

The optimal initial condition should be

$$\boldsymbol{x}_e(t_0) = \boldsymbol{m}_0 = \mathcal{E}[\boldsymbol{x}(t_0)] = \mathcal{E}[\boldsymbol{x}_0]. \tag{12.98}$$

If conditions (12.95) - (12.98) hold, then

$$\mathcal{E}\left\{[\boldsymbol{x}_e(t) - \boldsymbol{x}(t)][\boldsymbol{x}_e(t) - \boldsymbol{x}(t)]^T\right\} = \Pi(t), \tag{12.99}$$

while the mean square reconstruction error is

$$\begin{aligned}\mathcal{E}[\langle\Sigma(t)\boldsymbol{e}(t), \boldsymbol{e}(t)\rangle] &= \mathcal{E}\left\{[\boldsymbol{x}_e(t) - \boldsymbol{x}(t)]^T\Sigma(t)[\boldsymbol{x}_e(t) - \boldsymbol{x}(t)]\right\} \\ &= trace\left[\Pi(t)\Sigma(t)\right]. \end{aligned}\tag{12.100}$$

In addition to the properties given in Theorem 12.10, one can show that $\Pi(t)$ is the "minimal" nonnegative solution to (12.96) in the following sense. If $F(t)$ is any observer gain and $\Phi(t)$ is a solution to the Lyapunov equation

$$\begin{aligned}\dot{\Phi}(t) =& [A(t) - F(t)C(t)]\Phi(t) + \Phi(t)[A(t) - F(t)C(t)]^T \\ &+ F(t)V(t)F(t)^T + M(t)\end{aligned} \tag{12.101}$$

with initial data (12.97), then

$$0 \leq trace[\Pi(t)\Sigma(t)] \leq trace[\Phi(t)\Sigma(t)]. \tag{12.102}$$

Let $\Phi(t)$ denote the variance matrix of the error $\boldsymbol{e}(t)$ and let $\boldsymbol{e}_m(t)$ denote the mean $\boldsymbol{e}(t)$ so that

$$\mathcal{E}\left\{[\boldsymbol{e}(t) - \boldsymbol{e}_m(t)][\boldsymbol{e}(t) - \boldsymbol{e}_m(t)]^T\right\} = \Psi(t) \;\; \text{and} \;\; \mathcal{E}[\boldsymbol{e}(t)] = \boldsymbol{e}_m(t). \tag{12.103}$$

It can be shown (see [115], Section 4.3) that $\boldsymbol{e}_m(t)$ satisfies the ordinary differential equation

$$\dot{\boldsymbol{e}}_m(t) = [A(t) - F(t)C(t)]\boldsymbol{e}_m(t)$$

and $\Phi(t)$ is the solution to the Lyapunov equation (12.101) with initial condition $\Phi(t_0) = \Pi_0$. To see how this is related to the estimation problem, assume that

$$F(t) = \Phi(t)C^T(t)N(t)^{-1} \tag{12.104}$$

and $\Phi(t)$ satisfies the Lyapunov equation (12.101) above. Then, by substitution

$$\begin{aligned}
\dot{\Phi}(t) &= [A(t) - F(t)C(t)]\Phi(t) + \Phi(t)[A(t) - F(t)C(t)]^T \\
&\quad + F(t)N(t)F(t)^T + M(t) \\
&= A(t)\Phi(t) - [\Phi(t)C^T(t)N(t)^{-1}]C(t)\Phi(t) + \Phi(t)A(t)^T \\
&\quad - \Phi(t)[\Phi(t)C^T(t)N(t)^{-1}C(t)]^T \\
&\quad + [\Phi(t)C^T(t)N(t)^{-1}]W(t)[\Phi(t)C^T(t)N(t)^{-1}]^T + M(t) \\
&= A(t)\Phi(t) + \Phi(t)A(t)^T - \Phi(t)C^T(t)N(t)^{-1}C(t)\Phi(t) \\
&\quad - \Phi(t)C(t)^T N(t)^{-1}C(t)\Phi(t) \\
&\quad + [\Phi(t)C^T(t)N(t)^{-1}N(t)[N(t)^{-1}C(t)\Phi(t)] + M(t) \\
&= A(t)\Phi(t) + \Phi(t)A(t)^T - 2\Phi(t)C^T(t)N(t)^{-1}C(t)\Phi(t) \\
&\quad + \Phi(t)C^T(t)N(t)^{-1}C(t)\Phi(t) + M(t) \\
&= A(t)\Phi(t) + \Phi(t)A(t)^T - \Phi(t)C^T(t)V(t)^{-1}C(t)\Phi(t) + M(t)
\end{aligned}$$

and hence $\Phi(t)$ satisfies the observer Riccati differential equation (12.96). If $\Phi(t)$ satisfies the initial condition $\Phi(t_0) = \Pi_0$, then $\Phi(t) = \Pi(t)$ and we have the optimal filter.

Note that the time averaged error is given by

$$\frac{1}{T - t_0}\int_{t_0}^{T} \mathcal{E}[\langle \Sigma(s)\boldsymbol{e}(s), \boldsymbol{e}(s)\rangle]ds = \frac{1}{T - t_0}\int_{t_0}^{T} trace\,[\Pi(s)\Sigma(s)]\,ds. \tag{12.105}$$

This cost has been used for optimal sensor location problems (see [10], [39], [50], [51], [49], [69], [70], [78], [152], [181]). A similar approach has been used for optimal actuator location problems (see [66], [139]).

12.6 The Time Invariant Infinite Interval Problem

In the case where $t_1 = +\infty$ and the linear system is time invariant, the linear quadratic optimal control problem takes the form

$$\dot{\boldsymbol{x}}(t) = A\boldsymbol{x}(t) + B\boldsymbol{u}(t), \ \boldsymbol{x}(0) = \boldsymbol{x}_0, \tag{12.106}$$

with cost functional

$$J(\boldsymbol{u}(\cdot)) = \int_0^{+\infty} \{\langle Q\boldsymbol{x}(s), \boldsymbol{x}(s)\rangle + \langle R\boldsymbol{u}(s), \boldsymbol{u}(s)\rangle\} \, ds. \tag{12.107}$$

Here, $Q = D^T D \geq 0$, and $R = R^T > 0$. The linear quadratic optimal control problem defined by (12.106)-(12.107) is called the **Linear Quadratic Regulator Problem**, or **LQR Problem** for short. *If an optimal control exists*, then one can show (see [115]) that it has the form of state feedback

$$\boldsymbol{u}^*(t) = -K_{opt}\boldsymbol{x}^*(t), \tag{12.108}$$

where the constant gain matrix K_{opt} is given by

$$K_{opt} = R^{-1}B^T P_{opt} \tag{12.109}$$

and P_{opt} is a non-negative definite solution to the **Algebraic Riccati Equation (ARE)**

$$PA + A^T P - PBR^{-1}B^T P + Q = 0. \tag{12.110}$$

The proof of this statement follows by considering LQ optimal control problems on $0 \leq t_1 < +\infty$ and then taking the limit as $t_1 \to +\infty$ (see [115] and [119]). In particular, P_{opt} is the minimal nonnegative solution to (12.110) in the sense that if P is any non-negative definite solution of (12.110), then

$$P \geq P_{opt} \geq 0. \tag{12.111}$$

Moreover, the optimal cost is given by

$$\begin{aligned} J(\boldsymbol{u}^*(\cdot)) &= \int_0^{+\infty} \{\langle Q\boldsymbol{x}^*(s), \boldsymbol{x}^*(s)\rangle + \langle R\boldsymbol{u}^*(s), \boldsymbol{u}^*(s)\rangle\} \, dt \\ &= \langle P_{opt}\boldsymbol{x}_0, \boldsymbol{x}_0\rangle , \end{aligned} \tag{12.112}$$

where $\boldsymbol{x}(0) = \boldsymbol{x}_0$ is the initial state.
Some important questions are:

1. When does the LQR problem defined by (12.106)-(12.107) have a solution?

2. When is this solution unique?

3. How is this related to solutions of the ARE (12.110)?

As noted above we consider the case where the weighting matrix Q can be factored as

$$Q = D^T D \tag{12.113}$$

for a $q \times n$ matrix D. Hence, we define the **controlled output** by

$$\boldsymbol{z}(t) = D\boldsymbol{x}(t), \tag{12.114}$$

where the matrix D represents the mapping $D : \mathbb{R}^n \to \mathbb{R}^q$. Thus, the cost function (12.107) can be written as

$$\begin{aligned} J(\boldsymbol{u}(\cdot)) &= \int_0^{+\infty} \{\langle Q\boldsymbol{x}(s), \boldsymbol{x}(s)\rangle_{\mathbb{R}^n} + \langle R\boldsymbol{u}(s), \boldsymbol{u}(s)\rangle\}\, ds \\ &= \int_0^{+\infty} \{\left\langle D^T D\boldsymbol{x}(s), \boldsymbol{x}(s)\right\rangle + \langle R\boldsymbol{u}(s), \boldsymbol{u}(s)\rangle\}\, ds \\ &= \int_0^{+\infty} \{\langle D\boldsymbol{x}(s), D\boldsymbol{x}(s)\rangle + \langle R\boldsymbol{u}(s), \boldsymbol{u}(s)\rangle\}\, ds \\ &= \int_0^{+\infty} \left\{\langle \boldsymbol{z}(s), \boldsymbol{z}(s)\rangle_{\mathbb{R}^q} + \langle R\boldsymbol{u}(s), \boldsymbol{u}(s)\rangle\right\} ds. \end{aligned}$$

In order to guarantee the existence of an optimal controller we must place additional conditions on the system. The following result can be found many control theory books. In particular, see Section 3.4.3 in [115].

Theorem 12.11 (Existence for LQR Problem) *If (A, B) is stabilizable and (A, D) is detectable, then the optimal LQR control problem has a unique solution and is given by (12.108)-(12.110). Moreover, there is only one non-negative definite solution P_{opt} to (12.110) and (12.112) holds.*

Now assume that the full state is not available for feedback. Thus, we have a sensed output

$$\boldsymbol{y}(t) = C\boldsymbol{x}(t) + E\boldsymbol{v}(t) \in \mathbb{R}^p, \tag{12.115}$$

and we must build an state estimator (observer). We will employ a Kalman filter of the form

$$\dot{\boldsymbol{x}}_e(t) = A\boldsymbol{x}_e(t)+B\boldsymbol{u}(t)+F\left[\boldsymbol{y}(t) - C\boldsymbol{x}_e(t)\right], \quad \boldsymbol{x}_e(t_0) = \boldsymbol{x}_{e,0} \in \mathbb{R}^n, \tag{12.116}$$

where the disturbances $\boldsymbol{v}(t)$ and $\boldsymbol{w}(t)$ are white, gaussian, have zero mean and are independent. In particular, we assume that for all $t > 0$ and $\tau > 0$,

$$\mathcal{E}\left[\boldsymbol{v}(t)\boldsymbol{v}(\tau)^T\right] = I_{n\times n}\delta(t-\tau), \qquad \mathcal{E}\left[\boldsymbol{v}(t)\right] \equiv 0, \tag{12.117}$$

$$\mathcal{E}\left[\boldsymbol{w}(t)\boldsymbol{w}(\tau)^T\right] = I_{p\times p}\delta(t-\tau), \qquad \mathcal{E}\left[\boldsymbol{w}(t)\right] \equiv 0, \tag{12.118}$$

and

$$\mathcal{E}\left[\boldsymbol{v}(t)\boldsymbol{w}(\tau)^T\right] = 0. \tag{12.119}$$

Hence, $M = GG^T$ and condition (12.89) implies $N = NN^T \geq \alpha I_{p\times p} > 0$ so that the optimal filter is defined by the observer gain F and given by

$$F = \Pi C^T N^{-1}, \tag{12.120}$$

where

$$A\Pi + \Pi A^T - \Pi C^T N^{-1} C\Pi + M = 0. \tag{12.121}$$

Now consider the closed-loop system obtained by using the filter defined by (12.120) - (12.121). It follows that if one sets

$$\boldsymbol{u}(t) = -K\boldsymbol{x}_e(t), \tag{12.122}$$

then the controlled system becomes

$$\begin{aligned}\dot{\boldsymbol{x}}(t) &= A\boldsymbol{x}(t) + B\boldsymbol{u}(t) + G\boldsymbol{w}(t)\\ &= A\boldsymbol{x}(t) - BK\boldsymbol{x}_e(t) + G\boldsymbol{w}(t)\end{aligned}$$

and the observer equation (12.116) becomes

$$\begin{aligned}\dot{\boldsymbol{x}}_e(t) &= A\boldsymbol{x}_e(t) + B\boldsymbol{u}(t) + F\left[\boldsymbol{y}(t) - C\boldsymbol{x}_e(t)\right] \\ &= A\boldsymbol{x}_e(t) - BK\boldsymbol{x}_e(t) + F\boldsymbol{y}(t) - FC\boldsymbol{x}_e(t) \\ &= A\boldsymbol{x}_e(t) - BK\boldsymbol{x}_e(t) + F[C\boldsymbol{x}(t) + E\boldsymbol{v}(t)] - FC\boldsymbol{x}_e(t) \\ &= [A\boldsymbol{x}_e(t) - BK\boldsymbol{x}_e(t) - FC\boldsymbol{x}_e(t)] + FC\boldsymbol{x}(t) + FE\boldsymbol{v}(t) \\ &= [A - BK - FC]\boldsymbol{x}_e(t) + FC\boldsymbol{x}(t) + FE\boldsymbol{v}(t).\end{aligned}$$

Consequently, the closed-loop system has the form

$$\frac{d}{dt}\begin{bmatrix}\boldsymbol{x}(t)\\ \boldsymbol{x}_e(t)\end{bmatrix} = \begin{bmatrix}A & -BK\\ FC & A_e\end{bmatrix}\begin{bmatrix}\boldsymbol{x}(t)\\ \boldsymbol{x}_e(t)\end{bmatrix} + \begin{bmatrix}G\boldsymbol{w}(t)\\ FE\boldsymbol{v}(t)\end{bmatrix}, \tag{12.123}$$

where the observer matrix is given by

$$A_e = A - BK - FC. \tag{12.124}$$

The controller defined by the observer (12.116) with feedback law (12.122) is called the **Linear Quadratic Gaussian Regulator** or **LQG Controller** for short. The rigorous development of these concepts can be found in any modern control book. The closed-loop system defined by (12.123) is stable and for many years it was thought to be robust. However, it is well known that this feedback control is not robust (see the famous paper by John Doyle [72]) so there needs to be a way to deal with both performance and robustness. One approach is to apply the so-called Min-Max theory (see [16]) to obtain a more robust controller. This theory is based on a game theory approach to controller design and will not be discussed here. However, in the next section we summarize the basic approach for the linear regulator problem above.

12.7 The Time Invariant Min-Max Controller

Observe that the solution to the LQG problem involves solving two uncoupled Riccati equations (12.110) and (12.121). In particular, there is no relationship between the optimal controller determined by the solution of the Riccati equation (12.110) and the optimal observer determined by the Riccati equation (12.121). Thus, the controller design does not consider disturbance and sensor noise. This is one source of the lack of robustness. In the papers [155] and [175] an optimal control problem based on a game theoretic formulation is used to develop an optimal controller and estimator where the control, disturbances and noise are included in the formulation. The cost function is formulated as a quadratic function of the state history, the initial state, the system disturbances, sensor noise and the control. We shall not discuss the derivation of this control here, but briefly summarize the approach. The books by Basar and Bernhard [16] and [17] contain a very nice presentation of the theory of H^{∞} and Min-Max control.

Let $\theta^2 \geq 0$ and $Q = D^T D$, $M = GG^T$, $N = EE^T$ and $R = R^T > 0$ be as above. Consider the Riccati equations

$$PA + A^T P - P[BR^{-1}B^T - \theta^2 M]P + Q = 0 \tag{12.125}$$

and

$$A\Pi + \Pi A^T - \Pi[C^T N^{-1} C - \theta^2 Q]\Pi + M = 0 \tag{12.126}$$

and let P_θ and Π_θ be the minimal nonnegative definite solutions of (12.125) and (12.126), respectively. If $\Pi_\theta > 0$ is positive definite, then for θ^2 sufficiently small $[\Pi_\theta]^{-1} - \theta^2 P_\theta > 0$ and

$$\left[I - \theta^2 \Pi_\theta P_\theta\right]^{-1}$$

exists. Define

$$K_\theta = R^{-1} B^* P_\theta, \tag{12.127}$$

$$F_\theta = \left[I - \theta^2 \Pi_\theta P_\theta\right]^{-1} \Pi_\theta C^T N^{-1}, \tag{12.128}$$

$$A_\theta = A - BK_\theta - F_\theta C + \theta^2 M P_\theta \tag{12.129}$$

and consider the closed-loop system

$$\frac{d}{dt}\begin{bmatrix} \boldsymbol{x}(t) \\ \boldsymbol{x}_\theta(t) \end{bmatrix} = \begin{bmatrix} A & -BK_\theta \\ F_\theta C & A_\theta \end{bmatrix}\begin{bmatrix} \boldsymbol{x}(t) \\ \boldsymbol{x}_\theta(t) \end{bmatrix} + \begin{bmatrix} G\boldsymbol{v}(t) \\ F_\theta E\boldsymbol{w}(t) \end{bmatrix}. \tag{12.130}$$

This system is called the **Min-Max closed-loop system** and provides some robustness when it exists (see[16]). One can show that there is a $\hat{\theta}^2 > 0$ so that if $0 \leq \theta^2 < \hat{\theta}^2$, then there exist minimal positive definite solutions P_θ and Π_θ to the Ricatti equations (12.125) and (12.126), $[I - \theta^2 \Pi_\theta P_\theta]$ is non-singular and the closed-loop system (12.130) is stable (robustly). Roughly speaking, larger $\hat{\theta}^2$ implies more robustness. Note that when $\theta^2 = 0$, the Min-Max controller reduces to the LQG controller.

Remark 12.6 *The short treatment of LQ optimal control presented here is clearly incomplete. The primary purpose of discussing this topic was to illustrate how optimal control and the Maximum Principle can be used to address problems of this type. There are numerous excellent books devoted to these topics and to more advanced applications involving systems governed by delay and partial differential equations. Hopefully, this brief introduction will encourage and motivate the reader to dive more deeply into these exciting areas.*

12.8 Problem Set for Chapter 12

Consider the linear control system

$$\dot{\boldsymbol{x}}(t) = A(t)\boldsymbol{x}(t) + B(t)\boldsymbol{u}(t), \tag{12.131}$$

with measured output

$$\boldsymbol{y}(t) = C(t)\boldsymbol{x}(t) \tag{12.132}$$

and controlled output

$$\boldsymbol{z}(t) = D(t)\boldsymbol{x}(t). \tag{12.133}$$

Here, we assume that $A : [t_0, t_1] \to \mathbb{R}^{n\times n}$, $B : [t_0, t_1] \to \mathbb{R}^{n\times m}$, $C : [t_0, t_1] \to \mathbb{R}^{p\times n}$ and $D : [t_0, t_1] \to \mathbb{R}^{q\times n}$ are piecewise continuous matrix valued functions.

Let $Q(t) = [D(t)]^T D(t)$ and assume $R : [t_0, t_1] \to \mathbb{R}^{m\times m}$ is a piecewise continuous symmetric matrix valued function such that

$$0 < \alpha I_{m\times m} \leq R(t),$$

for some $\alpha > 0$. Finally, we assume that the controls belong to the space
$PWC(t_0, T; \mathbb{R}^m)$ and the quadratic cost functional is defined by

$$J(\boldsymbol{u}(\cdot)) = \langle S\boldsymbol{x}(t_1), \boldsymbol{x}(t_1)\rangle \tag{12.134}$$

$$+\frac{1}{2}\int_{t_0}^{t_1} \{\langle Q(s)\boldsymbol{x}(s), \boldsymbol{x}(s)\rangle + \langle R(s)\boldsymbol{u}(s), \boldsymbol{u}(s)\rangle\}\, ds, \tag{12.135}$$

where $S = S^T \geq 0$ is a constant symmetric matrix. All the following problems concern these LQ control problems.

Problem 12.1 *Consider the control system*

$$\frac{d}{dt}\begin{bmatrix} x_1(t) \\ x_2(t) \end{bmatrix} = \begin{bmatrix} -1 & R_e \\ 0 & -2 \end{bmatrix}\begin{bmatrix} x_1(t) \\ x_2(t) \end{bmatrix} + \begin{bmatrix} 0 \\ 1 \end{bmatrix} u(t),$$

with output

$$y(t) = x_1(t) + x_2(t) = \begin{bmatrix} 1 & 1 \end{bmatrix}\begin{bmatrix} x_1(t) \\ x_2(t) \end{bmatrix},$$

where $R_e \geq 0$ is a constant. (a) For what values of R_e is this system controllable? (b) For what values of R_e is the system observable?

Problem 12.2 *Consider the control system*

$$\frac{d}{dt}\begin{bmatrix} x_1(t) \\ x_2(t) \end{bmatrix} = \begin{bmatrix} -1 & R_e \\ 0 & -2 \end{bmatrix}\begin{bmatrix} x_1(t) \\ x_2(t) \end{bmatrix} + \begin{bmatrix} \alpha \\ \beta \end{bmatrix} u(t),$$

with output

$$y(t) = x_1(t) + x_2(t) = \begin{bmatrix} 1 & 1 \end{bmatrix}\begin{bmatrix} x_1(t) \\ x_2(t) \end{bmatrix}.$$

(a) For what values of R_e, α and β is this system controllable? (b) For what values of R_e, α and β is the system stabilizable?

Problem 12.3 *Consider the control system*

$$\frac{d}{dt}\begin{bmatrix} x_1(t) \\ x_2(t) \end{bmatrix} = \begin{bmatrix} -1 & 10 \\ 0 & 2 \end{bmatrix}\begin{bmatrix} x_1(t) \\ x_2(t) \end{bmatrix} + \begin{bmatrix} 0 \\ 1 \end{bmatrix} u(t)$$

with output

$$y(t) = \alpha x_1(t) + \beta x_2(t) = \begin{bmatrix} \alpha & \beta \end{bmatrix}\begin{bmatrix} x_1(t) \\ x_2(t) \end{bmatrix}.$$

(a) Is this system controllable? (b) For what values of α and β is the system observable? (c) Is this system stabilizable? (d) For what values of α and β is the system detectable?

Problem 12.4 *Consider the control system*

$$\frac{d}{dt}\begin{bmatrix} x_1(t) \\ x_2(t) \end{bmatrix} = \begin{bmatrix} 1 & 0 \\ 0 & -1 \end{bmatrix}\begin{bmatrix} x_1(t) \\ x_2(t) \end{bmatrix} + \begin{bmatrix} \delta \\ \epsilon \end{bmatrix} u(t)$$

with output

$$y(t) = x_1(t) + x_2(t) = \begin{bmatrix} 1 & 1 \end{bmatrix}\begin{bmatrix} x_1(t) \\ x_2(t) \end{bmatrix}.$$

For what values of δ and ϵ is this system (a) controllable, (b) observable, (c) stabilizable, (d) detectable?

Problem 12.5 *Consider the LQR control problem for the system*

$$\frac{d}{dt}\begin{bmatrix} x_1(t) \\ x_2(t) \end{bmatrix} = \begin{bmatrix} 1 & 0 \\ 0 & -1 \end{bmatrix}\begin{bmatrix} x_1(t) \\ x_2(t) \end{bmatrix} + \begin{bmatrix} \delta \\ \epsilon \end{bmatrix} u(t)$$

with quadratic cost

$$J(u(\cdot)) = \int_0^{+\infty} \left\{[x_1(s)]^2 + [x_2(s)]^2 + [u(s)]^2\right\} ds.$$

For each δ and ϵ, find the solution $P = P(\delta, \epsilon)$ of the algebraic Riccati equation

$$A^T P + PA - PBB^T P + Q = 0$$

and compute the optimal feedback gain

$$K(\delta, \epsilon) = B^T P(\delta, \epsilon).$$

Discuss these solutions as $\delta \longrightarrow 0$ *and* $\epsilon \longrightarrow 0$.

Problem 12.6 *Let* $R_e \geq 0$ *be a constant and consider the LQR control problem for the system*

$$\frac{d}{dt}\begin{bmatrix} x_1(t) \\ x_2(t) \end{bmatrix} = \begin{bmatrix} -1 & R_e \\ 0 & -2 \end{bmatrix}\begin{bmatrix} x_1(t) \\ x_2(t) \end{bmatrix} + \begin{bmatrix} 1 \\ 1 \end{bmatrix} u(t)$$

with quadratic cost

$$J(u(\cdot)) = \int_0^{+\infty} \left\{[x_1(s)]^2 + [x_2(s)]^2 + [u(s)]^2\right\} ds.$$

For each $R_e > 0$, *find the solution* $P = P(R_e)$ *of the algrbraic Riccati equation*

$$A^T P + PA - PBB^T P + Q = 0$$

and compute the optimal feedback gain

$$K(R_e) = B^T P(R_e).$$

Discuss these solutions as $R_e \longrightarrow 0$ *and as* $R_e \longrightarrow +\infty$.

Problem 12.7 *Consider the control system*

$$\frac{d}{dt}\begin{bmatrix} x_1(t) \\ x_2(t) \end{bmatrix} = \begin{bmatrix} 0 & 1 \\ 0 & 0 \end{bmatrix}\begin{bmatrix} x_1(t) \\ x_2(t) \end{bmatrix} + \begin{bmatrix} 0 \\ 1 \end{bmatrix} u(t).$$

Show that this system is controllable. Find the minimum energy control that transfers a state $\boldsymbol{x}_1$ *to zero in time* $t_1 = 2$. *Hint: the matrix exponential is given by*

$$e^{At} = \begin{bmatrix} 1 & t \\ 0 & 1 \end{bmatrix}.$$

Problem 12.8 *Consider the control system*

$$\frac{d}{dt}\begin{bmatrix} x_1(t) \\ x_2(t) \end{bmatrix} = \begin{bmatrix} -1 & e^{2t} \\ 0 & -1 \end{bmatrix}\begin{bmatrix} x_1(t) \\ x_2(t) \end{bmatrix} + \begin{bmatrix} 0 \\ e^{-t} \end{bmatrix} u(t).$$

Show that this system is completely controllable for all $t_1 > 0$. *Find the minimum energy control that transfers the state* $\boldsymbol{x}_1 = [1 \;\; 1]^T$ *to zero in time* $t_1 = 2$. *Hint: the state transition matrix is given by*

$$\Phi(t,s) = \begin{bmatrix} e^{-(t-s)} & \frac{1}{2}[e^{(t+s)} - e^{-(t--3s)}] \\ 0 & e^{-(t-s)} \end{bmatrix}.$$

Problem 12.9 *Consider the control system*

$$\frac{d}{dt}\begin{bmatrix} x_1(t) \\ x_2(t) \end{bmatrix} = \begin{bmatrix} 0 & 1 \\ 0 & 0 \end{bmatrix}\begin{bmatrix} x_1(t) \\ x_2(t) \end{bmatrix}$$

with measured output

$$\mathbf{y}(t) = \begin{bmatrix} 2x_1(t) \\ x_2(t) \end{bmatrix} = \begin{bmatrix} 2 & 0 \\ 0 & 1 \end{bmatrix}\begin{bmatrix} x_1(t) \\ x_2(t) \end{bmatrix}.$$

Construct a Luenberger observer for this system.

Problem 12.10 *Consider the control system*

$$\frac{d}{dt}\begin{bmatrix} x_1(t) \\ x_2(t) \end{bmatrix} = \begin{bmatrix} 0 & 1 \\ 0 & 0 \end{bmatrix}\begin{bmatrix} x_1(t) \\ x_2(t) \end{bmatrix} + \begin{bmatrix} 0 \\ 1 \end{bmatrix} u(t).$$

with measured output

$$\boldsymbol{y}(t) = \begin{bmatrix} 2x_1(t) \\ x_2(t) \end{bmatrix} = \begin{bmatrix} 2 & 0 \\ 0 & 1 \end{bmatrix}\begin{bmatrix} x_1(t) \\ x_2(t) \end{bmatrix}.$$

Let

$$Q = \begin{bmatrix} 2 & 0 \\ 0 & 1 \end{bmatrix}, \; R = 1, \; M = \begin{bmatrix} 1 & 0 \\ 0 & 1 \end{bmatrix}, \; \text{and } N = 1.$$

Construct the Kalman (LQG) observer for this system.

Advanced Problems

These problems require Matlab^{TM} or some other numerical toolbox.

Problem 12.11 *Consider the control system*

$$\frac{d}{dt}\begin{bmatrix} x_1(t) \\ x_2(t) \end{bmatrix} = \begin{bmatrix} -0.4 & R_e \\ 0 & -0.2 \end{bmatrix}\begin{bmatrix} x_1(t) \\ x_2(t) \end{bmatrix} + \begin{bmatrix} 0 \\ 1 \end{bmatrix} u(t),$$

with output

$$y(t) = x_1(t) + x_2(t) = \begin{bmatrix} 1 & 1 \end{bmatrix}\begin{bmatrix} x_1(t) \\ x_2(t) \end{bmatrix},$$

where $R_e \geq 0$ *is a constant. Let*

$$Q = \begin{bmatrix} 1 & 0 \\ 0 & 1 \end{bmatrix}, \; R = 1, \; M = \begin{bmatrix} 1 & 0 \\ 0 & 1 \end{bmatrix}, \; \text{and } N = 1.$$

For each $R_e = 1$ *and* 2, *compute the LQR gain* $K(R_e) = R^{-1}B^T P(R_e)$ *and the Kalman filter gain* $F(R_e) = \Pi C^T N^{-1}$, *where* $P(R_e)$ *and* $\Pi(R_e)$ *satisfy*

$$A^T P + PA - PBR^1 B^T P + Q = 0$$

and

$$A\Pi + \Pi A^T - \Pi C^T N^{-1} C\Pi + M = 0,$$

respectively. Construct the optimal LQG closed-loop system and simulate the response to the initial conditions

$$\boldsymbol{x}_0 = [1\ 0]^T, \; \boldsymbol{x}_0 = [0\ 0]^T \text{ and } \boldsymbol{x}_0 = [0\ 1]^T,$$

with initial state estimates $\boldsymbol{x}_{e,0} = (0.95) \cdot \boldsymbol{x}_0$.

Problem 12.12 *Consider the nonlinear control system*

$$\begin{aligned} \frac{d}{dt}\begin{bmatrix} x_1(t) \\ x_2(t) \end{bmatrix} &= \begin{bmatrix} -0.4 & R_e \\ 0 & -0.2 \end{bmatrix}\begin{bmatrix} x_1(t) \\ x_2(t) \end{bmatrix} \\ &\quad + \sqrt{[x_1(t)]^2 + [x_1(t)]^2}\begin{bmatrix} 0 & -1 \\ +1 & 0 \end{bmatrix}\begin{bmatrix} x_1(t) \\ x_2(t) \end{bmatrix} \\ &\quad + \begin{bmatrix} 0 \\ 1 \end{bmatrix} u(t). \end{aligned}$$

where $R_e = 2$. *Use the LQR optimal controller from* ***Problem*** *12.11 above and simulate the nonlinear response to the initial conditions*

$$\boldsymbol{x}_0 = [1\ 0]^T,\ \boldsymbol{x}_0 = [0\ 0]^T \text{ and } \boldsymbol{x}_0 = [0\ 1]^T.$$

Problem 12.13 *Consider the nonlinear control system*

$$\begin{aligned}\frac{d}{dt}\begin{bmatrix} x_1(t) \\ x_2(t) \end{bmatrix} = \begin{bmatrix} -0.4 & R_e \\ \varepsilon & -0.2 \end{bmatrix}\begin{bmatrix} x_1(t) \\ x_2(t) \end{bmatrix} \\ + \sqrt{[x_1(t)]^2 + [x_1(t)]^2}\begin{bmatrix} 0 & -1 \\ +1 & 0 \end{bmatrix}\begin{bmatrix} x_1(t) \\ x_2(t) \end{bmatrix} \\ + \begin{bmatrix} 0 \\ 1 \end{bmatrix} u(t).\end{aligned}$$

where $R_e = 2$ *and* $\varepsilon = 0.05$. *Use the LQR optimal controller from* ***Problem*** *12.11 above and simulate the nonlinear response to the initial conditions*

$$\boldsymbol{x}_0 = [1\ 0]^T,\ \boldsymbol{x}_0 = [0\ 0]^T \text{ and } \boldsymbol{x}_0 = [0\ 1]^T.$$

Problem 12.14 *Consider the system*

$$\dot{\boldsymbol{x}}_N(t) = A_N \boldsymbol{x}_N(t) + B_N \boldsymbol{u}(t),$$

where for each $N = 2, 4, 8, 16, \ldots$, *the* $N \times N$ *system matrix* A_n *is given by*

$$A_N = (N+1)^2 \begin{bmatrix} -2 & 1 & 0 & \cdots & 0 & 0 \\ 1 & -2 & 1 & 0 & \cdots & 0 \\ 0 & 1 & -2 & 1 & \ddots & \vdots \\ \vdots & \vdots & \ddots & \ddots & \ddots & 0 \\ 0 & 0 & \cdots & 1 & -2 & 1 \\ 0 & 0 & \cdots & 0 & 1 & -2 \end{bmatrix}_{N\times N}$$

and

$$B_N = \begin{bmatrix} 1/(N+1) \\ 2/(N+1) \\ 3/(N+1) \\ \vdots \\ (N-1)/(N+1) \\ N/(N+1) \end{bmatrix}$$

For $N = 2, 4, 8$, *use Matlab*^{TM} *to solve the LQR problem where* $Q = I_{N \times N}$ *is the* $N \times N$ *identity matrix and* $R = 1$. *(This system comes from approximating an optimal control problem for heat transfer in a rod.)*

Bibliography

[1] B. M. Adams, H. T. Banks, H. D. Kwon, and H. T. Tran. Dynamic multidrug therapies for hiv: Optimal and sti control approaches. *Mathematical Biosciences and Engineering*, 1(2):223–241, 2004.

[2] R. A. Adams. *Sobolev Spaces.* Academic Press, New York, 1975.

[3] J. H. Ahlberg, E. N. Nilson, and J. L. Walsh. *The Theory of Splines and Their Applications.* Academic Press, 1967.

[4] P. S. Aleksandrov, V. G. Boltyanskii, R. V. Gamkrelidze, and E. F. Mishchenko. Lev Semenovich Pontryagin (on his sixtieth birthday). *Uspekhi Mat. Nauk*, 23:187–196, 1968.

[5] B. D. Anderson and J. B. Moore. *Linear Optimal Control.* Prentice-Hall, Englewood Cliffs, 1971.

[6] B. D. O. Anderson and J. B. Moore. *Optimal Control: Linear Quadratic Methods.* Prentice-Hall, Inc., Upper Saddle River, 1990.

[7] P. J. Antsaklis and A. N. Michel. *Linear Systems.* Birkhauser, Boston, 2005.

[8] J. Appel, E. Cliff, M. Gunzburger, and A. Godfrey. Optimization-based design in high-speed flows. In *CFD for Design and Optimization, ASME, New York*, pages 61–68, 1995.

[9] E. Arian and S. Ta'asan. Multigrid one shot methods for optimal control problems: Infinite dimensional control. Technical Report ICASE Report No. 94–52, NASA, Langley Research Center, Hampton VA 23681–0001, 1994.

[10] A. Armaou and M. A. Demetriou. Optimal actuator/sensor placement for linear parabolic PDEs using spatial H2 norm. *Chemical Engineering Science*, 61(22):7351–7367, 2006.

[11] B. S. Attili and M. I. Syam. Efficient shooting method for solving two point boundary value problems. *Chaos, Solitons and Fractals*, 35(5):895–903, 2008.

[12] J. P. Aubin. *Approximation of Elliptic Boundary-Value Problems.* RE Krieger Pub. Co., 1972.

[13] S. Badrinarayanan and N. Zabaras. A sensitivity analysis for the optimal design of metal-forming processes. *Computer Methods in Applied Mechanics and Engineering*, 129(4):319–348, 1996.

[14] H. T. Banks, S. Hu, T. Jang, and H. D. Kwon. Modelling and optimal control of immune response of renal transplant recipients. *Journal of Biological Dynamics*, 6(2):539–567, 2012.

[15] R. G. Bartle. *The Elements of Real Analysis: Second Edition.* John Wiley & Sons, New York, 1976.

[16] T. Basar and P. Bernhard. *H^∞-Optimal Control and Related Minimax Design Problems: A Dynamic Game Approach.* Birkhauser, 1995.

[17] T. Başar and P. Bernhard. *H-Infinity Optimal Control and Related Minimax Design Problems: a Dynamic Game Approach.* Birkhäuser, Boston, 2008.

[18] J. Z. Ben-Asher. *Optimal Control Theory with Aerospace Applications.* AIAA, 2010.

[19] J. A. Bennett and M. E. Botkin. Structural shape optimization with geometric description and adaptive mesh refinement. *AIAA Journal*, 23(3):458–464, 1985.

[20] A. Bensoussan and A. Friedman. Nonlinear variational inequalities and differential games with stopping times. *Journal of Functional Analysis*, 16(3):305–352, 1974.

[21] A. Bensoussan and J. L. Lions. *Applications of Variational Inequalities in Stochastic Control*, volume 12. North-Holland, 1982.

[22] A. Bensoussan, G. Da Prato, M. C. Delfour, and S. K. Mitter. *Representation and Control of Infinite Dimensional Systems - I.* Systems & Control: Foundations & Applications. Birkhäuser, Boston, 1992.

[23] A. Bensoussan, G. Da Prato, M. C. Delfour, and S. K. Mitter. *Representation and Control of Infinite Dimensional Systems - II.* Systems & Control: Foundations & Applications. Birkhäuser, Boston, 1992.

[24] L. D. Berkovitz. *Optimal Control Theory*, volume 12 of *Applied Mathematical Sciences.* Springer-Verlag, New York, 1974.

[25] J. T. Betts. *Practical Methods for Optimal Control Using Nonlinear Programming.* Advances in Design and Control. Society for Industrial and Applied Mathematics, Philadelphia, 2001.

[26] F. Billy, J. Clairambault, and O. Fercoq. Optimisation of cancer drug treatments using cell population dynamics. *Mathematical Methods and Models in Biomedicine*, pages 263–299, 2012.

[27] G. A. Bliss. *Calculus of Variations.* Carus Mathematical Monographs, Chicago, 1925.

[28] G. A. Bliss. Normality and abnormality in the calculus of variations. *Trans. Amer. Math. Soc*, 43:365–376, 1938.

[29] G. A. Bliss. *Lectures on the Calculus of Variations.* Phoenix Science Series, University of Chicago Press, Chicago, 1963.

[30] V. G. Boltyanski. The maximum principle - how it came to be? report 526. Technical report, Mathematisches Institut, 1994.

[31] O. Bolza. *Lectures on the Calculus of Variations.* University of Chicago Press, Chicago, 1904.

[32] J. Borggaard. *The Sensitivity Equation Method for Optimal Design.* PhD thesis, Virginia Tech, Blacksburg, VA, December 1994.

[33] J. Borggaard. On the presence of shocks in domain optimization of Euler flows. In M. Gunzburger, editor, *Flow Control*, volume 68 of *Proceedings of the IMA*. Springer-Verlag, 1995.

[34] J. Borggaard and J. Burns. A sensitivity equation approach to optimal design of nozzles. In *Proc. 5th AIAA/USAF/NASA/ISSMO Symposium on Multidisciplinary Analysis and Optimization*, pages 232–241, 1994.

[35] J. Borggaard and J. Burns. A sensitivity equation approach to shape optimization in fluid flows. In M. Gunzburger, editor, *Flow Control*, volume 68 of *Proceedings of the IMA*. Springer-Verlag, 1995.

[36] J. Borggaard and J. Burns. Asymptotically consistent gradients in optimal design. In N. Alexandrov and M. Hussaini, editors, *Multidisciplinary Design Optimization: State of the Art*, pages 303–314, Philadelphia, 1997. SIAM Publications.

[37] J. Borggaard and J. Burns. A PDE sensitivity equation method for optimal aerodynamic design. *Journal of Computational Physics*, 136(2):366–384, 1997.

[38] J. Borggaard, J. Burns, E. Cliff, and M. Gunzburger. Sensitivity calculations for a 2D, inviscid, supersonic forebody

problem. In H.T. Banks, R. Fabiano, and K. Ito, editors, *Identification and Control of Systems Governed by Partial Differential Equations*, pages 14–24, Philadelphia, 1993. SIAM Publications.

[39] J. Borggaard, J. Burns, and L. Zietsman. On using lqg performance metrics for sensor placement. In *American Control Conference (ACC), 2011*, pages 2381–2386. IEEE, 2011.

[40] J. Borggaard, E. Cliff, J. Burkardt, M. Gunzburger, H. Kim, H. Lee, J. Peterson, A. Shenoy, and X. Wu. Algorithms for flow control and optimization. In J. Borggaard, J. Burkardt, M. Gunzburger, and J. Peterson, editors, *Optimal Design and Control*, pages 97–116. Birkhauser, 1995.

[41] J. Borggaard and D. Pelletier. Computing design sensitivities using an adaptive finite element method. In *Proceedings of the 27th AIAA Fluid Dynamics Conference*, 1996.

[42] J. Borggaard and D. Pelletier. Observations in adaptive refinement strategies for optimal design. In J. Borggaard, J. Burns, E. Cliff, and S. Schreck, editors, *Computational Methods for Optimal Design and Control*, pages 59–76. Birkhauser, 1998.

[43] J. Borggaard and D. Pelletier. Optimal shape design in forced convection using adaptive finite elements. In *Proc. 36th AIAA Aerospace Sciences Meeting and Exhibit*, 1998. AIAA Paper 98-0908.

[44] F. Brezzi and M. Fortin. *Mixed and Hybrid Finite Elements.* Springer-Verlag, New York, 1991.

[45] R. W. Brockett. *Finite Dimensional Linear Systems.* Wiley, New York, 1970.

[46] A. E. Bryson and Y. C. Ho. *Applied Optimal Control.* John Wiley & Sons, New York, 1975.

[47] A.E. Bryson Jr. Optimal control-1950 to 1985. *Control Systems Magazine, IEEE*, 16(3):26–33, 1996.

[48] C. Buck. *Advanced Calculus.* McGraw Hill, New York, 1978.

[49] J. A. Burns, E. M. Cliff, C. Rautenberg, and L. Zietsman. Optimal sensor design for estimation and optimization of pde systems. In *American Control Conference (ACC), 2010*, pages 4127–4132. IEEE, 2010.

[50] J. A. Burns and B. B. King. Optimal sensor location for robust control of distributed parameter systems. In *Proceedings of the 33rd IEEE Conference on Decision and Control*, volume 4, pages 3967–3972. IEEE, 1994.

[51] J. A. Burns, B. B. King, and Y. R. Ou. A computational approach to sensor/actuator location for feedback control of fluid flow systems. In J.D. Paduano, editor, *Sensing, Actuation, and Control in Aeropropulsion*, volume 2494 of *Proc. International Society for Optical Engineering*, pages 60–69, 1995.

[52] D. Bushaw. Optimal discontinuous forcing terms. In *Contributions to the Theory of Non-linear Oscillations*, pages 29–52, Princeton, 1958. Princeton University Press.

[53] C. Carathéodory. Die methode der geodätischen äquidistanten und das problem von lagrange. *Acta Mathematica*, 47(3):199–236, 1926.

[54] C. Carathéodory. *Calculus of Variations and Partial Differential Equations of First Order.* American Mathematical Society, Providence, 1999.

[55] P. G. Ciarlet. *The Finite Element Method for Elliptic Problems.* North-Holland, Amsterdam, 1978.

[56] P. G. Ciarlet. *Introduction to Numerical Linear Algebra and Optimisation.* Cambridge Press, Cambridge, 1989.

[57] F. H. Clarke. *Necessary Conditions for Nonsmooth Problems in Optimal Control and the Calculus of Variations.* PhD thesis, University of Washington, 1973.

[58] E. A. Coddington and N. Levinson. *Theory of Ordinary Differential Equations.* McGraw-Hill, New York, 1972.

[59] R. Courant and K. L. Friedrichs. *Supersonic Flow and Shock Waves.* Interscience Publishers, New York, 1948.

[60] M. G. Crandall, H. Ishii, and P. L. Lions. Users guide to viscosity solutions of second order partial differential equations. *Bull. Amer. Math. Soc*, 27(1):1–67, 1992.

[61] M. G. Crandall and P. L. Lions. Viscosity solutions of hamilton-jacobi equations. *Trans. Amer. Math. Soc*, 277(1):1–42, 1983.

[62] J. J. Crivelli, J. Földes, P. S. Kim, and J. R. Wares. A mathematical model for cell cycle-specific cancer virotherapy. *Journal of Biological Dynamics*, 6(sup1):104–120, 2012.

[63] R. F. Curtain and H. Zwart. *An Introduction to Infinite-Dimensional Linear Systems Theory.* Springer, New York-Berlin - Heidelberg, 1995.

[64] B. Dacorogna. *Direct Methods in the Calculus of Variations*, volume 78. Springer, 2007.

[65] B. Dacorogna. *Introduction to the Calculus of Variations.* Imperial College Press, London, 2nd edition, 2009.

[66] N. Darivandi, K. Morris, and A. Khajepour. Lq optimal actuator location in structures. In *American Control Conference (ACC), 2012*, pages 646–651. IEEE, 2012.

[67] M. C. Delfour and J. P. Zolesio. Shape sensitivity analysis via MinMax differentiability. *SIAM Journal of Control and Optimization*, 26(4), July 1988.

[68] M. C. Delfour and J. P. Zolesio. Velocity method and lagrangian formulation for the computation of the shape hessian. *SIAM Journal of Control and Optimization*, 29(6):1414–1442, November 1991.

[69] M. A. Demetriou. Integrated actuator-sensor placement and hybrid controller design of flexible structures under worst case spatiotemporal disturbance variations. *Journal of Intelligent Material Systems and Structures*, 15(12):901, 2004.

[70] M.A. Demetriou and J. Borggaard. Optimization of a joint sensor placement and robust estimation scheme for distributed parameter processes subject to worst case spatial disturbance distributions. In *American Control Conference, 2004. Proceedings of the 2004*, volume 3, pages 2239–2244. IEEE, 2004.

[71] P. Dorato, V. Cerone, and C. T. Abdallah. *Linear-Quadratic Control.* Prentice Hall, Englewood Cliffs, 1995.

[72] J. C. Doyle. Guaranteed margins for LQG regulators. *Transactions on Automatic Control*, 23:756–757, 1981.

[73] J. C. Doyle, B. A. Francis, and A. R. Tannenbaum. *Feedback Control Theory.* Macmillan Publishing Company, New York, 1992.

[74] M. Engelhart, D. Lebiedz, and S. Sager. Optimal control for selected cancer chemotherapy ode models: A view on the potential of optimal schedules and choice of objective function. *Mathematical Biosciences*, 229(1):123–134, 2011.

[75] L. C. Evans. On solving certain nonlinear partial differential equations by accretive operator methods. *Israel Journal of Mathematics*, 36(3):225–247, 1980.

[76] L. C. Evans. *Partial Differential Equations.* Graduate Studies in Mathematics: Vol. 19. American Mathematical Society, Providence, 1998.

[77] George M. Ewing. *Calculus of Variations with Applications.* Dover Publications, New York, 1985.

[78] A. L. Faulds and B. B. King. Sensor location in feedback control of partial differential equation systems. In *Proceedings of the 2000 IEEE International Conference on Control Applications*, pages 536–541. IEEE, 2000.

[79] W. H. Fleming and R. W. Rishel. *Deterministic and Stochastic Optimal Control.* Springer, New York, 1975.

[80] A. R. Forsyth. *Calculus of Variations.* Cambridge University Press, New York, 1927.

[81] B. A. Francis. *A Course in H^∞-Control Theory, Lecture Notes in Control, Vol. 88.* Springer-Verlag, New York, 1987.

[82] P. D. Frank and G. R. Shubin. A comparison of optimization-based approaches for a model computational aerodynamics design problem. *Journal of Computational Physics*, 98:74–89, 1992.

[83] R. H. Gallagher and O. C. Zienkiewicz, editors. *Optimum Structural Design: Theory and Application*, John Wiley & Sons, New York, 1973.

[84] R. V. Gamkrelidze. Discovery of the maximum principle. *Journal of Dynamical and Control Systems*, 5(4):437–451, 1999.

[85] R. V. Gamkrelidze. Discovery of the maximum principle. *Mathematical Events of the Twentieth Century*, pages 85–99, 2006.

[86] I. M. Gelfand and S. V. Fomin. *Calculus of Variations.* Prentice-Hall, Inc., Englewood Cliffs, 1963.

[87] I. M. Gelfand and S. V. Fomin. *Calculus of Variations.* Prentice-Hall, Inc., Englewood Cliffs, 1963.

[88] I. M. Gelfand, G. E. Shilov, E. Saletan, N. I. A. Vilenkin, and M. I. Graev. *Generalized Functions*, volume 1. Academic Press, New York, 1968.

[89] M. Giaquinta. *Multiple Integrals in the Calculus of Variations and Nonlinear Elliptic Systems.(AM-105)*, volume 105. Princeton University Press, 1983.

[90] A. Godfrey. Using sensitivities for flow analysis. In J. Borggaard, J. Burns, E. Cliff, and S. Schreck, editors, *Computational Methods for Optimal Design and Control*, pages 181–196. Birkhauser, 1998.

[91] H. H. Goldstine. *A History of the Calculus of Variations from the 17th through the 19th Century.* Springer-Verlag, 1980.

[92] G. H. Golub and C. F. Van Loan. *Matrix Computations: Third Edition.* Johns Hopkins University Press, Baltimore, 1996.

[93] L. M. Graves. *The Derivative as Independent Function in the Calculus of Variations.* PhD thesis, University of Chicago, Department of Mathematics, 1924.

[94] L. M. Graves. A transformation of the problem of Lagrange in the calculus of variations. *Transactions of the American Mathematical Society*, 35(3):675–682, 1933.

[95] L. M. Graves. *Calculus of Variations and Its Applications.* McGraw-Hill, New York, 1958.

[96] M. Gunzburger. *Finite Element Methods for Viscous Incompressible Flows.* Academic Press, 1989.

[97] H. Halkin. A generalization of LaSalle's bang-bang principle. *Journal of the Society for Industrial & Applied Mathematics, Series A: Control*, 2(2):199–202, 1964.

[98] J. Haslinger and P. Neittaanmaki. *Finite Element Approximation for Optimal Shape Design: Theory and Application.* John Wiley & Sons, 1988.

[99] E. J. Haug, K. K. Choi, and V. Komkov. *Design Sensitivity Analysis of Structural Systems*, volume 177 of *Mathematics in Science and Engineering*. Academic Press, Orlando, FL, 1986.

[100] H. Hermes and J. P. LaSalle. *Functional Analysis and Time Optimal Control. Number 56 in Mathematics in Science and Engineering*. Academic Press, New York, 1969.

[101] M. R. Hestenes. *Calculus of Variations and Optimal Control Theory*. John Wiley & Sons, New York, 1966.

[102] M. R. Hestenes. *Optimization Theory: The Finite Dimensional Case*. Wiley, 1975.

[103] R. Holsapple, R. Venkataraman, and D. Doman. A modified simple shooting method for solving two point boundary value problems. In *Proc. of the IEEE Aerospace Conference, Big Sky, MT*, 2003.

[104] R. Holsapple, R. Venkataraman, and D. Doman. New, fast numerical method for solving two-point boundary-value problems. *Journal of Guidance Control and Dynamics*, 27(2):301–303, 2004.

[105] A. Isidori. *Nonlinear Control Systems*. Springer, 1995.

[106] M. Itik, M. U. Salamci, and S. P. Banks. Optimal control of drug therapy in cancer treatment. *Nonlinear Analysis: Theory, Methods & Applications*, 71(12):e1473–e1486, 2009.

[107] V. G. Ivancevic and T. T. Ivancevic. *Applied Differential Geometry: A Modern Introduction*. World Scientific Publishing Company Incorporated, 2007.

[108] A. Jameson. Aerodynamic design via control theory. *Journal of Scientific Computing*, 3:233–260, 1988.

[109] A. Jameson and J. Reuther. Control theory based airfoil design using the Euler equations. In *Proc.*

5th AIAA/USAF/NASA/ISSMO Symposium on Multidisciplinary Analysis and Optimization, pages 206–222, 1994.

[110] T. Kailath. *Linear Systems.* Prentice-Hall, 1980.

[111] J. H. Kane and S. Saigal. Design sensitivity analysis of solids using BEM. *Journal of Engineering Mechanics*, 114:1703–1722, 1988.

[112] H. B. Keller. *Numerical Methods for Two-Point Boundary-Value Problems.* Blaisdell, 1968.

[113] C. T. Kelley and E. W. Sachs. Quasi-Newton methods and unconstrained optimal control problems. *SIAM Journal on Control and Optimization*, 25:1503, 1987.

[114] H. W. Knobloch, A. Isidori, and D. Flockerzi. *Topics in Control Theory.* Birkhäuser Verlag, 1993.

[115] H. Kwakernaak and R. Sivan. *Linear Optimal Control Systems.* Willey-Interscience, New York, 1972.

[116] S. Lang. *Fundamentals of Differential Geometry*, volume 160. Springer Verlag, 1999.

[117] U. Ledzewicz, M. Naghnaeian, and H. Schättler. Optimal response to chemotherapy for a mathematical model of tumor–immune dynamics. *Journal of Mathematical Biology*, 64(3):557–577, 2012.

[118] U. Ledzewicz and H. Schättler. Multi-input optimal control problems for combined tumor anti-angiogenic and radiotherapy treatments. *Journal of Optimization Theory and Applications*, pages 1–30, 2012.

[119] E. B. Lee and L. Markus. *Foundations of Optimal Control Theory, The SIAM Series in Applied Mathematics.* John Wiley & Sons, 1967.

[120] G. Leitmann. *The Calculus of Variations and Optimal Control.* Springer, Berlin, 1981.

[121] N. Levinson. Minimax, Liapunov and "bang-bang," *Journal of Differential Equations*, 2:218–241, 1966.

[122] D. Liberzon. *Calculus of Variations and Optimal Control Theory: A Concise Introduction.* Princeton University Press, Princeton, 2012.

[123] J. L. Lions. *Optimal Control of Systems Governed by Partial Differential Equations, vol. 170 of Grundlehren Math. Wiss.* Springer-Verlag, Berlin, 1971.

[124] J. L. Lions. *Optimal Control of Systems Governed by Partial Differential Equations, vol. 170 of Grundlehren Math. Wiss.* Springer-Verlag, 1971.

[125] J.-L. Lions. *Control of Distributed Singular Systems.* Gauthier-Villars, Kent, 1985.

[126] J. L. Lions and E. Magenes. *Non-homogeneous boundary value problems and applications, Vol. 1, 2.* Springer, 1972.

[127] J. L. Lions and E. Magenes. Non-homogeneous Boundary Value Problems and Applications. Volume 3 (Translation of Problemes aux limites non homogenes et applications). 1973.

[128] J. L. Lions, E. Magenes, and P. Kenneth. *Non-homogeneous Boundary Value Problems and Applications*, volume 1. Springer-Verlag Berlin, 1973.

[129] D. G. Luenberger. Observing the state of a linear system. *Military Electronics, IEEE Transactions on*, 8(2):74–80, 1964.

[130] D. G. Luenberger. Observers for multivariable systems. *Automatic Control, IEEE Transactions on*, 11(2):190–197, 1966.

[131] D. G. Luenberger. *Optimization by Vector Space Methods.* John Wiley & Sons, New York, 1969.

[132] D. G. Luenberger. An introduction to observers. *Automatic Control, IEEE Transactions on*, 16(6):596–602, 1971.

[133] E. J. McShane. The calculus of variations from the beginning through optimal control theory. In W. G. Kelley, A. B. Schwarzkopf and S. B. Eliason, editors, *Optimal Control and Differential Equations*, pages 3 –49. Academic Press, New York, 1978.

[134] E. J. McShane. The calculus of variations from the beginning through optimal control theory. *SIAM Journal on Control and Optimization*, 27:916, 1989.

[135] M. Mesterton-Gibbons. *A Primer on the Calculus of Variations and Optimal Control Theory*, volume 50. American Mathematical Society, Providence, 2009.

[136] A. Miele. *Theory of Optimum Aerodynamic Shapes.* Academic Press, New York, 1965.

[137] A. Miele and H.Y. Huang. Missile shapes of minimum ballistic factor. *Journal of Optimization Theory and Applications*, 1(2):151–164, 1967.

[138] R. K. Miller and A. N. Michel. *Ordinary Differential Equations.* Dover Publications, New York, 2007.

[139] K. Morris. Linear-quadratic optimal actuator location. *Automatic Control, IEEE Transactions on*, 56(1):113–124, 2011.

[140] J. M. Murray. Optimal control for a cancer chemotheraphy problem with general growth and loss functions. *Mathematical Biosciences*, 98(2):273–287, 1990.

[141] J. M. Murray. Some optimal control problems in cancer chemotherapy with a toxicity limit. *Mathematical Biosciences*, 100(1):49–67, 1990.

[142] R. Narducci, B. Grossman, and R. Haftka. Design sensitivity algorithms for an inverse design problem involving a shock wave. *Inverse Problems in Engineering*, 2:49–83, 1995.

[143] M. Z. Nashed. Some remarks on variations and differentials. *American Mathematical Monthly*, pages 63–76, 1966.

[144] L. W. Neustadt. *Optimization: A Theory of Necessary Conditions*. Princeton University Press, 1976.

[145] G. Newbury. *A Numerical Study of a Delay Differential Equation Model for Breast Cancer*. PhD thesis, Virginia Polytechnic Institute and State University, Blacksburg, Virginia, 2007.

[146] K. O. Okosun, O. D. Makinde, and I. Takaidza. Impact of optimal control on the treatment of hiv/aids and screening of unaware infectives. *Applied Mathematical Modelling*, 2012.

[147] B. O'neill. *Elementary Differential Geometry*. Academic Press, second edition, 1997.

[148] J. C. Panetta and K. R. Fister. Optimal control applied to cell-cycle-specific cancer chemotherapy. *SIAM Journal on Applied Mathematics*, 60(3):1059–1072, 2000.

[149] H. J. Pesch and M. Plail. The maximum principle of optimal control: A history of ingenious ideas and missed opportunities. *Control and Cybernetics*, 38(4A):973–995, 2009.

[150] O. Pironneau. *Optimal Shape Design for Elliptic Systems*. Springer-Verlag, 1983.

[151] BN Pshenichny. Convex analysis and optimization. *Moscow: Nauka*, 25:358, 1980.

[152] C. W. Ray and B. A. Batten. Sensor placement for flexible wing shape control. In *American Control Conference (ACC), 2012*, pages 1430–1435. IEEE, 2012.

[153] W. T. Reid. *Ordinary Differential Equations.* John Wiley and Sons, New York, 1971.

[154] W. T. Reid. A historical note on the maximum principle. *SIAM Review*, 20:580, 1978.

[155] I. Rhee and J. L. Speyer. A game theoretic controller and its relationship to H^∞ and linear-exponential-gaussian synthesis. In *Decision and Control, 1989, Proceedings of the 28th IEEE Conference on*, pages 909–915. IEEE, 1989.

[156] A. C. Robinson. A survey of optimal control of distributed-parameter systems. *Automatica*, 7(3):371–388, 1971.

[157] D. L. Russell. *Mathematics of Finite Dimensional Control Systems: Theory and Design.* M. Dekker, 1979.

[158] H. Sagan. *Introduction to the Calculus of Variations.* Courier Dover Publications, 1992.

[159] M. H. Schultz. *Spline Analysis.* Prentice-Hall, 1973.

[160] L. Schumaker. *Spline Functions: Basic Theory.* Cambridge University Press, 2007.

[161] L. Schwartz. Théorie des distributions. ii, hermann. *Paris, France*, 1959.

[162] L. F. Shampine, J. Kierzenka, and M. W. Reichelt. Solving boundary value problems for ordinary differential equations in Matlab with bvp4c. Manuscript, available at ftp://ftp. mathworks. com/pub/doc/papers/bvp, 2000.

[163] L. L. Sherman, A. C. Taylor III, L. L. Green, P.A. Newman, G.J.-W. Hou, and V.M. Korivi. First- and second-order aerodynamic sensitivity derivatives via automatic differentiation with incremental iterative methods. In *Proc. 5th AIAA/USAF/NASA/ISSMO Symposium on Multidisciplinary Analysis and Optimization*, pages 87–120, 1994.

[164] D. B. Silin. On the variation and riemann integrability of an optimal control in linear systems. *Dokl. Akad. Nauk SSSR*, 257(3):548–550, 1981. English Translation in *Soviet Math. Doklady*, 23(2): 309–311, 1981.

[165] M. H. N. Skandari, H. R. Erfanian, A. V. Kamyad, and S. Mohammadi. Optimal control of bone marrow in cancer chemotherapy. *European Journal of Experimental Biology*, 2(3):562–569, 2012.

[166] D. R. Smith. *Variational Methods in Optimization.* Prentice-Hall, Englewood Cliffs, 1974.

[167] D. R. Smith. *Variational Methods in Optimization.* Courier Dover Publications, 1998.

[168] D. Stewart. *Numerical Methods for Accurate Computation of Design Sensitivities.* PhD thesis, Virginia Tech, 1998.

[169] J. Stoer, R. Bulirsch, W. Gautschi, and C. Witzgall. *Introduction to Numerical Analysis.* Springer Verlag, 2002.

[170] G. Strang and G. Fix. *An Analysis of the Finite Element Method.* Prentice-Hall, Englewood Cliffs, 1973.

[171] G. Strang and G. Fix. *An Analysis of the Finite Element Method.* Wellesley-Cambridge Press, Wellesley, 1988.

[172] H. J. Sussmann and J. C. Willems. 300 years of optimal control: from the brachystochrone to the maximum principle. *IEEE Control Systems Magazine*, 17(3):32–44, 1997.

[173] G. W. Swan. Role of optimal control theory in cancer chemotherapy. *Mathematical Biosciences*, 101(2):237–284, 1990.

[174] S. Ta'asan. One shot methods for optimal control of distributed parameter systems I: Finite dimensional control. Technical Report ICASE Report No. 91-2, NASA Langley Research Center, Hampton, VA, 1991.

[175] M. Tahk and J. Speyer. Modeling of parameter variations and asymptotic LQG synthesis. *Automatic Control, IEEE Transactions on*, 32(9):793–801, 1987.

[176] A. E. Taylor. *General Theory of Functions and Integration.* Dover Publications, 2010.

[177] A. C. Taylor III, G. W. Hou, and V. M. Korivi. A methodology for determining aerodynamic sensitivity derivatives with respect to variation of geometric shape. In *Proc. AIAA/ASME/ASCE/AHS/ASC 32nd Structures, Structural Dynamics, and Materials Conference*, Baltimore, MD, April 1991. AIAA paper 91-1101.

[178] A. C. Taylor III, G. W. Hou, and V. M. Korivi. Sensitivity analysis, approximate analysis and design optimization for internal and external viscous flows. In *Proc. AIAA Aircraft Design Systems and Operations Meeting*, 1991. AIAA paper 91-3083.

[179] R. Temam. *Navier-Stokes Equations: Theory and Numerical Analysis*, volume 2 of *Studies in Mathematics and its Applications.* North-Holland Publishing Company, New York, 1979.

[180] J. L. Troutman. *Variational Calculus with Elementary Convexity.* Springer-Verlag, 1983.

[181] D. Uciński. Optimal sensor location for parameter estimation of distributed processes. *Int. J. Control*, 73:1235–1248, 2000.

[182] F.A. Valentine. *Convex Sets.* McGraw Hill, New York, 1964.

[183] F. Y. M. Wan. *Introduction to the Calculus of Variations and Its Applications.* Chapman & Hall/CRC, 1995.

[184] T. J. Willmore. *An Introduction to Differential Geometry.* Dover Publications, 2012.

[185] W. Yang, W. Y. Yang, W. Cao, T. S. Chung, and J. Morris. *Applied Numerical Methods Using MATLAB.* Wiley-Interscience, 2005.

[186] L. C. Young. *Lectures on the Calculus of Variations and Optimal Control Theory.* Chelsea, 1969.

[187] K. Zhou, J. C. Doyle, and K. Glover. *Robust and Optimal Control.* Prentice Hall, Englewood Cliffs, NJ, 1996.

Index

Milton Keynes UK
Ingram Content Group UK Ltd.
UKHW021928071024
449327UK00022B/1728

9 780367 379551